Neuromethods

Series Editor
Wolfgang Walz
University of Saskatchewan,
Saskatoon, SK, Canada

For further volumes:
http://www.springer.com/series/7657

Neuromethods publishes cutting-edge methods and protocols in all areas of neuroscience as well as translational neurological and mental research. Each volume in the series offers tested laboratory protocols, step-by-step methods for reproducible lab experiments and addresses methodological controversies and pitfalls in order to aid neuroscientists in experimentation. *Neuromethods* focuses on traditional and emerging topics with wide-ranging implications to brain function, such as electrophysiology, neuroimaging, behavioral analysis, genomics, neurodegeneration, translational research and clinical trials. *Neuromethods* provides investigators and trainees with highly useful compendiums of key strategies and approaches for successful research in animal and human brain function including translational "bench to bedside" approaches to mental and neurological diseases.

New Aspects in Analyzing the Synaptic Organization of the Brain

Edited by

Joachim H. R. Lübke

Institute of Neuroscience and Medicine INM-10, Research Centre Jülich GmbH, Jülich, Germany; Department of Psychiatry, Psychotherapy and Psychosomatics, Medical Faculty/RWTH University Hospital Aachen, Aachen, Germany; JARA Translational Brain Medicine, Jülich/Aachen, Germany

Astrid Rollenhagen

Institute of Neuroscience and Medicine INM-10, Research Centre Jülich GmbH, Jülich, Germany

Humana Press

Editors
Joachim H. R. Lübke
Institute of Neuroscience and Medicine
INM-10, Research Centre Jülich
GmbH
Jülich, Germany

Department of Psychiatry
Psychotherapy and Psychosomatics
Medical Faculty/RWTH
University Hospital Aachen
Aachen, Germany

JARA Translational Brain Medicine
Jülich/Aachen, Germany

Astrid Rollenhagen
Institute of Neuroscience and Medicine
INM-10, Research Centre Jülich GmbH
Jülich, Germany

ISSN 0893-2336 ISSN 1940-6045 (electronic)
Neuromethods
ISBN 978-1-0716-4018-0 ISBN 978-1-0716-4019-7 (eBook)
https://doi.org/10.1007/978-1-0716-4019-7

This Humana imprint is published by the registered company Springer Science+Business Media, LLC, part of Springer Nature.
The registered company address is: 1 New York Plaza, New York, NY 10004, U.S.A.

Foreword

Almost 10 years ago, Javier DeFelipe and I co-edited a Frontiers Special Topic on *Neuroanatomy for the XXIst Century*, including a follow-up General Discussion by multiple researchers on "The Anatomical Problem Posed by Brain Complexity and Size." Not surprisingly, many of the results and viewpoints presented then are still under vigorous discussion and still directly relevant to the state of the field broadly construed. This amounts no less than to the resurgence of neuroanatomy as a discipline and the correlative promise for a deeper understanding of the structural and functional organization of the nervous system. Central to this ongoing endeavor is the development of neuroanatomical methods and segmentation tools, which can convert qualitative observations to quantitative systematic data, which in turn might facilitate the formulation of testable hypotheses and principles.

Since 2016, a substantial research effort in cataloguing and mapping neuronal cell types has led to a major re-positioning in neuroscience; i.e., the recognition that the number of identifiable neuronal subtypes is large and their mix variable across species and areas. Large datasets relating to cell types are fast becoming available (e.g., Science, volume 382, 2023). These neuroanatomical data have wide relevance, from the study of development and disease, to the elucidation of substrates of behavior, including human cognition.

To some extent, comparable progress at the level of microanatomy and ultrastructure has been hindered by labor-intensive, slower protocols and smaller sample sizes. This is unfortunate, given the potentially unique and fundamental insights from work at the microstructural level in relation to general function, to area and species specializations, and to changes over the lifespan and in disease. Thanks to recent advances in hardware, software, and analysis techniques; however, this relative neglect is no longer the case. Representative examples of high-end electron microscopy include methods such as transmission, focused ion beam, electron microscopic tomography, and volume electron microscopy, which are increasingly used in the descriptions of brain connectomics, and in fact are recognized to have experienced a user-explosion due to rapid advances in volume, resolution, and ease of use ("quiet revolution," Peddie et al. 2023).

This book entitled: *Methods Unravelling the Synaptic Organization of the Brain* by Drs. Lübke and Rollenhagen represents a compelling state-of-the-field overview of a broad spectrum of methods, including structural, structural/functional, molecular and computational. It will serve as a how-to practicum and an accessible guide to experts as well as to a broad readership. Over the subsequent decades since the term "synapse" was first introduced, synaptology has become a subject of intensive research, appropriately enough since these structures are key elements in the normal and abnormal functioning of the brain, and a scaffolding intimately linked to a panoply of functional properties. The brain is in many aspects still a mystery even under normal conditions, and our understanding of the pathologically altered brain is notably limited. This book describes methods that will effectively advance our knowledge of this fascinating organ and is a welcome addition to the existing literature, extending even to the human brain.

Chapters in this book cover a broad range from detailed protocols to critical assessments and mini-reviews of what has been and can now be accomplished by microanatomical investigations at both the cellular and subcellular levels. Drs. Lübke and Rollenhagen as co-Editors speak from decades-long expertise in the field of neuroscience and have themselves contributed seminal articles on the distribution and phenomenology of synaptic boutons in multiple areas and species. They have meticulously quantified structural features and shown the correlation with functionally important parameters such as synaptic efficacy and plasticity. Notably, they have been early members of what up to now has been a small group of researchers addressing synaptic attributes and distributions in the human brain.

The promise of microanatomical data is great. The data are a necessary extension and validation of cellular-level spatial transcriptomics and a necessary bridge to understanding cellular interactions. Microanatomical spatial parameters are essential to further probing of molecular interactions among tissue components at different spatial proximities, and to better understanding the complex neuropil environment in normal and abnormal conditions. With all this in mind, the present book will be appreciated as an excellent summary of the state of the field and a helpful user's guide moving forward. Most importantly, the book may serve to spotlight the experimental results and controversies to date so as to attract new researchers to this increasingly relevant and fascinating research area. The crucial importance of "details" has been repeatedly emphasized (e.g., DeFelipe et al. 2016; Südhof 2017). As in the best detective stories, details can serve as vital clues, in this case, to what is going on in the brain.

Katheen S. Rockland

Research Professor, Department of Anatomy & Neurobiology, Chobanian & Avedisian School of Medicine, Boston University, Boston, MA, USA

References

1. DeFelipe J et al (2016) Comments and general discussion on "The anatomical problem posed by brain complexity and size: a potential solution." Front Neuroanat 10:70. https://doi.org/10.3389/fnana.2016.00070
2. Peddie CJ, Genoud C, Kreshuk A, Meechan K, Micheva KD, Narayan K, Pape C et al (2023) Volume electron microscopy. Nat Rev Methods Primers 2. https://doi.org/10.1038/s43586-022-00131-9
3. Südhof TC (2017) Molecular neuroscience in the 21st century: a personal perspective. Neuron 96. https://doi.org/10.1016/j.neuron.2017.10.005

Preface to the Series

Experimental life sciences have two basic foundations: concepts and tools. The Neuromethods series focuses on the tools and techniques unique to the investigation of the nervous system and excitable cells. It will not, however, shortchange the concept side of things as care has been taken to integrate these tools within the context of the concepts and questions under investigation. In this way, the series is unique in that it not only collects protocols but also includes theoretical background information and critiques which led to the methods and their development. Thus, it gives the reader a better understanding of the origin of the techniques and their potential future development. The Neuromethods publishing program strikes a balance between recent and exciting developments like those concerning new animal models of disease, imaging, in vivo methods, and more established techniques, including, for example, immunocytochemistry and electrophysiological technologies. New trainees in neurosciences still need a sound footing in these older methods in order to apply a critical approach to their results.

Under the guidance of its founders, Alan Boulton and Glen Baker, the Neuromethods series has been a success since its first volume published through Humana Press in 1985. The series continues to flourish through many changes over the years. It is now published under the umbrella of Springer Protocols. While methods involving brain research have changed a lot since the series started, the publishing environment and technology have changed even more radically. Neuromethods has the distinct layout and style of the Springer Protocols program, designed specifically for readability and ease of reference in a laboratory setting.

The careful application of methods is potentially the most important step in the process of scientific inquiry. In the past, new methodologies led the way in developing new disciplines in the biological and medical sciences. For example, Physiology emerged out of Anatomy in the nineteenth century by harnessing new methods based on the newly discovered phenomenon of electricity. Nowadays, the relationships between disciplines and methods are more complex. Methods are now widely shared between disciplines and research areas. New developments in electronic publishing make it possible for scientists that encounter new methods to quickly find sources of information electronically. The design of individual volumes and chapters in this series takes this new access technology into account. Springer Protocols makes it possible to download single protocols separately. In addition, Springer makes its print-on-demand technology available globally. A print copy can therefore be acquired quickly and for a competitive price anywhere in the world.

Saskatoon, SK, Canada ***Wolfgang Walz***

Preface

The aim of this book is to summarize and describe recent methodological developments that will help to investigate the synaptic organization of the brain from different viewpoints including structural, structural/functional, molecular, and computational methods. We believe that it will help students, post-docs, and even group leaders, in the discipline of cellular and synaptic neuroscience, who will find benefits from the content of this book. Moreover, without claiming to cover the entire spectrum of methods, we hope the several chapters, individually and as a whole, may inform ongoing discussions and decision-making as to which direction of neuroscience research will take.

The brain of higher mammals including humans contains a network of 100 Milliards of neurons and even 100 Billion of synapses. Each neuron itself receives input from approximately 20,000 synapses distributed over the entire cellular domain with a highly specialized pattern of innervation. Thus, neurons together with their synaptic network are the structural and functional building blocks that govern the computational properties of any given network in the brain.

The term "synapse", derived from the Greek words *syn* (together) and *haptein* (to grasp, to touch), was introduced more than 150 years ago first by the well-known English physiologist Charles Sherrington and was later adopted by the famous Spanish neuroanatomist Ramon y Cajal, the founder of the neural doctrine who together with Camillo Golgi received the Nobel Prize in Medicine in 1906 for their groundbreaking work on the nervous system. Ramon y Cajal's great achievement was his description of structures that connect neurons building individual networks without ever seeing them.

A major step forward in describing the synaptic organization of the brain was twofold: First, a tremendous increase in scientific technology that made it possible to directly record from neurons in vivo using sharp electrodes (Llyod Hodgkin and Fielding Huxley, Nobel prize in Medicine 1963) at the squid giant axon. This was later combined with intracellular horseradish peroxidase-filling to visualize the recorded neurons (see for example Hubel and Wiesel 1969, Nobel Prize in Medicine 1981). A further development in unraveling the structure/function relationships of neurons and their networks was in vitro recording using acute brain slice preparations with either sharp electrode or patch-clamp recordings (Erwin Neher and Bert Sakmann, Nobel prize in Medicine 1991), also combined with intracellular dye-filling. This approach allowed the study of the structural/function relationship of a given neuron in individual microcircuits, networks, or brain regions as exemplified in investigations of the neuronal organization of the cortical column, the fundamental building block of the neocortex in higher mammals.

Secondly, the invention of high-resolution microscopic techniques, in particular the introduction of electron microscopy by Ernst Ruska and Max Knoll (1933; post hum Nobel Prize in Physics to Ernst Ruska 1986). This opened a new window in the description of highly specialized synaptic structures by early investigators (Gray 1959, Beaulieu and Colonnier 1988). At the high-resolution fine-scale cellular and subcellular level, investigations identified synapses as excitatory or inhibitory by the shape of the active zone (equivalent to a functional neurotransmitter release site) and the shape and size of synaptic vesicles. Moreover, electron microscopy led to the visualization and description of the pre- and postsynaptic complex, including the pre- and postsynaptic density, and the synaptic cleft and synaptic vesicles, all fundamental structural elements that build a synapse.

The constant development of these techniques meanwhile resulted in paired, quadruple, or even multiple (up to 10 electrodes) simultaneous recordings in acute brain slices and in vivo recordings in different networks of the brain, the development of high-resolution fluorescent microscopy (confocal, STED, STEM microscopy), and further developments of transmission, scanning, focused ion beam, electron microscopic tomography, and cryo-electron microscopy. Sophisticated molecular approaches such as specific gene tagging and targeting combined, with green fluorescence protein or other labeling, have become common. The toolkit now includes highly specific single and multiple pre- and postembedding immunohistochemistry of synaptic proteins involved in synaptic transmission and plasticity, single and multiple receptor autoradiography, and finally freeze fracture replication combined with immunohistochemistry. This impressive repertoire of techniques is still advancing and can realistically be expected to help unravel the still open mysteries of the brain.

Finally, we would like to thank all authors for their participation and willingness to contribute to this book although of the burden to publish original articles. Without their effort and input, this book would not have been realized.

Julich, Germany
February, 2024

Joachim H. R. Lübke
Astrid Rollenhagen

Contents

Foreword .... *v*
Preface to the Series .... *vii*
Preface .... *ix*
Contributors .... *xiii*

PART I PREPARATION OF BRAIN TISSUE AND TECHNIQUES

1 Long-Distance Transport of Surgically Resected, Alive Human Brain Tissue 3
Franz X. Mittermaier, Henrik Alle, and Jörg R. P. Geiger

2 Modern Glyoxal Fixation for the Determination of the Copy Number of Synaptic Vesicle Proteins at Individual Synapses 19
Jailenne Quiñones-Rodríguez and Thomas Schikorski

3 Cell Labeling and FIB–SEM Imaging 31
Marta Turegano-Lopez, José-Rodrigo Rodríguez, Javier DeFelipe, and Angel Merchán-Pérez

4 Unbiased Sampling and Volume-Sorting of Origin-Specific Terminals Using SBEM Image Stacks 41
Alev Erisir, Alex Briegel, Erin E. Maher, and Francesca Sciaccotta

5 A Technique to Perform Structure–Function Analysis of Synapses 53
Carolin Wichmann and Susann Michanski

6 Large-Volume Electron Microscopy 81
Yoshiyuki Kubota, Takaaki Miyazaki, and Marc Takeno

7 Espina: A Software Tool for the Study of 3D Synaptic Organization 105
Marta Turegano-Lopez, Félix de las Pozas, José-Rodrigo Rodríguez, Javier DeFelipe, and Angel Merchán-Pérez

8 Automated Imaging and Analysis of Synapses in Freeze-Fracture Replica Samples with Deep Learning 123
David Kleindienst, Tommaso Costanzo, and Ryuichi Shigemoto

PART II CONNECTIVITY

9 Multiplexed *d*STORM Imaging of Neuronal Tissue and Cells 141
Marina Ruth Hesse, Thomas Kuner, and Maja Klevanski

10 Detailed Phenotyping of the Physiology, Structure, and Connectivity of Individual Neocortical Neurons 161
Maria Medalla and Jennifer I. Luebke

11 3D Quantitative Analysis of Synaptic Ultrastructure in Identified Long-Range Projection Axons 183
Pablo J. Martin-Correa, Javier Rodriguez-Moreno, Astrid Rollenhagen, Joachim H. R. Lübke, and Francisco Clascá

12 Quantitative Analysis of the Synaptic Organization of the Human Temporal Lobe Neocortex 201
Astrid Rollenhagen, Kurt Sätzler, and Joachim H. R. Lübke

PART III SYNAPTIC VESICLES

13 Counting Synaptic Vesicles: As Simple as That? 233
Bernd Walkenfort, Astrid Rollenhagen, Mike Hasenberg, and Joachim H. R. Lübke

14 Readily Releasable Pool 253
Sina Zobel and Christian Rosenmund

PART IV MODELLING

15 Phenomenological Modeling of Diverse and Heterogeneous Synaptic Dynamics at Natural Density 277
Agnes Korcsak-Gorzo, Charl Linssen, Jasper Albers, Stefan Dasbach, Renato Duarte, Susanne Kunkel, Abigail Morrison, Johanna Senk, Jonas Stapmanns, Tom Tetzlaff, Markus Diesmann, and Sacha J. van Albada

16 Computational Methods and Algorithms to Segment and Model Recurrent Bursting Events in Long-Time Series Springer Nature Book about Neuromethods 323
L. Zonca, E. Dossi, N. Rouach, and D. Holcman

PART V DENDRITIC SPINES

17 Probing Dendritic Function with Holographic Structured Illumination 373
Vincent R. Daria and Hans A. Bachor

18 Intravital Imaging and Analysis of the Structural Plasticity of Dendritic Spines in Multiple Brain Regions 395
Yuan Shi, Mochen Cui, Tanja Blume, and Jochen Herms

Index *431*

Contributors

JASPER ALBERS • *Institute for Advanced Simulation (IAS-6), Jülich Research Centre, Julich, Germany; Department of Physics, Faculty 1, RWTH Aachen University, Aachen, Germany*

HENRIK ALLE • *Charité-Universitätsmedizin Berlin Corporate Member of Freie Universität Berlin, Humboldt-Universität zu Berlin, Berlin Institute of Health, Berlin, Germany*

HANS A. BACHOR • *Research School of Physics, Australian National University, Canberra, ACT, Australia*

TANJA BLUME • *German Center for Neurodegenerative Diseases (DZNE), Munich, Germany*

ALEX BRIEGEL • *University of Virginia, Charlottesville, VA, USA*

FRANCISCO CLASCÁ • *Department of Anatomy & Neuroscience, School of Medicine, Autónoma de Madrid University, Madrid, Spain*

TOMMASO COSTANZO • *Institute of Science and Technology Austria (IST Austria), Klosterneuburg, Austria*

MOCHEN CUI • *Center for Neuropathology, Ludwig-Maximilians-University of Munich, Munich, Germany; German Center for Neurodegenerative Diseases (DZNE), Munich, Germany; Munich Medical Research School, Ludwig-Maximilians-University of Munich, Munich, Germany*

VINCENT R. DARIA • *Research School of Physics, Australian National University, Canberra, ACT, Australia; John Curtin School of Medical Research School, Australian National University, Canberra, ACT, Australia*

STEFAN DASBACH • *Institute for Advanced Simulation (IAS-6), Institute of Energy and Climate Research - Plasma Physics (IEK-4), Jülich Research Centre, Julich, Germany; Faculty of Mathematics and Natural Sciences, Heinrich Heine University Düsseldorf, Dusseldorf, Germany*

FÉLIX DE LAS POZAS • *Visualization & Graphics Lab (VG-Lab), Universidad Rey Juan Carlos, Madrid, Spain*

JAVIER DEFELIPE • *Laboratorio Cajal de Circuitos Corticales, Centro de Tecnología Biomé dica, Universidad Politécnica de Madrid, Madrid, Spain; Centro de Investigación Biomé dica en Red sobre Enfermedades Neurodegenerative (CIBERNED), ISCIII, Madrid, Spain; Instituto Cajal, Consejo Superior de Investigaciones Científicas (CSIC), Madrid, Spain*

MARKUS DIESMANN • *Institute for Advanced Simulation (IAS-6), JARA-Institute Brain Structure-Function Relationships (INM-10), Jülich Research Centre, Julich, Germany; Department of Physics, Faculty 1, RWTH Aachen University, Aachen, Germany; Department of Psychiatry, Psychotherapy, and Psychosomatics, Medical School, RWTH Aachen University, Aachen, Germany*

ELENA DOSSI • *Neuroglial Interactions in Cerebral Physiopathology, Center for Interdisciplinary Research in Biology, Collège de France, CNR UMR 7241, INSERM U1050, Labex Memolife, PSL Research University, Paris, France*

RENATO DUARTE • *Donders Institute for Brain, Cognition and Behaviour, Radboud University, Nijmegen, Netherlands; Institute for Advanced Simulation (IAS-6), Jülich Research Centre, Julich, Germany; Center for Neuroscience and Cell Biology (CNC), University of Coimbra, Coimbra, Portugal; Centre for Innovative Biomedicine and Biotechnology (CIBB), University of Coimbra, Coimbra, Portugal*

ALEV ERISIR • *University of Virginia, Charlottesville, VA, USA*
JÖRG R. P. GEIGER • *Charité-Universitätsmedizin Berlin Corporate Member of Freie Universität Berlin, Humboldt-Universität zu Berlin, Berlin Institute of Health, Berlin, Germany*
MIKE HASENBERG • *Imaging Center Essen (IMCES), Electron Microscopy Unit (EMU), Medical Faculty of the University of Duisburg-Essen, Essen, Germany*
JOCHEN HERMS • *Center for Neuropathology, Ludwig-Maximilians-University of Munich, Munich, Germany; German Center for Neurodegenerative Diseases (DZNE), Munich, Germany; Munich Cluster for Systems Neurology (SyNergy), Munich, Germany*
MARINA RUTH HESSE • *Department of Functional Neuroanatomy, Institute for Anatomy and Cell Biology, Heidelberg University, Heidelberg, Germany*
DAVID HOLCMAN • *Group of Applied Mathematics and Computational Biology, Ecole Normale Supérieure, PSL University, Paris, France; Churchill College, Cambridge University, Cambridge, UK*
DAVID KLEINDIENST • *Institute of Science and Technology Austria (IST Austria), Klosterneuburg, Austria*
MAJA KLEVANSKI • *Department of Functional Neuroanatomy, Institute for Anatomy and Cell Biology, Heidelberg University, Heidelberg, Germany*
AGNES KORCSAK-GORZO • *Institute for Advanced Simulation (IAS-6), Jülich Research Centre, Julich, Germany; Department of Physics, Faculty 1, RMTH Aachen, Aachen, Germany*
YOSHIYUKI KUBOTA • *Section of Electron Microscopy, National Institute for Physiological Sciences, Okazaki, Aichi, Japan; Department of Physiological Sciences, The Graduate University for Advanced Studies (SOKENDAI), Okazaki, Japan; Support Unit for Electron Microscopy Techniques, Research Resources Division, RIKEN Center for Brain Science, Wako, Japan*
THOMAS KUNER • *Department of Functional Neuroanatomy, Institute for Anatomy and Cell Biology, Heidelberg University, Heidelberg, Germany*
SUSANNE KUNKEL • *Neuromorphic Software Ecosystems (PGI-15), Jülich Research Centre, Julich, Germany; Faculty of Science and Technology, Norwegian University of Life Sciences, Oslo, Norway*
CHARL LINSSEN • *Simulation and Data Lab (SDL) Neuroscience, Institute for Advanced Simulation (IAS-6), Jülich Supercomputing Centre (JSC), Jülich Research Center, Julich, Germany; Department of Computer Science 3 – Software Engineering, RWTH Aachen University, Aachen, Germany*
JOACHIM H. R. LÜBKE • *Institute of Neuroscience and Medicine INM-10, Research Centre Jülich GmbH, Julich, Germany; JARA Translational Brain Medicine, Julich, Germany; Department of Psychiatry, Psychotherapy and Psychosomatics, Medical Faculty/RWTH University Hospital Aachen, Aachen, Germany*
JENNIFER I. LUEBKE • *Department of Anatomy and Neurobiology, Boston University Chobanian & Avedisian School of Medicine, Boston, MA, USA*
ERIN E. MAHER • *University of Virginia, Charlottesville, VA, USA*
PABLO J. MARTIN-CORREA • *Department of Anatomy & Neuroscience, School of Medicine, Autónoma de Madrid University, Madrid, Spain*
MARIA MEDALLA • *Department of Anatomy and Neurobiology, Boston University Chobanian & Avedisian School of Medicine, Boston, MA, USA*
ANGEL MERCHÁN-PÉREZ • *Laboratorio Cajal de Circuitos Corticales, Centro de Tecnología Biomédica, Universidad Politécnica de Madrid, Madrid, Spain; Centro de Investigación*

Biomédica en Red sobre Enfermedades Neurodegenerative (CIBERNED), ISCIII, Madrid, Spain; Departamento de Arquitectura y Tecnología de Sistemas Informáticos, Universidad Politécnica de Madrid, Pozuelo de Alarcon, Madrid, Spain

SUSANN MICHANSKI • *Institute of Auditory Neuroscience and InnerEarLab, Center for Biostructural Imaging of Neurodegeneration, University Medical Center Goettingen, Goettingen, Germany*

FRANZ X. MITTERMAIER • *Charité-Universitätsmedizin Berlin Corporate Member of Freie Universität Berlin, Humboldt-Universität zu Berlin, Berlin Institute of Health, Berlin, Germany*

TAKAAKI MIYAZAKI • *Section of Electron Microscopy, National Institute for Physiological Sciences, Okazaki, Aichi, Japan; Support Unit for Electron Microscopy Techniques, Research Resources Division, RIKEN Center for Brain Science, Wako, Saitama, Japan*

ABIGAIL MORRISON • *Institute for Advanced Simulation (IAS-6), Jülich Research Centre, Julich, Germany; Department of Computer Science 3 – Software Engineering, RWTH Aachen University, Aachen, Germany*

JAILENNE QUIÑONES-RODRÍGUEZ • *Universidad Central del Caribe, Bayamon, PR, USA*

JOSÉ-RODRIGO RODRÍGUEZ • *Laboratorio Cajal de Circuitos Corticales, Centro de Tecnología Biomédica, Universidad Politécnica de Madrid, Madrid, Spain; Instituto Cajal, Consejo Superior de Investigaciones Científicas (CSIC), Madrid, Spain; Centro de Investigación Biomédica en Red sobre Enfermedades Neurodegeneratives (CIBERNED), ISCIII, Madrid, Spain*

JAVIER RODRÍGUEZ-MORENO • *Department of Anatomy & Neuroscience, School of Medicine, Autónoma de Madrid University, Madrid, Spain*

ASTRID ROLLENHAGEN • *Institute of Neuroscience and Medicine INM-10, Research Centre Jülich GmbH, Julich, Germany*

CHRISTIAN ROSENMUND • *Charité-Universitätsmedizin Berlin Corporate Member of Freie Universität Berlin, Humboldt-Universität zu Berlin, Berlin Institute of Health, Berlin, Germany*

NATHALIE ROUACH • *Neuroglial Interactions in Cerebral Physiopathology, Center for Interdisciplinary Research in Biology, Collège de France, CNR UMR 7241, INSERM U1050, Labex Memolife, PSL Research University, Paris, France*

KURT SÄTZLER • *School of Biomedical Sciences, University of Ulster, Coleraine, UK*

THOMAS SCHIKORSKI • *Universidad Central de Caribe, Bayamon, PR, USA*

FRANCESCA SCIACCOTTA • *University of Virginia, Charlottesville, VA, USA*

JOHANNA SENK • *Institute for Advanced Simulation (IAS-6), Jülich Research Centre, Julich, Germany; Sussex AI, School of Engineering and Informatics, University of Sussex, Brighton, UK*

YUAN SHI • *Center for Neuropathology, Ludwig-Maximilians-University of Munich, Munich, Germany; German Center for Neurodegenerative Diseases (DZNE), Munich, Germany; Munich Cluster for Systems Neurology (SyNergy), Munich, Germany*

RYUICHI SHIGEMOTO • *Institute of Science and Technology Austria (IST Austria), Klosterneuburg, Austria*

JONAS STAPMANNS • *Institute for Advanced Simulation (IAS-6), Jülich Research Centre, Julich, Germany; Department of Physics, Faculty 1, RWTH Aachen University, Aachen, Germany; Department of Physiology, University of Bern, Bern, Switzerland*

MARC TAKENO • *Allen Institute for Brain Science, Seattle, WA, USA*

TOM TETZLAFF • *Institute for Advanced Simulation (IAS-6), Jülich Research Centre, Julich, Germany*

MARTA TUREGANO-LOPEZ • *Laboratorio Cajal de Circuitos Corticales, Centro de Tecnología Biomédica, Universidad Politécnica de Madrid, Pozuelo de Alarcon, Madrid, Spain; Centro de Investigación Biomédica en Red sobre Enfermedades Neurodegenerative (CIBERNED), ISCIII, Madrid, Spain*

SACHA J. VAN ALBADA • *Institute for Advanced Simulation (IAS-6), Jülich Research Centre, Julich, Germany; Institute of Zoology, University of Cologne, Cologne, Germany*

BERND WALKENFORT • *Imaging Center Essen (IMCES), Electron Microscopy Unit (EMU), Medical Faculty of the University of Duisburg-Essen, Essen, Germany*

CAROLIN WICHMANN • *Institute of Auditory Neuroscience and InnerEarLab, Center for Biostructural Imaging of Neurodegeneration, University Medical Center Goettingen, Goettingen, Germany*

SINA ZOBEL • *Charité-Universitätsmedizin Berlin Corporate Member of Freie Universität Berlin, Humboldt-Universität zu Berlin, Berlin Institute of Health, Berlin, Germany*

LOU ZONCA • *Group of Applied Mathematics and Computational Biology, Ecole Normale Supérieure, PSL University, Paris, France; Center for Brain and Cognition, University Pompeu Fabra, Barcelona, Spain*

Part I

Preparation of Brain Tissue and Techniques

Chapter 1

Long-Distance Transport of Surgically Resected, Alive Human Brain Tissue

Franz X. Mittermaier, Henrik Alle, and Jörg R. P. Geiger

Abstract

Tissue samples obtained during neurosurgical interventions offer a unique opportunity to study the alive human brain ex vivo on the level of individual neurons and synaptic networks. However, when sophisticated research projects are anticipated, the main challenge at hand is the scarcity of eligible surgeries. To counter this, an attempt can be made to collect samples from multiple hospitals, including remote ones. This creates the necessity to transport alive brain tissue over long distances. In this book chapter, we discuss general considerations that should be taken into account when establishing acute brain slices (i.e., a specific preparation where brain tissue is sliced into thin sections) from neurosurgical tissue. We then outline various methods for tissue transport over long distances (e.g., Bielefeld to Berlin: ~400 km; door-to-door 4.5–6 h by car, 3–4.5 h by train). We describe both a procedure for transporting a non-sliced specimen and options for transporting already sliced tissue. For the transport of brain slices, we describe the design of a simple and robust transport container (BrainCase). The BrainCase can be perfused with continuously gassed sucrose-based artificial cerebrospinal fluid or can be employed without perfusion. We share our experience regarding slice quality for the different transport methods, using tissue obtained from local surgeries as the reference material. Based on our experience, we argue that if moderately cooled and protected from vibration, BrainCases help secure human brain tissue and thus slice quality also without perfusion. Transport of cooled, non-sliced specimens is a valid option, although some inferiority concerning inducibility of network oscillations might result from this method. Taken together, long-distance transport of alive human brain tissue is feasible without quality cutbacks and thus a promising strategy to counter the problem of limited availability.

Key words Neurosurgery, Human brain slice, Sterile preparation and slicing procedure, Long-distance transport, Car vs. train, Multineuron patch clamp, Slice quality evaluation

1 Introduction

1.1 Motivation to Study the Human Brain on the Level of Neurons and Networks

Typically, fundamental neuroscientific research is conducted on laboratory animals like rodents. However, human cognitive abilities exceed those of common laboratory animals in many areas. These differences in intellectual capabilities are believed to stem from distinctions in anatomy, molecular and cellular architecture, and physiology. In fact, multiple studies have revealed important

Joachim H. R. Lübke and Astrid Rollenhagen (eds.), *New Aspects in Analyzing the Synaptic Organization of the Brain*, Neuromethods, vol. 212, https://doi.org/10.1007/978-1-0716-4019-7_1,

differences between human brains and, for example, rodent brains [1–8]. Such differences indicate that it is not always possible to directly extrapolate results from animal studies to humans. It becomes evident that in order to unravel the neural correlates of human cognitive abilities and susceptibility to neurologic and psychiatric diseases, we need to study the human brain directly.

1.2 Neurosurgical Resections Offer a Unique Opportunity to Study Alive Human Neurons and Networks

While certain anatomical and molecular questions can be addressed using post-mortem tissue samples, functional studies can only be performed on alive samples taken from living brains. For human brains, the utilization of such samples for scientific purposes is ethically limited to neurosurgical procedures, during which tissue is removed for medical reasons. The resected samples typically contain the disease focus as well as access tissue. For example, patients with epilepsy that cannot be controlled with medication may undergo surgical therapy involving the removal of the epileptic focus and surrounding cortex [9]. While most of the tissue is used for neuropathological examination, it is both medically and ethically justifiable to allocate a portion of the resected tissue to scientific investigations. The research questions that can be addressed in this tissue will ultimately lead to a deeper understanding of how the human brain works, what differentiates our brains from that of other species, and how certain diseases affect the brain.

1.3 Ethical Considerations

In the context of experiments performed on the tissue from the human brains, it is important to state that the well-being of the patient is of utmost priority. It is of great importance that the planned scientific projects are discussed in an interdisciplinary team, considering ethical aspects. Furthermore, approval of the projects by an independent ethics committee is always necessary. All individuals involved in the implementation of such projects must conduct in a manner that ensures the quality of patient care is not compromised.

1.4 Limitations of Surgically Resected Human Brain Specimens and Long-Distance Transportation as a Possible Solution

There are some important limitations of studies on surgically resected human brain tissue as compared to classic studies in laboratory animals. One of the main limitations is that the human tissue inevitably comes from patients. It is likely that certain diseases, such as refractory epilepsy and long-term medication affect neuron physiology and synaptic networks. Consequently, the results must be interpreted in the context of the specific disease history. However, researchers often want to know if a certain result also applies to individuals who don't have a specific disease. To address whether a finding is likely to be disease independent, one could repeat experiments in multiple different disease entities (i.e., hippocampal sclerosis, focal cortical dysplasia, brain tumors favorably without a history of seizures, etc.). If a result is found in tissue samples irrespective of the disease entity, then this is suggestive of a more general finding that could potentially even apply to healthy individuals. However, this approach requires a large number of samples from different patients, which brings us to the next shortcoming. This is the limited availability of neurosurgical resections suitable for scientific purposes. Only a few procedures yield tissue samples that can be used for research. These include certain epilepsy surgeries as well as procedures for the removal of brain tumors of various types, which involve resection of neural tissue along the access path. Overall, however, eligible surgeries are rare. In this book chapter, we will discuss an approach to transport alive human brain tissue over large distances (>4 h) to utilize samples from different hospitals to counter the limitation of tissue availability.

2 Methods

2.1 General Considerations

When aiming to utilize neurosurgical resections for functional studies, there are several considerations that, based on our experience, should be taken into account. We will systematically address these general considerations in the following paragraphs, focusing on neocortical tissue samples and acute brain slices (Fig. 1). Acute brain slices are preparations in which the brain tissue has been sliced into 300–500-μm-thick sections to favor cellular viability through diffusion-based nutrient supply and to enable visually guided patch-clamp experiments using video-microscopy.

Considerations for surgery/resection technique Over the course of a neurosurgical operation, various instruments and devices are employed. Electrocautery is frequently used, particularly for hemostasis, and can result in localized intense heat generation. Additionally, ultrasound-based devices are often used for tissue fragmentation. Brain tissue located in the immediate vicinity of sites where these invasive techniques are used may be subject to damage. Such thermal or mechanical damage may not necessarily be visible macroscopically. Ideally, an experienced surgeon can attempt to spare the tissue designated for experimental purposes. A restrictive use of electrocautery, ultrasound-based devices, and other traumatic actions is desirable. In some cases, it may be possible to perform the resection using a sharp instrument [10] (e.g., scalpel), which is significantly less detrimental to the tissue near the resection margin compared to, for example, ultrasound-based methods. Decisions regarding the details of the resection can only be made by the treating neurosurgeon, with the greatest priority given to the well-being of the patient.

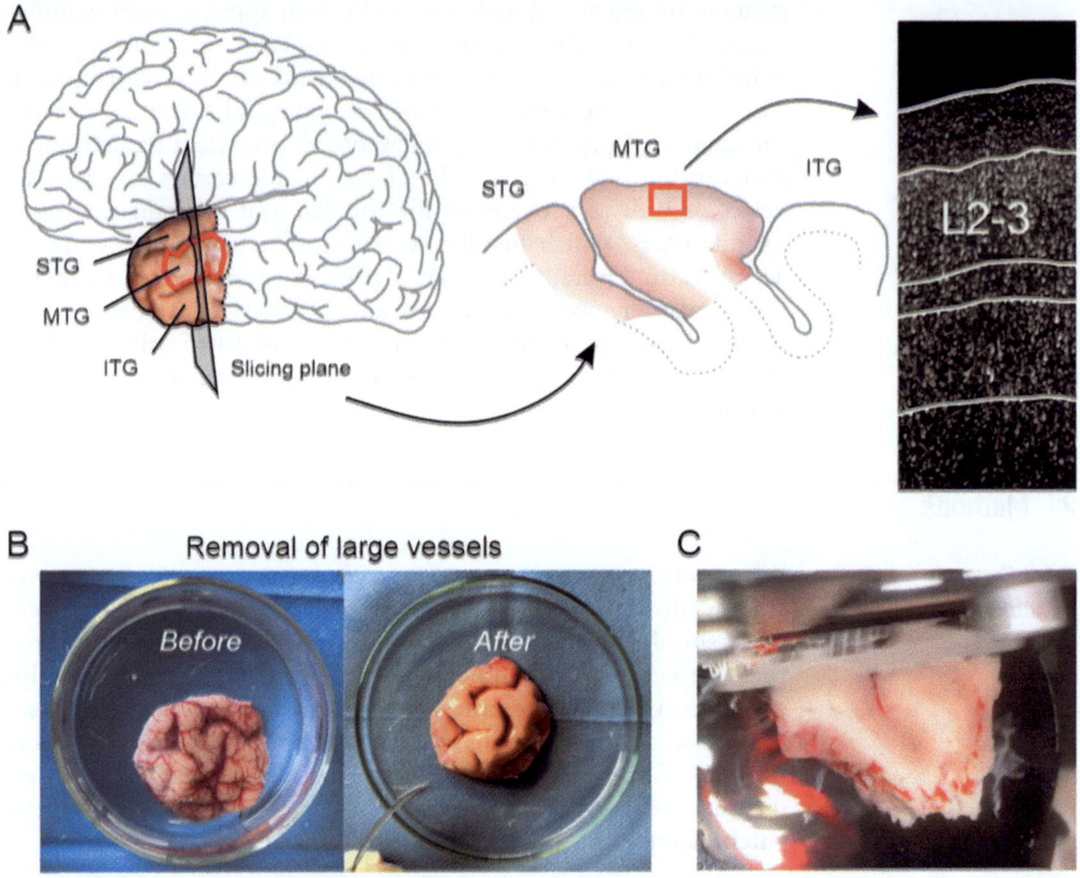

Fig. 1 Acute cortical brain slices from human tissue samples. (**a**) Left: Schematic showing a left human temporal pole (colored) that is resected during epilepsy surgery (dotted line: resection border), on the lateral side consisting of superior (STG), medial (MTG), and inferior (ITG) temporal gyrus. Middle: Neocortex of MTG and adjacent STG colored: dotted line indicates the border of the cortex to the subcortical white matter. Red box schematically reflects the area from which staining is done (right). In the stained slice, cell bodies are discernible as white dots, and borders between layers are indicated by white lines. (Adapted from Ref. 16). (**b**) An example of a resected temporal pole, before (left; see general considerations, preparation procedure) and after removal of large vessels (right); note the dense vessel distribution on the surface (left) and that the entire tissue block was slightly turned counterclockwise during the removal procedure (right). (**c**) A specimen section glued into a slicing chamber; the cortical (rosy) and subcortical (white) parts are clearly visible. At closer inspection layer 4 is usually distinguishable from other layers. In the upper part, the horizontally moving blade is cutting into the tissue generating a macro-slice

Sterile conditions We strictly maintain sterile conditions from the handover of the resected tissue in the operating room (OR) all the way to the storage container for slices. The rationale behind this practice is that the solutions, in which the acute brain slices are stored, provide ideal conditions for the growth of bacteria. Even the slightest contamination can lead to the proliferation of microbiological organisms that significantly impact neuron vitality. To adhere to these sterile conditions, all our glassware, preparation instruments, and storage containers are autoclaved (20 min at 121 °C, 200 kPa, 100% humidity). The physiological solutions are filtered through sterile filters (pore size 0.2 μm) before use. The participating scientific personnel wear sterile gloves and face masks, and those concerned with slicing the tissue additionally wear sterile gowns.

Transport from OR to laboratory After the resection, the specimen is handed over to the scientist in the OR using a sterile Petri dish. Without delay, the specimen is transferred into a wide-neck bottle. The wide-neck bottle contains ice-cold (~4 °C), carbogen-gassed (95/5% O_2/CO_2) sucrose-containing artificial cerebrospinal fluid (sACSF, in mM, 87 NaCl, 1.25 NaH_2PO_4, 2.5 KCl, 0.5 $CaCl_2$, 3 $MgCl_2$, 10 glucose, 25 $NaHCO_3$, 75 sucrose). The bottle is hermetically sealed for transport and placed in a styrofoam box surrounded by ice. In this styrofoam box, the tissue is transported to the designated laboratory facilities. In our experience, it is crucial to consistently maintain the cooling chain from the OR to the laboratory.

Preparation procedure The resected tissue is gently removed from the wide-neck bottle and transferred to a large glass Petri dish filled with ice cold, carbogen-gassed sACSF. The tissue specimen is then inspected, and photographic documentation is conducted. Subsequently a portion of the specimen, defined by diagnostic requirements, is separated and processed by neuropathology. The remaining piece is then processed into acute brain slices. To achieve this, an experienced scientific staff member initially removes larger blood vessels from the cortical surface, using sterile instruments (Fig. 1b). In our experience, this is achieved most efficiently using two fine but blunt tweezers. Great care must be taken during this process to avoid damaging the cortex. However, it is necessary to remove the blood vessels, as their fibrous components would impede the slicing of the tissue. As the next step, the specimen is sectioned into pieces that can be glued onto a flat surface for the slicing procedure. It is important to glue the specimen sections in a way which ensures a perpendicular orientation of the slicing plane to the cortical surface and to the longitudinal axis of gyri or sulci (Fig. 1a). This leads to a cutting angle that preserves apical dendrites and the axon of cortical neurons.

Slicing procedure A vibratome (VT1200, Leica, Germany) is utilized to cut the specimen sections into acute brain slices, which we call "macro-slices" at this stage, to distinguish them from the final slices resulting from subdividing these macro-slices (Fig. 1c). During slicing the tissue is submerged in ice-cold sACSF, and the solution is continuously gassed with carbogen [11]. We use a slice thickness of 300 μm for patch-clamp experiments and 450–500 μm for field recordings. Slices are transferred into sterile storage containers filled with sACSF which is heated to 34 °C for 30 min to support resealing of the cut membranes (recovery period). The storage containers are made in-house and consist of a plastic box, a sterile net (medical gauze) that is held by a pair of concentrically clampable plastic rings, and a micro-filter candle (VitraPOR™, ROBU, Germany) to allow for gassing of the solution during storage. All components of the storage container can be autoclaved to ensure sterile conditions (however, the sterile net is a one-time use component). After recovery, the slices are kept at room temperature. Experiments are started 2–4 h after the end of recovery, as we have occasionally observed that the membrane potential of neurons is slightly depolarized early after recovery. Typically, slice viability is of sufficient quality up to 30–48 h after recovery.

Methods for slice quality evaluation Acute brain slices are evaluated with multiple methods. The visual aspect of the slices is explored with our standard optical method for performing patch-clamp recordings (differential interference contrast video-microscopy, DIC; Nikon Eclipse FN1 microscope; Nikon water immersion objective ×40; Hamamatsu ORCA-Flash 4.0 camera). Cellular and synaptic function is assessed with high-resolution multineuron patch-clamp recordings [12] (Fig. 3). Network function is evaluated through local field potential recordings with up to 16 electrodes on 8 setups [13], recording kainate-induced gamma-oscillations (kainate concentrations 400–600 nM).

2.2 Reference Procedure: Local Neurosurgery (<30 min Transport Time)

Adequacy of any human brain transportation method must be evaluated by comparing tissue quality to that of the tissue obtained from the closest site available—in our case from the Department of Neurosurgery, Charité Campus Mitte, right on the same campus where our institute is located. The medial temporal gyrus of either hemisphere, obtained from patients with medication-resistant temporal lobe epilepsy, served as our standard or reference tissue for method comparisons. In addition to the general considerations listed above, some particularities of this reference procedure are important to note:

1. *Transport from OR to laboratory* (general considerations see above): The time needed for the transfer from the OR to our Berlin laboratory is between 15 and 20 min. The tissue is transported from the OR to the Berlin laboratory by foot.
2. *Preparation procedure* (general considerations see above): The preparation takes place in the rooms of the Berlin laboratory.

3. *Slicing procedure* (general considerations see above): Due to the short distance from the OR to the Berlin laboratory, the entire procedure, including slicing and recovery period, is typically completed within 2.5 h of the timepoint of handover to the scientist in the OR.
4. *Slice quality evaluation* (general considerations see above): This tissue is considered as the current gold standard that the other transport conditions are compared to.

2.3 Remote Neurosurgery (Transport Time > 3 h)

During the initial projects with the tissue from our reference neurosurgery (Charité Neurosurgery Department in Berlin), we made the observation that human cortical neurons in the medial temporal gyrus are generally more resilient than what we were used to from rodent ones. Even 24 h after slicing, we typically observed a healthy visual appearance in DIC microscopy and stable resting membrane potentials when recording from human neurons. Conversely, these parameters decay only hours after the preparation in rodent slices. The general robustness of the human tissue prompted us to test whether long-distance transport of such samples might be feasible. For this purpose, we established collaborations with Bielefeld (Department of Neurosurgery, Evangelisches Klinikum Bethel, Bielefeld University; roughly 400 km from our Berlin laboratory) and Hamburg-Eppendorf (Department of Neurosurgery, Universitätsklinikum Hamburg Eppendorf (UKE); roughly 300 km from our Berlin laboratory). Different options for long-distance transportation are to transport a non-sliced specimen or to transport brain slices. Transporting slices can then be further divided into methods with perfused slice containers (using carbogenated, sucrose-based solution; by car) or without perfusion (by car or train). Continuous carbogenation of perfusion solutions over long distances necessitates a pressurized gas bottle, which was the reason why we could only use a modified car for this transportation option, but not the train.

2.3.1 Non-sliced Specimen Transport

Transporting a non-sliced specimen from a remote neurosurgery is basically the extension of the method of local tissue pick-up but over longer distances and thus time. The advantages of this method are its simplicity and its convenience; however, it might lead to shortage of oxygen and nutrient supply within the tissue over time, especially in deeper layers of the neocortex. We found that, when using this transport version, it is crucial to keep the temperature of the transport solution at sufficiently low values (5–10 °C) for the entire time:

1. *Transport from OR to Laboratory* (general considerations see above): After the handover in the remote OR (Klinikum Bethel, Bielefeld; UKE, Hamburg), the specimen is transported to a laboratory (15–20 min transport time) within the

same respective city ("Waldlabor" in Bielefeld; ZMNH in Hamburg) to remove the pieces designated for diagnostic evaluation by neuropathology. Afterward the piece dedicated for experiments is transferred back into a glass bottle containing precooled and gassed sACSF. The bottle is hermetically sealed and placed on ice within a styrofoam box. The transport to the Berlin laboratory is conducted by train or car.

2. *Preparation procedure* (general considerations see above): After the long-distance transport of the non-sliced specimen, the preparation then takes place in the rooms of the Berlin laboratory.
3. *Slicing procedure* (general considerations see above): The slicing in the Berlin laboratory is performed as described under the section "General Considerations." The procedure, including slicing and recovery, is typically completed within 5–6 h of the handover in the OR.
4. *Slice quality evaluation* (general considerations see above): Assessing the quality of the slices with several methods showed no obvious differences with respect to visual appearance of the slices, single neuron, and synaptic properties (Fig. 3). However, some inferiority regarding inducibility of extracellularly recorded network activity and oscillations compared to the tissue obtained from the Department of Neurosurgery at Charité Campus Mitte was observed.

2.3.2 Slices in BrainCase with Perfusion: Car

To prevent potential tissue hypo-oxygenation, lack of glucose, and shifts in pH of extracellular and cellular compartments during the transport of human brain slices over long distances, we developed a method that allows for continuous perfusion of slices with a carbogenated, sucrose-based solution. This is considered our maximum variant with respect to tissue preservation measures (continuous perfusion and carbogenation). The container for the slices is composed of several different parts that will be explained below and is shown in a technical drawing (Fig. 2a–c). This container, called "BrainCase", contains a bottom part with an inlet for the gassed solution and a top part with an outlet. The slices are placed between two layers of sterile gauze that are mounted on two double-clamp ring pairs (coverframes), respectively. A spacer sheet corresponding to the thickness of slices (300 μm) is placed in between the two opposing gauzes to hold the slices in place (Fig. 2d–e). The entire system is constructed such that a stepwise and sterile assembly is possible. Sterile loading and unloading of slices into and out of the BrainCase is straightforward and requires only conventional tools and parts of daily lab routine; however, some practice is advisable before first use which can be acquired within half an hour. The material used to build the BrainCase is PEEK (polyether ether ketone); all other materials we tested are not stable in form when they are sterilized in an autoclave.

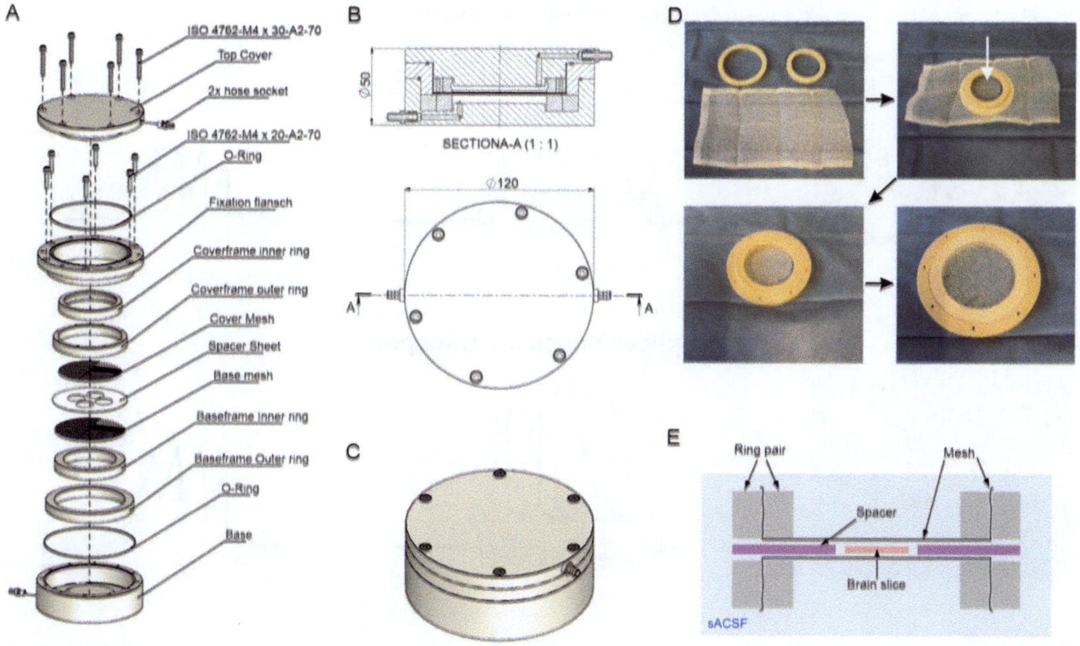

Fig. 2 BrainCase transport system for acute brain slices. (**a**) Explosion assembly drawing showing all the single parts of a BrainCase. (**b**) Upper part: Cross-sectional, lateral view drawing including the outlet for sACSF of the top cover and the inlet of the base. Total height of assembled BrainCase is 50 mm. Lower part: top view, total diameter is 120 mm excluding the hose sockets. (**c**) Isometric view of the assembled BrainCase. Note the compact design that allows for stacking of multiple BrainCases. (**d**) Assembly procedure of a clamp ring pair and sterile medical gauze (mesh). Picture on the lower right shows the completed baseframe including tensioned mesh. Small holes in the outer ring facilitate insertion and removal of frame into and out of the BrainCase; similar for coverframe. (**e**) Principle of placement of spacer sheet and macro-slices in between meshes of baseframe and coverframe, see (a) for spacer sheet with four cutouts, each for one macro-slice, respectively

Each BrainCase is perfused employing a miniaturized peristaltic pump (RP-QX1.5S-1Z, 3 V; Takasago Fluidic Systems, USA) working at a rate of perfusion of 0.3 to 0.7 mL/min. The volume inside a BrainCase amounts to roughly 28 mL; thus about every 40 to 90 min, the entire solution will be exchanged. The flow of sucrose-based, carbogenated solution in mL/min is measured after the outlet of the BrainCase using light beams, and the pump rates of the BrainCases can thus be adjusted individually if necessary. Flow of carbogen into the large glass bottle containing the perfusion solution is monitored using a gas flow sensor (FS 2012—1020NG, 3 V; Renesas, Japan). The entire electronic system is custom-made and based on an Arduino-Mega (Arduino S. r. l., Italy).

Special care must be taken regarding the installation and locking of the highly pressurized gas bottles (carbogen; initial pressure of new bottles is about 20 MPa) in the fond of the modified transporter car, which is provided by our facility management.

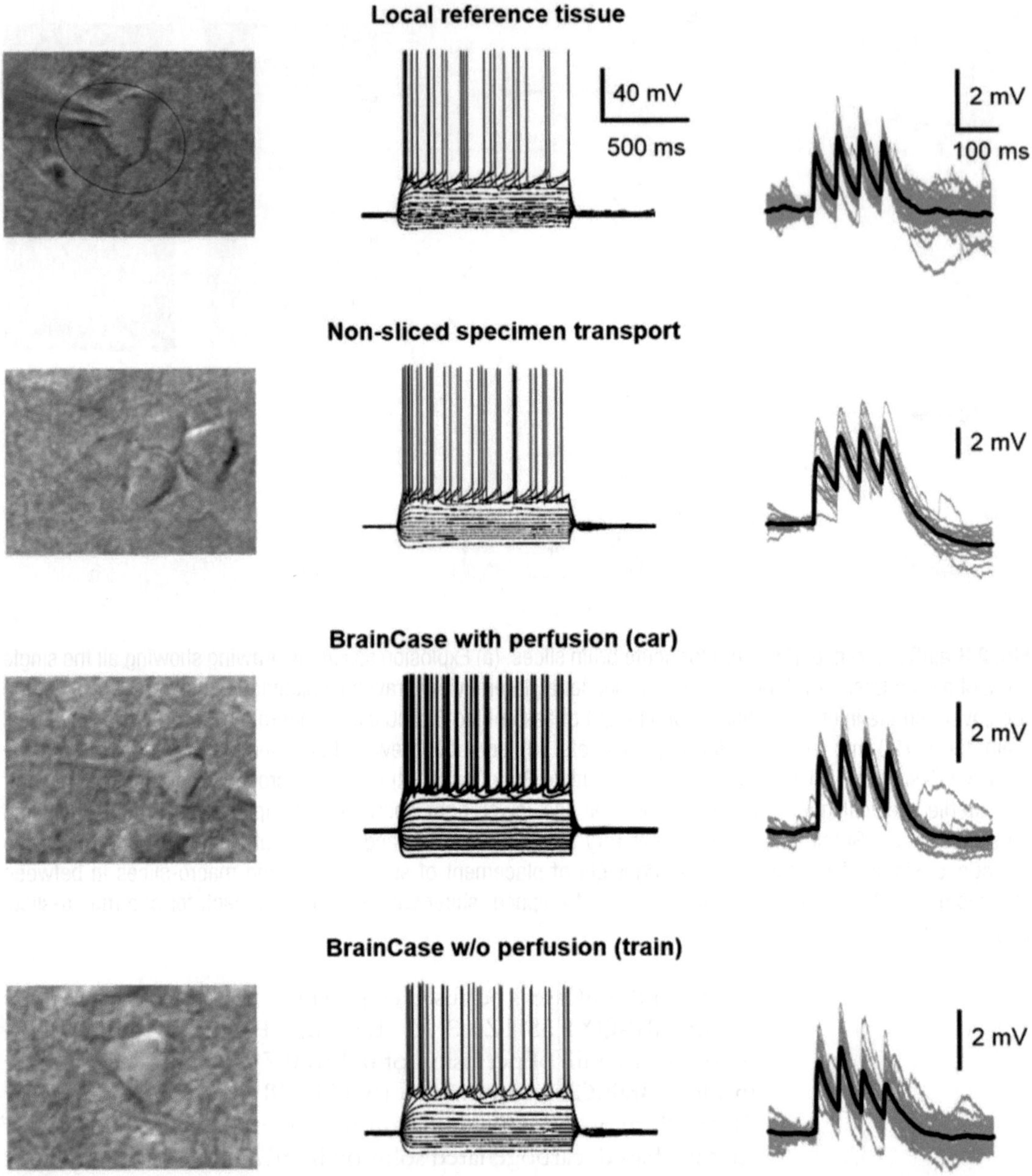

Fig. 3 Qualitative examples of slice appearance in DIC-video-microscopy and cellular and synaptic patch-clamp recordings of neurons from different transport versions. Rows from top to bottom: Examples of slice surface, membrane potential changes in response to current injections in single neurons (L2–L3 pyramidal neurons of MTG), and unitary excitatory postsynaptic potentials (single traces in gray, average trace in black) from paired recordings, from reference procedure (top row), from transport of a non-sliced specimen (second row), and from transport of macro-slices in BrainCases either with perfusion of gassed sACSF (third row; car) or without perfusion (bottom row; train)

Instructions regarding the gas manometer had to be provided by an external specialist to all participants in our car transport trips. In addition, we built devices to securely put and place all the necessary components of the transportation system into the car (container box for solution bottle(s), up to four BrainCases in parallel, waste bottles, electronics), minimizing the transmission of unavoidable vibrations from the car onto the different parts. For more detailed descriptions of the entire transport system, please contact the authors:

1. *Transport from OR to laboratory* (general considerations see above): After the handover in the remote OR (Klinikum Bethel, Bielefeld; UKE Hamburg), the specimen is transported to a laboratory (15–20 min transport time) within the same city ("Waldlabor" in Bielefeld; ZMNH in Hamburg) to remove the piece designated for diagnostic evaluation by neuropathology.
2. *Preparation procedure* (general considerations see above): The preparation takes place in the remote laboratories ("Waldlabor" Bielefeld, ZMNH Hamburg).
3. *Slicing procedure* (general considerations see above): The slicing is performed in the remote laboratories ("Waldlabor" Bielefeld, ZMNH Hamburg).
4. *Transfer into/out of the BrainCases and car transport*: After slicing and recovery, the brain macro slices are transferred into the pre-assembled BrainCases by a scientific staff member. The BrainCases are then fully assembled and tightly closed, and perfusion with gassed sACSF is started. The BrainCases are put into the container box, which is then loaded into the car. After arrival at the Berlin laboratory, the BrainCases are disassembled, and the slices are transferred into our normal storage containers (see slicing procedure under general considerations). Macro-slices are then kept at room temperature until subdivided, and experiments are started. Typically, we wait for 2–3 h upon arrival at the laboratory in Berlin before starting the experiments.
5. *Slice quality evaluation* (general considerations see above): With regard to quality evaluation of slices transported, this way we find the slices being of the same quality as compared to local tissue-derived slices when visualized in DIC and analyzed using multi-patch-clamp recording technique (Fig. 3), and we observed an almost complete rescue of network oscillation inducibility, which suggested that this method is presumably superior to the variant of non-sliced specimen transportation.

Taken together, it is a comparatively large effort to provide both oxygenation and perfusion to macro-slices during the transfer from our remote neurosurgical partner sites to our Berlin laboratory by car. Furthermore, the collecting trip itself is quite straining and exhausting for the scientific staff members, especially if they are also those performing experiments upon arrival. We did not follow previously described methods [14–15] as those would not have been easily scalable to larger numbers of slices, which is very well possible with the closed BrainCase container box solution: they simply can be stacked on top of each other (Fig. 2c). Other drawbacks of this variant (car transport over several hundred kilometers) are time (as using a car is usually slower and more prone to unforeseen complications like traffic jams along the route than taking a train) and finally uncontrollable vibrations and shocks resulting from the interaction of the car and the road surface.

2.3.3 Slices in BrainCase Without Perfusion

Since the BrainCase car transport version with continuous oxygenation and perfusion requires large amounts of resources in materials and personnel, we decided to test a further transport version to reduce the complexity. For this version we use BrainCases to increase practicability and robustness; however we transport them by train and without continuous oxygenation and perfusion. This variant saves time, is less exhausting, and very likely confers less mechanical stress (vibrations) to the slices:

1. *Transport from OR to laboratory* (general considerations see above): After the handover in the remote OR (Klinikum Bethel, Bielefeld; UKE Hamburg), the specimen is transported to a laboratory (15–20 min transport time) within the same city ("Waldlabor" in Bielefeld, ZMNH Hamburg) to remove the piece designated for diagnostic evaluation by neuropathology.
2. *Preparation procedure* (general considerations see above): The preparation takes place in the remote laboratories ("Waldlabor" Bielefeld, ZMNH Hamburg).
3. *Slicing procedure* (general considerations see above): The slicing is performed in the remote laboratories ("Waldlabor" in Bielefeld, ZMNH Hamburg).
4. *Transfer into/out of the BrainCases and car transport*: After slicing and recovery, the brain macro-slices are transferred into the BrainCases by a scientific staff member. Just before leaving the remote lab at our neurosurgical partners' locations, we replace the sACSF inside the BrainCases with fresh, cold, and carbogenated sACSF and then seal them. Subsequently, we stack the BrainCases into a cool box. Using cool packs inside the box, we keep the temperature below 15 °C, also in the warm months of the year. While we did not measure the temperature within the BrainCases directly, we estimate it to

also be in the range of 10–15 °C. After arrival at the laboratory in Berlin, the BrainCases are disassembled, and the macro-slices are transferred into our normal storage containers (see slicing procedure in general considerations). Macro-slices are then kept at room temperature until subdivided, and experiments are started. Typically, we wait for 2–3 h upon arrival at the laboratory in Berlin before starting the experiments.

5. *Slice quality evaluation* (general considerations see above): With regard to quality evaluation of slices transported, this way we find the slices to be of the same quality as compared to local tissue-derived slices when visualized in DIC and analyzed using multi-patch-clamp recording technique (Fig. 3). Currently we do not have sufficient data to draw any safe conclusions about network activity with this transport method.

3 Discussion and Conclusions

To unravel the neural correlates of human cognitive abilities and susceptibility to neurological and psychiatric diseases, it is essential to attain a fundamental understanding of how the brain functions. Neurosurgical resections for medical reasons represent a unique opportunity to investigate a functionally intact human brain tissue ex vivo. However, such surgeries are rare. A way to counter the problem of limited availability is to establish collaborations with multiple hospitals, thereby increasing the pool of eligible procedures. However, this typically entails the transportation of the tissue over extended distances from remote locations. In this book chapter, we have outlined three approaches for realizing long-distance transport of alive human tissue. These are as follows: (1) The tissue is transported as a non-sliced specimen from the remote location to our laboratory in Berlin, where it is subsequently processed into brain slices. (2) The tissue is processed into brain slices at the remote location and then transported in our BrainCase, which is perfused with a gassed solution (by car). (3) The tissue is processed into brain slices at the remote location and transported in the BrainCase, but without perfusion. When establishing such transportation methods, it is crucial that the quality of the tissue transported from remote sites is not inferior compared to the local reference tissue. We used multiple methods to address whether the transported tissue represents an equally good alternative. For instance, we used our multineuron patch-clamp method, which consists of up to ten simultaneous whole-cell recordings, to study tissue samples from different transport versions. This method enables the experimenter to record passive and active electrophysiological properties of neurons, as well as unitary synaptic connections. While we do not yet have a rigorous quantitative proof, we can say with a fair amount of confidence that the brain slices obtained from remote neurosurgeries are not inferior to those from local tissue samples. Regarding the different transport versions, we do not find obvious differences in visual appearance of slice surface integrity, as well as cellular and synaptic physiology (Fig. 3). We find that patient-specific differences (such as age, disease and reason for surgery, differences in surgical procedure) have a substantially larger impact on tissue quality than the transport method itself. This inter-patient variability is also the reason why large amounts of samples from different patients would have to be compared in a very systematic way to quantitatively detect subtle differences between transport methods. Furthermore, we performed local field potential recordings to explore whether there are any differences in inducibility and expression of large network activity between the different transport versions. We attempted to induce gamma-oscillations; while we have not obtained enough data to come to a definitive conclusion, we found a trend of a lower responder rate (percentage of slices incubated with kainate that exhibited oscillations of all incubated slices) if the tissue was transported as a non-sliced specimen from the remote site compared to the BrainCase transport version and the local reference tissue. Frequency range and amplitude of oscillations, however, were quite similar. Further and more rigorous experiments are needed to come to a more conclusive result. While the BrainCase transport version (continuous carbogenation of sACSF and BrainCase perfusion) should be considered the maximum variant over long distances, we argue that many experiments can also be performed with the considerably less demanding transport either of the non-sliced specimen or the sliced specimen in BrainCases without continuous gassing and perfusion. In conclusion, we successfully developed different strategies that allowed for long-distance transport of alive human cortical tissue samples from two remote sites, namely, Bielefeld (~4 h by car, 3.5 h by train) and Hamburg (~3 h by train). These transport strategies should encourage researchers to utilize the human brain tissue from remote sites and ultimately boost their research.

Acknowledgments

We thank our collaborators from the Neurosurgery Departments in Bielefeld (T. Kalbhenn and M. Simon; Evangelisches Klinikum Bethel, Bielefeld University, Medical School) and Hamburg (T. Sauvigny, Universitätsklinikum Hamburg-Eppendorf; O. Ohana, ZMNH); as well as colleagues of Charité—Universitätsmedizin Berlin: Department of Neurosurgery (J. Onken and R. Xu), NWFZ (P. Fidzinski, V. Metodieva), and Institute of Neurophysiology (M. Hadler and A. Podesta; E. Tütüncü for sharing preliminary results of oscillatory network behavior; A. Wilke for excellent technical assistance); we also thank M. Schebsdat from our mechanical workshop, and R. Mittermaier for technical drawings of the BrainCase.

This work was supported by the NeuroCure Cluster of Excellence, Berlin, Germany.

References

1. DeFelipe J, Alonso-Nanclares L, Arellano JI (2002) Microstructure of the neocortex: comparative aspects. J Neurocytol 31(3-5): 299–316
2. Molnár G, Oláh S, Komlósi G, Füle M, Szabadics J, Varga C et al (2008) Complex events initiated by individual spikes in the human cerebral cortex. PLoS Biol 6(9):e222
3. Boldog E, Bakken TE, Hodge RD, Novotny M, Aevermann BD, Baka J et al (2018) Transcriptomic and morphophysiological evidence for a specialized human cortical GABAergic cell type. Nat Neurosci 21(9): 1185–1195
4. Hodge RD, Bakken TE, Miller JA, Smith KA, Barkan ER, Graybuck LT et al (2019) Conserved cell types with divergent features in human versus mouse cortex. Nature 573(7772):61–68
5. Berg J, Sorensen SA, Ting JT, Miller JA, Chartrand T, Buchin A et al (2021) Human neocortical expansion involves glutamatergic neuron diversification. Nature 598(7879): 151–158
6. Testa-Silva G, Verhoog MB, Linaro D, de Kock CP, Baayen JC, Meredith RM et al (2014) High bandwith synaptic communication and frequency tracking in human neocortex. PLoS Biol 12(11):e1002007
7. Hunt S, Leibner Y, Mertens EJ, Barros-Zulaica N, Kanari L, Heistek TS et al (2023) Strong and reliable synaptic communication between pyramidal neurons in adult human cerebral cortex. Cereb Cortex 33(6): 2857–2878
8. Gidon A, Zolnik TA, Fidzinski P, Bolduan F, Papoutsi A, Poirazi P et al (2020) Dendritic action potentials and computation in human layer 2/3 cortical neurons. Science 367(6473):83–87
9. Wiebe S, Blume WT, Girvin JP, Eliasziw M (2001) Effectiveness and Efficiency of Surgery for Temporal Lobe Epilepsy Study Group. A randomized, controlled trial of surgery for temporal-lobe epilepsy. N Engl J Med 345(5): 311–318
10. Straehle J, Ravi VM, Heiland DH, Galanis C, Lenz M, Zhang J et al (2023) Technical report: surgical preparation of human brain tissue for clinical and basic research. Acta Neurochir 165(6):1461–1471
11. Geiger JRP, Bischofberger J, Vida I, Fröbe U, Pfitzinger S, Weber HJ et al (2002) Patch-clamp recording in brain slices with improved slicer technology. Pflügers Arch 443(3): 491–501
12. Peng Y, Mittermaier FX, Planert H, Schneider UC, Alle H, Geiger JRP (2019) elife 19: 8e48178
13. Meier MA, Lemercier CE, Kulisch C, Kiss B, Lendvai B, Adham N et al (2020) The novel antipsychotic cariprazine stabilizes gamma oscillations in rat hippocampal slices. Br J Pharmacol 177(7):1622–1634
14. Knowles WD (1985 Aug) A portable in vitro brain slice chamber. J Neurosci Methods 14(3):187–192
15. Köhling R, Lücke A, Straub H, Speckmann EJ (1996) A portable chamber for long-distance transport of surviving human brain slice preparations. J Neurosci Methods 67(2):233–236
16. Planert H, Mittermaier FX, Grosser S, Fidzinski P, Schneider UC, Radbruch H, et al. Cellular and synaptic diversity of layer 2-3 pyramidal neurons in human individuals. bioRxiv 2021.11.08.467668; https://doi.org/10.1101/2021.11.08.467668

Chapter 2

Modern Glyoxal Fixation for the Determination of the Copy Number of Synaptic Vesicle Proteins at Individual Synapses

Jailenne Quiñones-Rodríguez and Thomas Schikorski

Abstract

This protocol introduces the use of acid-free glyoxal as a fixative for immunogold localization of proteins. Acid-free glyoxal is a glyoxal solution that was purified from glyoxylic acid. The composition of the presented fixative was created to optimize acid-free glyoxal fixation for increased sensitivity of antibody labeling while preserving the fine structure at the highest possible level. The protocol describes the use of acid-free glyoxal for indirect immunofluorescence and pre-embedding immunogold localization of neuronal antigens.

Key words Acid-free glyoxal fixation, Electron microscopy, Pre-embedding immunocytochemistry, Quantitative immunogold protein localization, Quadruple immunofluorescence

1 Introduction

Immunogold localization of various proteins on the ultrastructural level has been a pillar of our understanding of molecular and structural processes in all areas of cell biology. In particular, protein localization plays a decisive role when testing hypothesis about the role of a protein in a certain molecular pathway. The validity of any hypothesis stands and falls with the necessity that a protein must be present at the location of the molecular process.

Twenty years ago, the interest in the determination of how many protein copies are actually involved in a molecular process or are present at a given cell organelle arose [1]. Since then immunocytochemistry has become an important tool to measure the copy number of proteins at synaptic vesicles [2], synaptic terminals [3–5], and postsynaptic structures [6]. Such quantitative measurements demand new pre-requisites on immunocytochemical techniques in terms of sensitivity of antigen detection and structural preservation. In other words, fixation should not interfere with the specific binding of antibodies and at the same time preserve

Joachim H. R. Lübke and Astrid Rollenhagen (eds.), *New Aspects in Analyzing the Synaptic Organization of the Brain*, Neuromethods, vol. 212, https://doi.org/10.1007/978-1-0716-4019-7_2,

the structure so that the antigen can be precisely located within the context of subcellular structures. For immunogold localization of antibodies at the fine-scale electron microscopic (EM) level, this is particularly challenging, because in principle rather harsh fixatives are needed for the preservation of the fine structure.

Glyoxal was first proposed as a fixative for biological specimens by Wicks and Suntzeff [7]. Later, in a study centered on EM, Sabatini et al. [8] tested a variety of aldehydes but found that glyoxal was not suitable for EM. However, in recent years, a renewed interest in glyoxal fixation was driven by its better health profile compared with the known toxicity of formalin. Today, a variety of fixatives based on glyoxal have been developed [9–11], and the application, in particular, in general histology has been successful and thus promising. Also, the superiority of glyoxal for retaining immunoreactivity and decreasing the need for antigen retrieval was demonstrated by [12]. The next breakthrough was the introduction of acid-free glyoxal as a fixative [13]. Commercial glyoxal stored at room temperature (RT) slowly oxidizes to glyoxylic acid which then accumulates in the solution over time. Glyoxylic acid has detrimental effects on the fine structure; thus removing glyoxylic acid improves the quality of fixation. This acid-free glyoxal (AFG) also improves staining of histological specimens, immunochemical labeling, and in situ hybridization.

Here, we describe a technique that applies AFG to immunofluorescence quadrolabeling and to immunogold localization of proteins at the subcellular level. Perhaps surprisingly, both procedures are identical up to the point when processing for EM starts. The technique leads to more intense labeling with less background staining when compared with paraformaldehyde (PFA)-based fixatives and achieves outstanding fine structural preservation.

2 Materials

2.1 Cell Culture

For our experiments, primary hippocampal cell cultures from newborn rat pups (Sprague–Dawley strain) were used. For the preparation of neuronal cultures, a step-by-step protocol was carried out as described previously [14]. However, the original medium was slightly modified. Minimum Essential Medium (MEM, Sigma-Aldrich, St. Louis, MO, USA) was used instead of Neurobasal-A Medium. Due to this switch, sodium pyruvate, serum, and GlutaMax (all available from Sigma-Aldrich, St. Louis, MO, USA) were added. The final medium was as follows:

Fetal bovine serum	5 mL
Glucose (1 M stock)	2 mL
Pen/Strep	1 mL

(continued)

GlutaMax (1:100, 200 mM stock)	0.25 mL
Fungizone (optional)	100 μL
Sodium pyruvate (100 mM; 11 g/l; 100×)	1 mL
B27 Supplement (50×)	2 mL

Minimum Essential Medium (MEM) was added to make 100 mL final volume

This medium simplified our cell culture preparation procedure because the addition of serum allowed us to skip pre-culturing astrocytes (astrocytes do not grow in Neurobasal serum-free medium). Increasing the number of cells plated was sufficient to grow neurons and astrocytes simultaneously. Adding FUDR or Ara-C at 14 days in vitro (DIV) prevented the overgrowth of astrocytes.

2.2 Preparation of Acid-Free Glyoxal

Glyoxal was purchased as a 40% aqueous stock solution (Sigma-Aldrich, St. Louis, MO, USA) and stored at room temperature (RT). For the removal of acids that form in commercial glyoxal solution over time, a basic ion-exchange resin was applied (AmberLite® IRA-67; Alfa Aesar, Ward Hill, Massachusetts, USA) following the protocol by Bussolati et al. [13]. To prepare the solution, a typical 250 mL filter unit, commonly used to sterilize media, was modified. The 0.22 um membrane was replaced by a gaze with fine nylon mesh width (smaller than the size of the resin beads) and glued with cyanoacrylic glue which is stable at high concentration of sodium hydroxide (NaOH). The flow throw of the filter was blocked with a common rubber stopper. This simple construct can be re-used many times. About 20 g of AmberLite IRA 67 was added to the filter and was mixed with distilled water to fully moisten the resin. After 1 min, the distilled water was replaced with 50 mL of 2 N NaOH to activate the resin for 5 min. The NaOH was drained and captured (e.g., for re-use), and many washes with large amounts of distilled water were applied to remove excessive NaOH until a pH between 7.4 and 8.0 was reached. The flow through was stopped, and 50 mL of commercial 40% glyoxal, which has a pH of around 3, was added and left for 30 min. The basic resin beads quickly captured the acids, but depending of the volume between the membrane and the flow out of the filter system, some swirling is recommended to free the entire solution of acids. Once the reaction is completed, the now acid-free glyoxal solution, with a pH of 7.0 is captured. The final stock solution was prepared by adding 20% ethanol to the 40% glyoxal resulting in 20% acid-free glyoxal and 10% ethanol. This 10× stock solution is stable at room temperature for about 3 months. However, it is more feasible to store the stock solution in aliquots at −80 °C. These aliquots are stable for more than 1 year without losing quality.

Our final fixative mixture was 2% acid-free glyoxal, 0.0125% glutaraldehyde, 1% ethanol, and 100 mM sucrose in McIlvaine buffer at pH 6.0.

McIlvaine Buffer

McIlvaine buffer is a biological buffer system that has a stable buffer capacity below pH 7.4 and thus is well suited for this protocol [15]. The McIlvaine buffer has two components: dibasic phosphate (200 mM) (Sigma-Aldrich, St. Louis, MO, USA) and citric acid (100 mM) (Sigma-Aldrich, St. Louis, MO, USA). To make a 100 mL final volume, 63 mL dibasic phosphate and 35 mL of citric acid were combined and then adjusted to pH 6.0 with additional citric acid.

2.3 Immunoreagents and Detergents

Permeabilization is a critical and often a deleterious step for the ultrastructure. In our immunocytochemical protocol, the detergent Triton X was used in all our experiments. Triton X has strong permeabilization properties and thus enhances immunolabeling. However, it is important to use purified Triton X that is peroxide- and carbonyl-free. The peroxides, and not Triton X itself, cause severe damage to the fine structure which sometimes can even be seen in light microscopy.

In our protocol, both Triton X-100 (10% proteomic grade, Roche, Indianapolis, IN, USA) and Triton X-114 (10% proteomic grade, Cepham Life Sciences, Fulton, MD, USA) were used at a concentration of 0.03–0.1%.

We have successfully used this protocol with a variety of antibodies against common cell and neuron-specific proteins. First antibodies were purchased from various manufactures like Developmental Studies Hybridoma Bank (Iowa City, IA, USA) and Synaptic Systems (SYSY, Göttingen, Germany). For fluorescence microscopy, secondary antibodies conjugated with Alexa 488, Alexa 647, or Cy-3 directed against rabbit or mouse antibodies were purchased from Jackson ImmunoResearch (Cambridgeshire, UK). For EM, we relied on anti-mouse and anti-rabbit fluoronanogold antibodies conjugated with Alexa 488 and Alexa 647 by Nanoprobes (Yaphank, NY, USA).

For the visualization of the 1.4 nm nanogold particles, particle size was increased with the gold enhancement reagent sold by Nanoprobes (Yaphank, NY, USA).

2.4 Fluorescence Microscopy

For imaging, an upright Nikon FN1 epifluorescence microscope, equipped with a water immersion lens with 25× magnification and high numerical aperture 1.1, was used.

As a light source served a spectral light (Lumencor) and fluorescent channels were selected with Nikon filter sets for blue (395 nm), green (488), red (555 nm), and dark red (647 nm), respectively. A sCMOS camera (Andor Zyla 5.5 megapixel, 11 bit monochrome) was used for image acquisition. The camera and the Lumencor were controlled by Nikon Instrumentation Software (NIS, Melville, NY, USA). Each monochrome image was automatically converted to a pseudo-color corresponding to the dye and filter set and composed to multicolor images. For publication, selected images were further processed using Adobe Photoshop™ (Adobe, Hamburg, Germany).

2.5 EM Processing and Imaging

Standard procedures as published previously [16] were carried out. In short, post-fixation was done with 1% osmium tetroxide (EMS, Hatfield, PA, USA) and 1% potassium ferrocyanide (SPI) in 150 mM cacodylate buffer (pH 7.2) for 30 min followed by en bloc contrasting with aqueous 2% uranyl acetate (EMS, Hatfield, PA, USA) for 1 h, followed by dehydration in an ascending ethanol, and final embedding in "EMbed 812"/Spurr's mixture (1 + 1) (both EMS, Hatfield, PA, USA). Specimens were flat embedded between Aclar® sheets (SPI supplies, West Chester, PA, USA) and polymerized at 60 °C overnight.

Selected areas were cut at 60 nm thickness with a Leica Ultracut UC6. Afterward ultrathin sections were transferred to Pioloform-coated slot copper grids (Ted Pella INC., Redding, CA, USA).

EM imaging, for the purpose of a first look and selection of specimens, was done with a JEOL 100CX. Selected specimens were then imaged for analysis and publication at high resolution with a Zeiss Libra 120 TEM (Fa. Zeiss, Oberkochen, Germany) equipped with a bottom-mounted Proscan 2 K digital camera using the Image SP software (Fa. Tröndle, Moorenweis, Germany) at a magnification of 5000× to 20,000× with panorama frames of up to nine images.

3 Methods

3.1 Fixation

After 17–20 DIV, cells were fixed via transferring coverslips into a new 24-well culture plate containing the fixative at RT. As soon as the coverslip was immersed, we agitated the whole culture plate to ensure quick and even mixing of the remaining culture medium with the fixative. This step is important for fast and even fixation. Once all coverslips for a particular experiment were transferred, the culture plate was placed on a shaker (Fig. 1b) for immersion fixation at RT for 1 h and followed by additional fixation in the same fixative at 4 °C for at least 18 h. Washing with 25 mM Tris-buffered saline (TBS, three times) ended the fixation, and cells were ready for immunolabeling.

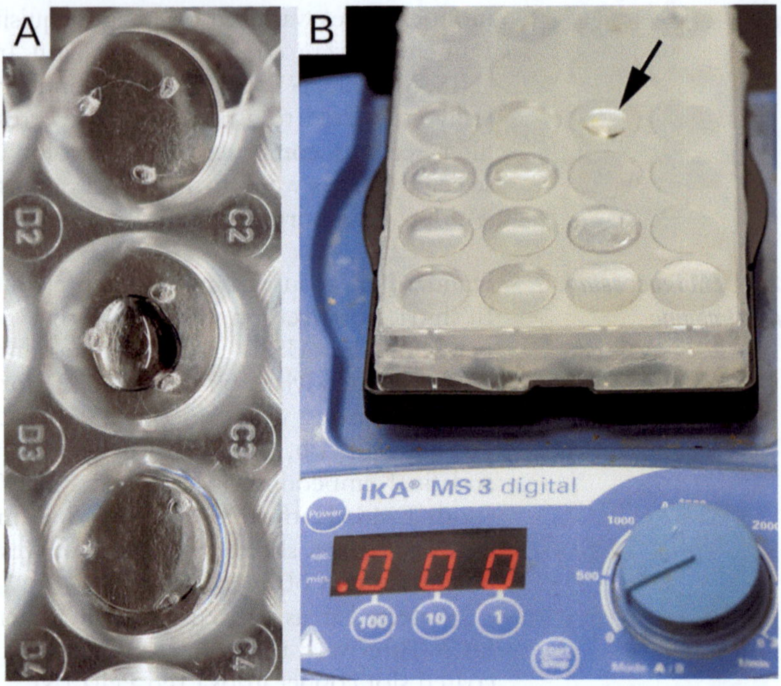

Fig. 1 The use of 24-well culture plates for immunolabeling. (**a**) Three wells are shown. Each well has three mounts on which a coverslip can be incubated. The top well is empty, the middle well contains 100 μL solution, and the bottom well shows a coverslip placed in the solution. (**b**) A 24-well plate covered with parafilm and placed on a shaker. The arrow points at a coverslip incubated in gold enhancement solution

3.2 Immunolabeling

We used indirect immunofluorescence either for fluorescent labeling alone or for co-labeling with fluoronanogold. In the first step, cells were permeabilized with detergent and treated with serum to suppress unspecific binding sites. This step was carried out with 0.03–0.1% Triton, 2% normal goat serum in TBS for 30–60 min while shaking. In the meantime, the first antibody solutions were prepared. All antibodies were dissolved in 0.2% NGS and 0.005% Trition in TBS. The exact antibody concentration was determined experimentally for each antibody. For incubating coverslips with the first antibody solutions, we prepared a 24-well plate similar to the Banker culture system. With a heated glass rod, three small plastic mounts were created on which a single coverslip could be placed. Once the quenching and permeabilization step was completed, a small amount of antibody solution (100 μL) was placed in the center of the three mounts. Without washing off the permeabilization solution, individual coverslips were placed—cells facing down—on the solution and the mount (Fig. 1a). To allow for quick mixing and even distribution, coverslips should be lifted a few times. After transferring all coverslips, the 24-well plate was covered with parafilm and placed on a shaker for incubation at 4 °C overnight.

Coverslips containing neurons and astrocytes were transferred to a clean culture plate, cells facing up, and were washed with TBS three times (10 min total). For incubation with the secondary antibody (either fluorescent or fluoronanogold), coverslips were again placed cells facing down on 100 μL of antibody anti-mouse or anti-rabbit solution into the washed culture plate containing the mounts and incubated on a shaker at 4 °C overnight.

For quadrolabeling, we used two first antibodies (one mouse monoclonal and one rabbit polyclonal) against two different antigens at the same time. After the simultaneous detection with secondary antibody (with anti-mouse and anti-rabbit antibodies conjugated with different fluorophores), we repeated these steps with a new pair of monoclonal and polyclonal antibodies against a new set of antigens.

After washing with TBS, cells are ready for imaging. Usually, we did not "coverslip" the specimens in an anti-fading reagent because most of our specimens were further processed for EM.

3.3 Imaging of Fluorescent Labeling

Specimens were imaged in TBS with a water immersion lens. Images were taken after optimizing the illumination intensity and exposure time for each dye. Once the exposure parameters were set, multilabeled specimens were automatically acquired with the NIS and converted to multicolor images of up to four channels (Fig. 2).

Up to this point, the protocol for fluorescent multilabeling and immunogold labeling are identical. For immunogold labeling, we proceeded with a gold enhancement procedure.

3.4 Gold Enhancement

Because nanogold particles are too small to be easily visible even at the EM level, an enhancement procedure is required to increase nanoparticle size. Here, an easy-to-use and very reliable gold enhancement technique was applied which is commercially available from Nanoprobes Inc. (Yaphank, NY, USA). In contrast to the manufacturer's directions, gold enhancement was done at 4 °C. First, we washed in distilled water and kept the cells in the same solution in the 24-well plate on wet ice. Gold enhancement solution was mixed as directed by the manufacturer and cooled to 0–4 °C. A clean culture plate was covered with parafilm and placed on a shaker. Single coverslips were placed on the parafilm, and 100 μL of the enhancement solution was pipetted on the coverslip (Fig. 1b); the shaker was turned on immediately, and enhancement was continued for ~30 s (longer enhancement times increase gold particle size). Coverslips were picked up with tweezers and dipped into ice-cold distilled water which stopped the enhancement procedure by dilution. The washed cells were placed back into a 24-well plate containing TBS. Finally, all specimens were post-fixed with 2.5% glutaraldehyde in 150 mM cacodylate buffer at pH 7.4 at RT for 1 h, and after a single wash, the cells were placed in cacodylate buffer at 4 °C until further processing for EM.

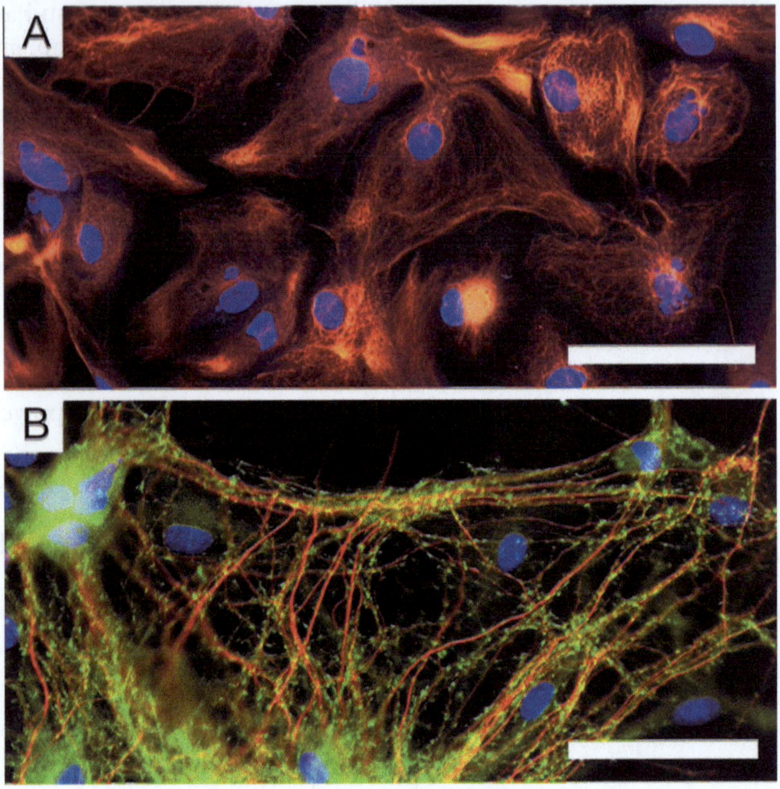

Fig. 2 Immunofluorescent labeling of cultured hippocampal astrocytes (**a**) and neurons (**b**). (**a**) Astrocytes are brightly labeled with anti-glial fibrillary acidic protein (GFAP, orange), and the nuclei are counterstained with DAPI. (**b**) Dendrites of neurons are labeled against MAP-2 (red), and synaptic contacts are labeled against synaptobrevin 2 (green). Nuclei are stained with DAPI. In the background some astrocytes are visible (labeled with anti-GFAP) in dark green. Scale bars = 100 μm

3.5 EM Processing, Sectioning, and Imaging

From here on, we followed standard embedding procedures which have been published previously in much detail [14, 16]. The steps were as follows: post-fixation with aqueous 1% osmium tetroxide including 1% potassium ferrocyanide in 150 mM cacodylate buffer for 30 min. En bloc contrasting was with 2% aqueous uranyl acetate for at least 1 h or longer if more contrast is needed. After dehydration in an ascending ethanol series, specimens were infiltrated in a 1 + 1 mixture of Spurr and Embed 812. Finally, coverslips with their cells facing up were flat embedded between Aclar foils and polymerized for 18 h at 60 °C. The top Aclar sheet was removed the next day, and a resin-filled BEEM capsule was placed up-side down on a selected region and polymerized for another 18 h at 60 °C. The BEEM capsule serves as a mounting block for sectioning.

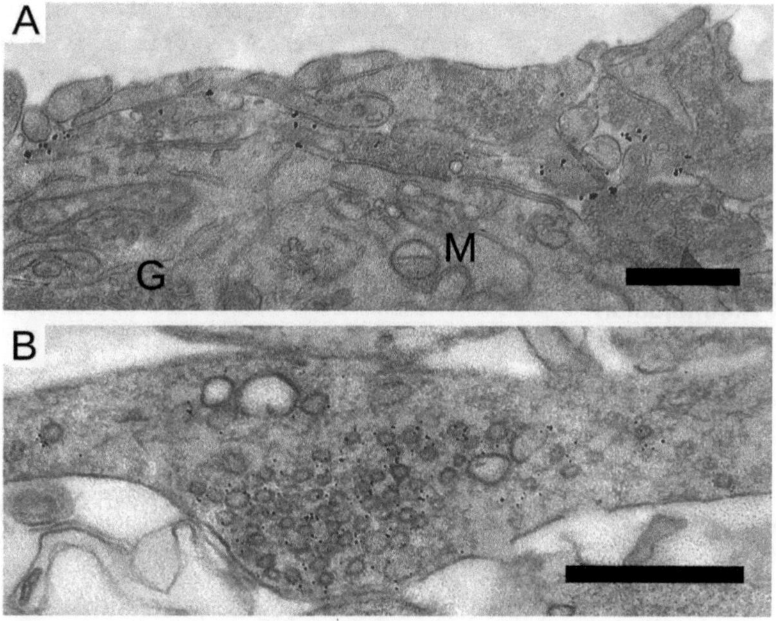

Fig. 3 EM micrographs of immunogold localization of VAMP4 and synaptobrevin 2. (**a**) Anti-VAMP4 distribution at presynaptic structures is visualized with gold particles. The image depicts several presynaptic boutons and axons in the upper half. The gold particles are located close to SVs and along the presynaptic membrane. The lower half of the image shows a cell body with cell organelles, mitochondria (M), Golgi apparatus (G), and others. Scale bar = 900 nm. (**b**) Immunogold localization of anti-synaptobrevin 2 antibodies. A single synaptic bouton is shown. In this image gold particles are small which allow their precise location at individual SVs. In these specimens, on-section staining was omitted to not obscure the fine gold particles, but resulted in overall low contrast images. Scale bar = 500 nm

A region of interest was selected and trimmed down to a trapezoid of about 800 μm by 200 μm. The surface of a block was aligned precisely with the knife, and serial sections (60 nm in thickness) were cut right from the start. The sections were collected on Pioloform-coated 1 × 2 mm slot copper grids.

Arrays of neighboring single frames were taken that were stitched together to one final image. The magnification chosen was high enough to resolve the fine structure of synapses (Fig. 3) in sufficient detail. These composed images were ideal for our unbiased analysis of gold particle number of SVs per synaptic bouton.

3.6 Gold Particle Quantification

The motivation to develop this protocol was to advance the quantification of the protein copy number by using immunogold localization at individual synaptic boutons. Synaptic boutons with their SVs are ideal for such endeavor because one can compare the SV number with the number of gold particles in individual boutons. To show that immunolabeling was increased after AFG fixation, labeling after fixation with AFG with standard PFA fixation was compared. For each experiment, both the SV number and gold particle number of individual SV clusters were counted. The ratio of gold particles over SV number is then a measure for the intensity of the labeling. Statistical analysis was done by calculating the average, standard error, and standard deviation and by exploiting the Kolmogorov–Smirnov test to determine significance.

4 Conclusion

Original or classical glyoxal fixatives often contained high amounts of ethanol and acetic acid. These reagents were commonly used as a standard fixative in the nineteenth century and were abandoned with the introduction of formalin. Both reagents, ethanol and acetic acid, were never used in EM preparations of biological samples because they led to a bad ultrastructural preservation, and the low pH of 4 is detrimental. In our protocol, we, therefore, used the McIlvaine buffer, a phosphate buffer system that provides an excellent buffer capacity in a wide pH range of 4–8. The pH was settled at 6 because it increases immunoreactivity but does not interfere with fine structural preservation. At lower pH the effects of the high acidity become visible at the EM level. Our AFG fixative contains small amounts of glutaraldehyde to support some cross-linking among proteins. Finally, added sucrose raises the osmolarity which contributes to stabilize structural elements during the fixation process.

This fixation protocol is capable of stabilizing the fine structure during fixation, after treatment with Triton X, and long incubation times for immunocytochemical labeling. The excellent fine structure achieved is visible in the undisturbed morphology of cell organelles. Mitochondria show a continuous matrix and cisternae. The stacks of the Golgi apparatus and the endoplasmic reticulum showed no sign of swelling. The cytoskeleton is not disrupted (Fig. 3). Judged by our extensive experience with EM and with pre-embedding immunogold techniques, the presented result surpasses any other pre-embedding protocol published so far.

Our AFG fixation protocol enhances immunofluorescence compared to PFA fixation as was previously shown for other glyoxal fixation methods [9, 13, 17] (see also Fig. 2). Obviously more intense labeling is visible by the eye, and unspecific background labeling is practically non-existing. These characteristics form the basis for studies that exploit multicolor immunolabeling.

Figure 3a shows that the increased specificity observed at the fluorescence microscopic level is maintained in immunogold labeling experiments. In this anti-VAMP4 labeling, the immunoreactivity is restricted to specific sites at presynaptic structures and is completely absent at the postsynaptic site and in cell bodies. Figure 3b shows intense immunogold labeling against the synaptic

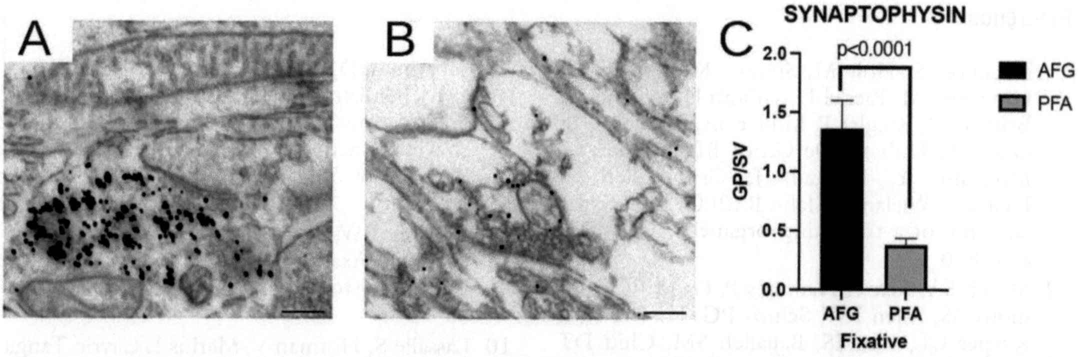

Fig. 4 Quantification of anti-synaptophysin labeling. In (**a**) numerous gold particles are located within a SV cluster after AFG fixation. (**b**) shows a parallel experiment but after using standard PFA fixation. Note that only a few SVs are labeled and that many gold particles appear at non-synaptic structures. This background disappears when fixing with AFG. From those images the number of SVs and number of gold particles of individual boutons were counted. The total number of gold particles divided by the total number of SVs (GP/SV) within an individual bouton is the average gold particle number (copy number of Synaptophysin) per SV. The bar graph in (**c**) shows a threefold increase in labeling intensity after AFG fixation in comparison with PFA. A significance test (Kolmogorov–Smirnov test) revealed *a p* value of <0.0001. Scale bars = 250 nm

vesicle protein synaptobrevin 2. At this high magnification, the superb preservation of cell organelles and the cytoskeleton is recognizable. In all experiments carried out so far and independent of antibodies used, a more intense labeling paired with undistorted fine structure was reliably observed.

Figure 4 shows an example of the quantification of immunogold labeling per cell organelle. Here, we measured the number of gold particles and the number of SVs in individual synaptic boutons which results in the average gold particle number per SV. This quantitative measure can be extended to other cell organelles such as mitochondria, Golgi apparatus, etc. and to specific sites of a given organelle. In addition, even immunogold labeling throughout synaptic structures were observed. This characteristic would make it possible to 3D reconstruct the entire synaptic boutons (or other cell organelles) from serial thin sections. Such 3D views are likely to provide a much more detailed understanding of the distribution of a particular protein and its potential interactions within the architecture of neuronal subcellular structure of the central nervous system.

Acknowledgments

The authors are indebted to Drs. Joachim Lübke and Astrid Rollenhagen for their outstanding cooperation and contributions to the acquisition of electron microscopy data. We thank Brigitte Marshallsay for her truly excellent electron microscopic imaging. This work was supported by the National Institute of Health (grant no. R15NS095318).

References

1. Takamori S, Holt M, Stenius K, Lemke EA, Gronborg M, Riedel D, Urlaub H, Schenck S, Brugger B, Ringler P, Muller SA, Rammner B, Grater F, Hub JS, De Groot BL, Mieskes G, Moriyama Y, Klingauf J, Grubmuller H, Heuser J, Wieland F, Jahn R (2006) Molecular anatomy of a trafficking organelle. Cell 127: 831–846
2. Mutch SA, Kensel-Hammes P, Gadd JC, Fujimoto BS, Allen RW, Schiro PG, Lorenz RM, Kuyper CL, Kuo JS, Bajjalieh SM, Chiu DT (2011) Protein quantification at the single vesicle level reveals that a subset of synaptic vesicle proteins are trafficked with high precision. J Neurosci 31:1461–1470
3. Wilhelm BG, Mandad S, Truckenbrodt S, Krohnert K, Schafer C, Rammner B, Koo SJ, Classen GA, Krauss M, Haucke V, Urlaub H, Rizzoli SO (2014) Composition of isolated synaptic boutons reveals the amounts of vesicle trafficking proteins. Science 344:1023–1028
4. Richter KN, Revelo NH, Seitz KJ, Helm MS, Sarkar D, Saleeb RS, D'Este E, Eberle J, Wagner E, Vogl C, Lazaro DF, Richter F, Coy-Vergara J, Coceano G, Boyden ES, Duncan RR, Hell SW, Lauterbach MA, Lehnart SE, Moser T, Outeiro TF, Rehling P, Schwappach B, Testa I, Zapiec B, Rizzoli SO (2018) Glyoxal as an alternative fixative to formaldehyde in immunostaining and super-resolution microscopy. EMBO J 37:139–159
5. Reshetniak S, Ussling JE, Perego E, Rammner B, Schikorski T, Fornasiero EF, Truckenbrodt S, Koster S, Rizzoli SO (2020) A comparative analysis of the mobility of 45 proteins in the synaptic bouton. EMBO J 39:e104596
6. Helm MS, Dankovich TM, Mandad S, Rammner B, Jahne S, Salimi V, Koerbs C, Leibrandt R, Urlaub H, Schikorski T, Rizzoli SO (2021) A large-scale nanoscopy and biochemistry analysis of postsynaptic dendritic spines. Nat Neurosci 24:1151–1162
7. Wicks LF, Suntzeff V (1943) Glyoxal, a non-irritating aldehyde suggested as substitute for formalin in histological fixations. Science 98:204. Epub 1943/08/27. PMID: 17843715. https://doi.org/10.1126/science.98.2539.204
8. Sabatini DD, Bensch K, Barrnett RJ (1963) Cytochemistry and electron microscopy. The preservation of cellular ultrastructure and enzymatic activity by aldehyde fixation. J Cell Biol 17:19–58. Epub 1963/04/01. PMID: 13975866
9. Dapson RW, Feldman AT, Wolfe D (2006) Glyoxal Fixation and Its Relationship to Immunohistochemistry. J Histotechnol 29: 65–76
10. Lassalle S, Hofman V, Marius I, Gavric-Tanga-V, Brest P, Havet K, Butori C, Selva E, Santini J, Mograbi B, Hofman P (2009) Assessment of morphology, antigenicity, and nucleic acid integrity for diagnostic thyroid pathology using formalin substitute fixatives. Thyroid 19: 1239–1248
11. Marcon N, Bressenot A, Montagne K, Bastien C, Champigneulle J, Monhoven N, Albuisson E, Plenat F (2009) Glyoxal: a possible polyvalent substitute for formaldehyde in pathology? Ann Pathol 29:460–467
12. Dapson RW (2007) Glyoxal fixation: how it works and why it only occasionally needs antigen retrieval. Biotech Histochem 82:161–166
13. Bussolati G, Annaratone L, Berrino E, Miglio U, Panero M, Cupo M, Gugliotta P, Venesio T, Sapino A, Marchio C (2007) Acid-free glyoxal as a substitute of formalin for structural and molecular preservation in tissue samples. PLoS One 12:e0182965
14. Schikorski T (2010) Pre-embedding immunogold localization of antigens in mammalian brain slices. Methods Mol Biol 657:133–144
15. McIlvaine TC (1921) A buffer solution for colorimetric comparison. J Biol Chem 49: 183–186
16. Schikorski T (2016) Monitoring synaptic vesicle protein sorting with enhanced horseradish peroxidase in the electron microscope. Methods Mol Biol 1474:327–341
17. Richter KN, Wildhagen H, Helm MS, Ussling JE, Schikorski T, Rizzoli SO (2018) Comparative synaptosome imaging: a semi-quantitative method to obtain copy numbers for synaptic and neuronal proteins. Sci Rep 8:14838

Chapter 3

Cell Labeling and FIB–SEM Imaging

Marta Turegano-Lopez, José-Rodrigo Rodríguez, Javier DeFelipe, and Angel Merchán-Pérez

Abstract

Volume electron microscopy is a powerful method for the study of brain connectivity at the ultrastructural level. Synapses and their parent axons and dendrites can be traced within the three-dimensional tissue sample, so their numbers and distribution can be quantified in different brain regions and in different conditions of health or disease. Cell labeling methods allow us to identify the nature and origin of nerve fibers. Here we briefly review several labeling methods that have been combined with Focused Ion Beam–Scanning Electron Microscopy (FIB–SEM). We describe the different methods used and the technical difficulties that they pose.

Key words Volume electron microscopy, Synapses, Brain connectivity, FIB–SEM, Tracing, Labeling

1 Introduction

Volume electron microscopy is a highly valuable tool for obtaining information about the structure of the brain, particularly in relation to synapses and their connections. With Focused Ion Beam milling and Scanning Electron Microcopy (FIB–SEM), synapses are identified on the basis of morphological features. The identification criteria include the presence of synaptic vesicles in the presynaptic element, the electron-dense thickening of the postsynaptic membrane (postsynaptic density, PSD), and the presence of the synaptic cleft, although this is only visible if the synaptic junction is sectioned perpendicularly [1]. Once synapses are identified, they can be classified mostly according to the thickness of the PSD: asymmetric synapses (AS) have a prominent PSD, while symmetric synapses (SS) have a thin PSD whose thickness is very similar to the presynaptic membrane [2]. This distinction is relevant since asymmetric synapses correspond to excitatory, mainly glutamatergic synapses, while symmetric ones correspond to inhibitory GABAergic synapses [2–7].

Joachim H. R. Lübke and Astrid Rollenhagen (eds.), *New Aspects in Analyzing the Synaptic Organization of the Brain*, Neuromethods, vol. 212, https://doi.org/10.1007/978-1-0716-4019-7_3,

Upon identifying and categorizing synapses, it becomes straightforward to differentiate axons (the presynaptic elements) from dendrites (the postsynaptic elements). Axons can be further sub-classified into excitatory and inhibitory axons, according to the morphology of their corresponding synapses, as mentioned above. Dendrites may or may not have dendritic spines, and they establish synapses on their shafts and spines. Quantitative information regarding the number of synapses, the proportion of excitatory and inhibitory synapses, and their distribution on dendritic shafts and spines is crucial. This information allows us to identify differences between brain regions and conditions, even if such differences would have gone unnoticed in a qualitative analysis.

Cell labeling techniques, in combination with volume electron microscopy, provide additional information that is otherwise not possible to obtain. For instance, when using FIB–SEM to acquire a series of sequential sections from any layer of the cerebral cortex, it is challenging to distinguish between dendritic branches originating from nearby local neurons and those originating from neurons situated at a distance, whether in the same or different layers. In the case of axons, categorizing them based on their origin and target is even more challenging, given that they can extend across microns, millimeters, or even centimeters beyond the sample. In these cases, various labeling techniques, such as anterograde tracing or immunocytochemistry, can be employed to identify the origin of specific axons or determine the cell type to which certain dendrites belong.

2 General Considerations

From a technical point of view, the great variety of tracing and labeling methods available for electron microscopy renders it impractical to formulate a single protocol adaptable to FIB–SEM. However, since the tissue preparation methods for FIB–SEM and transmission electron microscopy (TEM) are very similar, it is reasonable to start with the premise that any labeling method proven effective in TEM should be applicable to FIB–SEM. Nevertheless, several adjustments are necessary, prompting the need for preliminary considerations.

Firstly, preservation of the ultrastructure is crucial for FIB–SEM, which is also the case for conventional TEM. The choice of the fixation method is pivotal for the preservation of the ultrastructure, and it must be compatible with the chosen labeling method. Additionally, the procedure to develop the labeling agent to make it visible under the electron microscope may also impact the ultrastructure. This is because, typically, tissue permeabilization is required to facilitate the access of reagents to the localization of the chromogen. Although some degree of alteration of tissue ultrastructure cannot be avoided, it is usually possible to achieve a compromise solution both for fixation and development of the labeling agent, as we will see below. Secondly, post-embedding labeling methods are not applicable for FIB–SEM since the tissue sample is milled in steps of fixed thickness. Therefore, we do not obtain actual sections but rather sequential images of the block face. Thirdly, given that we do not obtain actual sections, any staining must be performed prior to embedding. In this respect, it is important to stain the tissue with the adequate selection of heavy metals to ensure good resolution and contrast under the FIB–SEM.

In the following paragraphs, we will review the main steps of tissue preparation and briefly comment on the dos and don'ts of different methods of cell labeling.

3 Fixation

The method of choice is perfusion fixation with 4% paraformaldehyde (PFA) (Fig. 1a). If necessary (as in the case of human brain tissue), fixation by immersion is also possible, but it must be performed immediately after surgery (biopsies)—or minimizing postmortem times when the tissue is obtained from autopsies. PFA is adequate as a fixative since it preserves both the ultrastructure and the labeling agent. A high concentration of glutaraldehyde (GA) should be avoided since this can compromise antigenicity or the enzymatic activity of many different labeling agents. However, low doses of GA (0.05–0.1%) can be used without any noticeable loss of labeling [8–11]. Additionally, a high concentration of GA will cause intense autofluorescence of the tissue, while a low concentration (0.125%) is compatible with methods that require detection of fluorescent tracers [12]. After perfusion, the brain is extracted (Fig. 1b) and left for 1–16 h (overnight) in fresh fixative at 4 °C.

4 Vibratome Sectioning

The brain is washed for 1 h in 0.1 M phosphate buffer (PB) at 4 °C, and 50- to 250-μm-thick sections are obtained with a vibratome (Fig. 1c). The free-floating sections are serially collected in 24-well plates with cold PB. The sections containing the region of interest are selected for further processing.

5 Tissue Permeabilization

If immunocytochemistry is intended as the primary labeling method, or if it is necessary for the identification of the labeling agent, tissue permeabilization is required so that reagents can

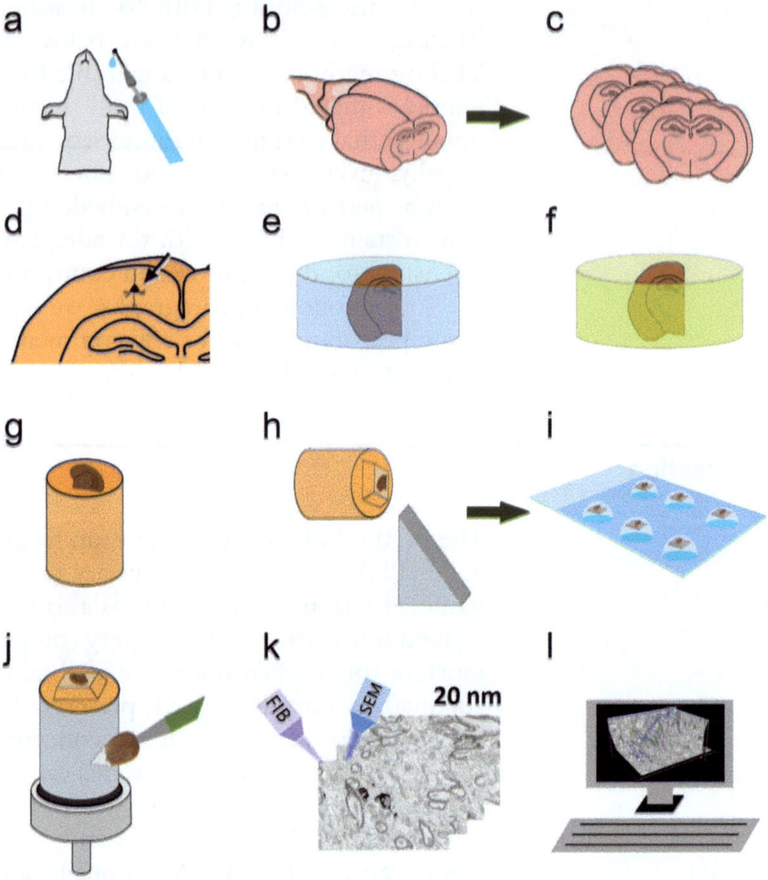

Fig. 1 Schematic representation of the preparation of samples for FIB–SEM imaging. (**a**) Fixation. (**b**) Extraction of the brain. (**c**) Vibratome sectioning. (**d**) Immunocytochemistry and/or development of the labeling agent. (**e**) Postfixation and osmication. (**f**) En bloc staining with uranyl acetate, dehydration, and flat embedding. (**g**) The flat-embedded section is mounted on a blank resin block. (**h**) Excess resin is removed from the surface of the block. (**i**) Semithin sections are obtained and examined. (**j**) The block is painted with conductive silver paint and sputter coated with gold/palladium to facilitate charge dissipation. (**k**) FIB–SEM imaging. (**l**) Analysis of the stacks of serial sections

penetrate the sample as much as possible. However, this process alters the fine structure of the tissue, so there needs to be a trade-off between permeabilization and preservation of ultrastructure. In our experience, the freeze-thawing method yields the best results. The samples are first cryo-protected by placing them in PB with 10% sucrose until they sink and then in PB with 30% sucrose until they sink again. Then, one to three freeze-thawing cycles in liquid nitrogen are performed [13].

With other labeling methods, such as the intracellular injection of horseradish peroxidase (HRP), freeze-thawing is not necessary.

In this case, dimethyl sulfoxide (DMSO) can be used for a milder permeabilization of the tissue [12]. In fact, DMSO may also be used in combination with freezing and thawing, since in that case it helps preserve the integrity of membranes [14]. Detergents such as Triton X-100 or saponin, which are commonly used in light and fluorescence microscopy, should be avoided for electron microscopy, since they cause severe disruption of biological membranes.

6 Immunocytochemistry and/or Label Development

Immunocytochemical procedures and the development of the labeling agents are both highly variable, but they usually involve precipitation of a substance that is rendered electron-dense after processing (Fig. 1d). For example, 3,3′-diaminobenzidine (DAB) is a commonly used chromogen for light and electron microscopy. Metal enhancement of this and other chromogens with nickel or cobalt, for example, which are commonly used for light microscopy, must be avoided for electron microscopy, since they are not necessary for the visualization of the labeling product and may cause unwanted precipitates. A special case is silver enhancement of gold nanoparticles, which is usually performed in both light and electron microscopy immunogold techniques with excellent results [9].

7 Postfixation and Osmication

After development of the labeling agent, we use a postfixation solution with a mixture of PFA and GA (2% PFA, 2.5% GA, and 3 mM $CaCl_2$ in 0.1 M cacodylate buffer, for 2 h) [12, 13]. Wash three times, 10 min each time, in 0.1 M cacodylate buffer, and then osmicate for 1 h in 1% OsO_4 in 0.1 M cacodylate buffer (Fig. 1e).

Osmication may darken the preparation to the extent that it makes it impossible to visualize the labeled cell or fibers against the black background. Adding 7% glucose to the 1% osmium solution will result in a much lighter staining of the section, so the labeled elements can be easily identified.

8 Uranyl Acetate Staining, Dehydration, and Flat Embedding

The vibratome sections are stained with 1% uranyl acetate in 50% ethanol for 30 min at 37 °C [12, 13] (Fig. 1f). We then proceed with dehydration with a graded series of ethanol.

We usually perform flat embedding in an epoxy resin [15]. Flat embedding allows us to process large vibratome sections, which can include a whole mouse hemisphere. These large preparations can later be trimmed, if necessary, to the appropriate size.

Once the sections are embedded, they are glued to a blank resin block with cyanoacrylate glue and trimmed for further processing (Fig. 1g).

9 Semithin Sectioning

We first need to remove the excess resin from the surface of the block, using a glass knife to remove the majority of the resin and then switching to a diamond knife when we are already close to the surface of the tissue (Fig. 1h). Then we obtain semithin sections (0.5–2-μm-thick) (Fig. 1i). Given that antibodies and reactants usually penetrate only a few microns into the tissue, it is very important to examine these sections with a light microscope as they are being obtained. As soon as the surface of the labeled tissue is reached, sectioning must be stopped. In this way, the thin layer of the tissue where penetration of reactants is optimal will remain just below the block face and will be easily accessible for FIB–SEM imaging. Note that the last semithin section obtained will match the surface of the block, so we can collate an image of the semithin section and an SEM image of the block surface. This allows us to select the region of interest with a precision of only a few microns. Semithin sections can also be examined under the SEM [16], so the location of the region of interest can be even more precise.

Alternatively, we can first obtain thin sections (up to 7 μm thick) that are serially collected. This is only necessary when we have labeled long cell processes or whole cells. These sections are observed and photographed to locate the region of interest, and then they are re-mounted on blank resin blocks and processed individually, as described above [12].

10 Preparation for FIB–SEM

Once the semithin sections have been obtained, the block is glued to an aluminum stub with a conductive carbon sticker. To allow charge dissipation, the block is painted with conductive silver paint, except for the block face where our sample is located (Fig. 1j). To further facilitate dissipation, the block is sputter-coated with gold or gold/palladium. The samples are then ready to be studied with FIB–SEM (Fig. 1k) and analyzed as serial sections (Fig. 1l).

11 FIB–SEM Milling and Imaging

FIB–SEM milling and imaging can be performed in many different ways, depending on the specific needs of each study. Here we will summarize the conditions that we routinely use for immunocytochemistry or intracellularly labeled cells [12, 13], as in the example shown in Fig. 2. The region of interest is previously located in the flat-embedded vibratome section (Fig. 2a) and in the corresponding photographs that have been taken from the semithin or thin section (Fig. 2b), as mentioned above. Then we take a low-magnification panoramic view of the block face. We use secondary electrons and a relatively high acceleration voltage (3–5 kV) to obtain it. Even though fine details may not be visible, large structures such as blood vessels and cell somata can be clearly identified, so we can collate this image with the previous semithin or thin sections to select the region that must be imaged with the FIB–SEM (Fig. 2c). We then mill a trench in the selected region using a 7 nA FIB current (30 kV accelerating voltage) (Fig. 2d). We will later acquire serial images from the front face of this viewing trench, so it has to be large enough for the intended field of view. For example, if the plan is to acquire images of 12 × 10 μm, the front face of the trench must be at least 24 μm wide and 20 μm deep to avoid shading effects.

An image of the front face of the trench is then taken using the back-scattered electron detector (Fig. 2e). This image is used to identify the labeled structure and to further determine the region where fine milling and high-resolution imaging will be performed (Fig. 2f). Finally, the milling/imaging cycle is started to obtain serial sections of the chosen region (Fig. 2g). We routinely use a milling thickness of 20 nm, although thinner or thicker milling cycles can be performed, according to the needs of the study. Regarding image resolution, we use 5 nm per pixel as our routine resolution, but, again, this can be modified as per requirements.

In Fig. 3, we show another example, but in this case, immunocytochemistry was used to identify GABAergic terminals, using anti-vesicular GABA transporter (VGAT) antibodies [13]. Although the labeling is lighter than in the HRP-injected cell of the previous example (Fig. 2), the labeled terminals are clearly identified. Additionally, unlabeled terminals can be seen to establish AS, often on the same postsynaptic targets as SS.

12 Conclusion

FIB–SEM can be utilized to examine the brain tissue that has been labeled with the techniques typically applied to TEM. However, only pre-embedding methods are feasible with FIB–SEM. A positive point is that automated sectioning and imaging enable the acquisition and analysis of hundreds or thousands of labeled serial images. These serial images can be stacked to represent a three-dimensional sample of the tissue, which can be analyzed and quantified with dedicated 3D software.

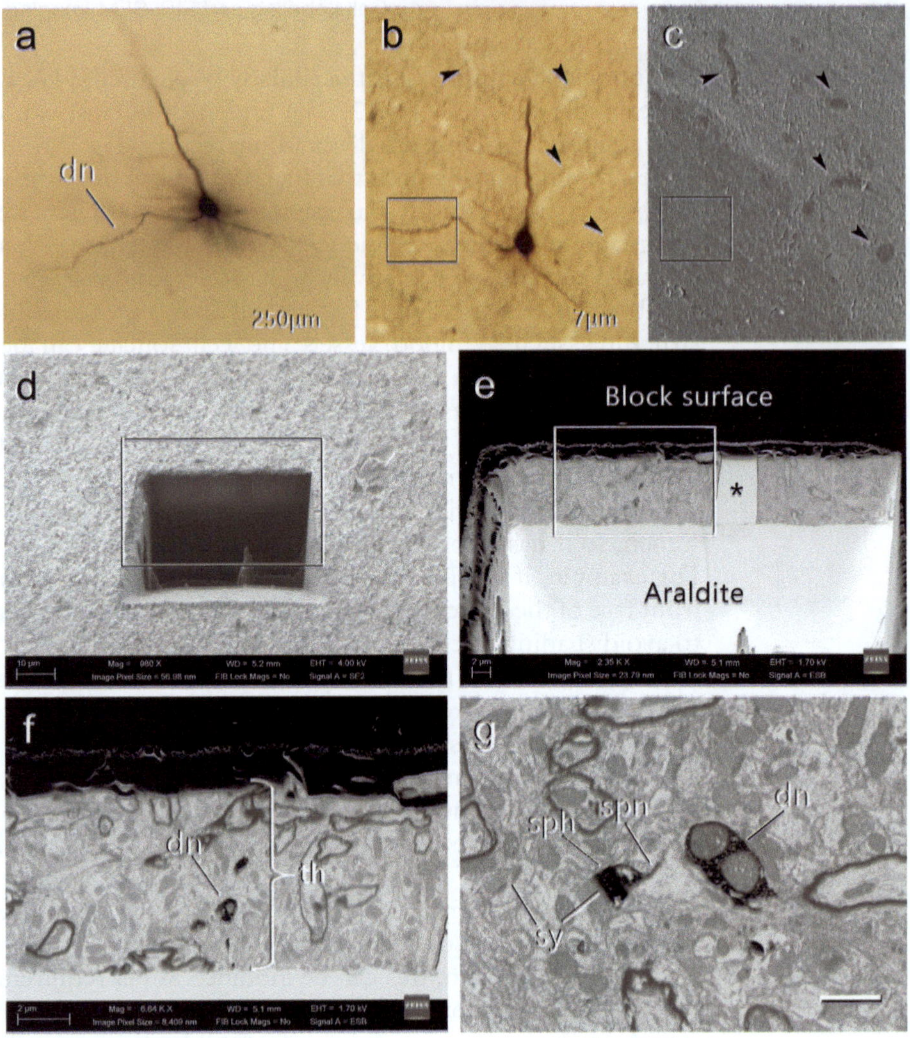

Fig. 2 FIB–SEM imaging of a labeled cell. (**a**) This example shows a flat-embedded 250-μm-thick vibratome section with a pyramidal neuron from layer 3 of the mouse motor cortex. The labeled cell was injected with HRP and developed with DAB. One dendritic segment (dn) was selected for further study. (**b**) One of the 7-μm-thick serial sections obtained from the 250-μm-thick vibratome section. This section contains the selected dendritic segment (inset). Blood vessels (arrowheads) were clearly identifiable and were later used as landmarks. Once this section was photographed, it was mounted on a blank Araldite block and observed with the SEM. (**c**) The surface of the 7 μm section photographed with the SEM using the secondary electron detector. The same blood vessels were visible (arrowheads), so they were used as landmarks to identify the region where the labeled dendritic segment was located (inset). (**d**) A viewing trench was excavated with the FIB to gain visual access to the target dendrite. This image was taken with the secondary electron detector. The inset indicates the region that is further magnified in the next panel. (**e**) Backscattered electron image showing the trench that was excavated with the FIB to identify the selected dendritic segment. The grey scale has been inverted so the block surface appears dark and the resin bed (Araldite) appears light. The visible portion of brain tissue is crossed vertically by a blood vessel (asterisk). The region within the inset is further magnified in the next panel. (**f**) Region of interest where the labeled dendrite (dn) is visible due to the dark, electron-dense precipitate. The thickness of the section (th) is approximately 7 μm. (**g**) Detail of one of the serial images acquired from the dendrite (dn). One of its dendritic spines is visible, showing both the spine head (sph) and spine neck (spn). Synaptic junctions (sy) can be identified on the dendritic spine and in the surrounding tissue. Calibration bar in (**g**): 50 μm for (**a**), (**b**), and (**c**); 16 μm for (**d**); 6.9 μm for (**e**); 2.4 μm for (**f**); and 1 μm for (**g**). (Figure taken from Ref. (12), with permission)

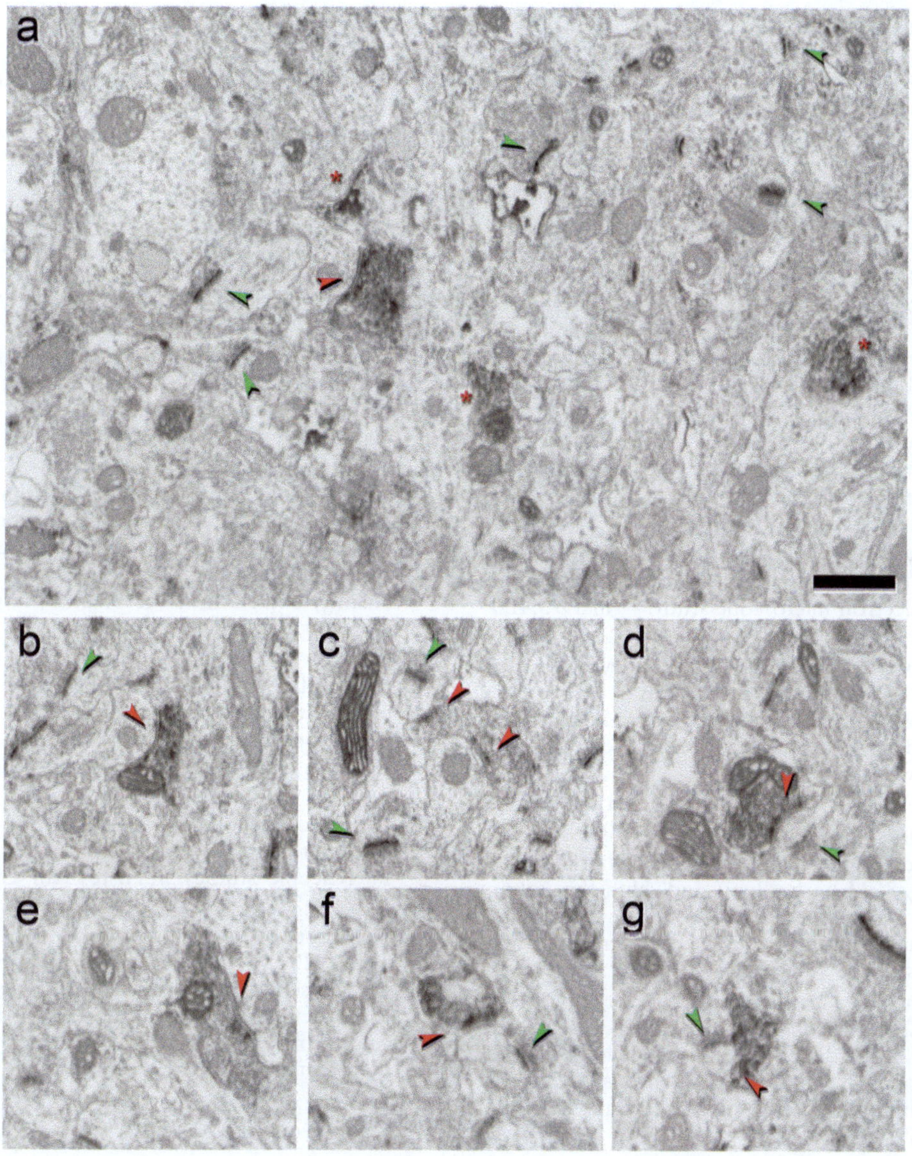

Fig. 3 FIB–SEM images of VGAT-immunostained tissue. (**a**) AS (green arrows) and SS (red arrows) in a panoramic view from the mouse somatosensory cortex. The classification of synapses is based on the thickness of the PSD and on the presence of VGAT-positive labeling. Green arrows indicate AS, and red arrows indicate SS. Note the VGAT-positive presynaptic staining on the SS (red asterisks). (**b**–**g**) Examples of AS and SS synapses. In (**c**), (**d**), (**f**), and (**g**), AS and SS synapses share the same postsynaptic target. (**c**) The VGAT staining is less intense compared to other terminals. The immunolabeling can be non-uniform throughout the tissue due to the differential penetration of reagents. Scale bar indicates 755 nm (**a**), 920 nm (**b**), 900 nm (**c**), 850 nm (**d**), 750 nm (**e**), 580 nm (**f**), and 630 nm (**g**)

References

1. Mayhew TM (1996) How to count synapses unbiasedly and efficiently at the ultrastructural level: proposal for a standard sampling and counting protocol. J Neurocytol 25:793–804
2. Colonnier M (1968) Synaptic patterns on different cell types in the different laminae of the cat visual cortex. An electron microscope study. Brain Res 9:268–287
3. Ascoli GA, Alonso-Nanclares L, Anderson SA et al (2008) Petilla terminology: nomenclature of features of GABAergic interneurons of the cerebral cortex. Nat Rev Neurosci 9:557–568
4. Colonnier M (1981) The electron-microscopic analysis of the neuronal organization of the cerebral cortex. In: The Organizaton of the Cerebral Cortex. MIT Press, Cambridge, MA, pp 125–152
5. Gray EG (1959) Axo-somatic and axo-dendritic synapses of the cerebral cortex: an electron microscope study. J Anat 93:420–433
6. Houser CR, Vaughn JE, Hendry SHC et al (1984) GABA neurons in the cerebral cortex. In: Cerebral cortex, vol. 2, functional properties of cortical cells. Plenum Press, New York, pp 63–89
7. Peters A, Palay SL (1996) The morphology of synapses. J Neurocytol 25:687–700
8. Bosch C, Martínez A, Masachs N et al (2015) FIB/SEM technology and high-throughput 3D reconstruction of dendritic spines and synapses in GFP-labeled adult-generated neurons. Front Neuroanat 9:60
9. Luján R, Merchán-Pérez A, Soriano J et al (2021) Neuron class and target variability in the three-dimensional localization of SK2 channels in hippocampal neurons as detected by immunogold FIB-SEM. Front Neuroanat 15:781314
10. Rodriguez-Moreno J, Porrero C, Rollenhagen A et al (2020) Area-specific synapse structure in branched posterior nucleus axons reveals a new level of complexity in thalamocortical networks. J Neurosci 40:2663–2679
11. Rodriguez-Moreno J, Rollenhagen A, Arlandis J et al (2018) Quantitative 3D ultrastructure of thalamocortical synapses from the "Lemniscal" ventral posteromedial nucleus in mouse barrel cortex. Cereb Cortex 28:3159–3175
12. Turegano-Lopez M, Santuy A, Kastanauskaite A et al (2022) Single-neuron labeling in fixed tissue and targeted volume electron microscopy. Front Neuroanat 16:852057
13. Turégano-López M, Rodríguez J-R, Alonso-Nanclares L et al (2021) Pre-embedding immunostaining of brain tissue and three-dimensional imaging with FIB-SEM. In: Lujan R, Ciruela F (eds) Receptor and ion channel detection in the brain. Springer, US, New York, NY, pp 285–302
14. Sydykov B, Oldenhof H, Oliveira Barros L d et al (2018) Membrane permeabilization of phosphatidylcholine liposomes induced by cryopreservation and vitrification solutions. Biochim Biophys Acta Biomembr 1860:467–474
15. DeFelipe J, Fairén A (1993) A simple and reliable method for correlative light and electron microscopic studies. J Histochem Cytochem 41:769–772
16. Rodríguez J-R, Turégano-López M, DeFelipe J et al (2018) Neuroanatomy from mesoscopic to nanoscopic scales: an improved method for the observation of semithin sections by high-resolution scanning electron microscopy. Front Neuroanat 12

Chapter 4

Unbiased Sampling and Volume-Sorting of Origin-Specific Terminals Using SBEM Image Stacks

Alev Erisir, Alex Briegel, Erin E. Maher, and Francesca Sciaccotta

Abstract

While massive sets of serially imaged electron microscopy stacks are required for large-scale brain circuitry reconstructions, smaller and more easily attainable image stacks can be used for studying the circuitry within the given nuclei and the contributions of different synaptic boutons whose origins can be identified by morphological or morphometric criteria using random sampling strategies. Here we describe an unbiased axon terminal sampling procedure for reconstructing and identifying morphological and morphometric properties of terminal boutons that contribute to the circuitry in a sample volume of the tissue. We also describe a statistical modeling tool for volume-based sorting of terminal boutons using the cutoff criteria obtained with the unbiased terminal sampling (UTS) approach and an independent dataset. The UTS approach, as described, yields a variety of measures, including bouton volume, synaptic zone area, and estimated synapse density, and may allow researchers to obtain a quantification of the synaptic circuitry in a given brain region in a relatively short time and then compare across any given variable such as age, development, or disease progression.

Key words Connectomics, 3D reconstruction, Circuitry, Electron microscopy

1 Introduction

Recent advances in techniques to obtain serial electron microscopy images have led to high-throughput connectomics approaches that are capable of reconstructing local neuronal circuitries and obtaining detailed maps of neuronal connectivity [1–6]. Compared to TEM, serial EM material is more reliable for the morphometric classifications of synaptic inputs, spines, dendrites, and the complex relationships among converging synaptic inputs on neurons. Furthermore, 3D reconstructions can yield quantitative information on the circuit element's volume, length, surface area, and frequency, allowing for network modeling of neuronal connectivity. The advantage and the precision of serial EM offer the potential to replace unbiased estimation approaches used with 2D images, which have been the gold standard for quantitative methods in

Joachim H. R. Lübke and Astrid Rollenhagen (eds.), *New Aspects in Analyzing the Synaptic Organization of the Brain*, Neuromethods, vol. 212, https://doi.org/10.1007/978-1-0716-4019-7_4,

studying ultrastructural morphology and connectivity [2, 7]. The application of estimation-based approaches to 3D-EM image stack datasets opens new possibilities for hypothesis-based research without reconstructing large volumes of the brain tissue in multiple ages, experimental conditions, or species.

In this chapter, we describe an unbiased terminal sampling (UTS) approach for collecting quantitative data representative of the composition of various axonal inputs that converge on neurons. In developing the approach, we chose the dorsal lateral geniculate nucleus (dLGN) of the thalamus, where the morphological and quantitative properties of origin-specific inputs were described in detailed TEM studies [8–12]. We also describe a statistical modeling approach to determine the cutoff points for subdividing the range of terminal volumes measured via UTS and a subsequent application of the cutoff criteria for analyzing the input selectivity on geniculate cell types in a non-UTS dataset [13].

2 Materials

1. Serial Electron Microscopy (SEM) Image Stack

   The materials used in our study were prepared to facilitate SBEM scanning (*see* Subheading 4.1 for notes on SEM stack preparation). However, the methodology described below can be suitable for implementing EM image stacks obtained with other instruments.

2. Computer

   A computer that matches the specifications (processor, REM, and graphic card) required by the segmentation software should be used. We used a Surface Studio computer (*see* Subheading 4.2) for Reconstruct software with 200–600 images.

3. Segementation Software Reconstruct Software (https://synapseweb.clm.utexas.edu/software-0 [14]; *see* Subheading 4.3 for Notes on Reconstruct Software).
4. R/RStudio (*see* Subheading 4.4 for Notes on R/RStudio).
5. R Packages Mclust, Arules, and Qgraph (*see* Subheading 4.4.1 for Necessary R Packages).

3 Methods

3.1 Opening the SEM Stack on the Segmentation Software

Import the SEM stack at a full resolution in Reconstruct, following the instructions in Reconstruct User Manual.

3.2 Stereology Grid

Generate a 3D stereology grid on the SEM stack using Series Options tool from the main menu bar: On the Grid tab, select the Sampling Frame option as the element type (see Reconstruct User Manual Creating a Grid of Traces [p. 73]). Set grid parameters, press OK, and select "place grid" arrow from the toolbar. The left mouse button places the upper left corner of the first grid element. For determining the X/Y/Z parameters of the 3D stereology grid, *see* Subheading 4.5 for notes on Generating the 3D Stereology Grid. In the current example (Fig. 1a), the following

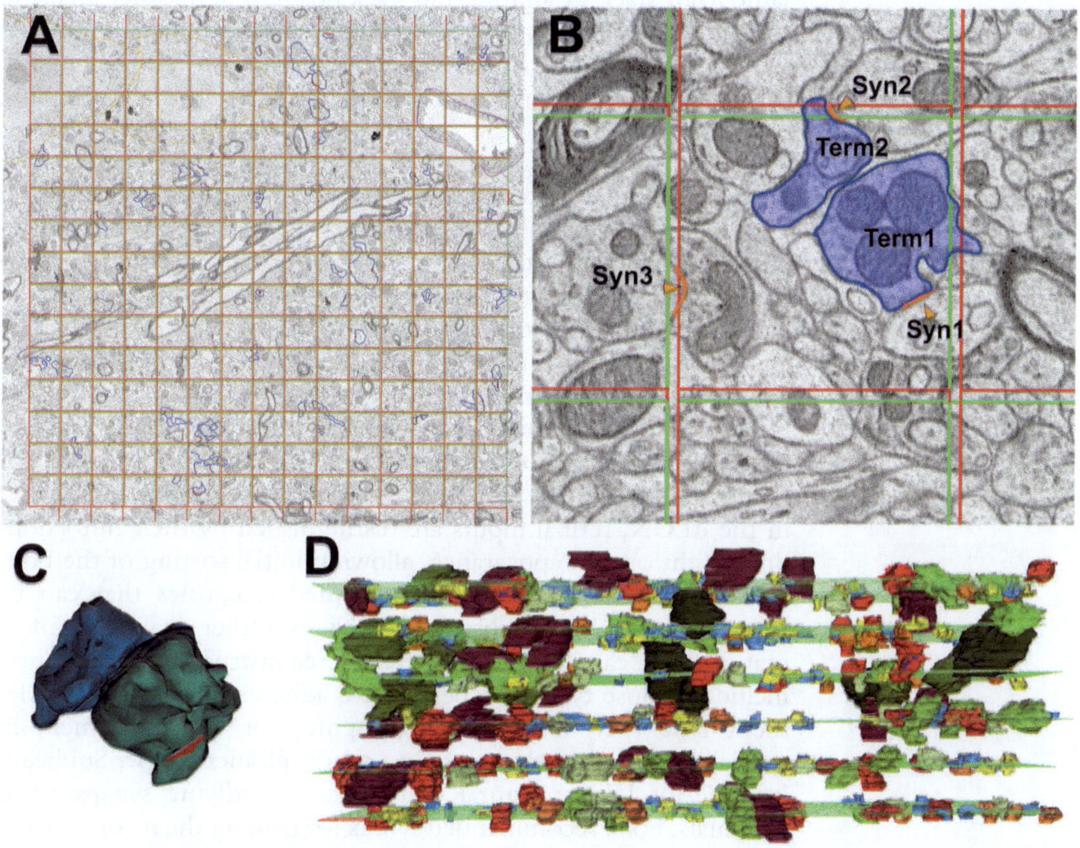

Fig. 1 (**a**) Placement of a stereology grid (red and green lines) on one image of the SBEM stack. The grid has 225 (15 × 15) locations, each outlines a 2.5 um^2 region of the tissue. (**b**) Sampling strategy from each location selected by a random number generator. Every synapse (orange lines; orange arrow points to the synapse from the postsynaptic side) that is within the location lines (Syn 1) or intersecting a green (inclusion) line of the location (Syn 2) are drawn. Synapses that intersect a red line (exclusion line; Syn 3) are ignored. All terminals (blue outlines and pseudo-coloring; term 1 and term 2) that are associated with the drawn synapses are reconstructed. (**c**) The reconstruction of two boutons from panel B. The reconstruction of the boutons was terminated at the transition from a bouton to an axon stalk. (**d**) Sparse reconstruction of dLGN boutons from one 3D stereology grid placement. Six grids colored light green are oriented at the near-z aspect of the image stacks. Reconstructed boutons were subsequently color-coded based on their terminal subpopulation as ascertained via mclust analysis. Four shades of green mark four distinct retinal terminal subpopulations. Blue, yellow, orange, red, and burgundy boutons are non-retinal in origin

parameters were used for placing grids on six images: element size = 2.4 microns, distance = 2.5 for X and Y, and number = 15 for both X and Y, yielding 1350 sequentially numbered grid "locations."

3.3 Sampling Strategy

Using a random number generator, designate a list for the order number for each grid location to be visited. We recommend sampling no more than 50% of the grid locations for each stereology grid placement in a single SEM stack. If a predetermined sample size is aimed, multiple runs of UTS can be performed from the same SEM stack by shifting the stack position to another image in the stack. This strategy also has the added benefit of sampling across any potential modularity in the brain structure (*see* Subheading 4.6 for Proof-of-Concept Experiments validating the sampling strategy).

3.4 Tracing

For each location visited, an experimenter traces each synapse that crosses no lines or only the green lines. The synapses that cross a redline are not traced. Then, the experimenter traces and reconstructs terminal boutons that make each synapse. Care should be taken to capture the entire presynaptic bouton in successive images and to exclude the thinner inter-bouton axon segments that are devoid of vesicles from the reconstructions (Fig. 1a, b).

Reconstructed terminals can be further categorized in brain nuclei where morphological criteria were previously defined for identifying boutons originating from distinct sources. For example, in the dLGN, retinal inputs are distinguished by their mitochondria's light or pale appearance, allowing initial sorting of the bouton dataset into retinal and non-retinal categories that can be followed in parallel in subsequent analyses. Other qualitative information that can be noted for each reconstructed bouton may include synapse type (i.e., perforated, adherent), postsynaptic element type (i.e., dendritic shaft, spine), or its engagement in specialized circuitry elements (i.e., triads, glomeruli) (*see* Subheading 4.7 for Ultrastructural Criteria for identifying synapses and terminals. For a recommended syntax for naming the reconstructed objects, *see* Subheading 4.8 Nomenclature section).

3.5 Analysis of UTS Datasets

The Reconstruct software yields several measures of reconstructed objects, including bouton volumes, centroid coordinates of objects, and surface area. In the volume-sorting approach described below, the bouton volumes were used for categorizing inputs from distinct origins based on demonstrated differences among the size range and mean size of individual input types in the dLGN that is the inputs from the retina, cortex, brainstem, thalamic reticular nucleus, and the inhibitory interneurons. An estimate of synapse density in neuropil regions can be calculated by subtracting the volume of cell bodies and capillaries from the total stack volume.

3.5.1 Volume-Sorting Analysis

The frequency distribution of randomly sampled boutons from and brain region yields a multi-peaked distribution due to size differences among different inputs converging on the same region. For example, primary excitatory axons in the sensory pathways have large terminal boutons, whereas the terminals that bring the feedback excitation are small in volume. The volume-sorting analysis assumes that each peak in the total UTS data represents input from a distinct origin, and the bouton volumes of each input are normally distributed. In our study that aimed to determine the number of normally distributed clusters within the larger distribution of axon terminal volumes, Gaussian normal mixture modeling through the R Package "mclust" [15] was used. The mclust models the data based on the Bayesian information criteria (BIC) and identifies the model that yields the highest number of BIC clusters, that is, the highest possible number of distinct inputs that make up the multi-peaked distribution of the bouton volume data. The selected model yields a sample size, mean, and standard deviation for each cluster, which can be used to model each cluster. The mclust algorithm also yields the marginal values, that is, the smallest and the largest bouton volume for each cluster (*see* Subheading 4.9 for R Code Used for Mclust Analysis).

3.5.2 Monte Carlo Analysis to Reveal the Cutoff Criteria Between Input Types with Adjacent Bouton Volume Distributions

The mixture modeling in mclust returns cluster classification for each data point; however, it does not provide generalizable cutoff values to sort any future data points or sets. We wanted to know specific volume cutoffs that could be used to sort future data. To that aim, we used the mclust subpopulation parameters (mean and standard deviation) to simulate larger sets of data for mixture modeling. The simulated dataset sizes can be scaled up in proportion to the original size of each subpopulation. In each of 10,000 replications, the average volume between adjacent simulated terminal clusters was used as a generalizable cutoff value for the particular brain region (*see* Subheading 4.10 for the R Code Used for the Monte Carlo Simulations Analysis).

3.6 Application of dLGN Volume-Sorting Cutoff Criteria in Analyzing Clustering of Origin-Sorted Bouton Synapsing on Origin-Identified Dendrites

For this analysis, we used a second dataset of terminal boutons that synapsed on select dendrite segments [13]. Terminals in this dataset were not sampled in an unbiased fashion by design: only the terminals synapsing on reconstructed geniculate dendrite segments constituted this dataset. Applying the volume-sorting cutoff values from the UTS dataset onto the second, independent dataset helped classify the boutons as from distinct origins on a subset of cell types.

3.6.1 Association Rules and Network Analysis

Following the classification of terminals, we aimed to identify the patterns of dendrite co-innervation by different bouton types using an association rule network modeling approach. First, to determine patterns of connectivity between input types, association rule learning was employed with the R package "arules" [16].

In its most common applications in commerce, the association rule approach is often used to determine what products consumers are likely to buy together at any "transaction." Rules are defined by two item sets, A and B, which each can include more than one actual item or product. Rules can be interpreted as if A and then B, meaning B is present if A is present. In some contexts, this relationship is represented with an arrow: {A → B}; A is referred to as the antecedent, and B is referred to as the consequent.

In our dataset for determining the tendency of inputs from a distinct origin to synapse on the same dendrite segment as another input, we treated each dendrite segment as a transaction and the terminals present on those segments as the purchases. The a priori algorithm to determine associations takes into account the net presence of an item in a given transaction, not the amount of that item present (i.e., treats item presence as a binary). For example, if a dendrite segment received 30, 10, 15, 8, 0, and 0 synapses from six bouton types identified in our dataset, the arules data for that dendrite segment were expressed as {1,1,1,1,0,0}. The confidence and lift values returned by the algorithm are used in interpreting the conditional probability of co-innervation by input pairs (*see* Subheading 4.11 for Interpretation of Arules Outcomes).

4 Notes and Considerations on Methodology

4.1 Serial Electron Microscopy Image Stack

The material used in the current study came from mouse dorsal lateral geniculate nucleus. Our material was prepared following the SBEM embedding protocol provided by Renovo Neural Inc., which subsequently prepared the image stacks. For preparation of new material, we recommend referring to more recent versions of the SBEM embedding protocol at https://www.protocols.io/view/preparation-of-biological-tissues-forserial-block-36wgq7je5vk5/v2 or the protocols that may be provided by the user's SEM Core Facility. The SBEM stacks used in the study are publicly available at https://bossdb.org/project/maher_briegel2023.

4.2 Computer

The minimum computer configuration requirements for the Reconstruct software are indicated as Microsoft Windows (XP/Vista/7/8/10), Pentium processor, 10 Mb of disk space for program files, and at least 128 Mb of RAM. Our reconstruction setup includes two.

Surface Studio 2 computers with 32GB RAM and 2 TB hard drive; operating systems Windows 10 and 11; Graphic adapter NVIDIA GeForce GTX 1070; Monitor is 28 inch 4500 × 3000 resolution. This setup was sufficient for manipulating SBEM image tracks at a full resolution (5 nm × 5 nm pixel size, 100 um × 100 um tile size, up to 600 images in stack).

4.3 Reconstruct Software

All image tracing was done using manual segmentation software Reconstruct [14], which is downloaded from ttps://synapseweb.clm.utexas.edu/software-0. The user manual for Reconstruct is available at https://synapseweb.clm.utexas.edu/sites/default/files/synapseweb/files/reconstructusermanualv 1.1.0.0.pdf.

In the current study, we used Reconstruct software and a manual segmentation approach, where a team of research assistants traced all sampled boutons and synapses. Reconstruct software was preferred because it allowed handling of a full stack of SBEM images without downsampling of resolution (pixel size: 5 nm × 5 nm) or tiling (40 um × 40 um image dimensions; 200–600 images in the stack). A manual segmentation approach allowed the experimenters to decide where a terminal bouton region ends and where the axon stalk begins. The manual approach also allowed the experimenter to directly identify and categorize the morphological features of terminal boutons, including mitochondria contrast (a morphological criterion for retinal terminals in the thalamus), the presence of protrusions within the bouton structure, synapse type (e.g., macular, perforated, adherentia), and the structure of the postsynaptic site (e.g., spine, dendritic shaft, soma).

4.4 R/RStudio

R is a programming language that can be used across multiple different graphic user interfaces. One of the most common (and recommended) interfaces is RStudio. R language and RStudio can be downloaded from the Comprehensive R Archive Network (CRAN) website https://cran.r-project.org/.

4.4.1 Necessary R Packages

We recommend using R packages mclust, arules, and qgraph for Gaussian mixture modeling, association rule learning, and network modeling, respectively.

To check if a package is already downloaded on your base R or RStudio, consult the "packages" window. If these packages are not present in base R or RStudio, use the *install.packages()* function to install and *library()* function to load before use.

4.5 Setting the X/Y/Z Parameters of the 3D Stereology Grid

In order to obtain an unbiased sample of synaptic terminal boutons in our tissue, we used a systematic random sampling approach as described in optical fractionator approaches (references here). To form a 3D grid of stereology cubes, we placed multiple 2D stereology matrices in the SEM stacks originating at the same X–Y coordinates of each Z-level image. Each 2D matrix contained rows and columns of counting frames that were drawn as two green inclusion lines and two red exclusion lines (Fig. 1). The X/Y/Z dimensions of the 3D cube (i.e., the X and Y dimensions of the counting frames and the distance between the 2D grids placed in Z dimension) were set to be larger than the largest synapse length that could be encountered in a particular tissue. This aimed to prevent oversampling of large boutons by ensuring that a synapse did not intersect both a green and red line that were parallel to each other in a 2D matrix and did not span across two subsequent matrices placed in Z dimension. In occasions when a synapse crossed a vertical and horizontal counting frame line, the nature of the vertical line (i.e., inclusion or exclusion) determined if the synapse would be included in our sample. For our sample from dLGN, where some very large boutons forming large synaptic zones are encountered, the X/Y/Z dimensions of the 3D cube was set as 2.5μm each. This yielded six 2D matrices with 15 × 15 counting frames, placed every 34 sections in the SEM stack of 200 images that were scanned at 5 nm × 5 nm pixel resolution and 75 nm section thickness. This yields 1350 cubes.

4.6 Sampling Strategy Validation Experiments

(1) The UTS was implemented on three different tissue samples, from mouse dLGN and vLGN, and the rat visual cortex layer 4, yielding 400–500 terminals in each dataset with 30–50% of the locations sampled in a single placement of the grid. The volume frequency distribution of all datasets displayed multiple peaks, suggesting multiple inputs with unique bouton sizes make up the circuitry in each brain region. On the other hand, the range of the volumes and the location of peaks in each brain region were different than in others (non-parametric tests, unpublished data), consistent with the presence of unique circuitry elements in each brain region. (2) The UTS was implemented twice on a dLGN SEM stack by placing the first grid on a different image in the sequence. In both sampling rounds, 50% of the locations were sampled, yielding over 500 terminals in each round. The distributions of terminal volumes obtaining at each round of data collection were not statistically different (non-parametric tests [13], suggesting that a single UTS grid placement was sufficient to capture samples from all inputs that contribute to the dLGN circuitry. This analysis was repeated using an SEM from V1 to layer 4, yielding over 200 boutons up to 30% of locations sampled in each stereology grid placement. Similarly, the cumulative volume distributions in each dataset were statistically indistinguishable (Mann–Whitney U, $p = 0.09$, unpublished data).

4.7 Criteria for Identifying Morphological Features of the Neuropil

Synapse A synapse is identified by the presence of a presynaptic cleft, a postsynaptic cleft, and vesicles in the presynaptic terminal. Frequently, vesicles can be seen docking to the presynaptic cleft. In some cases, the postsynaptic side will display a postsynaptic density, which consists of scaffolding proteins and appears as a dark smudge.

Axon Terminal An axon terminal is identified by the presence of vesicles and the formation of a synapse at the synaptic cleft. Some terminals are filled with vesicles, whereas some have only sparse vesicles.

Dendrite Neuron dendrites can be identified as such when they are postsynaptic to an axon terminal. Dendrites may contain mitochondria and scaffolding proteins.

Interneuron Dendrites Interneuron dendrites in the sensory thalamus are unique in that they are both postsynaptic to an axon terminal and presynaptic to another dendrite. In an EM image, these interneuron dendrites can be identified by the presence of a postsynaptic zone, as well as a presynaptic zone with vesicles.

Cell Body (Soma) Somata, or cell bodies, can be identified by the presence of a nucleus and, in some sections, a dense, darkly stained nucleolus. Somas also contain organelles such as endoplasmic reticulum and ribosomes.

4.8 Nomenclature

We recommend naming the synapses and the terminals encountered in each location in a way to facilitate further sorting by the type (synapse or bouton) or morphological properties of the bouton. In the current study, the following nomenclature scheme was suitable for our purpose:

Syn [location number, in 4 digits] [a, b, c. . ., if more than one synapse in a location] [on_d, on_F2, or on_soma to indicate the postsynaptic site is a dendrite, a presynaptic profile or a cell body, respectively].

Term [location number in 4 digits] [a, b,c. . ., matching the associated synapse's name] [LM, DM or NM, for boutons containing light, dark or no mitochondria] [F2, if the bouton is postsynaptic to a synapse anywhere along the reconstructed volume].

Examples: Syn0367 is a synapse that is found within the inclusion zones of location 367. This was the only synapse in this location. The postsynaptic profile was a dendrite.

Term0367_NM is a bouton that forms Syn0367. It does not contain a mitochondrion within the reconstructed volume.

Syn0026b_on_F2 is a synapse that is found within the inclusion zones of location 26, which displays more than one synapse. The postsynaptic profile was both post- and presynaptic, a feature of triadic arrangements in the mammalian thalamus.

Term0026b_LM is the bouton that forms Syn026b. It contains light-contrast mitochondria, a property of retinal terminals in the LGN.

Once retrieved from Reconstruct, the data were sorted to organize synapses and terminals associated with each other on the same rows, and the categorical information in the object names are converted to codes for analysis.

4.9 MClust Analysis

The R script *MClust_Volume_Sorting.R* (University of Virginia Dataverse https://doi.org/10.18130/V3/9VXAZ7) is an example of implementing mclust package for the analysis of a dataset of about 1000 terminal bouton volumes (the sample data is available at doi.org/10.18130/V3/9VXAZ7). The boutons are initially categorized into two categories based on morphological criteria to distinguish inputs from the retina (with light mitochondria) vs from other sources (with dark or no mitochondria). The R script yields the parameters (mean, n, and SD) of subpopulations within each category, identified via Gaussian normal mixture modeling.

4.10 Monte Carlo Simulations

The R script *MonteCarlo_VolumeSortingCutoffs.R* (University of Virginia Dataverse https://doi.org/10.18130/V3/9VXAZ7) is an example for implementing the subpopulation parameters obtained by the mclust analysis in Monte Carlo simulations to compute generalizable cutoff points between each subpopulation with overlapping distributions. The example is based on the population of retinal terminals, the distribution of which reveals four subpopulations.

4.11 Interpretation of Arules Outcomes

Three values that the algorithm returns are of specific note: support, confidence, and lift (Hornik et al. 2005). Support is defined as the proportion of total transactions that contain a given item set. Support is used to calculate both confidence and lift:

$$supp = (A \cup B).$$

The confidence of a rule can be interpreted as the conditional probability of finding item A when item B is given as present:

$$Confidence(A \rightarrow B) = supp(A \cup B)/supp(A).$$

Lift is the ratio of occurrence of a rule to the occurrence of each item individually, if they were independent. Lift is related to confidence but has a greater focus on controlling the popularity of an item:

$$Lift(A \rightarrow B) = supp(A \cup B)/(supp(A) * supp(B)).$$

While confidence can be a useful tool to identify relationships, it is highly dependent on the popularity of an item set. Lift is an additional filter to select only the strongest association rules. An additional difference between lift and confidence is that lift is symmetric, while confidence is not (i.e., reversing the antecedent and consequent does change the value of lift) [17].

Interpretation of Lift Values If the lift is equal to 1, there is no relationship between the two items in a rule. If the lift is greater than 1, then the presence of one item positively impacts the presence of the other. If the lift is less than 1, then the presence of one item negatively affects the presence of the other.

Interpretation of Confidence Values The confidence values of terminal associations can be plotted as a network model of the relationships between terminal clusters using the R package "qgraph" [18]. Networks involve nodes, which represent individual elements, and edges, which represent relationships between those elements [19]. Two measures of centrality are used to measure relationships: out-flow and in-flow. Out-flow is a measure of how much one node influences other nodes. In-flow is a measure of how much one node is influenced by other nodes. The larger the size of the node, the larger the measure of centrality (see Maher et al. (2023) [13] for implementation of the method and visualization of lift and confidence values).

5 Summary

Here, we outlined a method to collect a random sample of terminal volumes from an SBEM image stack, representing input heterogeneity of all terminal populations that make up the synaptic circuitry. This method may allow for addressing hypothesis-based comparisons across multiple experimental conditions, ages, or brain regions using SBEM stacks. As a further application of the quantitative measures yielded from UTS data, a statistical volume-sorting approach is developed and implemented to reveal the co-innervation of neurons by a set of inputs from distinct origins. The UTS manual segmentation is a feasible approach that is suitable for quantifications of input morphology and synapse density using small SEM stacks from multiple experimental conditions.

References

1. Denk W, Horstmann H (2004) Serial block-face scanning electron microscopy to reconstruct three-dimensional tissue nanostructure. PLoS Biol 2(11):e329
2. Merchán-Pérez A, Rodriguez JR, Alonso-Nanclares L, Schertel A, DeFelipe J (2009) Counting synapses using FIB/SEM microscopy: A true revolution for ultrastructural volume reconstruction. Front Neuroanat 3:3. https://doi.org/10.3389/neuro.05.018.2009
3. Kasthuri N, Hayworth KJ, Berger DR et al (2015) Saturated Reconstruction of a Volume of Neocortex. Cell 162(3):648–661. https://doi.org/10.1016/j.cell.2015.06.054
4. Joesch M, Mankus D, Yamagata M et al (2016) Reconstruction of genetically identified neurons imaged by serial-section electron microscopy. elife 5:5. https://doi.org/10.7554/ELIFE.15015
5. Morgan JL, Berger DR, Wetzel AW, Lichtman JW (2016) The fuzzy logic of network

connectivity in mouse visual thalamus. Cell 165(1):192–206. https://doi.org/10.1016/j.cell.2016.02.033

6. Scheffer LK, Xu CS, Januszewski M et al (2020) A connectome and analysis of the adult drosophila central brain. elife 9:1–74. https://doi.org/10.7554/ELIFE.57443
7. DeFelipe J, Marco P, Busturia I, Merchán-Pérez A. Estimation of the number of synapses in the cerebral cortex: methodological considerations. Cereb cortex (New York 1991. 9(7): 722–732. http://cercor.oxfordjournals.org/content/9/7/722.long
8. Guillery RW (1969) The organization of synaptic interconnections in the laminae of the dorsal lateral geniculate nucleus of the cat. Zeitschrift für Zellforsch und Mikroskopische Anat 96(1):1–38. https://doi.org/10.1007/BF00321474
9. Jones EG, Powell TP (1969) Electron microscopy of synaptic glomeruli in the thalamic relay nuclei of the cat. Proc R Soc Lond B Biol Sci 172(1027):153–171. http://eutils.ncbi.nlm.nih.gov/entrez/eutils/elink.fcgi?dbfrom=pubmed&id=4388107&retmode=ref&cmd=prlinks
10. Fitzpatrick D, Diamond IT, Raczkowski D (1989) Cholinergic and monoaminergic innervation of the cat's thalamus: comparison of the lateral geniculate nucleus with other principal sensory nuclei. J Comp Neurol 288(4): 647–675. https://doi.org/10.1002/cne.902880411
11. Erisir A, Van Horn SC, Sherman SM (1997) Relative numbers of cortical and brainstem inputs to the lateral geniculate nucleus. Proc Natl Acad Sci USA 94(4):1517–1520. https://doi.org/10.1073/pnas.94.4.1517
12. Erisir A, Van Horn SC, Sherman SM (1998) Distribution of synapses in the lateral geniculate nucleus of the cat: differences between laminae A and A1 and between relay cells and interneurons. J Comp Neurol 390(2): 247–255. https://doi.org/10.1002/(SICI)1096-9861(19980112)390:2<247::AID-CNE7>3.0.CO;2-1
13. Maher EE, Briegel AC, Imtiaz S, Fox MA, Golino H, Erisir A (2023) 3D electron microscopy and volume-based bouton sorting reveal the selectivity of inputs onto geniculate relay cell and interneuron dendrite segments. Front Neuroanat 17:17. https://doi.org/10.3389/fnana.2023.1150747
14. Fiala JC (2005) Reconstruct: a free editor for serial section microscopy. J Microsc 218(1): 52–61. https://doi.org/10.1111/j.1365-2818.2005.01466.x
15. Scrucca L, Fop M, Murphy TB, Raftery AE (2016) mclust 5: clustering, classification and density estimation using Gaussian finite mixture models. R J 8(1):289. https://doi.org/10.32614/rj-2016-021
16. Hahsler M, Grun B, Hornik K (2005) arules—a computational environment for mining association rules and frequent item sets. J Stat Softw 14(15):1–6
17. Azevedo PJ, Jorge AM (2007) Comparing rule measures for predictive association rules. Lect Notes Comput Sci (including Subser Lect Notes Artif Intell Lect Notes Bioinformatics). 4701 LNAI:510–517. https://doi.org/10.1007/978-3-540-74958-5_47/COVER
18. Epskamp S, Cramer AOJ, Waldorp LJ, Schmittmann VD, Borsboom D (2012) qgraph: network visualizations of relationships in psychometric data. J Stat Softw 48:1–18. https://doi.org/10.18637/JSS.V048.I04
19. Harary F (1969) Graph theory (on demand printing of 02787). CRC Press. https://doi.org/10.1201/9780429493768

Chapter 5

A Technique to Perform Structure–Function Analysis of Synapses

Carolin Wichmann and Susann Michanski

Abstract

In this chapter, we will present a detailed methodology for nanoscale structure–function analysis of an ultrafast and high-performing synapse, the auditory ribbon synapse. Ribbon synapses are found in sensory systems and are important to convey sensory information without synaptic fatigue. These synapses are characterized by a highly specialized structure, the so-called synaptic ribbon. This ribbon tethers a large number of synaptic vesicles to facilitate constant neurotransmission. Auditory ribbon synapses, located in the cochlear inner hair cells, furthermore display a specific set of synaptic proteins, that are partially distinct from those found in conventional central synapses or even other ribbon synapses. In order to understand how such specialized structures mediate indefatigable neurotransmission, we perform structure–function analysis of hair cell synapses by combining a simple stimulation method with electron microscopy. We describe in detail how to use chemical solutions in order to depolarize inner hair cell ribbon synapses and then rapidly immobilize the tissue utilizing cryofixation by high-pressure freezing followed by freeze-substitution. In this chapter, we will put emphasis on the sample depolarization as well as further processing steps. Finally, we provide a short description of how to analyze chemically induced changes of morphological vesicle pools and active zone parameters to obtain information about the fine structures of individual synapses.

Key words Structure–function relationship, Synapse, Chemical stimulation, High-pressure freezing, Freeze-substitution, Cryofixation

1 Introduction

In neuroscience, the correlation of structure and function of synapses is key to understanding brain function, memory formation, sensory processing, and disease mechanisms. Synaptic neurotransmission is a highly coordinated process that requires the interplay of a large number of proteins on the pre- and postsynaptic side. Upon a stimulus, rapid electrical and chemical processes take place to convey the incoming signal to the adjacent neuron. Already more than a century ago, methods were developed to determine the structure of neurons and correlate them to different

Joachim H. R. Lübke and Astrid Rollenhagen (eds.), *New Aspects in Analyzing the Synaptic Organization of the Brain*, Neuromethods, vol. 212, https://doi.org/10.1007/978-1-0716-4019-7_5,

functionalities. Heinrich Wilhelm Gottfried von Waldeyer-Hartz (1836–1921) established the term *Neuron*, while different compartments of neurons were already discovered early on by Otto Deiters (1834–1863), who described the axon for the first time. The anatomist Deiters investigated brain tissue as well as muscles, the spinal cord, and peripheral organs. He mainly teased out the finer structures, and this way revealed many so far unknown details of nervous cells [1]. Wilhelm His (1831–1904) developed the microtomy, and this way enabled the investigation of thin sections to provide lasting insight into the development of the central and peripheral nervous system (CNS and PNS) [2]. In order to investigate neuronal structures in finer detail, the invention of the silver impregnation method by Camillo Golgi (1843–1926) was groundbreaking. This staining method was used and improved by Santiago Ramón y Cajal (1852–1934). Cajal traced neurons in a plethora of organs and tissues, and his beautiful drawings substantially increased our knowledge about the anatomy of the brain but also peripheral organs such as the cochlea. Based on these studies, Cajal postulated that the neuronal network of the brain consists of individual cells—a hypothesis that challenged the general believe at these times that neural circuits are organized of neurons with physical continuity, allowing an information transfer in each direction [3]. Based on the morphology of neurons, Cajal further postulated a unidirectional information transfer [3]. This hypothesis set the ground for neuroscience research. Charles Scott Sherrington (1857–1952) finally elucidated the synapse, a nexus for reflex-arc function, and synaptic transmission [4].

The synapse as the basic communication unit in the CNS and PNS is since then subject of intense research, and a vast body of literature exists that deals with combined structure–function analysis of synapses. The term "structure–function" applies for a broad range of experimental set-ups and research questions. The correlation of structural changes to functional manipulations spans from the investigation of synapse morphology of mutants lacking important synaptic proteins, during synapse formation or neurodegeneration, in disease models or during acute or chronic activation or inhibition. Structural changes can be determined using different microscopic techniques such as confocal or electron microscopy after cell-physiological manipulations [5–12]. Similar to Cajal's studies, neurons can be traced back using different filling methods after cell-physiological recordings. This way, neuronal morphology and their target cells can be identified in direct correlation to their function. For example, Zandt et al. determined the cellular morphology after whole-cell recordings from AII amacrine cells in rat retinal slices. Using fluorescent dye filling, subsequent multiphoton excitation microscopy, and quantitative morphological reconstruction by computer-aided manual tracing, evidence for morphological subtypes of AII amacrines could not be found

[13]. The correlation of cell-physiological recordings and the subsequent imaging on the level of synapses of a previously characterized neuron using electron microscopy is one of the most challenging tasks. Such experiments were, for example, performed by Holler and co-workers. The researchers recorded from neocortical synapses and subsequently used biocytin fillings of dendrites and axons in order to trace back the synapses of recorded neurons on the light and finally electron microscopic level [14]. Furthermore, horseradish peroxidase injections were used to receive a direct correlation of recorded neurons and labeled boutons [15]. Horseradish peroxidase is also frequently used to label endosomal structures upon stimulation-dependent internalization. Labeled vesicular structures could this way be identified on the ultrastructural level after activity induction [16]. Finally, in a complex workflow, Scholl and co-workers combined visually driven activity, two-photon microscopy and ultrastructural anatomy of synapses using serial block face scanning electron microscopy on single cortical neurons from the ferret visual cortex 1 [17]. These are only very few examples of many beautiful studies demonstrating a direct correlation of structure and function in regard to active zone size (AZ), number of synaptic vesicles, or cleft dimensions. However, these methods are technically extremely challenging, and the structural preservation of the synapse is often limited due to the use of biocytin, horseradish peroxidase, or fluorescent dye staining in combination with chemical fixation reagents such as paraformaldehyde, glutaraldehyde, or picric acid.

Even though these methods are of utmost importance to deliver direct structural and functional information, there are other—more simplistic—ways to circumvent the need to stimulate a specific neuron and find this neuron and its synapses back in an ultrastructural preparation. Such methods allow for a broader quantification, are easier to perform, and give the opportunity to make use of excellent tissue preservation methods in order to uncover so far hidden structural information.

Heuser and Reese performed pioneering experiments by applying electrical stimulation on frog neuromuscular junctions of isolated muscles [18]. They conducted different stimulation protocols with durations and recovery periods and immobilized the samples using immersion fixation and later quick-freezing [19]. Subsequently, they determined the vesicular structures within junctional boutons on the ultrastructural level. Step by step, Heuser and Reese put the puzzle together to finally identify the pathway underlying synaptic vesicle recycling at neuronal synapses [18].

The arrival of cryopreservation methods for a rapid tissue immobilization lifted the ultrastructural investigation to a new level. However, ultra-rapid freezing relies on fast cooling by slamming the sample onto a cryogen-cooled metal block or directly plunging the sample itself into the cryogen, which is typically ethane or propane [20]. Yet, such approaches can only be applied to single cell layers. The development of ultra-rapid cryofixation using high-pressure freezing (HPF) opened a new window for researchers, since this method represents a way to freeze the tissue in a near-to-native state up to 200 μm thickness [20–23]. During HPF, the immobilization of the sample occurs within milliseconds under a pressure above 1700 bar and liquid nitrogen (LN_2), gaining a temporal "snapshot" of the sample. Hereby, ice-crystal formation is prevented due to the high pressure, and the cell water is frozen in an amorphous stage. The actual fixation (such as the chemical cross-linking) takes place during a process called freeze-substitution (FS) [24–27]. FS involves substituting organic solvents for the cellular water at low temperatures, typically −78° to −90 °C. For this purpose, the most common solvent is acetone [20]. Therefore, brain slices, organotypic cultures, or whole tissues like the organ of Corti can be cryopreserved without ice-crystal formation and subsequently processed using HPF/FS [28–50]. HPF with subsequent FS requires a high-pressure freezing machine, which is available from different companies and ideally a FS machine. However, the latter is not necessarily required due to the existing protocols that are based on normal laboratory tools together with LN_2 [51].

One key feature for a direct correlation of morphology and function is the close temporal coupling of the stimulation with the onset of freezing. However, direct stimulation methods of neurons are not easy to combine with ultra-rapid cryofixation—especially when aiming for a close coupling between stimulation and freezing. Most freezing devices are not equipped with any stimulation unit. In principle, a field stimulation could also be performed outside the freezing machine, but mounting fresh tissue into these specialized machines might take seconds or even minutes. Therefore, either the stimulation has to take place in the freezing chamber such as in high-pressure freezers that are equipped with a light source in order to perform optogenetic activation of neurons or electrical stimulation units. Finally, one can use chemical depolarizations, which can be combined easily with a longer mounting procedure.

Chemical depolarization of neurons is widely performed using high-potassium solutions (hereafter referred to as "high-K^+"). This method has been applied on many different synapse types with a variety of protocols, for example, to stimulate *Drosophila melanogaster* neuromuscular junctions (NMJs) to drive activity-dependent FM1–43 dye uptake/release. Bathing synaptic preparations in high concentrations of K^+ is often used to induce and study synaptic vesicle (SV) cycling; as done for frog cholinergic NMJs [52], cultured rodent brain hippocampal neurons [53], and *Drosophila* glutamatergic NMJ models [54, 55].

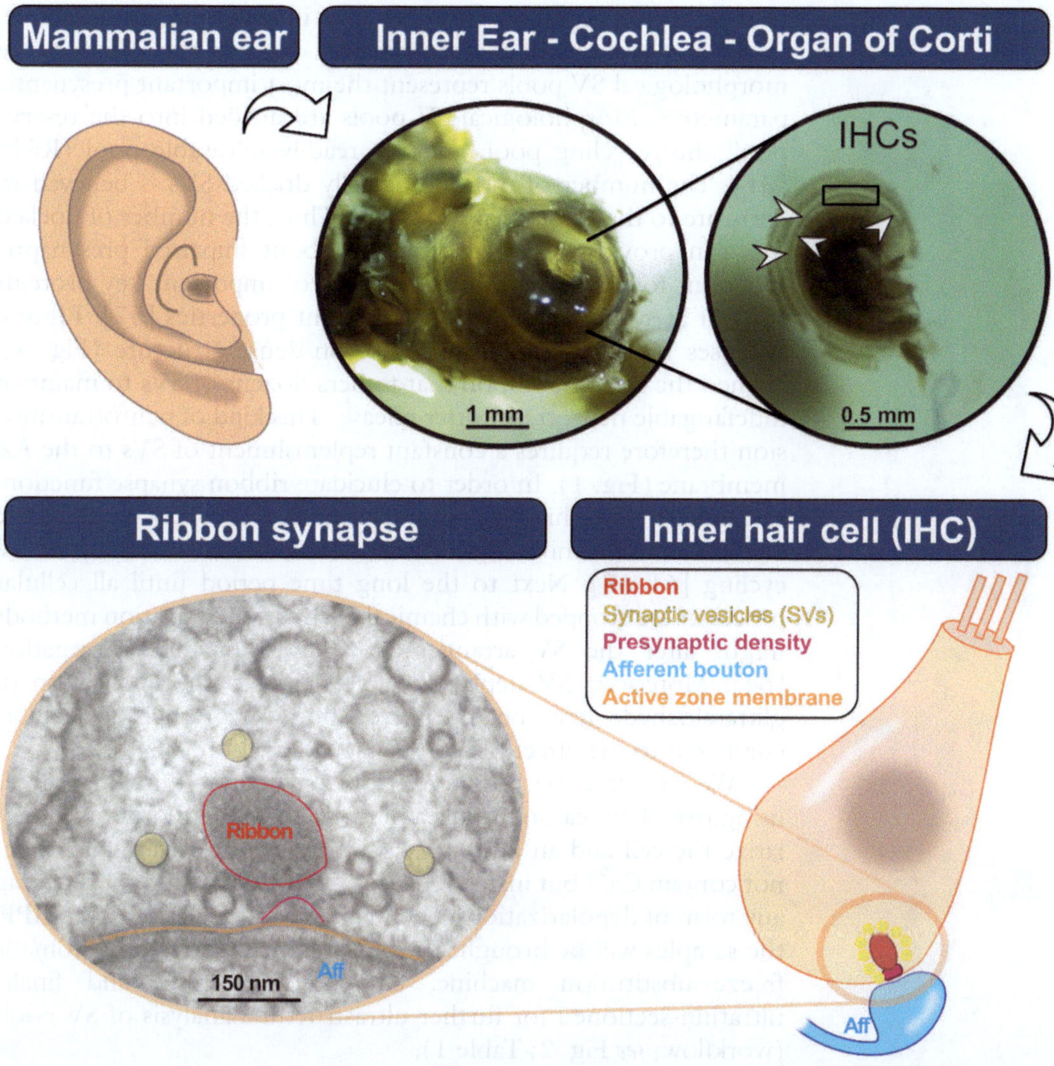

Fig. 1 Ribbon synapses within the cochlear inner hair cells (IHCs). The mammalian ear consists of the outer (upper left panel), middle, and inner ear. The inner ear harbors the cochlea (upper middle panel, fixed and osmium-stained for better visibility). Inside the cochlea the sensory tissue, the organ of Corti is found (upper right panel, apical 1.5 turns of the organ of Corti, osmium-stained for better visibility). The IHC row is visible as a thin line (box and arrowheads). Lower right panel: Schematic drawing of an IHC harboring a ribbon synapse, innervated by an afferent bouton (Aff). The lower left panel shows an exemplary HPF/FS murine ribbon synapse tethering SVs close to the AZ (60 nm ultrathin section). (Figure was created with Adobe Illustrator and Affinity Designer)

In this chapter, we present how to perform this simple chemical stimulation combined with HPF/FS on the cochlear inner hair cell (IHC) ribbon synapse, which are located in the sensory organ of Corti (Fig. 1). Ultra-rapid freezing approaches immobilize cellular processes within milliseconds [20] and have therefore become a standard tool for synapse research [28–32, 41, 43–45, 49, 56–59]. Structural components at individual AZs can be captured in a near-to-native state and this way give information, for example, about the strength of a given synapse. Next to the AZ size, the morphological SV pools represent the most important presynaptic parameters. Morphological SV pools are divided into the reserve pool, the recycling pool, and the readily releasable pool (RRP) [31]. The number of morphologically docked SVs is believed to correlate to the functional RRP [60]. Thus, the number of docked SVs can provide direct information about impaired presynaptic function, for instance, due to the lack of important key proteins [30] or at synapses that exhibit different properties [57]. Ribbon synapses harbor a specialized electron-dense structure (Fig. 1), termed the synaptic ribbon that tethers dozens of SVs to maintain indefatigable neurotransmitter release. This kind of neurotransmission therefore requires a constant replenishment of SVs to the AZ membrane (Fig. 1). In order to elucidate ribbon synapse function, researchers used high-K^+ stimulation and subsequent aldehyde fixation to determine the different structural features upon SV cycling [61–63]. Next to the long time period until all cellular processes are stopped with chemical fixatives, such fixation methods might alter the SV arrangement due to protein aggregation [64]. Moreover, SV depletion effects due to the application of glutaraldehyde have been reported but were rather neglectable compared to the effect of high-K^+ depolarization [65].

We will describe how to perform structure–function analysis using two chemical solutions: a high-K^+ solution in order to depolarize the cell and an inhibitory solution as a control, which does not contain Ca^{2+} but instead the Ca^{2+}-chelator EGTA, for blocking any form of depolarization-dependent synaptic release. After HPF, the samples will be brought to room temperature in an automatic freeze-substitution machine, embedded in resin and finally ultrathin-sectioned for further ultrastructural analysis of SV pools (workflow; *see* Fig. 2; Table 1).

2 Materials

2.1 Preparation of the Organ of Corti and High-Pressure Freezing (HPF)

The dissection of the organ of Corti is carried out in Sylgard-coated glass-bottom dishes filled with ice-cold HEPES-HANKS solution (pH 7.2, osmolarity 300 mmol/kg). In addition to scissors and forceps for tissue preparation, a micro curette is needed for the transfer of the organ of Corti from the glass dish into the inhibitory or stimulation solution containing petri dish (Please note that animal handling and experimental procedure for organ of Corti preparation (Fig. 1) were performed in accordance with the national animal care guidelines issued by the animal welfare committees of the University of Göttingen and the Animal Welfare Office of the State of Lower Saxony). Alternatively, a small piece of filter paper can be used for the transfer step. For the HPF sample

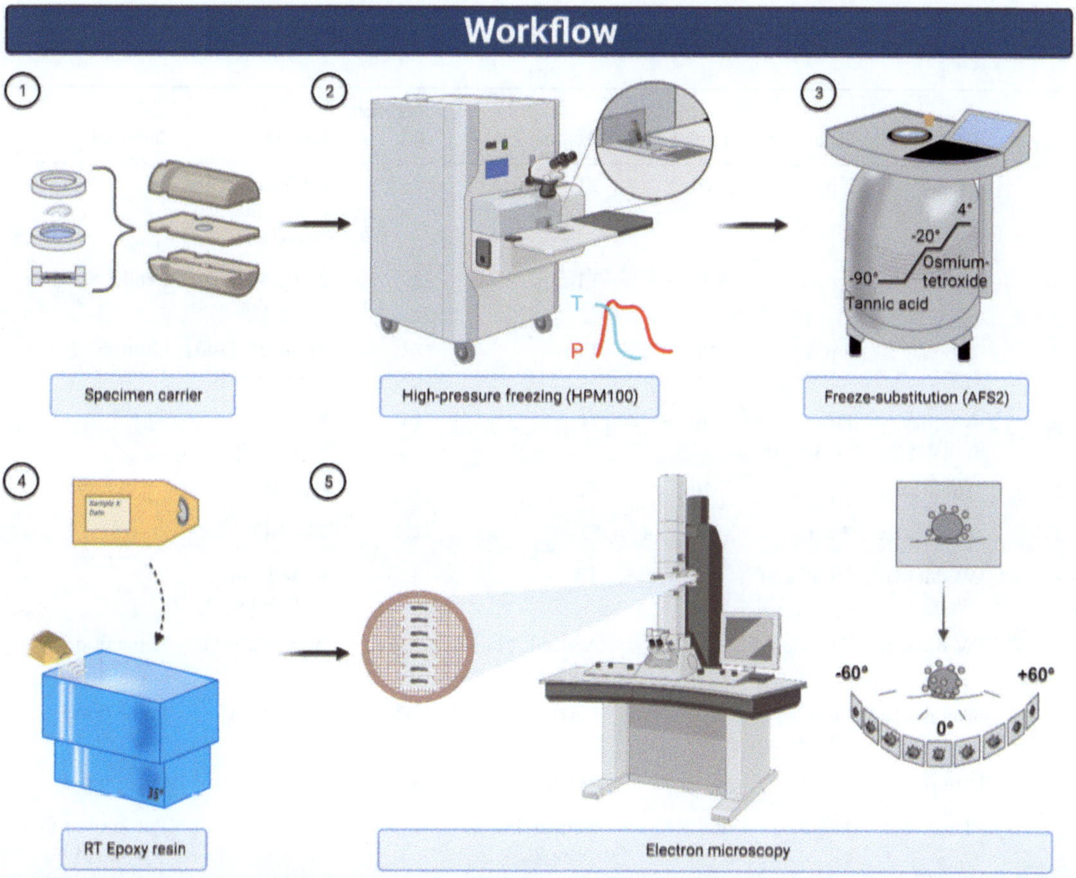

Fig. 2 Summary of the workflow from specimen assembly to image acquisition. (1) After dissection of the tissue followed by a chemical stimulation/inhibition step, the sample is placed into the 3 mm aluminum specimen carrier sandwich, which goes into the middle plate of the cartridge system for the high-pressure freezer (HPM100). (2) Next, the cartridge system is assembled, and the cryoimmobilization process is started using the HPM100. (3) The high-pressure frozen sample can now be stored in liquid nitrogen (LN_2) until further processing in the freeze-substitution machine (EM AFS2) where the sample is gradually warmed up to 4 °C and the frozen water is exchanged with an organic solvent. (4) Subsequently, the sample is embedded and polymerized in epoxy resin from which ultrathin or semithin sections are cut at room temperature (RT). (5) Finally, on formvar-coated copper grids, the collected sections are subjected to electron microscopy for image acquisition. (Figure was created with BioRender.com (1, 2, 5) and Adobe Illustrator)

assembly system, specimen carrier type A and B, sample holder (cartridge system), and 1-hexadecane as a filler to prevent air bubble formation are required. For more information about the individual steps, products, and solutions, see Fig. 3, Subheading 3.1, Table 2.

2.2 Freeze-Substitution (FS)

FS is performed in a Leica EM AFS2 (Leica Microsystems) machine. Next to forceps for the transfer, aluminum containers, plastic lids, plastic bottles, plastic baskets, falcons, a petri dish lid, disposable pipettes, razor blade, styrofoam box, metal stage, tannic

Table 1
Freeze-substitution protocol program

Day #	Day	Time	Temp [°C]	Program	Duration [h]	Solution	Comments
1	Friday[a]	2 p.m.	−90	T1	73	0.1% tannic acid in acetone	
4	Monday[a]	3 p.m.	−90	T1	2	3× wash in acetone	Refill LN_2
4	Monday[a]	5 p.m.	−90	T1	7	Apply 2% OsO_4 in acetone	Caution!
4	Monday/ Tuesday	From 12 a.m. on	−90 → −20	S1	14	5 °C/h increment	
5	Tuesday	2 p.m.	−20	T2	17	2% OsO_4	
6	Wednesday	7 a.m.	-20 → +4	S2	2.4	2% OsO_4	
6	Wednesday[a]	~9:30 a.m.	+4	T3	3	3× wash in acetone	
6	Wednesday[a]	~1:30 p.m.	RT	Acetone/resin 1:1	3–6	Acetone/resin	Fume hood
6	Wednesday[a]	From ~3:30 p.m.	RT	Pure resin	ON	100% resin	Fume hood
7	Thursday[a]	~9 a.m.	RT	3–6 h	3–6	100% resin	Fume hood
7	Thursday[a]	Noon	RT	Embedding mold, positioning			Put in oven for at least 48 h (70 °C) for polymerization

[a]You need to be present/be active

acid, acetone, osmium tetroxide (OsO_4), Na-thiosulfate, special nitril gloves, safety goggles, FFP3 mask, and epoxy resin are required for performing the FS (*see* also Fig. 4). A detailed description of each step and respective material can be found in Tables 2 and 3 as well as in Subheading 3.2. The program is depicted in Table 1 and was previously published in [32].

2.3 Trimming and Ultrathin Sectioning

After embedding, specimen bloc trimming and ultrathin sectioning need to be performed. These steps require a set of diamond knives (DiATOME, trimming 45° and ultra 35° knife), a file, an ultramicrotome, formvar-coated grids (e.g., copper slot grids and mesh grids), grid boxes, and forceps.

2.4 Electron Microscopy and Analysis: Brief

Electron microscopy can be performed with any transmission electron microscope (TEM); for specific electron tomography, any kind of TEM or scanning (S)TEM tomography microscope is suitable.

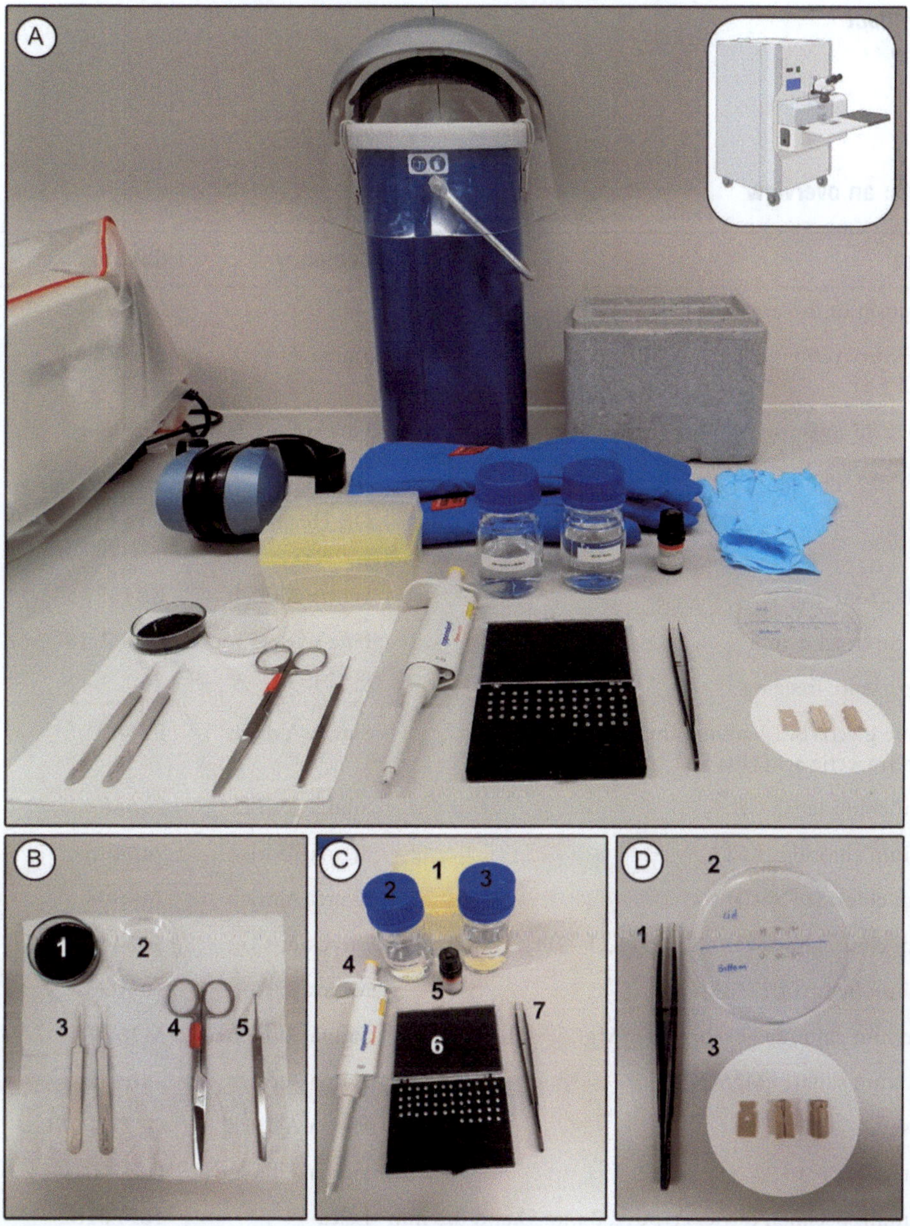

Fig. 3 Material for sample preparation and high-pressure freezing. (**a**) Summary of all requirements for the specimen preparation and HPF procedure. (**b**) Detailed illustration of the preparation material for the organ of Corti dissection, including: (1) Sylgard-coated glass-bottom dish, (2) plastic petri dish, (3) Dumont #5 forceps, (4) scissors, and (5) micro-curette. (**c**) After sample preparation the following tools are used: (1) pipette tips, (2) stimulatory/inhibitory solution, (3) HEPES-HANKS solution, (4) pipette, (5) 1-hexadecane, (6) specimen carrier type A and B, and (7) coarse forceps. (**d**) For the final sample assembly step, coarse forceps (1), the presorted specimen carriers (Bottom = type A, Lid = type B) (2), the cartridge system with two half cylinders, and one middle plate and filter paper (3) are required. Please note that for the handling with LN_2 special protection equipment is essential. As shown in (**a**) besides the lab coat and normal gloves for the tissue preparation also cryogloves, a protective face shield or safety goggles and noise cancelling ear protection are necessary. To keep the sample during the preparation time at ice-cold temperature, an ice-filled styrofoam box needs to be prepared in addition to the LN_2 dewar, which is needed for sample storage after the HPF protocol (**a**). (Figure was created with BioRender.com and Affinity Designer)

2.5 Product Overview

Table 2
Materials: an overview

Product	Company	Cat. no.
Preparation of the organ of Corti		
4-(2-Hydroxyethyl)-1-piperazineethanesulfonic acid (HEPES)	Sigma-Aldrich	H3375
Dumont #5 forceps	Fine science tools	11295-10
Petri dishes (plastic)	Greiner bio-one	633180
Sylgard	GE Healthcare Bayer	RTV615-A + B-1LBS-2-Silikon
Glass-bottom dish	Sigma-Aldrich	BR455717
Micro curette 1.5 mm	Fine science tools	10082-15
High-pressure freezing		
Ethylene glycol-bis(β-aminoethyl ether)-N,N,N′,N′-tetraacetic acid (EGTA)	Sigma-Aldrich	E3889
Cesium chloride (CsCl)	Sigma-Aldrich	746487
Potassium chloride (KCl)	Sigma-Aldrich	746436
Sodium chloride (NaCl)	Sigma-Aldrich	746398
Sodium citrate dihydrate ($Na_3C_6H_5O_7$-$2H_2O$)	Merck & Co.	567446
Potassium hydroxide (KOH)	Merck & Co.	105033
Magnesium chloride hexahydrate ($MgCl_2$-$6H_2O$)	Sigma-Aldrich	M2670
Magnesium sulfate ($MgSO_4$)	Sigma-Aldrich	746452
Glucose	Sigma-Aldrich	G8270-1KG
L-Glutamine	Sigma-Aldrich	G3126-100 g
Aluminum specimen carriers type A, 3 mm Ø and 0.2 mm depth	Leica microsystems	16770152
Aluminum specimen carriers type B, 3 mm Ø and 0.1 mm depth	Leica microsystems	16770153
1-Hexadecene	Sigma-Aldrich	52276.5 ml
Sample holder half cylinder cartridge	Leica microsystems	16770135
Sample holder middle plate 3 mm Ø	Leica microsystems	16770136
Freeze-substitution		
Tannic acid	Sigma-Aldrich	403040

(continued)

Table 2
(continued)

Product	Company	Cat. no.
Acetone, glass distilled	Carl Roth	5025.5
Osmium tetroxide (OsO_4) solid	Sigma-Aldrich	201030.250 g.D
Embedding molds	Agar scientific	G3533
EPON pre-mix kit	Agar scientific	R1140
Aluminum container	Leica microsystems	16702735
Plastic basket with gaze	Leica microsystems	16702734
Plastic lid	Pöppelmann	GPN600 B 300
Plastic bottles	Thermo Fisher Scientific	2006-9025
SHIELDskin CHEM™ NEO NITRILE™ 300	SHIELDskin	67-64-1
Trimming and ultrathin sectioning		
Vinylec E (Formvar®) Resin	TED Pella, Inc.	19222
Chloroform	Merck & Co.	1.024.451.000
Copper 100 mesh grids	Plano	G2410C
Copper slot grids 3.05 mm	Plano	G2500C
Uranyl acetate replacement	EMS	22405
Grid boxes	Plano	B801003080
Protein A 10 nm gold	CMC Utrecht	PAG 10 nm/S
Equipment		
EM AFS2	Leica microsystems	
EM HPM100	Leica microsystems	
JEM-1011 (80 kV)	JEOL	
JEM-2100 (200 kV)	JEOL	
Orius SC1000 CCD camera	GATAN	
Ultra 35° diamond knife	DiATOME	
UC7	Leica microsystems	
Osmometer VAPRO 5520	Kreienbaum	
Cryotrim 45°	DiATOME	

2.6 Solutions

Table 3
Solution details

Solution	Concentrations and using and storing conditions
Formvar solution	1% in chloroform (w/v), light protected, used under the fume hood at RT
HEPES-HANKS balanced salt solution	NaCl 141.7 mM, HEPES 10 mM, KCl 5.36 mM, $MgCl_2$-$6H_2O$ 1 mM, $MgSO_4$-$7H_2O$ 0.5 mM, glucose 2 mg/ml, L-glutamine 0.5 mg/ml. pH was adjusted to 7.2 with NaOH. Osmolarity: 300 mmol/kg *Prepare stock solution which can be stored at 4 °C for several weeks. Take the amount needed for the experiment, and add glucose as well as L-glutamine freshly*
Inhibitory solution	NaCl 130 mM, HEPES 10 mM, KCl 5 mM, $MgCl_2$-$6H_2O$ 3 mM, EGTA 5 mM, glucose 2 mg/ml. pH was adjusted to 7.2 with NaOH. Osmolarity: 300 mmol/kg *Prepare stock solution which can be stored at 4 °C for several weeks. Take the amount needed for the experiment, and add glucose freshly*
Tannic acid solution	1% tannic acid in acetone (w/v) *CAUTION with acetone: Wear special gloves (that protect against acetone; see* Table 2 *under "Freeze-substitution")*
Solid OsO_4	2% in acetone (w/v) *CAUTION! Toxic! Never open or use outside the fume hood. Wear special gloves (that protect against acetone; see* Table 2 *under "Freeze-substitution"), lab coat, and safety goggles. Prepared freshly on the day of the experiment under the fume hood*
Stimulatory solution	NaCl 95 mM, HEPES 10 mM, KCl 50 mM, $MgCl_2$-$6H_2O$ 1 mM, $CaCl_2$ 5 mM, glucose 2 mg/ml. pH was adjusted to 7.2 with NaOH. Osmolarity: 300 mmol/kg *Prepare stock solution which can be stored at 4 °C for several weeks. Take the amount needed for the experiment, and add glucose freshly*

3 Methods

For the best cryofixation results, the following in detail described technical steps are essential as depicted in Fig. 2. We use HPF in combination with FS, which allows for a rapid sample immobilization with a near-to-native structural preservation and a subsequent analysis at room temperature. Details are described in the legend of Fig. 2, and respective material can be found in Tables 2 and 3.

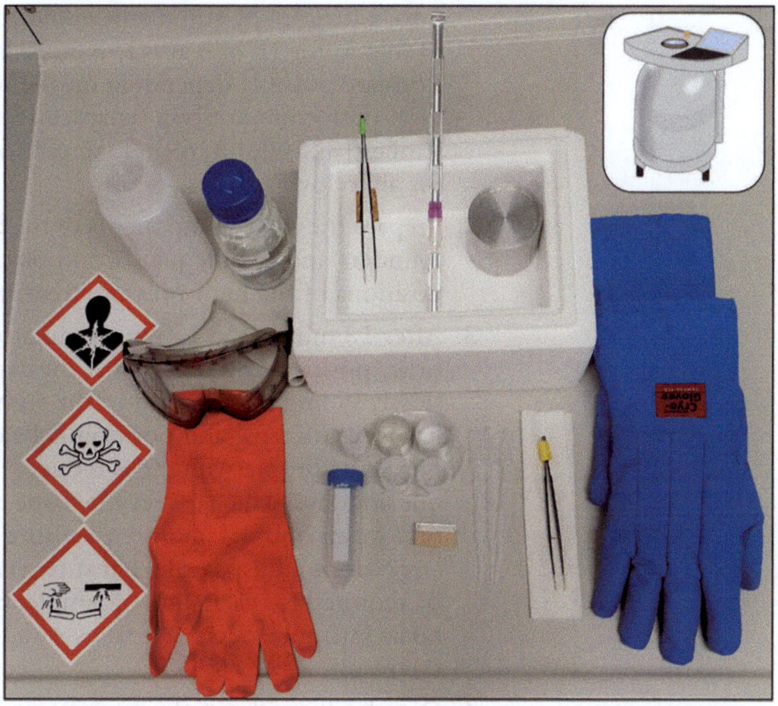

Fig. 4 Material for freeze-substitution. Overview of the above-listed materials that are required for the freeze-substitution protocol. For further description see main text. (Figure was created with Adobe Illustrator and Affinity Designer)

3.1 Preparation and High-Pressure Freezing (HPF)

As described previously, HPF enables the stabilization of whole cells and thick tissue preparations by cryofixation within milliseconds, thereby reducing fixation artifacts, tissue shrinkage, and extraction of cellular components [66–68].

Before the actual start of such a process, the high-pressure freezing machine (HPM100, please note that also all other high-pressure freezing machines are suitable) needs to be prepared by starting a bake-out overnight or at least for 2 h. Thereby, the remaining humidity is removed. Afterwards, the machine can be filled with LN_2, and when the display shows "ready for freezing," a few test freezing cycles should first be performed to verify the quality of the freezing procedure by checking the pressure and temperature indications (*see* **Note 1**). If all values are in a good range, the experiment can be started by following the subsequent steps:

1. Use only freshly dissected tissue (*see* **Note 3**). This could be freshly prepared brain slices (*see*, e.g., [31]), organotypic slice cultures [30, 56, 57], or other organs. In this example, the apical turn of the murine organ of Corti is dissected in an ice-cold HEPES-HANKS solution containing Sylgard-coated glass-bottom dish and transferred with a micro-curette either

into a droplet of stimulatory solution or inhibitory solution (Table 3) for 15 min at RT. Note: Also other incubation durations are possible depending on the research question. For the high-performing ribbon synapses 15 min were chosen in many studies [32, 62, 63, 69]. Take care of all required protections for all following steps.

2. Prepare the three components of the cartridge system (two half cylinders and middle plate), and place them in the designated positions of the high-pressure freezing machine (Fig. 5a,b).
3. After 13 min of the 15 min stimulatory/inhibitory incubation time, position the tissue into the type A specimen carrier by slowly sliding the carrier into the stimulatory/inhibitory droplet using forceps, and position it below the tissue (Fig. 5c,e). Next, the droplet volume is reduced, and the carrier is pushed out sideways, which causes the tissue to sink down. Take care to remove air bubbles in order to ensure an optimal freezing result (*see* **Note 4**). Please note that such a careful handling might not be required for each kind of sample, but in general it should be taken into consideration that every additional touch might alter the result.
4. Dip the flat side of the type B specimen carrier into a filler (in this example 1-hexadecene) before placing it on top of the tissue containing type A specimen carrier (Fig. 5c).
5. Transfer the specimen carrier sandwich with forceps onto the filter paper, and remove excessive liquid in order to ensure proper removal from the cartridge system after freezing (Fig. 5d; *see* **Note 2**).
6. Position the specimen carrier sandwich with forceps into the hole of the middle plate (Fig. 5f).
7. Turn the handle on the left side of the machine to close the cartridge system (Fig. 5g).
8. Check the correct alignment of the assembled cartridge system with all three components (Fig. 5h; *see* **Note 2**).
9. Turn the handle backward, push the slider to insert the specimen cartridge system into the machine, and initiate the freezing process (Fig. 5i). Use proper ear protection for this step.
10. Take out all pieces from the small LN_2 dewar on the left side of the machine, and transfer them in a LN_2-filled styrofoam box. (Fig. 6a).
11. Place the specimen carrier sandwich containing the middle plate into the designated release device to punch out the carrier sandwich (Fig. 6a,b).
12. Transfer the specimen carrier sandwich with LN_2-precooled forceps into a prepared cryo-vial (Fig. 6c). Therefore, label

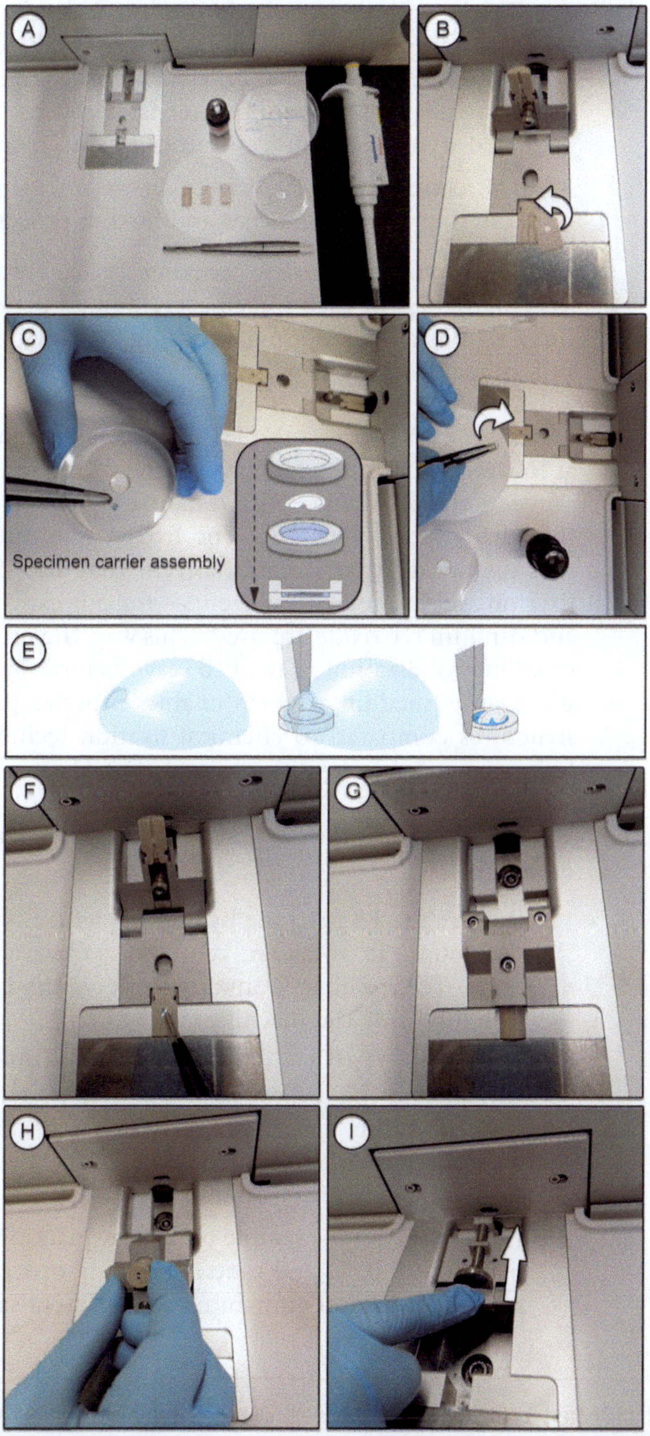

Fig. 5 Sample mounting and high-pressure freezing with the HPM100. (**a**) Summary of the required equipment for the HPF process (forceps, pipette, petri dish with presorted type A and B aluminum specimen carriers, 1-hexadecene, prepared drop of stimulatory or inhibitory solution in a petri dish, filter paper, and the three components of the cartridge system). (**b**) Positioning of the cartridge half cylinders and the middle plate, which

the vial with the required experiment details (freezing number, date, sample information), and bore two holes (one up and one a bit lower in the side of the vial in order to ensure LN_2 access) using coarse forceps. Put the vial with opened lid in the LN_2-filled styrofoam box for precooling and filling with LN_2.

13. The closed cryo-vial, with the specimen inside, can be now stored in LN_2 (Fig. 6d,e) and further processed for the FS.

Before starting a new freezing cycle, always document the freezing details of each freezing cycle (e.g., pressure and temperature) to ensure an optimal freezing or document samples with suboptimal freezing details.

3.2 Freeze-Substitution (FS)

During the FS process, the HPF samples get gradually dehydrated at low temperatures, which avoids the formation of ice crystals and prevents the previous reported damaging effects of room temperature dehydration [20, 26]. The frozen water inside the specimen gets dissolved by an organic solvent that also contains additional fixation and impregnation agents—for our experiments tannic acid and osmium tetroxide are used. This way, the process combines the rapid immobilization of the cell (cryofixation by HPF) with plastic RT resin embedding, which enables a better preservation of fine structures compared to chemical fixation techniques [24]. In the following step by step list, we provide a FS protocol optimized for synapses:

Day 1 (Thursday or Friday): Advised starting time, ~11: 30 a.m.:

1. Prepare the FS machine by filling it with LN_2 (for the protocol ~40–60 L in total are required). Wear appropriate protections. To circumvent any kind of contamination, place a petri dish lid inside the machine (Fig. 7).
2. Fill 6–10 mL 0.1% tannic acid in acetone in an aluminum container using a disposable plastic pipette, and seal it with a corresponding plastic lid. Transfer the aluminum container with forceps onto the petri dish inside the FS machine, and let it cool down for ~30 min.
3. Prepare the plastic sample baskets by marking them with a razor blade (e.g., 0, 1, 2, 3 indentations) in order to distinguish the individual baskets containing the different samples. Place them

Fig. 5 (continued) is highlighted by the arrow. (**c**–**f**) Assembly of the 3 mm specimen carrier sandwich and its transfer into the middle plate (white arrow). (**e**) Detailed illustration of the individual sample mounting steps. (**g** and **h**) View on the fully completed cartridge system followed by a last quality control of the assembly. (**i**) Pushing the slider finally starts the freezing process. (Figure was created with Adobe Illustrator and Affinity Designer)

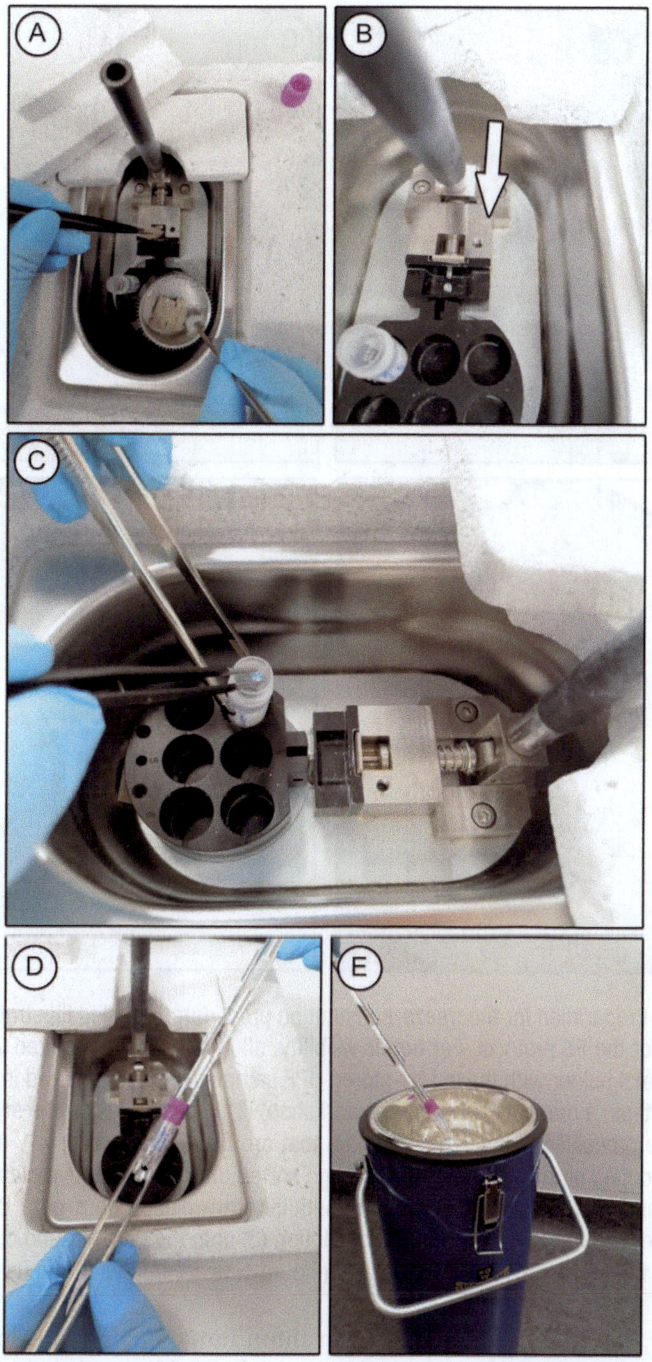

Fig. 6 Sample handling and storing after the high-pressure freezing process. (**a**) After freezing, the individual parts of the cartridge system are transferred into a LN_2-filled styrofoam box, where the aluminum specimen carrier is punched out (**b**). Please note that for better visibility all images were acquired without LN_2. Subsequently, the specimen carrier is placed into a LN_2-filled cryo-vial (**c**) and is ready for a transfer and storage in a LN_2 container (**d** and **e**). (Figure was created with Affinity Designer)

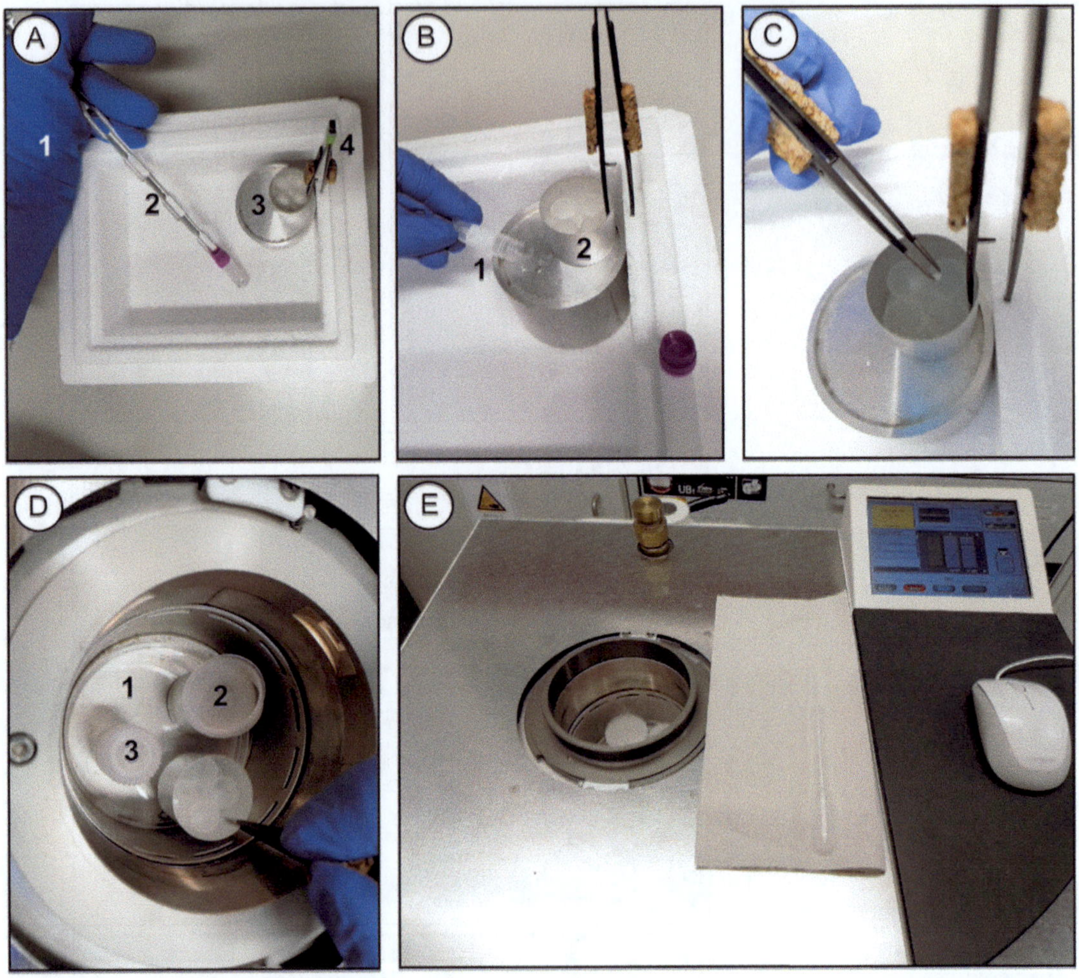

Fig. 7 Preparation for the freeze-substitution procedure. Detailed illustration of the below-described individual steps of the FS protocol. For better visibility, all images were acquired without LN_2. (**a**) Using cryogloves (1), the metal holder with the in LN_2-stored HPF samples (2) is transferred into a LN_2-filled styrofoam box. For the next steps, a metal stage (made in workshop) (3) is placed inside the styrofoam box, which serves as a base for the precooled aluminum containers that are fixed with forceps (4). (**b**) Afterwards, the specimen carriers are dropped onto the metal stage (1) and sorted into the respective plastic baskets (**2** and **c**). (**d** and **e**) Now, the specimen-containing aluminum containers are placed on a petri dish lid (1) inside the FS machine. If required, put more 0.1% tannic acid solution (3) onto the carriers, and close the aluminum vials with a plastic lid (2) before starting the FS program. (Figure was created with Affinity Designer)

in an aluminum container inside the FS machine, and fill it with the precooled 0.1% tannic acid.

4. Fill a dewar and styrofoam box with LN_2. The HPF samples, stored with a metal holder in a LN_2 storage dewar, are transferred to the styrofoam box using the dewar. Please note that until step 7 the samples must always remain in LN_2 (*see* **Note 5**).

5. Place the metal holder inside the styrofoam box, and remove the individual cryo-vials (Fig. 7a). Please take care to protect yourself with a lab coat, appropriate gloves, and safety goggles.
6. Take out the precooled aluminum container from the FS machine, and transfer it onto the metal stage inside the styrofoam box by fixing it with forceps (Fig. 7a).
7. Take out the specimen carriers inside the cryo-vials by dropping them out onto the metal stage, and transfer them on top of the frozen tannic acid solution of the corresponding plastic basket using precooled forceps (Fig. 7b,c). Ensure that the carriers are open to guarantee the access of chemicals.
8. Close the aluminum vials with the lid, and place them inside the FS machine using precooled forceps (Fig. 7d,e). Wait a few minutes, and check if the carriers are completely surrounded by the 0.1% tannic acid solution.
9. Refill the FS machine to 100% with LN_2 if required, and start the program (Table 1) with −90° until Monday incubation with 0.1% tannic acid solution.

Day 2 (Monday): Advised starting time ~ 9.00 a.m.:

1. Refill the FS machine up to 75% with LN_2, and precool an aluminum container filled with acetone for ~10 min inside the chamber.
2. Perform three washing steps with acetone for 30–60 min, respectively.
3. Add freshly prepared osmium tetroxide solution (2% solid OsO_4 in acetone, Table 3).

   Caution: Carefully handle OsO_4 steps under the fume hood with appropriate protection (laboratory coat, nitril gloves, safety goggles, FFP3 mask) as well as waste bottles, forceps, and pipettes dedicated for OsO_4 steps, and always have Na-thiosulfate in place for neutralization. In detail, open the OsO_4 glass ampule, and place it inside a falcon tube that contains the prepared amount of acetone. Mix carefully with the lid always up. Transfer the solution in a properly sealed plastic bottle, and let it cool down inside the FS machine before usage. Remaining solution, containers, etc. that were in contact with OsO_4 are neutralized with a few drops of Na-thiosulfate. Dispose all plastic bottles and pipettes immediately into the OsO_4 waste.
4. Next, the temperature is raised in the automatic FS protocol first to −20 °C and kept until it further increases up to 4 °C (Table 1).

Day 3 (Wednesday): Advised starting time ~ 9:00 a.m.:

1. Wash three times for 30–60 min with precooled acetone. To do so, open the aluminum containers with the OsO_4 forceps. Remove OsO_4 solution into the small waste plastic bottle inside the FS machine, and add pure acetone.
2. Prepare a 1:1 epoxy/acetone solution, and place the aluminum containers from the FS machine under the hood at RT in order to transfer the tissue first into acetone-filled glass vials. Dispose empty carriers using OsO_4 forceps into the OsO_4 waste. Remove excessive acetone, and add 1–2 mL of the epoxy/acetone solution. Incubate the samples for 3–6 h on a shaker at low speed (~20 rpm) at RT.
3. After the incubation time, remove as much solution as possible using a disposable pipette, and infiltrate in fresh 100% epoxy resin. Once for 2 h and subsequent overnight incubation on a shaker (~20 rpm) at RT.
4. Prepare 100% epoxy resin half-filled embedding molds for each sample, and incubate them overnight at 70 °C.

Day 4 (Thursday): Advised starting time ~ 9:00 a.m.:

1. Incubate the samples for additional 3–6 h with fresh 100% epoxy resin on a shaker (~20 rpm) at RT.
2. Place printed labels with the sample details on top of the half-filled and by now polymerized embedding molds. Fill them up with fresh 100% epoxy resin.
3. Transfer the tissue with a toothpick into the embedding molds, and place them in the correct orientation. Ensure to remove any air bubbles.
4. Let the samples polymerize at 70 °C for at least 48 h.

3.3 Trimming and Ultrathin Sectioning

After sample polymerization, the blocks are trimmed first with a fine file (DiATOME) to get closer to the region of interest. In this case, till the tunnel of Corti (located abneural from the single row of IHCs), which serve as a reference point. Thereafter, excess of polymerized resin is removed with a razor blade to create a small trapeze-shaped block face (Fig. 2). Final trimming of the block face is performed with a 45° trimming knife.

Next, 60–70 nm ultrathin sections are obtained from the trimmed blocks using an UC7 (Leica Microsystems) equipped with a 35° diamond knife (DiATOME) and collected on formvar coated (Table 3) slot grids for 2D TEM imaging. For electron tomography, semithin 250 nm sections are obtained and collected on formvar-coated mesh grids. Afterwards, the grids are contrasted with UranyLess at RT for 20–30 min followed by several brief disH_2O washing steps. The mesh grids for electron tomography, in addition, need to be incubated on 10 nm gold fiducial droplets for 5–10 min on each side. Finally, the dried slot and mesh grids can be acquired.

3.4 Data Acquisition

The data acquisition can be performed on the basis of different electron microscopic techniques. Here, very briefly, we will introduce a quantification for IHC ribbon SV pools on the basis of HPF/FS and electron tomography (3D). Typically, our tomograms are acquired from −60 to +60° using a JEM2100 (JEOL, Germany) TEM. Beforehand, we recommend to pre-screen the samples for quality (70 nm ultrathin sections) as well as the presence of IHC ribbons (250 nm semithin sections) using a conventional TEM (JEM1011, JEOL Germany). Afterwards, tomogram acquisition and model generation will be performed as it was published in several publications by our group [29, 31, 32, 43, 49, 61, 69] and others [30, 57, 70]. A possible analysis workflow is depicted in Fig. 8. The pool definition for analysis strongly relies on the type of synapses [72]. For central synapses, SV pool analysis after HPF/FS and electron tomography as well as 2D analysis is done, for example, in [28, 30, 31, 33–35, 56, 57]. Here, the determination of morphological vesicle pools is largely performed in different bins, assuming that the distances from SVs to the AZ membrane correlates with their availability for neurotransmission.

However, for ribbon synapses, often the separation in membrane-proximal (plus specifically docked) and ribbon-associated SVs was chosen. The parameters used by our group are depicted in Fig. 8. Please note that depending on the used embedding method (usage of fixatives (dehydration and cross-linking artifacts are possible) vs. cryofixation and FS (rather near-to-native structural preservation)) and the type of analysis (2D, random sections (limited axially by ultrathin section thickness vs. 3D (tomography with an axial resolution of 1–2 nm)), we employed different quantification parameters. In detail, morphological SV pools in random 2D sections are analyzed in size and number by categorizing them into (i) membrane-proximal synaptic vesicles (MP-SVs, ≤25 nm distance of the SV membrane to the AZ membrane and ≤80 nm distance from the presynaptic density) and (ii) ribbon-associated synaptic vesicles (RA-SVs, first layer of vesicles around the ribbon with a maximum distance of 80 nm). For tomograms we further modified the quantification criteria. Here, MP-SVs are defined as vesicles with ≤50 nm distance from the AZ membrane and ≤100 nm from the presynaptic density. Similar to the 2D analysis, RA-SVs in tomograms are defined as vesicles with a ≤80 nm distance from the ribbon surface to the vesicle membrane. For 2D data analysis, we use the program Image J [73], while tomograms are generated from their respective tilt series using the IMOD package *etomo*, and 3D models are generated using *3dmod* of the IMOD software package [74, 75].

Furthermore, any other parameters such as postsynaptic density (PSD) extent, mitochondria, ribbon size, and shape can be analyzed with this method. It is important to be aware of the limitations of the specific approaches and how they influence or

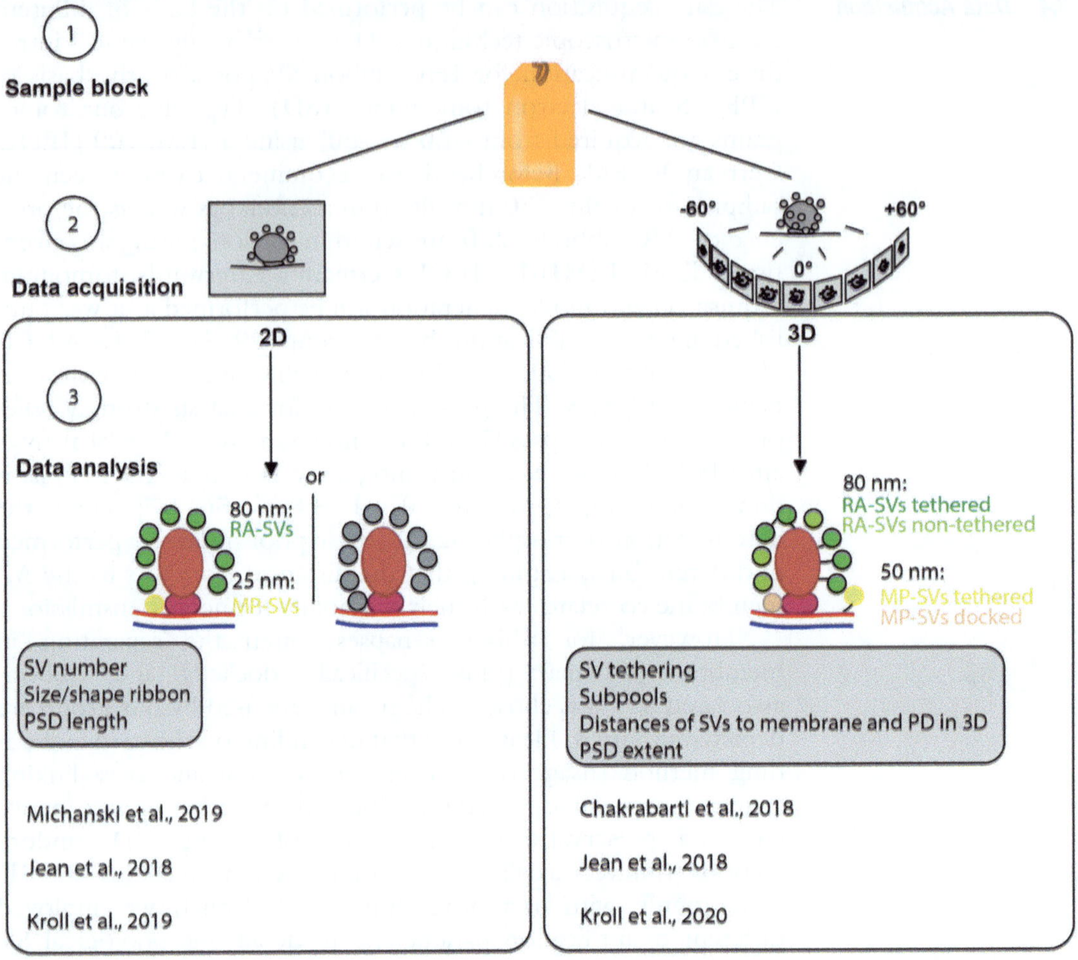

Fig. 8 Suggested data analysis for ribbon synapses in 2D and 3D. After generation of the sample block (1), data can be acquired either as 2D or 3D (2). The data analysis is depicted in (3) with some example publications [29, 32, 49, 69, 71] performing the analysis according to these criteria. (Figure was created with Adobe Illustrator)

support your research question. 2D data acquisition and post processing are faster in comparison to the generation of tomograms, but it is limited in the specimen information by the section thickness of 50–70 nm. With serial sections larger areas can be imaged and 3D reconstructed as well; however, part of the structural information always gets lost from section to section. Data from tomograms are limited by the area of 250 nm semithin sections that enables the quantification of larger areas in 3D per section. In addition, the tilt series image acquisition with 1° increment results in a higher z-resolution, which allows the quantification of finer structures like the filamentous and proteinaceous structures that associate SVs with the AZ (Fig. 8). These filaments are described in different species and synapses [72] and can be divided into three categories: (i) interconnectors that link SVs to each other within the presynaptic terminal; (ii) tethers, which connect SVs to the AZ membrane; and (iii) in ribbon-type synapses, tethers that connect SVs with the ribbon.

4 Outlook

High-K^+ stimulation is the easiest way to induce a widespread depolarization and this way activate neurotransmitter release at synapses. It was heavily used to study bulk endocytic events, for example, at neuromuscular junctions of *Drosophila* larvae [76, 77]. However, several limitations of this method need to be considered when applying it. High-K^+ stimulation is not timely precise or can be performed in a millisecond range or even in trains of stimulation. Therefore, it is suggested to be the least physiological approach in comparison to electrical or optogenetically evoked activity of synapses [76, 77]. However, here specific devices equipped for this type of stimulations are required. Despite this difficulty, combining optogenetic light stimulation, high-pressure freezing, and electron microscopy for a millisecond precise stimulation and a close coupling of stimulation and freezing became more and more prominent in the past years [45, 59, 70, 78, 79]. Using chemical solution-based depolarization of cell, the time course is more in a minute than in a second range. Hence, for studying rapid cellular processes such as early exocytosis events that require a more precise and shorter stimulation or stimulation sequence, optogenetic stimulation is the method of choice, as it was also recently described for IHC ribbon synapses. Only this method is suitable to investigate early events of exocytosis on the level of electron microscopy in a distinct manner. This way, different protocols can be applied with different stimulation paradigms and resting durations, and thus distinct phases of vesicle release can be approached [80].

5 Notes

This method flow comes with a number of difficulties that should be mentioned. First of all, it should be taken into consideration how well the stimulation paradigm we are using for the specific ribbon synapse also suits for other synapses. High-K^+ stimulation was frequently applied to ribbon-type synapses [32, 62, 63, 69] as well as to conventional neuronal synapses [52–55], but it might not be physiological for each kind of synapse. However, when taking into account this limitation as well as formulating a suitable research question, the application of chemical solutions for depolarization can be a very meaningful tool.

Next to these principle problems, there are a few points that should be taken special care of:

1. This applies to all steps of the described workflow. Control for the condition of your equipment beforehand. Specifically: HPM (test freezing cycles), aluminum specimen carriers, cartridge system, aluminum container, solutions (prepare freshly). All materials must be clean and intact in order to prevent contaminations and technical problems with the machines.
2. The cartridge system for mounting the samples for HPF is very sensitive to errors (*see* Subheading 3.1). If the cartridges are not properly aligned or wet, the specimen carrier can get stuck inside the HPM. The sample assembly should be precise for proper freezing results, and excessive liquid at the specimen carrier needs to be removed for ensuring an easy disassembly of the two carrier types and avoid them freezing together.
3. Another important point is the quality of the tissue and sample assembly. Always use fresh tissue, and try to be fast with the preparation to prevent structural changes and possible artifacts (*see* Subheading 3.1).
4. During the tissue positioning inside the carrier and carrier assembly, remove any air bubbles (*see* Subheading 3.1, **step 3**). The usage of hexadecane prevents air bubble formation. Air bubbles can affect the freezing quality by acting as insulators, deforming the tissue or tearing apart the carriers [81].
5. Furthermore, ensure that the samples always stay inside LN_2 at the corresponding steps of the protocols. Any exposure to air will damage the cell structure as a result of ice-crystal formation.

Acknowledgments

We thank Wiebke Möbius and Christian Vogl for the critical reading of the chapter.
This work was supported by funding from the Deutsche Forschungsgemeinschaft via the Collaborative Research Center 889 (project A07 to CW) and the Collaborative Research Center 1286 (project A04 to CW). CW is further supported by the Deutsche Forschungsgemeinschaft under Germany's Excellence Strategy–EXC 2067/1-390729940.

References

1. Deiters VS, Guillery RW (2013) Otto Friedrich Karl Deiters (1834–1863). J Comp Neurol 521(9):1929–1953. https://doi.org/10.1002/cne.23316
2. Glover JC, Elliott KL, Erives A, Chizhikov VV, Fritzsch B (2018) Wilhelm His' lasting insights into hindbrain and cranial ganglia development and evolution. Dev Biol 444:S14–S24.

https://doi.org/10.1016/j.ydbio.2018.02.001

3. Swanson LW, Lichtman JW (2016) From Cajal to Connectome and beyond. Annu Rev Neurosci 39(1):197–216. https://doi.org/10.1146/annurev-neuro-071714-033954
4. Pearce JMS (2004) Sir Charles Scott Sherrington (1857–1952) and the synapse. J Neurol Neurosurg Psychiatry 75(4):544
5. Deuchars J, West DC, Thomson AM (1994) Relationships between morphology and physiology of pyramid-pyramid single axon connections in rat neocortex in vitro. J Physiol 478 (Pt 3):423–435
6. Markram H, Lubke J, Frotscher M, Roth A, Sakmann B (1997) Physiology and anatomy of synaptic connections between thick tufted pyramidal neurones in the developing rat neocortex. J Physiol 500(Pt 2):409–440
7. Feldmeyer D, Lübke J, Silver RA, Sakmann B (2002) Synaptic connections between layer 4 spiny neurone-layer 2/3 pyramidal cell pairs in juvenile rat barrel cortex: Physiology and anatomy of interlaminar signalling within a cortical column. J Physiol 538(3):803–822
8. Tamás G, Simon Anna AL, Szabadics J (2003) Identified Sources and Targets of Slow Inhibition in the Neocortex. Science 299(5614): 1902–1905. https://doi.org/10.1126/science.1082053
9. Feldmeyer D, Lübke J, Sakmann B (2006) Efficacy and connectivity of intracolumnar pairs of layer 2/3 pyramidal cells in the barrel cortex of juvenile rats: Layer 2/3 pyramidal cell synapses. J Physiol 575(2):583–602. https://doi.org/10.1113/jphysiol.2006.105106
10. Silberberg G, Markram H (2007) Disynaptic inhibition between neocortical pyramidal cells mediated by Martinotti cells. Neuron 53(5): 735–746. https://doi.org/10.1016/j.neuron.2007.02.012
11. Helmstaedter M, Briggman KL, Denk W (2008) 3D structural imaging of the brain with photons and electrons. Curr Opin Neurobiol 18(6):633–641. https://doi.org/10.1016/j.conb.2009.03.005
12. Kapfer C, Glickfeld LL, Atallah BV, Scanziani M (2007) Supralinear increase of recurrent inhibition during sparse activity in the somatosensory cortex. Nat Neurosci 10(6):743–753. https://doi.org/10.1038/nn1909
13. Zandt B-J, Liu JH, Veruki ML, Hartveit E (2017) AII amacrine cells: quantitative reconstruction and morphometric analysis of electrophysiologically identified cells in live rat retinal slices imaged with multi-photon excitation microscopy. Brain Struct Funct 222(1): 151–182. https://doi.org/10.1007/s00429-016-1206-0
14. Holler S, Köstinger G, Martin KAC, Schuhknecht GFP, Stratford KJ (2021) Structure and function of a neocortical synapse. Nature 591(7848):111–116. https://doi.org/10.1038/s41586-020-03134-2
15. Neale EA, MacDonald RL, Nelson PG (1978) Intracellular horseradish peroxidase injection for correlation of light and electron microscopic anatomy with synaptic physiology of cultured mouse spinal cord neurons. Brain Res 152(2):265–282. https://doi.org/10.1016/0006-8993(78)90255-X
16. Lentz TL, Chester J (1982) Synaptic vesicle recycling at the neuromuscular junction in the presence of a presynaptic membrane marker. Neuroscience 7(1):9–20. https://doi.org/10.1016/0306-4522(82)90148-8
17. Scholl B, Thomas CI, Ryan MA, Kamasawa N, Fitzpatrick D (2021) Cortical response selectivity derives from strength in numbers of synapses. Nature 590(7844) Art. no. 7844. https://doi.org/10.1038/s41586-020-03044-3
18. Heuser JE, Reese TS (1973) Evidence for recycling of synaptic vesicle membrane during transmitter release at the frog neuromuscular junction. J Cell Biol 57(2):315–344. https://doi.org/10.1083/jcb.57.2.315
19. Heuser JE, Reese TS, Dennis MJ, Jan Y, Jan L, Evans L (1979) Synaptic vesicle exocytosis captured by quick freezing and correlated with quantal transmitter release. J Cell Biol 81(2): 275–300. https://doi.org/10.1083/jcb.81.2.275
20. McDonald KL, Auer M (2006) High-pressure freezing, cellular tomography, and structural cell biology. BioTechniques 41(2):137–143. https://doi.org/10.2144/000112226
21. Moor H (1987) Theory and practice of high pressure freezing. In: Steinbrecht PDRA, Zierold DK (eds) Cryotechniques in biological electron microscopy. Springer, Berlin Heidelberg, pp 175–191. https://doi.org/10.1007/978-3-642-72815-0_8
22. Dubochet J (1995) High-pressure freezing for cryoelectron microscopy. Trends Cell Biol 5(9):366–368
23. Leunissen JLM, Yi H (2009) Self-pressurized rapid freezing (SPRF): a novel cryofixation method for specimen preparation in electron microscopy. J Microsc 235(1):25–35. https://doi.org/10.1111/j.1365-2818.2009.03178.x
24. Steinbrecht RA, Müller M (1987) Freeze-Substitution and Freeze-Drying. In:

Steinbrecht RA, Zierold K (eds) Cryotechniques in biological electron microscopy. Springer Berlin Heidelberg, Berlin, Heidelberg, pp 149–172. https://doi.org/10.1007/978-3-642-72815-0_7

25. Feder N, Sidman RL (1958) Methods and principles of fixation by freeze-substitution. J Biophys Biochem Cytol 4(5):593–600
26. Hobot JA, Villiger W, Escaig J, Maeder M, Ryter A, Kellenberger E (1985) Shape and fine structure of nucleoids observed on sections of ultrarapidly frozen and cryosubstituted bacteria. J Bacteriol 162(3):960–971
27. Kellenberger E (1991) The potential of cryofixation and freeze substitution: observations and theoretical considerations. J Microsc 161(2):183–203. https://doi.org/10.1111/j.1365-2818.1991.tb03083.x
28. Cole AA, Chen X, Reese TS (2016) A network of three types of filaments organizes synaptic vesicles for storage, mobilization, and docking. J Neurosci 36(11):3222–3230. https://doi.org/10.1523/JNEUROSCI.2939-15.2016
29. Jean P et al (2018) The synaptic ribbon is critical for sound encoding at high rates and with temporal precision. elife 7:e29275. https://doi.org/10.7554/eLife.29275
30. Imig C et al (2014) The morphological and molecular nature of synaptic vesicle priming at presynaptic active zones. Neuron 84(2):416–431. https://doi.org/10.1016/j.neuron.2014.10.009
31. Hintze A, Gültas M, Semmelhack EA, Wichmann C (2021) Ultrastructural maturation of the endbulb of Held active zones comparing wild-type and otoferlin-deficient mice. iScience 24(4):102282. https://doi.org/10.1016/j.isci.2021.102282
32. Chakrabarti R, Michanski S, Wichmann C (2018) Vesicle sub-pool organization at inner hair cell ribbon synapses. EMBO Rep:e44937. https://doi.org/10.15252/embr.201744937
33. Siksou L, Triller A, Marty S (2011) Ultrastructural organization of presynaptic terminals. Curr Opin Neurobiol 21(2):261–268. https://doi.org/10.1016/j.conb.2010.12.003
34. Siksou L et al (2007) Three-dimensional architecture of presynaptic terminal cytomatrix. J Neurosci 27(26):6868–6877. https://doi.org/10.1523/JNEUROSCI.1773-07.2007
35. Siksou L, Varoqueaux F, Pascual O, Triller A, Brose N, Marty S (2009) A common molecular basis for membrane docking and functional priming of synaptic vesicles. Eur J Neurosci 30(1):49–56. https://doi.org/10.1111/j.1460-9568.2009.06811.x
36. Fernández-Busnadiego R et al (2013) Cryo-electron tomography reveals a critical role of RIM1α in synaptic vesicle tethering. J Cell Biol 201(5):725–740. https://doi.org/10.1083/jcb.201206063
37. Fernández-Busnadiego R, Zuber B, Maurer UE, Cyrklaff M, Baumeister W, Lučić V (2010) Quantitative analysis of the native presynaptic cytomatrix by cryoelectron tomography. J Cell Biol 188(1):145–156. https://doi.org/10.1083/jcb.200908082
38. Tao C-L et al (2018) Differentiation and characterization of excitatory and inhibitory synapses by cryo-electron tomography and correlative microscopy. J Neurosci 38(6):1493–1510. https://doi.org/10.1523/JNEUROSCI.1548-17.2017
39. Fernández-Busnadiego R et al (2011) Insights into the molecular organization of the neuron by cryo-electron tomography. J Electron Microsc 60(suppl 1):S137–S148. https://doi.org/10.1093/jmicro/dfr018
40. Lucić V, Kossel AH, Yang T, Bonhoeffer T, Baumeister W, Sartori A (2007) Multiscale imaging of neurons grown in culture: from light microscopy to cryo-electron tomography. J Struct Biol 160(2):146–156. https://doi.org/10.1016/j.jsb.2007.08.014
41. Vogl C et al (2015) Unconventional molecular regulation of synaptic vesicle replenishment in cochlear inner hair cells. J Cell Sci 128(4):638–644. https://doi.org/10.1242/jcs.162099
42. Bullen A et al (2015) Association of intracellular and synaptic organization in cochlear inner hair cells revealed by 3D electron microscopy. J Cell Sci 128(14):2529–2540. https://doi.org/10.1242/jcs.170761
43. Jung S et al (2015) Rab3-interacting molecules 2α and 2β promote the abundance of voltage-gated CaV1.3 Ca2+ channels at hair cell active zones. Proc Natl Acad Sci USA 112(24):E3141–E3149. https://doi.org/10.1073/pnas.1417207112
44. Cooper B et al (2012) Munc13-independent vesicle priming at mouse photoreceptor ribbon synapses. J Neurosci 32(23):8040–8052. https://doi.org/10.1523/JNEUROSCI.4240-11.2012
45. Watanabe S et al (2013) Ultrafast endocytosis at mouse hippocampal synapses. Nature 504(7479):242–247. https://doi.org/10.1038/nature12809
46. Rostaing P et al (2006) Analysis of synaptic ultrastructure without fixative using high-pressure freezing and tomography. Eur J Neurosci 24(12):3463–3474. https://doi.org/10.1111/j.1460-9568.2006.05234.x

47. Imig C, Cooper BH (2017) 3D analysis of synaptic ultrastructure in organotypic hippocampal slice culture by high-pressure freezing and electron tomography. In: Poulopoulos A (ed) Synapse development: methods and protocols. Springer, New York, pp 215–231. https://doi.org/10.1007/978-1-4939-6688-2_15
48. Möbius W et al (2010) Chapter 20—electron microscopy of the mouse central nervous system. In: Müller-Reichert T (ed) Methods in cell biology, vol 96. Academic Press, pp 475–512. https://doi.org/10.1016/S0091-679X(10)96020-2
49. Kroll J et al (2020) AP180 promotes release site clearance and clathrin-dependent vesicle reformation in mouse cochlear inner hair cells. J Cell Sci 133(2). https://doi.org/10.1242/jcs.236737
50. Wong AB et al (2014) Developmental refinement of hair cell synapses tightens the coupling of Ca2+ influx to exocytosis. EMBO J 33(3): 247–264. https://doi.org/10.1002/embj.201387110
51. McDonald KL, Webb RI (2011) Freeze substitution in 3 hours or less. J Microsc 243(3): 227–233. https://doi.org/10.1111/j.1365-2818.2011.03526.x
52. Betz W, Mao F, Bewick G (1992) Activity-dependent fluorescent staining and destaining of living vertebrate motor nerve terminals. J Neurosci 12(2):363–375. https://doi.org/10.1523/JNEUROSCI.12-02-00363.1992
53. Ryan TA, Reuter H, Wendland B, Schweizer FE, Tsien RW, Smith SJ (1993) The kinetics of synaptic vesicle recycling measured at single presynaptic boutons. Neuron 11(4):713–724. https://doi.org/10.1016/0896-6273(93)90081-2
54. Ramaswami M, Krishnan KS, Kelly RB (1994) Intermediates in synaptic vesicle recycling revealed by optical imaging of Drosophila neuromuscular junctions. Neuron 13(2):363–375. https://doi.org/10.1016/0896-6273(94)90353-0
55. Trotta N, Rodesch CK, Fergestad T, Broadie K (2004) Cellular bases of activity-dependent paralysis in Drosophila stress-sensitive mutants. J Neurobiol 60(3):328–347. https://doi.org/10.1002/neu.20017
56. Imig C (2013) Molecular and morphological correlates of synaptic vesicle priming. Georg August University Göttingen. Accessed: 18 Jan 2016. [Online]. Available: http://ediss.uni-goettingen.de/bitstream/handle/11858/00-1735-0000-0023-9917-A/Imig.pdf?sequence=1
57. Maus L et al (2020) Ultrastructural correlates of presynaptic functional heterogeneity in hippocampal synapses. Cell Rep 30(11): 3632–3643.e8. https://doi.org/10.1016/j.celrep.2020.02.083
58. Becker L et al (2018) The presynaptic ribbon maintains vesicle populations at the hair cell afferent fiber synapse. eLife Sciences 7: e30241. https://doi.org/10.7554/eLife.30241
59. Watanabe S et al (2013) Ultrafast endocytosis at Caenorhabditis elegans neuromuscular junctions. elife 2:e00723. https://doi.org/10.7554/eLife.00723
60. Schikorski T, Stevens CF (1997) Quantitative ultrastructural analysis of hippocampal excitatory synapses. J Neurosci 17(15):5858–5867
61. Strenzke N et al (2016) Hair cell synaptic dysfunction, auditory fatigue and thermal sensitivity in otoferlin Ile515Thr mutants. EMBO J 35(23):e201694564. https://doi.org/10.15252/embj.201694564
62. Lenzi D, Runyeon JW, Crum J, Ellisman MH, Roberts WM (1999) Synaptic vesicle populations in saccular hair cells reconstructed by electron tomography. J Neurosci 19(1): 119–132
63. Pangrsic T et al (2010) Hearing requires otoferlin-dependent efficient replenishment of synaptic vesicles in hair cells. Nat Neurosci 13(7):869–876. https://doi.org/10.1038/nn.2578
64. Fouquet W et al (2009) Maturation of active zone assembly by Drosophila Bruchpilot. J Cell Biol 186(1):129–145. https://doi.org/10.1083/jcb.200812150
65. Rosenmund C, Stevens CF (1997) The rate of aldehyde fixation of the exocytotic machinery in ultured hippocampal synapses. J Neurosci Methods 76(1):1–5
66. Weibull C, Villiger W, Carlemalm E (1984) Extraction of lipids during freeze-substitution of Acholeplasma laidlawii-cells for electron microscopy. J Microsc 134(2):213–216. https://doi.org/10.1111/j.1365-2818.1984.tb02513.x
67. Humbel B, Schwarz H (1989) Freeze-substitution for immunochemistry. In: Immuno-gold labeling in cell biology, p 115
68. Hunziker EB, Herrmann W (1987) In situ localization of cartilage extracellular matrix components by immunoelectron microscopy after cryotechnical tissue processing. J Histochem Cytochem 35(6):647–655. https://doi.org/10.1177/35.6.3553318
69. Kroll J et al (2019) Endophilin-A regulates presynaptic Ca2+ influx and synaptic vesicle recycling in auditory hair cells. EMBO J 38(5). https://doi.org/10.15252/embj.2018100116

70. Imig C et al (2020) Ultrastructural imaging of activity-dependent synaptic membrane-trafficking events in cultured brain slices. Neuron 108(5):843–860.e8. https://doi.org/10.1016/j.neuron.2020.09.004
71. Michanski S et al (2019) Mapping developmental maturation of inner hair cell ribbon synapses in the apical mouse cochlea. PNAS 116(13):6415–6424. https://doi.org/10.1073/pnas.1812029116
72. Chakrabarti R, Wichmann C (2019) Nanomachinery organizing release at neuronal and ribbon synapses. Int J Mol Sci 20(9):2147. https://doi.org/10.3390/ijms20092147
73. Schneider CA, Rasband WS, Eliceiri KW (2012) NIH Image to ImageJ: 25 years of image analysis. Nat Methods 9:671–675
74. Kremer JR, Mastronarde DN, McIntosh JR (1996) Computer visualization of three-dimensional image data using IMOD. J Struct Biol 116(1):71–76. https://doi.org/10.1006/jsbi.1996.0013
75. Mastronarde DN (2005) Automated electron microscope tomography using robust prediction of specimen movements. J Struct Biol 152(1):36–51. https://doi.org/10.1016/j.jsb.2005.07.007
76. Kopke DL, Broadie K (2018) FM Dye cycling at the synapse: comparing high potassium depolarization, electrical and channelrhodopsin stimulation. J Vis Exp no. 135:57765. https://doi.org/10.3791/57765
77. Vijayakrishnan N, Woodruff EA, Broadie K (2009) Rolling blackout is required for bulk endocytosis in non-neuronal cells and neuronal synapses. J Cell Sci 122(1):114–125. https://doi.org/10.1242/jcs.036673
78. Borges-Merjane C, Kim O, Jonas P (2020) Functional electron microscopy, 'flash and freeze,' of identified cortical synapses in acute brain slices. Neuron 105(6):992–1006.e6. https://doi.org/10.1016/j.neuron.2019.12.022
79. Watanabe S (2016) Flash-and-freeze: coordinating optogenetic stimulation with rapid freezing to visualize membrane dynamics at synapses with millisecond resolution. Front Synaptic Neurosci 8:24. https://doi.org/10.3389/fnsyn.2016.00024
80. Chakrabarti R, Jaime Tobón LM, Slitin L, Redondo Canales M, Hoch G, Slashcheva M, Fritsch E, Bodensiek K, Özçete ÖD, Gültas M, Michanski S, Opazo F, Neef J, Pangrsic T, Moser T, Wichmann C (2022) Optogenetics and electron tomography for structurefunction analysis of cochlear ribbon synapses. Elife. 11: e79494. https://doi.org/10.7554/eLife.79494
81. Dahl R, Staehelin LA (1989) High-pressure freezing for the preservation of biological structure: theory and practice. J Electron Microsc Tech 13(3):165–174. https://doi.org/10.1002/jemt.1060130305

Chapter 6

Large-Volume Electron Microscopy

Yoshiyuki Kubota, Takaaki Miyazaki, and Marc Takeno

Abstract

A large volume electron microscopy method was developed in association with serial electron microscopy technologies, which have improved quite significantly during the past two decades. These methods include focused ion beam (FIB)-scanning electron microscopy (SEM), serial block face (SBF)-SEM, automated tape-collecting ultramicrotome (ATUM)-SEM, and ATUM-transmission electron microscopy (TEM). Here, an outline of these imaging technologies and the latest advancements in ATUM-SEM and ATUM-TEM imaging are described. One of the distinguishing features of these approaches is the capability of large-volume EM imaging. This state-of-the-art technology enables the acquisition of data from one cubic millimeter or larger volumes of brain tissue for high-magnification electron micrographic and three-dimensional (3D) reconstruction of neural elements via AI-assisted automated segmentation, thereby allowing the analysis of synaptic connections in brain microcircuitry.

Key words FIB-SEM, ATUM, SBF-SEM, Dense segmentation, Serial EM, Connectome, Synapse

1 Introduction

The cerebral cortex controls higher brain functions. The neural network or connectome in the cerebral cortex is extremely complex, and we still do not know or understand this cerebral architecture fully. Electron microscopy is an essential tool with which to investigate synaptic connections, which form the basis of signal conduction among neurons in the neural microcircuit. Here, we describe the development of image acquisition and image processing techniques for the three-dimensional reconstruction of neural elements from serial electron micrographs. This cutting-edge technology has become popular over the past two decades as a means to understand the cerebral connectome.

1.1 Serial Electron Micrograph Method: Initial Era (1980–)

In 1980, White and Rock first introduced a three-dimensional reconstruction analysis method of neural structures [1]. Spiny stellate cells of mouse cerebral cortex in 80-nm-thick serial ultrathin sections were imaged using transmission electron microscopy

Joachim H. R. Lübke and Astrid Rollenhagen (eds.), *New Aspects in Analyzing the Synaptic Organization of the Brain*, Neuromethods, vol. 212, https://doi.org/10.1007/978-1-0716-4019-7_6,

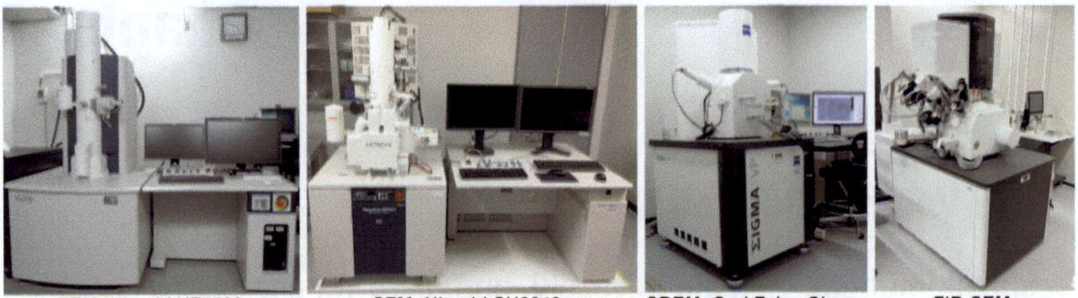

Fig. 1 Electron microscopes

(TEM) (Fig. 1), and the spiny stellate cells in each of the serial electron micrographs were modeled on a 3.2-mm-thick wooden plate. Then they were reconstructed three-dimensionally by stacking the wooden plates serially. Back then, the ultramicrotome that created ultrathin sections was a rotary manual cutting device and the quality of ultrathin sections was greatly affected by the skill of the operator.

Later, in the 1980s, motor-driven ultramicrotomes provided better performance, precision and ease of use and allowed more stable ultrathin sectioning. The serial ultrathin sections were carefully collected on one-hole grids, instead of mesh grids, to preserve neural element continuity. This EM technology made it possible to collect 1000 serial ultrathin sections, although it required an operator with significant skill, concentration and dexterity [1].

In the 1990s and early 2000s, three-dimensional (3D) reconstruction methods from serial electron micrographs using computer applications were developed, and they were adapted gradually into neuroscience research. However, the 3D reconstruction method was used in only a few EM laboratories, because it still required unique technical skills which were not easily acquired [2–5].

1.2 Serial Electron Micrograph Method: Second Era (2000–)

In the 2000s, a number of automated serial ultrathin sectioning imaging methods were developed following the recognition of the significant importance of this sort of analysis to understand brain microcircuits. Such new EM devices include focused ion beam (FIB)-SEM, serial block face-SEM (SBF-SEM or SBEM), automated tape-collecting ultramicrotome (ATUMtome)-SEM, and TEM camera array (TEMCA) (Fig. 1) [6].

FIB-SEM is a device which was invented to examine the structure of semiconductor chips. It has been applied to neuroscience research by Graham Knott [7]. A neural tissue block surface is milled by FIB (>4 nm), and the new surface revealed is imaged using SEM. Conveniently, the device automatically repeats this process and acquires serial ultrastructural image data. The effective

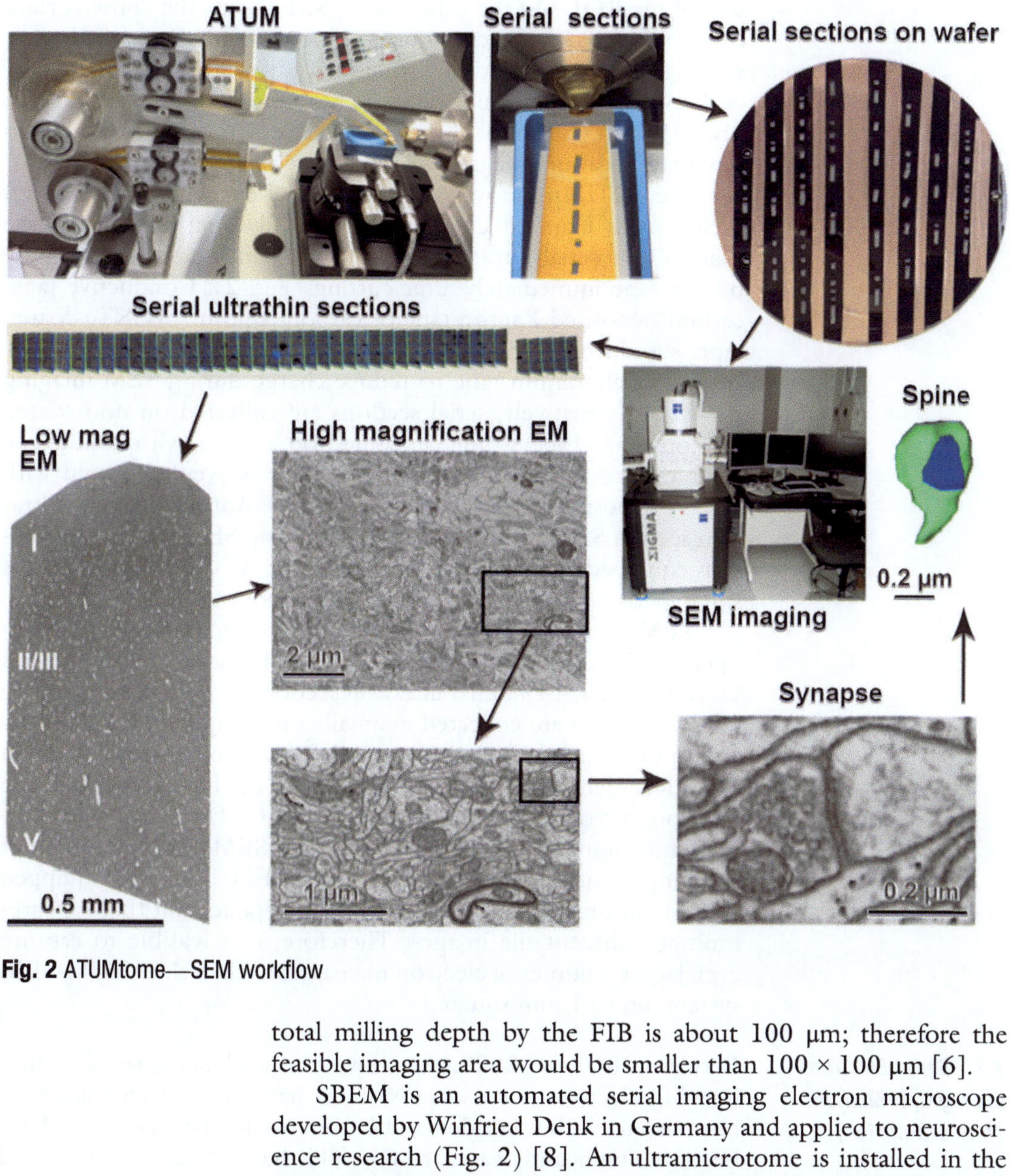

Fig. 2 ATUMtome—SEM workflow

total milling depth by the FIB is about 100 μm; therefore the feasible imaging area would be smaller than 100 × 100 μm [6].

SBEM is an automated serial imaging electron microscope developed by Winfried Denk in Germany and applied to neuroscience research (Fig. 2) [8]. An ultramicrotome is installed in the SEM chamber for sectioning a tissue block with a diamond knife (cutting thickness >20 nm), and the fresh block surface exposed is imaged using the SEM. This automated process allows the collection of more than several hundred serial electron micrographs overnight. While imaging with the FIB-SEM and SBEM, electric charging occurs due to electron retention. To reduce/prevent the charging, the entire block surface must be covered by a layer of silver paste to secure a path for electrons. The charging problem in the FIB-SEM is less than the SBEM, because only a small area is required to ablate for imaging, which is only slightly larger than the size of the ROI. On the other hand, SBEM cuts the entire surface of the block, and the diamond knife can remove the conductive coating metal layer entirely, so significant charging occurs, especially with high-magnification (<5 nm/pixel) imaging. Thus, SBEM is currently only suitable for medium- to low-magnification imaging (>7 nm/pixel) [9].

The ATUMtome was developed by Ken Hayworth and Jeff Lichtman in Harvard University [10]. This is an ultramicrotome that collects serial ultrathin sections while maintaining continuity on the tape immediately after cutting (Fig. 2). Conductive tape, carbon-deposited Kapton tape or carbon nanotube (CNT)-coated tape, should be used, and the tape surface must be grounded by copper or aluminum tape to reduce charge during SEM imaging [6, 11]. Alternatively, serial sections are collected on non-coated Kapton tape. Tape strings are glued on a wafer with conductive double-sided tape, and then their surface is evenly covered with carbon deposition in a carbon coater [12]. Automatic serial ultrathin section SEM imaging is processed using SEM control application equipped in FE-SEM. Therefore, the ATUM-SEM system is suitable for a relatively large area (up to about 0.5 mm square).

TEMCA is based on a transmission electron microscope developed by Clay Reid, Davi Bock, and Wei-Chung A. Lee of Harvard University [13, 14]. Serial ultrathin sections are cut using an ultramicrotome and are collected manually on a single-hole grid. They are captured using four 4-megapixel CCD cameras installed in an extended vacuum chamber camera room of the TEM. Since the electron micrographs are captured using a CCD camera, the imaging time is significantly faster than with the SEM, which uses a point scanning image capturing system. The TEMCA is also equipped with an automatic serial section imaging application that captures multiple adjacent tile images. Therefore, it is feasible to capture even larger volumes of electron micrographs than the ATUM-SEM system, up to 1 mm square.

1.3 Serial Electron Micrograph Method: Third Era (2015–)

Between 2010 and 2020, the above four automated serial section imaging electron microscope systems have evolved technically in response to an intense demand by neuroscientists, and they have become significantly easier to use. This has dramatically increased the use of the 3D reconstruction of serial EM methods. A few significant developments have occurred. One is the introduction of multi-beam SEM (MSEM: Carl Zeiss Microscopy, Inc.) that splits the SEM beam into 61 or 91 beams and enables image acquisition in each beam channel in parallel [15]. Because SEM imaging is based on a secondary electron detection system, the time required for image capturing is dramatically shortened (100–800 nsec/pixel). Imaging time is about 100 times faster than that of a single-beam FE-SEM. Therefore, this approach is useful for large-volume data acquisition. Resolution of the image is

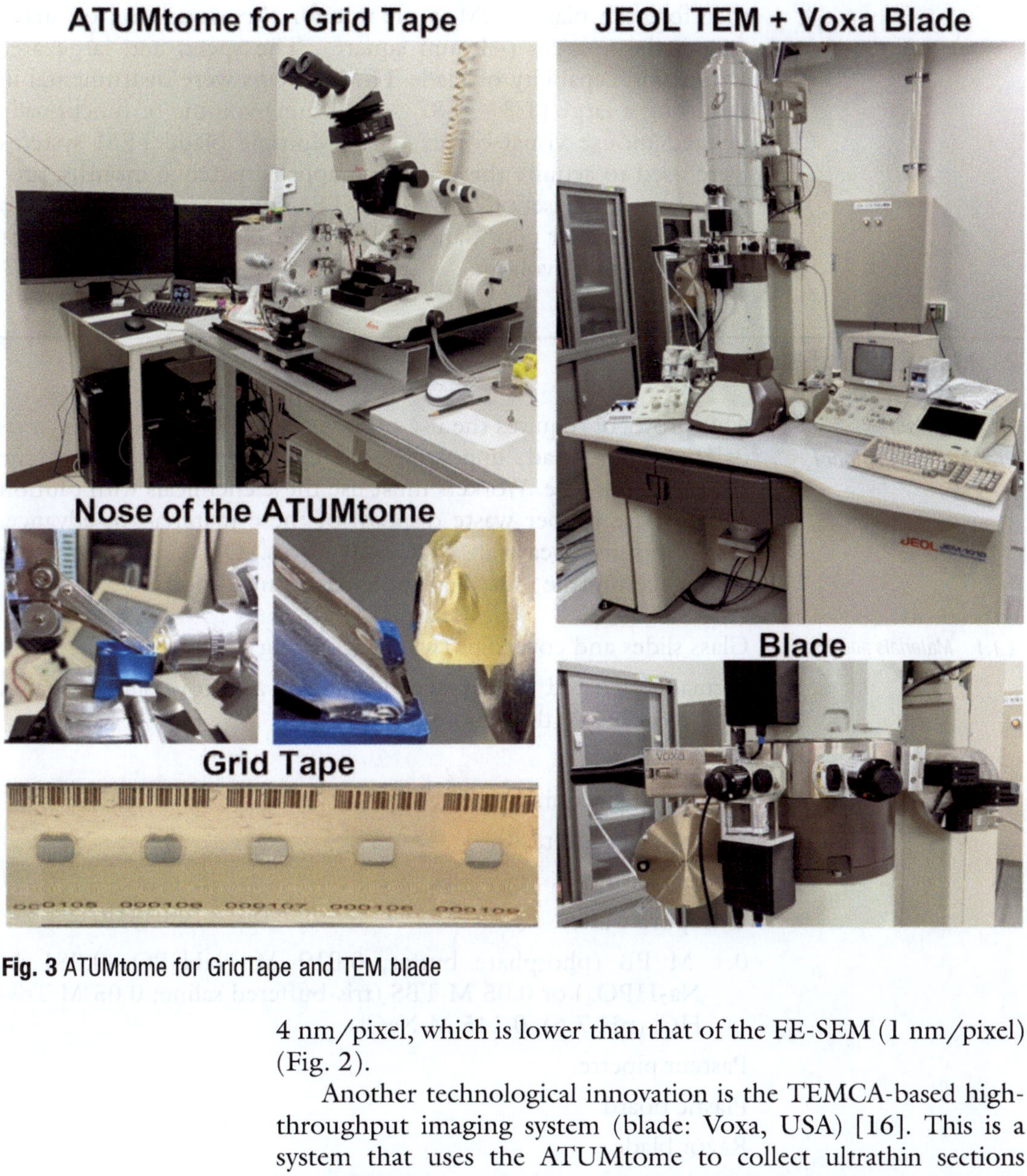

Fig. 3 ATUMtome for GridTape and TEM blade

4 nm/pixel, which is lower than that of the FE-SEM (1 nm/pixel) (Fig. 2).

Another technological innovation is the TEMCA-based high-throughput imaging system (blade: Voxa, USA) [16]. This is a system that uses the ATUMtome to collect ultrathin sections onto a GridTape, which is an 8-mm-width aluminum-coated Kapton tape with 1.5 × 2 mm slots with 50-nm polyimide thin film at every 6 mm length. The GridTape with the serial ultrathin sections is put in the blade-mounted TEM, and serial electron micrographs are captured using an automatic image capturing application (Fig. 3). The SEM, which captures images by the point scanning method, is much slower in imaging speed than the blade-TEM, which uses a 50-megapixel cMOS camera to capture images at 30 msec/image. So, the imaging speed of the blade-TEM is extremely fast. In addition, large images can be tiled and imaged;

therefore the blade-TEM is suitable for the acquisition of image data in large areas (~1 mm square). The speed and large area acquisition capability of blade-TEM systems were instrumental in acquiring a large (1.3 × 0.87 × 0.82 mm) volume of functionally imaged mouse visual cortex [17]. Multiple blade-TEM systems were used to acquire this dataset in approximately 6 months, producing a raw dataset of 2 PB. Blade-TEM systems were also used to produce a smaller pilot dataset [18] in development for the larger MICrONS mm^3 volume.

2 Materials and Methods

2.1 Sample Preparation Protocol for EM

This protocol requires the use of harmful substances such as cacodylate, OsO_4, lead, unpolymerized epoxy resin, and propylene oxide/acetonitrile. Workers must use these chemicals with caution and prepare proper waste disposal of these materials in advance. Workers must wear lab coat and gloves. OsO_4 and propylene oxide/acetonitrile must be handled in a fume hood.

2.1.1 Materials and Tools

Glass slides and coverslips (without coating).

Sigmacote-coated (siliconized) glass slide and coverslip (prepared according to the procedures below).

Glass dish.

Sigmacote siliconizing reagent.

Staining jars and racks.

Six-well dish.

Ultrapure water.

0.1 M PB (phosphate buffer; 0.019 M NaH_2PO_4–0.081 M Na_2HPO_4) or 0.05 M TBS (tris-buffered saline; 0.05 M Tris-HCl (pH 7.6), 0.145 M NaCl).

Pasteur pipette.

Plastic board.

Razor blade.

Glass vials.

Fine-point brush.

0.4 M cacodylate buffer ($Na(CH_3)_2AsO_2$-HCl, pH 7.4).

Disposable plastic pipette.

Specified waste bottles for cacodylate/OsO_4-stained LabWare, cacodylate/OsO_4 waste solution, TCH waste solution, samarium waste solution, lead waste solution, and ethanol waste.

Pipette tip with carbon filter.

File (for opening a glass ampul).

Tough glass bottle with a screw cap (for opening a glass ampul).

Beakers and tubes (some steps require centrifugable tubes).

4% osmium tetroxide (OsO_4) (in water solution).

Potassium ferrocyanide ($K_4[Fe(CN)_6]$).

TCH (thiocarbohydrazide).

Incubator (to keep materials at 65 °C).

Samarium(III) chloride ($SmCl_3$).

Parafilm.

Shaker and refrigerator different from the one being used for antibody labeling.

Oven which can be set at 50/70 °C.

Syringe.

0.22-μm filter.

Water bath.

Weight to prevent the bottle from floating up in the water bath.

pH meter.

L-Aspartic acid.

NaOH.

KOH.

Lead(II) nitrate ($Pb(NO_3)_2$).

EtOH (50, 70, 90, 95, 99.5%).

100% EtOH (prepared by putting copper(II) sulfate anhydrate ($CuSO_4$) or molecular sieve in 99.5% EtOH).

Epoxy resin (Durcupan, Hard-Plus Resin-812 or LX112).

Analytic balance.

Aluminum foil.

Centrifuge.

Propylene oxide or acetonitrile.

Silicone containers (or glass vials and plastic mesh).

Wooden stick (disposable chopstick).

Spacer (250-μm-thick insulating tape or 70-μm-thick punch label).

Twisted small pieces of Kimwipes.

Weights.

Buckets.

2.1.2 Preparation of Sigmacote-Coated Glass Slide and Coverslip

Wear gloves to prevent contamination on glass:

1. Put Sigmacote siliconizing reagent in a glass dish. Soak glass slides and coverslips in the Sigmacote for 5 min.

2. Set the glass slides and coverslips in a staining rack. Wash the slides and coverslips with running tap water for 30 min.
3. Wash with ultrapure water:
   (i) Put the rack with the slides and coverslips in the staining jar filled with ultrapure water.
   (ii) Repeat two to three times.
4. Air-dry the slides and coverslips.

   The Sigmacote-coated glass slides and coverslips should be stored in the rack in a container (e.g., staining jar) with a lid.

2.1.3 First Day

Sections of fixed brain samples in anti-freezing solution

1. Move the samples into a six-well dish by decantation. Keep on ice.

   Rinse the samples with buffer solution. The following buffers can be used for this step: 0.1 M PB (phosphate buffer; 0.019 M NaH_2PO_4–0.081 M Na_2HPO_4) or 0.05 M TBS (tris-buffered saline; 0.05 M Tris-HCl (pH 7.6), 0.145 M NaCl). Remove the solution with a Pasteur pipette, and add buffer solution. Repeat two to three times.
2. If the samples are larger than 2–3 mm square, cut them into smaller pieces, and keep the region of interest (ROI) intact. Move the samples onto a plastic board with a fine-point brush, and then cut them using a razor blade. Avoid cutting into too small pieces because such pieces are hard to handle.
3. Move the samples into glass vials using a fine-point brush. Keep on ice:

   Check if workers wear lab coats and gloves.

   Preparation for **step 4**:

   Prepare 0.1 M cacodylate buffer ($Na(CH_3)_2AsO_2$-HCl, pH 7.4) from stock of 0.4 M cacodylate buffer (pH 7.4) and ultrapure water.
4. Rinse the samples with 0.1 M cacodylate buffer twice. Remove the solution from the glass vials using a plastic pipette, and then add 0.1 M cacodylate buffer using a plastic pipette. Buffer solution used in **step 1** can be discarded into the sink, but the cacodylate buffer must be discarded into a specified cacodylate/OsO_4 waste bottle. Use a single plastic pipette to handle 0.1 M cacodylate/osmium solution in order to reduce plastic pipette waste.

   Preparation for **step 5**:

   Prepare the first osmium reaction solution (2% OsO_4, 0.1 M cacodylate buffer).

Handle osmium compounds in the fume hood. To measure out osmium compound solution, use a pipette tip with carbon filter. Use a minimal amount of osmium.

When discarding tools and containers used with osmium, put them into a designated garbage bottle in the fume hood.

To open an OsO_4 ampule, wash the outer surface of the ampule, scratch the neck of the ampul with a file, put the ampule into a tough glass bottle with a screw cap, close the cap of the bottle, and then agitate the bottle. Measure out the OsO_4 solution from this bottle.

For 1 mL of the first osmium reaction solution:

4% osmium tetroxide (OsO_4)	500 μL (final: 2%).
0.4 M cacodylate buffer	250 μL (final: 0.1 M).
Ultrapure water	250 μL

This solution is required 1 mL per vial. The excess 4% OsO_4 solution is used in **step 12**.

5. First osmium reaction. Remove the cacodylate buffer, and add the first osmium reaction solution. Gently shake. RT for 1 h.

   Preparation for **step 6**:

   Prepare potassium ferrocyanide reaction solution (3% $K_4[Fe(CN)_6]$, 0.1 M cacodylate buffer).

   For 1 mL of the potassium ferrocyanide reaction solution:

Potassium ferrocyanide ($K_4[Fe(CN)_6]$)	30 mg (final: 3%)
0.4 M cacodylate buffer	250 μL (final: 0.1 M)
Ultrapure water	750 μL

This solution is required 1 mL per vial.

6. Potassium ferrocyanide reaction. Remove the osmium solution, and add the potassium ferrocyanide reaction solution. Incubation time is 1 h or 1.5 h. Discard the osmium reaction solution into the cacodylate/OsO_4 waste bottle. Handle gently because the OsO_4-processed samples have got fragile.

   Preparation for **step 9**:

Prepare 1% TCH (thiocarbohydrazide) solution:

TCH	0.1 g (final: 1% w/v)
Ultrapure water	10 mL

Put the solution into 65 °C incubator for ~1 h to dissolve TCH. Periodically agitate the solution using a vortex mixer.

7. Rinse the samples with ultrapure water three times. Remove the solution, and add ultrapure water. Discard the potassium ferrocyanide reaction solution and the first rinse solution into the cacodylate/OsO_4 waste bottle. Handle gently because the samples are fragile.
8. Flatten the samples. Using a fine-point brush, move the sample onto a glass slide, and put a coverslip on the sample. Gently push the coverslip with the fine-point brush. After a 10-min incubation, add ultrapure water on the glass slide to float the coverslip. Then remove the coverslip, and move the sample back into a glass vial using the fine-point brush. Add ultrapure water.

   Preparation for **step 9**:

   Cool the TCH solution with ice to RT. Put the TCH solution into a syringe. Put a 0.22-μm filter on the syringe.
9. TCH reaction. Add filtered TCH solution into samples. Gently shake. RT for 30 min.

   Preparation for **step 12**:

   Prepare the second osmium reaction solution (2% OsO_4, no ferrocyanide or cacodylate).

   For 1 mL of the second osmium reaction solution:

4% osmium tetroxides (OsO_4)	500 μL (final: 2%)
Ultrapure water	500 μL

This solution is required 1 mL per vial.

10. Rinse the samples with ultrapure water three times. Removed TCH solution must be discarded into a TCH waste bottle.
11. Flatten the samples as **step 8**.
12. Second osmium reaction. Remove the ultrapure water, and add the second osmium reaction solution. Gently shake. RT for 30 min (for ~30-μm-thick section) or 2 h (for ~250-μm-thick section).

    Preparation for **step 15**:

Prepare samarium(III) chloride ($SmCl_3$) solution:

$SmCl_3 \cdot 6H_2O$	250 mg (final: 2.5% w/v as hexahydrate)
Ultrapure water	10 mL

13. Rinse the samples with ultrapure water three times. Remove the solution, and add ultrapure water. Discard the osmium reaction solution and the first and second rinse solution into the cacodylate/OsO_4 waste bottle. Put the samples on ice.

    If you did not use all of the 4% OsO_4 solution in the ampul, seal the bottle, which was used to open the ampul, using parafilm. This unused 4% OsO_4 solution can be stored in 4 °C for months. Use a refrigerator different from the one being used for antibody labeling.

    After finishing the osmium work, close the caps of the cacodylate/OsO_4 waste bottle and the osmium garbage bottle.
14. Flatten the samples as **step 8**.
15. Remove the solution, and add 2.5% $SmCl_3$. Gently shake at 4 °C overnight. Cover from light. Use a shaker/refrigerator different from the one being used for antibody labeling.

2.1.4 Second Day

16. Remove the 2.5% $SmCl_3$ and add ultrapure water. The removed $SmCl_3$ solution must be discarded into a samarium waste bottle. Rinse the sample with ultrapure water twice more.
17. Flatten the samples as **step 8**.
18. Put the samples into a 50 °C oven.

    Preparation for **step 19**:

    (i) Put a syringe, a 0.22-μm filter, and ultrapure water into a 50 °C oven.

    (ii) Prepare and calibrate a pH meter.

    (iii) Prepare and warm L-aspartic acid solution:

L-Aspartic acid	0.998 g (final: 30 mM)
Ultrapure water	250 mL

Keep the solution at 50 °C using a water bath and stir.

To dissolve the L-aspartic acid quickly, add 1 M NaOH until the solution pH goes up to pH 3.8.

(*) This stock solution of 30 mM L-aspartic acid is stable for 1–2 months if refrigerated.

(iv) Prepare and warm 1 M KOH solution.

Keep the solution at 50 °C using the water bath.

(v) Prepare lead aspartate solutions at 50 °C.

Lead(II) nitrate ($Pb(NO_3)_2$)	0.066 g (final: 20 mM)
30 mM L-aspartic acid	10 mL

Put $Pb(NO_3)_2$ into a bottle (or beaker). Gently place L-aspartic acid solution into the bottle, running it along the wall. Avoid bubble formation because lead aspartate solution makes precipitation when it takes in CO_2 in the air. To prevent the bottle from floating up in the water bath, some weights are tied to the bottle. Put a thermometer's probe in the solution. Stir gently.

Record pH and temperature of the solution.

Add 1 M KOH until pH goes up to 5.0 (or ≥4.95). 400 μL + α of 1 M KOH will be required. Check if the temperature is kept at 50 °C.

(vi) Suck the lead aspartate solution into the syringe, and keep at 50 °C. Put the 0.22-μm filter on the syringe.

19. Lead reaction. Remove the solution from the sample, and add filtered lead aspartate solution into the sample. Add the solution on the wall of the vial to avoid bubble formation. Keep at 50 °C in the oven for 1 h.

    Preparation for **step 23** (only needed if you use LX112 as epoxy resin):

    Start **step 23** and proceed concurrently with **steps 20–22**. This is because the preparation of LX112 takes time (3 h or more) to thoroughly mix components.

20. Remove the lead aspartate solution from the samples, and add ultrapure water at 50 °C. Discard the lead aspartate solution into a lead waste bottle. Rinse the samples with ultrapure water twice more at RT.
21. Flatten the samples as **step 8**.
22. Dehydration. 100% EtOH is prepared by putting copper (II) sulfate anhydrate ($CuSO_4$) or molecular sieve in 99.5% EtOH. Removed EtOH must be discarded into an ethanol waste tank:

(i)	50% EtOH on ice (4 °C)	5 min (for ~30-μm-thick) or 10 min (for ~250-μm-thick)
	After incubation in 50% EtOH, flatten the samples as **step 8**. When floating the coverslip to remove it, add 50% EtOH on the glass slide instead of ultrapure water	
(ii)	70% EtOH on ice (4 °C)	5 min (for ~30-μm-thick) or 10 min (for ~250-μm-thick)

(continued)

	After incubation in 70% EtOH, flatten the samples as **step 8**. When floating the coverslip to remove it, add 70% EtOH on the glass slide instead of ultrapure water	
(iii)	80% EtOH on ice (4 °C)	10 min (for ~30-μm-thick) or 20 min (for ~250-μm-thick)
(iv)	90% EtOH on ice (4 °C)	10 min (for ~30-μm-thick) or 20 min (for ~250-μm-thick)
(v)	95% EtOH on ice (4 °C)	10 min (for ~30-μm-thick) or 20 min (for ~250-μm-thick)
	After this step, leave a small amount of EtOH when removing the EtOH to prevent the sample from drying up	
(vi)	99.5% EtOH at RT	10 min (for ~30-μm-thick) or 20 min (for ~250-μm-thick)
	Do not put $CuSO_4$ (or molecular sieve) powder into the sample when adding 100% EtOH	
(vii)	100% EtOH at RT	10 min (for ~30-μm-thick) or 20 min (for ~250-μm-thick)
(viii)	100% EtOH at RT	10 min (for ~30-μm-thick), 20 min (for ~250-μm-thick) or 30 min (for ~500-μm-thick)

23. Prepare epoxy resin (Durcupan, Hard-Plus Resin-812 or LX112). Handle mixtures and components of epoxy resin in a fume hood. Check if the room humidity is kept low (<40%). You can use epoxy resin mixture which was previously made and stored at −20 °C (see also **step 25**) (*see* **Note *1**). In this case, bring the mixture to RT thoroughly before opening, to prevent moisture absorption. Avoid spilling the mixtures or components of the epoxy resin. Collect tools and Kimwipes stained with epoxy resin onto a sheet of aluminum foil.

    For Durcupan:

Component A	11.4 g
Component B	10.0 g (21.4 g together with component A)
Component C	0.3 g (21.7 g together with components A and B)
Component D	0.05–0.1 g

(i) Put a plastic 50-mL tube on an analytic balance. Put component A into the tube by decantation or using a disposable plastic pipette until the balance reads 11.4 g.

(ii) Add component B until the balance reads 21.4 g.

(iii) Add component C until the balance reads 21.7 g.

(iv) Stir with disposable plastic microspatula until the color of the mixture gets dark orange.

(v) Reset the scale of the balance to 0.
Add component D until the balance reads 0.05–0.1 g.

(vi) Stir the mixture.

(vii) Centrifuge the mixture (3000 rpm, 5 min) to remove bubbles. Keep balance.

(viii) Using the aluminum foil, wrap the tools and Kimwipes stained with Durcupan, and put them into a 60 °C oven.

For Hard-Plus Resin-812:

Hard-Plus Resin-812	50 g
Hardener Hard-Plus	50 g
Accelerator	2.5 mL

(i) Put Hard-Plus Resin-812 into a bottle of Hardener Hard-Plus. Gently tumble the bottle until the two liquids are well mixed.

(ii) Add Accelerator into the bottle. Handle the Accelerator in a fume hood. Gently tumble the bottle until the liquids are well mixed.

(iii) Move the mixture into centrifugal tubes, and then centrifuge it (3000 rpm, 5 min) to remove bubbles. Keep balance.

(iv) Using the aluminum foil, wrap the tools and Kimwipes stained with Hard-Plus Resin-812, and put them into a 60 °C oven.

For LX112:

Mixture A

LX112	10 g
NSA (nonenyl succinic anhydride)	11.2 g

Mixture B

LX112	30 g
NMA (nadic methyl anhydride)	26.4 g

Resin

Mixture A	18 g
Mixture B	42 g
BDMA (benzyl dimethyl amine)	1.334 mL

(*) Since the volumes of mixture B and resin listed above do not fit into a 50-mL tube, these mixtures are prepared in two separate tubes if necessary.

(i) Put a plastic tube on an analytic balance. Put LX112 into the tube by decantation or using a disposable plastic pipette until the balance reads 10 g.

(ii) Reset the scale of the balance to 0. Add NSA until the balance reads 11.2 g. This mixture is mixture A.

(iii) Put a new plastic tube on an analytic balance. Put LX112 into the tube by decantation or using a disposable plastic pipette until the balance reads 30 g.

(iv) Reset the scale of the balance to 0. Add NMA until the balance reads 26.4 g. This mixture is mixture B.

(v) Put mixtures A and B, which are prepared in steps (ii) and (iv), on a rotator, and stir well for 1 h.

(vi) Put a new plastic tube on an analytic balance. Put mixture A into the tube by decantation or using a disposable plastic pipette until the balance reads 18 g.

(vii) Reset the scale of the balance to 0. Add mixture B into the tube by decantation or using a disposable plastic pipette until the balance reads 42 g.

(viii) Put the mixtures of resins on a rotator, and stir well for 1 h.

(ix) Add 1.334 mL BDMA to the resin prepared in (viii). Handle the BDMA in a fume hood.

(x) Gently tumble the tube, put it on a rotator, and stir well for 1 h.

(xi) Centrifuge the mixtures (3000 rpm, 5 min) to remove bubbles.

(xii) Using the aluminum foil, wrap the tools and Kimwipes stained with LX112, and put them into a 60 °C oven.

24. Remove the EtOH from the samples, and add propylene oxide. Propylene oxide may be substituted by acetonitrile. Handle the propylene oxide/acetonitrile in the fume hood.

    Preparation for **step 25**:

    Mix the epoxy resin and propylene oxide (or acetonitrile) by 1:1 volume ratio.

25. Remove the propylene oxide (or acetonitrile) from the samples, and add the 1:1 mixture of the epoxy resin and the organic solvent. Gently shake. RT overnight.

    Collect tools and Kimwipes stained with epoxy resin onto a sheet of aluminum foil. Using the aluminum foil, wrap the tools and Kimwipes stained with epoxy resin, and put them into a 60 °C oven.

    Seal the tube of unused mixture of epoxy resin using parafilm. This mixture can be stored at −20 °C for months (*see* **Note *1**).

2.1.5 Third Day

26. Prepare epoxy resin (Durcupan, Hard-Plus Resin-812, or LX112) as **step 23**.

    Preparation for **step 27**:

    Put the epoxy resin into silicone containers (or glass vials) using a disposable plastic pipette with a cut tip. (When you use glass vials, put a piece of plastic mesh on the bottom of the glass vial to prevent the sample from sticking on the bottom).

    Collect tools and Kimwipes stained with epoxy resin onto a sheet of aluminum foil.

27. Put the samples into the epoxy resin. In the fume hood, using a wooden stick (disposable chopstick) with a sharpened tip, pick the samples in the 1:1 mixture, and put them into the epoxy resin in the silicone containers (or glass vials).

    Collect the waste liquid containing organic solvent (propylene oxide/acetonitrile) into a disposable plastic ware, and leave it in the fume hood for 24 h (*see* **Note *2**).

28. Leave the samples in the epoxy resin for 2–3 h at RT. Keep the humidity around the samples low (<40%; if necessary, take the samples into a low humidity room), and leave the cap open to evaporate organic solvent. Cover the containers to prevent dust contamination.

    Preparation for **step 29**:

(i) Set temperature of an oven to 70 °C.

(ii) Put pieces of a spacer (250-μm-thick insulating tape or 70-μm-thick punch label) on a Sigmacote-coated (siliconized) glass slide.

29. Put the glass slides on the aluminum foil-covered tray of the oven.

Put a drop of epoxy resin on the glass slide using a disposable plastic pipette with a cut tip.

Put the samples on the glass slides using a wooden stick with a sharpened tip. Be careful because the samples are fragile. If necessary, collect and move the small fragments of the samples onto the glass slides.

Place a sigma-coated coverslip on the sample.

30. Put twisted small pieces of Kimwipes on the edges of the coverslip. The Kimwipes are used to absorb excess epoxy resin.

Then put a weight on the coverslip.

Before taking the samples into the 70 °C oven, put aside the weights. Take the tray on which the samples are placed into the 70 °C oven. Then put the weights back on the coverslip.

Leave the samples at 70 °C for 48–72 h (2–3 days).

Tidying up:

Seal the tube of unused mixture of epoxy resin using parafilm. This mixture can be stored at −20 °C for months.

Remove epoxy resin from the glass/silicone wares, and collect it into a disposable plastic ware. Wipe the glass/silicone wares with Kimwipes to remove epoxy resin. Move the glass/silicone wares into EtOH in a plastic bucket marked "2." Use forceps to immerse the glass/silicone wares thoroughly in the EtOH (*see* **Note *3**).

Using the aluminum foil, wrap the tools and Kimwipes stained with epoxy resin, and put them into the 70 °C oven. Also, put the tube containing the epoxy resin into the oven.

2.1.6 Notes

*1. (**steps 23** and **25**) Epoxy resin (Durcupan, Hard-Plus Resin-812, or LX112) becomes much more viscous the next day when stored at room temperature overnight. It must be stored in the freezer.

*2. (**step 27**) After evaporating propylene oxide in the fume hood, collect the waste liquid (containing epoxy resin) into a disposable plastic ware, and put it into the 60 °C/70 °C oven.

*3. (**step 30**) After 2 days, move the glass/silicone wares into EtOH in a plastic bucket marked "2." After ~1 month, pick them up, and wash them with water and detergent as general lab wares. Sometimes discard the EtOH in the bucket "1" into an Epson/ethanol waste tank, move the EtOH from the bucket "1" to the bucket "2," and fill the bucket "1" with new EtOH.

2.2 Serial Ultrathin Section Imaging with FE-SEM Using CNT Tape

1. A typical tissue block size for ATUM (automated tape ultramicrotome) sectioning using carbon nanotube-coated (CNT)-PET tape is 0.7–0.3 mm length × 1.0–0.5 mm width. Design the trimming geometry to yield a modified trapezoid with shorter upper line parallel to longer lower line (Fig. 4). The block should be asymmetrical. This makes orientation easily identifiable in the electron microscope which will aid determination of imaging area of a region of interest (ROI).
2. Serial ultrathin sections are cut with the ATUM, and they are collected on the CNT-PET tape (Fig. 2).
3. The tape with the serial ultrathin sections is cut into strings of 10 cm or shorter length, and they are then put on 10 cm silicon wafer with double-sided adhesive conductive tape. Side edge of the CNT tape must be grounded by copper or aluminum tape to prevent charging problem.
4. Put the wafer into the chamber of FE-SEM to observe the tissue ultrastructure. Use automated array tomography imaging control application equipped in the FE-SEM to capture the

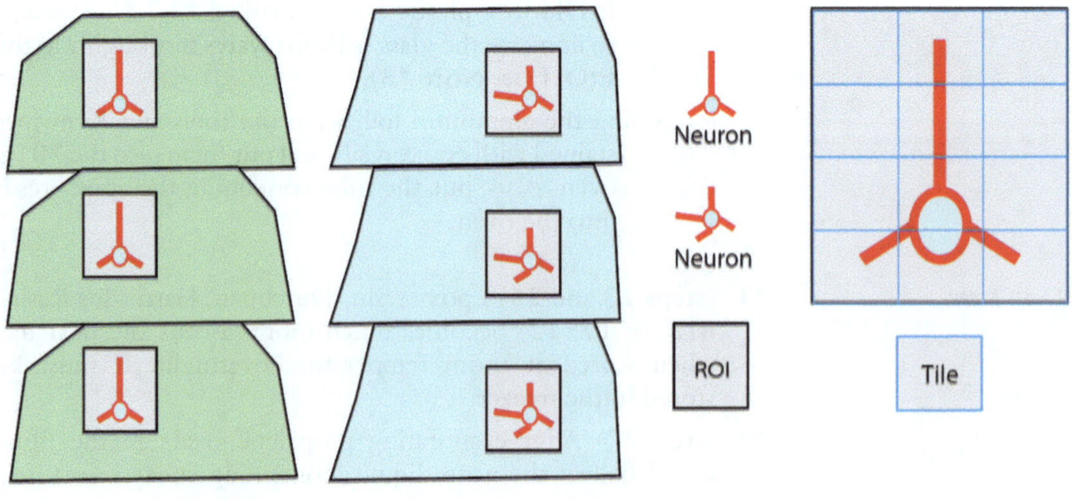

Fig. 4 Serial ultrathin tissue sections

electron micrographs of the ROI in the serial ultrathin sections. The ROI image can be tiled and imaged.

5. Montage may be generated with stitching the tiles using the TrakEM2 in NIH image (https://imagej.nih.gov/nih-image/index.html) with a PC that has a high-power GPU and large RAM.
6. The serial electron micrographs are aligned using the TrakEM2 in the NIH image.

   Segmentation of the neuronal elements is processed manually using VAST Lite (https://software.rc.fas.harvard.edu/lichtman/vast/) [19].

2.3 Serial Ultrathin Section Imaging with Blade/TEMCA Using GridTape

Preparing a block of tissue for fully automated, or semi-automated, sectioning with a robotic device requires that the block be prepared in such a way as to maximize consistency. This includes positional accuracy and consistency in section thickness:

1. Identify a region of interest (ROI) from the osmicated tissue sections.

   A typical tissue block sizes for ATUM (automated tape ultramicrotome) sectioning using GridTape is 4.0–5.5 mm length × 1.0 mm width. If possible, design the trimming geometry to yield a visible mark on the EM section (e.g., a corner) that is easily identifiable in the electron microscope. This will aid the determination of the imaging area of the ROI.

   Useful tools in the preparation of such a block are a diamond trim tool (e.g., EM Trim Tool (or EM RAPID milling device), razor blades, and diamond trimming knives.
2. Examine a few ultrathin (e.g., 40 nm) sections with a diamond knife to assess ultrastructural quality and suitability for further sectioning. One caveat is that often with en bloc staining methods, the outermost tissue is often of lesser quality than deeper in the tissue block. However, even with heavily staining osmium methods as described above, the tissue quality improves rapidly after progressing 5–10 μm in depth from the surface.
3. When the ultrastructure is determined to be suitable for large-scale sectioning, plan the ROI of the block.

Fig. 5 Generalized hexagon shape for ATUM-style block. Sectioning direction would be in the horizontal direction, e.g., left to right

Fig. 6 Block with vertical sides for automated sectioning

4. A shape that is useful for automated sectioning is roughly hexagonal, with longer parallel sides that are perpendicular to the sectioning direction (Fig. 5). In our experience, this shape helps decrease the immediate changes in cutting pressure of the diamond microtome blade on the leading and trailing edges of the tissue block as the block meets and leaves the knife.
5. Another useful block property is to have vertical sides extending away from the block face, so that the overall section size does not change as blocks are sectioned (Fig. 6). This step is not strictly necessary, but not doing so could add additional work to align ROI with the tape. Some considerations in preparing a block for serial sections are given in [20].
6. Prepare the microtome and ATUM tape collection device for sectioning. Typically, this involves aligning the block face, loading tape, and cutting a few test sections. Specific instructions for how to do so vary depending on the model of the microtome and the tape collection hardware and software used; in our experience, we have used either a modified RMC microtome or Leica UC7 with serial control and a modified ATUM device from RMC. For more considerations on ATUM-based serial section collection, see [21, 22].
7. The first few sections collected can be used to synchronize the ATUM tape collection machine and the microtome. Generally speaking, it is best to position the section so as to provide a buffer zone of ~50–100 μm from the edges of the aperture (Fig. 7). The milled aperture in this example is 2.0 × 1.5 mm.
8. If the intended imaging volume is large, it may be worth performing a short tape collection and EM imaging as a quality check. This would help predict the sectioning behavior of the block over multiple sections and help indicate if the ultrastructural quality and contrast of the tissue are sufficient for the large-scale volume collection.
9. It may be necessary to clean the knife edge of debris or stuck section fragments; in that case, follow the diamond knife manufacturer's guidelines for cleaning. In our experience, stopping section collection generally causes the epoxy tissue block to contract slightly, and thus it is necessary to reposition the knife

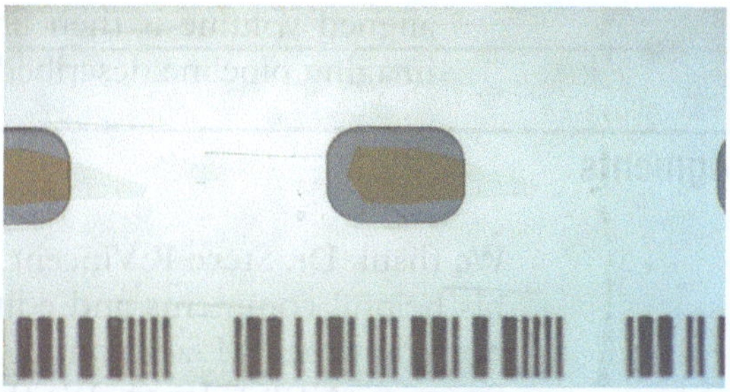

Fig. 7 Close-up of GridTape™ showing sequential section pickup and good placement of section (light brown hexagonal shape) on TEM film-coated aperture (rounded rectangle shape)

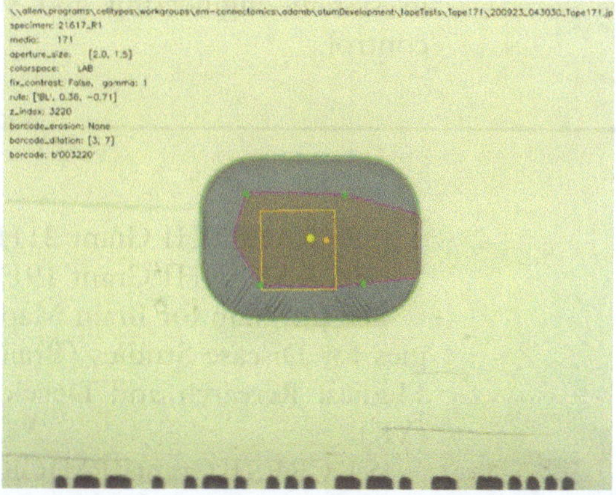

Fig. 8 Automated image analysis identifies an ROI (yellow rectangle) from an optical inspection image

to start up again. Ideally there are no lost sections between the start and stop of the section collection for knife cleaning.

10. After tape collection, the tape is run through an inspection machine to acquire high-resolution optical images (Fig. 8). These images are then uploaded to a TEM imaging database for ROI extraction based on image analysis and population of an imaging database. The database contains metadata and imaging information for that section as reported by the electron microscopy control software.

11. The GridTape is then loaded into the EM and prepared for imaging using the imaging database.

12. As imaging progresses, quality control metrics in software determine if a section needs to be re-imaged.

13. An image series is now ready for stitching and alignment phase of the pipeline to create a rough-aligned volume. This rough-

aligned volume is then fine-aligned and segmented with the imaging pipeline described in [23].

Acknowledgments

We thank Dr. Steve R Vincent, University of British Columbia, for his helpful comments and editing. Electron microscopy was supported by the EM facility in the National Institute for Physiological Sciences (NIPS), Japan. We thank the Hitachi High-Tech Corporation team and the Allen Institute Connectomics team for the development of software and hardware for automated section collection and imaging. We also thank the Allen Institute Manufacturing and Process Engineering team for their expertise in system engineering and software development for TEMCA control.

Funding

JSPS KAKENHI Grant 21H00217 (YK).

JSPS KAKENHI Grant 19H03336 (YK).

The program for Brain Mapping by Integrated Neurotechnologies for Disease Studies (Brain/MINDS) from Japan Agency for Medical Research and Development (AMED) JP22dm0207084 (YK).

JST CREST Grant JPMJCR21E2 to (YK).

US Intelligence Advanced Research Projects Activity (IARPA) via DoI/IBC contract number D16PC00004 (MMT).

References

1. White EL, Rock MP (1980) Three-dimensional aspects and synaptic relationships of a Golgi-impregnated spiny stellate cell reconstructed from serial thin sections. J Neurocytol 9:615–636
2. Fiala JC (2005) Reconstruct: a free editor for serial section microscopy. J Microsc 218:52–61
3. Fiala JC, Allwardt B, Harris KM (2002) Dendritic spines do not split during hippocampal LTP or maturation. Nat Neurosci 5:297–298
4. Kubota Y et al (2007) Neocortical inhibitory terminals innervate dendritic spines targeted by thalamocortical afferents. J Neurosci 27:1139–1150
5. White EL, Weinfeld E, Lev DL (2004) Quantitative analysis of synaptic distribution along thalamocortical axons in adult mouse barrels. J Comp Neurol 479:56–69
6. Kubota Y (2015) New developments in electron microscopy for serial image acquisition of neuronal profiles. Microscopy (Oxf) 64:27–36
7. Knott G et al (2008) Serial section scanning electron microscopy of adult brain tissue using focused ion beam milling. J Neurosci 28:2959–2964
8. Denk W, Horstmann H (2004) Serial block-face scanning electron microscopy to reconstruct three-dimensional tissue nanostructure. PLoS Biol 2:e329
9. Haruwaka K et al (2019) Dual microglia effects on blood brain barrier permeability induced by systemic inflammation. Nat Commun 10:5816
10. Hayworth KJ et al (2014) Imaging ATUM ultrathin section libraries with Wafer Mapper: a multi-scale approach to EM reconstruction of neural circuits. Front Neural Circuits 8

11. Kubota Y et al (2018) A carbon nanotube tape for serial-section electron microscopy of brain ultrastructure. Nat Commun 9:437
12. Mikuni T et al (2016) High-throughput, high-resolution mapping of protein localization in mammalian brain by in vivo genome editing. Cell 165:1803–1817
13. Bock DD et al (2011) Network anatomy and in vivo physiology of visual cortical neurons. Nature 471:177–182
14. Lee WC et al (2016) Anatomy and function of an excitatory network in the visual cortex. Nature 532:370–374
15. Eberle AL, Zeidler D (2018) Multi-beam scanning electron microscopy for high-throughput imaging in connectomics research. Front Neuroanat 12:112
16. Yin W et al (2020) A Petascale automated imaging pipeline for mapping neuronal circuits with high-throughput transmission electron microscopy. Nat Commun 11:4949
17. MICrONS Consortium, Bae JA, Baptiste M et al (2021) Functional connectomics spanning multiple areas of mouse visual cortex. bioRxiv. [Online]. https://doi.org/10.1101/2021.07.28.454025
18. Turner NL et al (2022) Reconstruction of neocortex: organelles, compartments, cells, circuits, and activity. Cell 185:1082–1100.e24
19. Berger DR, Seung HS, Lichtman JW (2018) VAST (volume annotation and segmentation tool): efficient manual and semi-automatic labeling of large 3D image stacks. Front Neural Circuits 12:88
20. Harris KM et al (2006) Uniform serial sectioning for transmission electron microscopy. J Neurosci 26:12101–12103
21. Graham BJ et al (2019) High-throughput transmission electron microscopy with automated serial sectioning. bioRxiv. [Online]. https://doi.org/10.1101/657346
22. Baena V et al (2019) Serial-section electron microscopy using automated tape-collecting ultramicrotome (ATUM). Methods Cell Biol 152:41–67
23. Mahalingam G et al (2022) A scalable and modular automated pipeline for stitching of large electron microscopy datasets. elife 11: e7653

Chapter 7

Espina: A Software Tool for the Study of 3D Synaptic Organization

Marta Turegano-Lopez, Félix de las Pozas, José-Rodrigo Rodríguez, Javier DeFelipe, and Angel Merchán-Pérez

Abstract

Automated electron microscopy methods produce a large number of serial sections that can be stacked to reconstruct a three-dimensional (3D) sample of brain tissue. The visualization, navigation, and analysis of these stacks of serial sections require dedicated software. Here we describe Espina, a software tool which helps to perform the segmentation and analysis of subcellular elements in a reconstructed 3D volume of the brain. Espina can navigate the stack in 3D, allowing the visualization of images through the original plane of section and through the other two orthogonal planes. Synaptic junctions can be identified and segmented in 3D. The software provides the spatial position of synapses, as well as their size and other geometrical features. The shape of synapses can also be analyzed, and the postsynaptic targets (dendritic spines or dendritic shafts) can be identified and tagged. Axons and dendrites can be traced, together with their connections, so the linear density of synapses established by individual fibers can be calculated. Finally, the relationship of synapses with other cellular and subcellular elements such as astrocytes, mitochondria, or multivesicular bodies can be analyzed.

Key words Volume electron microscopy, Synapses, Brain connectivity, FIB SEM, 3D software

1 Introduction

Electron microscopy (EM) is the method of choice for the study of the structure of brain tissue at the synaptic and subsynaptic level. If 3D studies are required, the most common approach is reconstruction from serial sections. However, manual serial sectioning for EM is a time-consuming and technically difficult task, and the number of sections that can be obtained is usually limited. Automated serial sectioning methods such as serial block-face scanning electron microscopy (SB-SEM), Focusing Ion Beam Milling-Scanning Electron Microscopy (FIB-SEM), or other serial sectioning SEM approaches overcome these difficulties and yield hundreds or thousands of serial sections in a relatively short period of time

Joachim H. R. Lübke and Astrid Rollenhagen (eds.), *New Aspects in Analyzing the Synaptic Organization of the Brain*, Neuromethods, vol. 212, https://doi.org/10.1007/978-1-0716-4019-7_7,

[1–4]. The availability of these methods has displaced the bottleneck from the difficulty in obtaining serial sections to the difficulty in performing the 3D reconstruction, visualization, and analysis of a vast number of serial sections.

A major aim of electron microscopy is to determine the number of synapses in a given brain area. This number can be estimated from collections of single electron micrographs or from series of EM images, using stereological methods such as the size frequency method or the disector method. However, these methods show a high statistical variability, which is strongly dependent on sample size, and they tend to underestimate the number of objects that are present in low numbers, such as inhibitory synapses [5]. Ideally, these drawbacks would be avoided if you could directly count the number of synapses within a known volume of tissue [3]. If you have a long-enough stack of serial sections, you can reconstruct a block of tissue, and a tool is then needed to identify the synapses within the stack.

Several software tools are freely available to perform this task [some of them reviewed in [6]]. However, in our experience, they are often difficult to use or do not provide certain data that we consider important to study the synaptic organization (see below). Thus, we designed a software package, named Espina, which was originally planned as a user-friendly tool for the visualization, segmentation, and quantification of synapses in stacks of FIB-SEM serial sections [7]. With further development, new functionalities have been added to Espina and, in its current form, it also helps determine the size and shape of synapses, their spatial distribution, and their position on dendritic spines or shafts. Axons and dendrites can also be traced within the stack of serial sections, so the linear densities of synapses established by different types of neuronal processes (dendrites and axons) can be calculated. The relationship of synapses with other cellular and subcellular elements, such as astrocytes, mitochondria, or multivesicular bodies, can also be investigated in 3D. Furthermore, as we will discuss below, Espina can also be used to analyze stacks of confocal images. We will describe these capabilities in the following sections, with reference to research work that has been performed with this tool. Espina is freely available for download at https://cajalbbp.csic.es/espina-2/.

2 Serial Image Acquisition and Preprocessing

Serial images can be obtained with different methods of light and electron microcopy, and any of them can be visualized and analyzed with Espina. We will use as an example a series of images that have been acquired with FIB-SEM. With this or any other method to obtain serial images, once the full series of images has been acquired, some drift is commonly noticeable, especially after long milling/imaging sessions. Thus, alignment of serial images—also known as registration—is necessary. To do this, we use freely available software such as ImageJ [8, 9] or FIJI, which is a version of ImageJ with a collection of preinstalled plug-ins for microscopy [10, 11]. We recommend using the "Register Virtual Stack Slices" plug-in. Among the several options available, we use a registration algorithm that only allows translation of individual images, with no rotation, to avoid any deformation of the original images, size changes, or other artifacts. After correcting brightness and contrast (if needed), we save the registered series of images as an uncompressed stack in TIFF format.

An additional preprocessing step is required for the segmentation of synaptic junctions. The segmentation algorithms that Espina uses for synapses rely on the fact that the pre- and postsynaptic densities are strongly electron-dense, so they appear darker than the surrounding structures. However, since EM images are noisy, the pixels corresponding to the synaptic densities have very heterogeneous gray levels, and light pixels appear among the generally darker ones. To eliminate these noisy pixels, we apply a Gaussian blur filter on the previously registered stack and save it as another TIFF file. Pixel gray-level heterogeneity is reduced, but the images are obviously blurred, so it is important to also keep the original registered stack. Notice that both stacks will have exactly the same dimensions.

3 Loading Stacks of Images into Espina and Spatial Calibration

One or more TIFF stacks of images can be loaded simultaneously into Espina. If only one stack is loaded, this stack will be both the *Active* stack (the one on which any operations are performed) and the *Visible* stack (the one that the user actually sees) (Fig. 1a). When two or more stacks are loaded, the user can choose which one is the *Active* stack and which one is the *Visible* stack. For example, a Gaussian blur stack is recommended to segment synaptic junctions, as mentioned above. If only this stack is loaded, the user will have to work with blurred images, which is uncomfortable and tiring. To overcome this difficulty, the original, unprocessed stack can also be loaded and made the *Visible* stack while keeping the blurred stack as the *Active* stack. In this way, the segmentation of synapses (or any other operation) will be performed on the blurred stack, while the user is actually viewing the original stack. In this and similar cases, it is of course essential that both stacks are exactly the same size so that they are perfectly superimposed.

Several stacks can be made *Visible* at the same time. This is common, for example, when using different channels in fluorescence microscopy. Suppose, for example, that cell nuclei and mitochondria have been labeled with different fluorophores and imaged

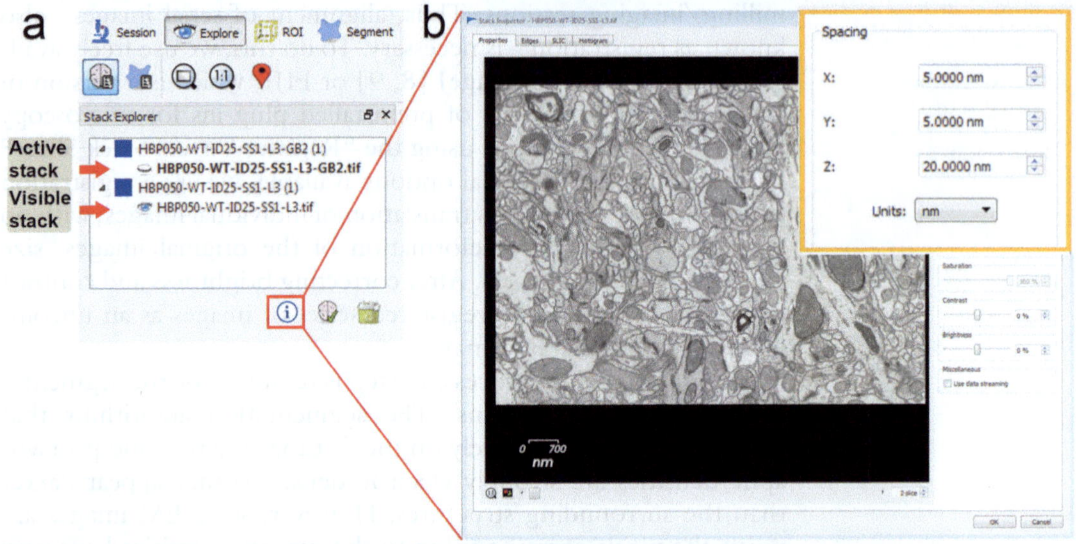

Fig. 1 Working with stacks of images in Espina. (**a**) Two stacks have been loaded simultaneously. The *Active* stack (the one used for segmentation) is highlighted in bold typeface, but it is not visible. The *Visible* stack is superimposed on the active stack and is the one that the user actually sees. (**b**) Spatial calibration in the stack of images by setting pixel size (X–Y planes) and section thickness (Z plane)

in two different channels. We can visualize any of the channels independently or a merged version of both. Then, cell nuclei and mitochondria can be segmented by making *Active* the corresponding channel.

Spatial calibration must be completed for any of the stacks that are loaded into Espina. In this step, we enter the pixel size of the images and the section thickness in the units chosen by the user (Fig. 1b). The voxels do not need to be isotropic (cubic), so the resolution of images in the XY plane (e.g., 5 nm per pixel) can be different from the section thickness in the Z axis (e.g., 20 nm). Thus, the voxel size must be entered independently in the X, Y, and Z axes. Spatial calibration affects the way stacks are visualized and navigated in 3D, as well as—very importantly—any measurements and calculations of length, surface, and volume performed by the software. Therefore, if you need, for example, two stacks to be perfectly superimposed, their original sizes and spatial calibrations must be identical.

4 Navigating the Stack of Images

The stack can be navigated through three windows that show the series of images using the three orthogonal planes of section. A fourth window can also be opened to have a 3D rendering of the stack (Fig. 2). By default, the stack of images will be shown in the

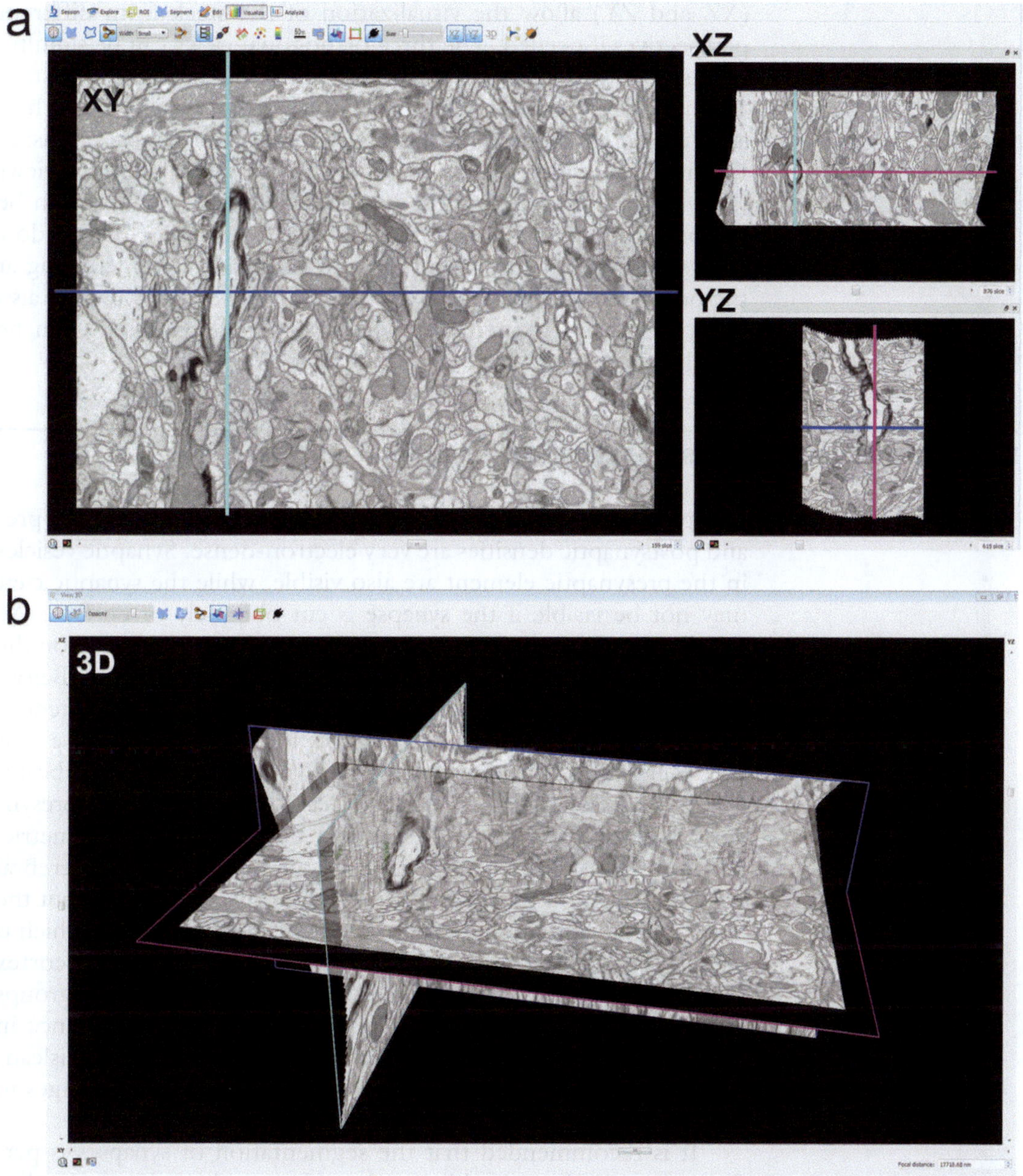

Fig. 2 Navigation of the stack of images in Espina. (**a**) Espina offers views from the three orthogonal planes (XY, XZ, YZ). The crosshair (pink and blue lines) aids the identification of different structures in each view. In this example, the crosshair is pointing to a myelinated axon. (**b**) 3D rendering of the three orthogonal planes, crossing at the myelinated axon shown in (a)

main window, in the original plane of the section (the XY plane), which shows the view that was used to acquire the images with the FIB-SEM. The user can go back and forth through the stack, along the Z axis, using the mouse wheel or the slider located at the bottom of the window. The two remaining orthogonal planes

(XZ and ZY) allow the visualization of the stack from different perspectives if needed. The three orthogonal views can be simultaneously shown, and the user can zoom and navigate the stack through any of these three windows, using the mouse wheel and/or the corresponding slider. Any object can be marked using the crosshair, which will show the same location on the three views (Fig. 2a). Segmentation, and any other editing operation, can be performed on any of the three orthogonal views. The 3D window shows a 3D rendering of the three orthogonal planes, crossing at the point marked by the crosshair (Fig. 2b). The 3D view can also be zoomed and rotated. Segmented objects (see below) can be visualized on any of the views of the stack.

5 Segmentation of Synapses in Serial EM Images

Synapses are identified by the user mainly on the basis that the pre- and postsynaptic densities are very electron-dense. Synaptic vesicles in the presynaptic element are also visible, while the synaptic cleft may not be visible if the synapse is cut obliquely or en face. We usually classify synapses as asymmetric and symmetric, based on the appearance of the postsynaptic density (PSD) [12, 13]. Before beginning the segmentation process, the user creates the category and subcategories that will be needed, for example, "synapse" for the category and "asymmetric" and "symmetric" for the subcategories (Fig. 3a). When the PSD is clearly thicker than the presynaptic density, the synapse is classified by the user as asymmetric, which is the usual appearance of excitatory synapses in the cerebral cortex [14]. If the presynaptic density and the PSD have about the same thickness, then the synapse is classified as symmetric, which is the usual appearance of inhibitory synapses in the cerebral cortex [15, 16]. The classification of synapses into these two groups (asymmetric and symmetric) is not based on their appearance in single images and types of synaptic vesicles, which is often misleading; instead, it is based on the examination of all serial images in which a given synapse is visible.

It is recommended that the segmentation of synapses is performed on the Gaussian blur stacks, although the user can visualize the superimposed unprocessed stack, as outlined above (Fig. 3b). To begin the segmentation process, the user only needs to click on the PSD, which is the most prominent and darkest part of the synaptic junction. The clicked pixel then acts as a seed for a 3D *flood fill* algorithm that extends the segmentation to the neighboring dark pixels. For the algorithm to work properly, the user must select a threshold value for the gray level of the segmentation (Fig. 3b). If the threshold is set to a gray level that is lighter (high range of color value) than the synaptic densities, the segmented objects will be too large, since they will include pixels that are

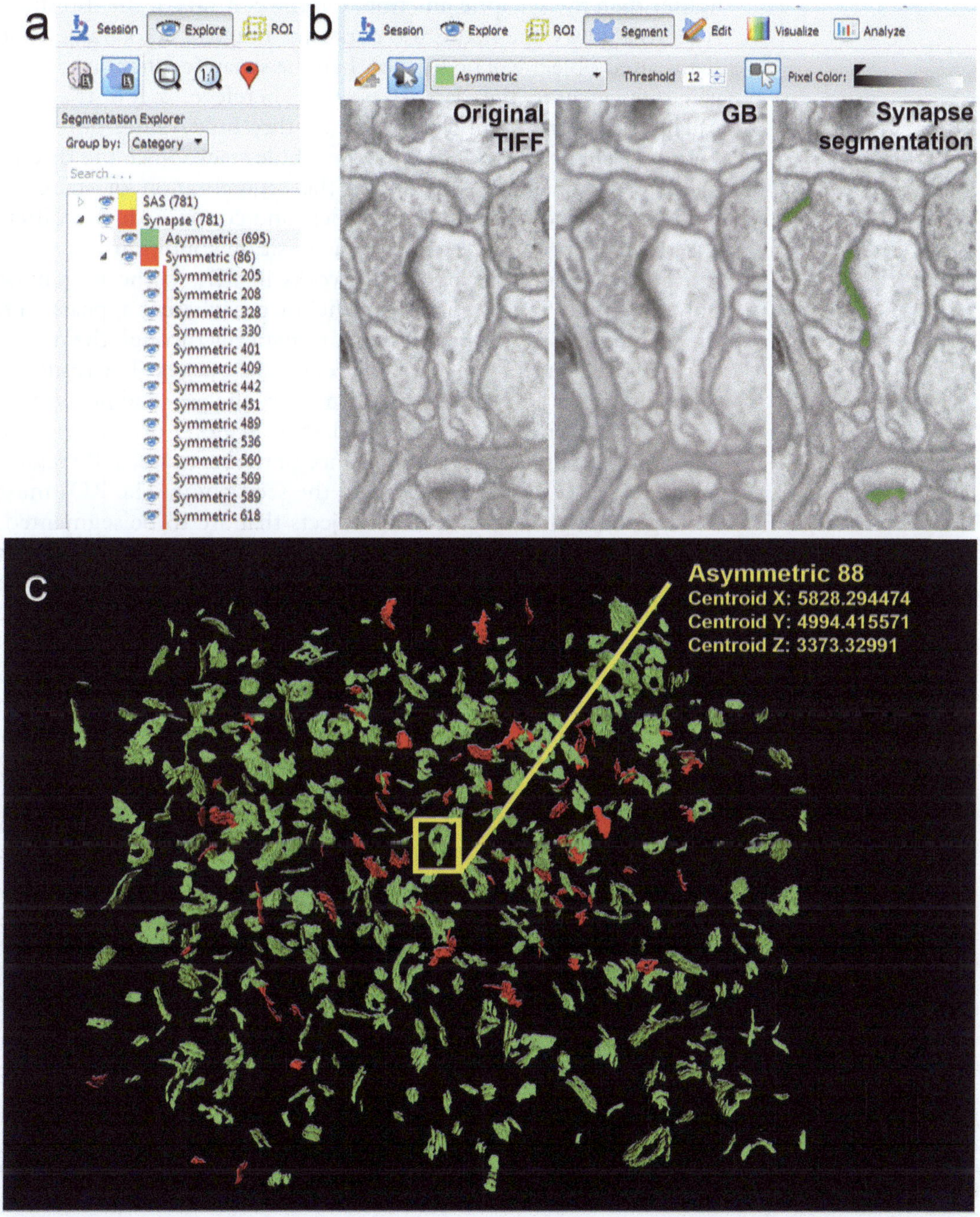

Fig. 3 Segmentation of synapses in Espina. (**a**) The user creates the "synapse" category which is subcategorized into "asymmetric" (green) and "symmetric" synapses (red). All the segmented objects are named with a unique numeric ID. (**b**) The grayscale threshold can be adapted to the EM sample, which facilitates the segmentation of synapses in stacks of sections that have been processed or acquired under different conditions. The original version of an electron microscopy image is shown, followed by the same image with a Gaussian blur (GB) filter and the same image with three segmented synapses (green). The stack used for visualization of synapses is usually the original stack; however, the stack with a Gaussian blur (GB) filter is commonly used to perform the segmentation of synapses. (**c**) The synapse cloud shows the positions of all synaptic junctions within the stack. In this example, the spatial coordinates of one of the synapses are shown

lighter than the synaptic junction. On the contrary, if the threshold is set to a gray value that is too dark (low range of color value), the segmented object will be smaller than the synaptic junction, since it will only include the darkest pixels. Therefore, a few trials are usually necessary to find the optimal threshold value, but, once found, it will usually be valid for the whole stack. Since the thresholds can be changed, this allows the segmentation of stacks of sections that show different brightness and contrast, due to different processing, staining, or image-acquisition conditions.

The segmentation algorithm works in 3D, so the user must only click the PSD once in any of the images where it appears, and the segmentation then progresses automatically through the neighboring voxels in the same and adjacent serial images. The resulting segmentation is a 3D object that comprises the pre- and postsynaptic densities, since these are the darkest regions of the synaptic junction. To improve the performance of the software, a 3D region of interest (ROI) is created around the seed point. The ROI must be large enough to contain the objects that are to be segmented, but it will typically be much smaller than the whole stack. In this way, when the user clicks on any given synaptic junction, the algorithm will only operate inside the ROI, instead of inside the whole stack. This saves computational resources and time but may result in a truncated segmentation when the object is larger than the ROI. In such cases, the software warns the user that the segmented object is larger than the ROI, so the user can modify the threshold if the segmentation is wrong or enlarge the ROI if the segmented object is larger than expected.

The final result of the segmentation of all synapses within a stack is a 3D cloud of synaptic junctions. Each individual segmentation is annotated and color-coded as asymmetric or symmetric and is given a unique ID number (Fig. 3c). The software provides the spatial coordinates of the centers of mass (centroids) of the segmentations (Figs. 3c and 8b). This information can be used offline to study the distribution of synapses in space [17, 18], to identify possible deviations from the normal pattern under pathological conditions [19]—or to compare between different brain regions or species.

6 Calculating the Density of Synapses with the Unbiased Counting Frame

To quantify the synapses, or any other segmented objects, per unit volume, a 3D stereological unbiased counting frame must be built within the stack. In its simplest form, the counting frame is a rectangular prism bounded by three acceptance planes and three exclusion planes (color-coded as green and red planes, respectively) (Fig. 4). Any segmented object located inside the counting frame or touching any of the acceptance planes is counted, while any

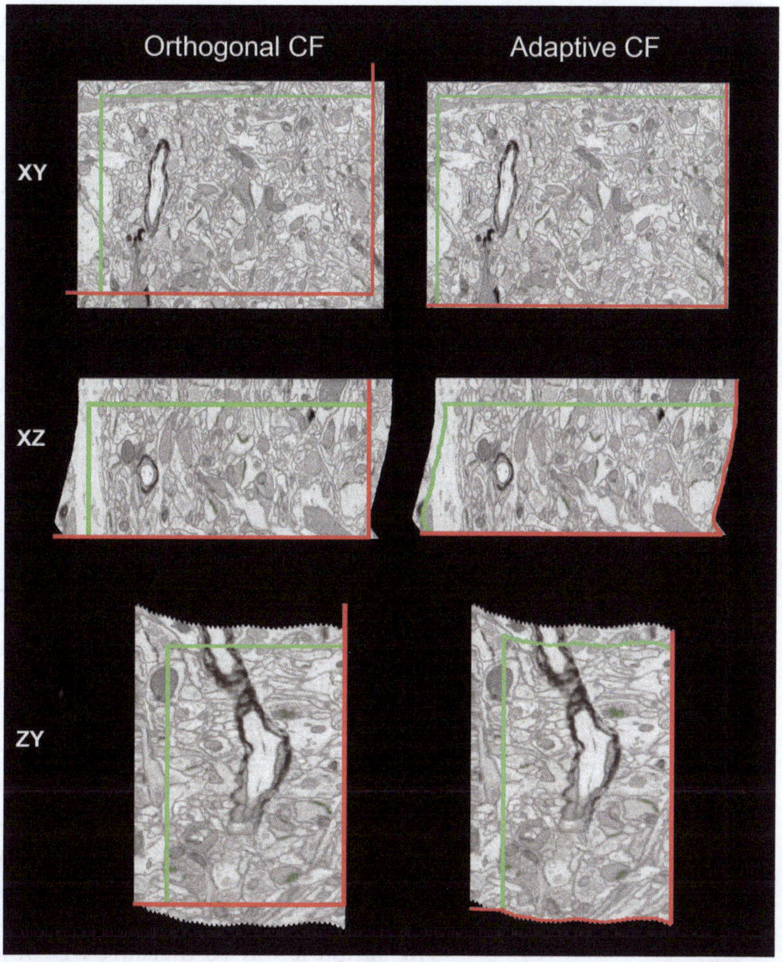

Fig. 4 The stereological counting frame (CF) in Espina assists in the unbiased quantification of synapses or any other segmented structures. Espina provides the option of an orthogonal CF (a regular rectangular prism) or an adaptive CF (whose edges adapt to the shape of the stack of images). In both cases the CF has three acceptance planes or faces (green) and three exclusion planes or faces (red)

object outside the counting frame or intersecting any of the exclusion planes is not [20]. Since Espina provides both the number of segmented objects inside the counting frame and its volume, the densities of synapses (or any other segmented structures) can be calculated directly.

The boundaries of the counting frame can be set manually by the user, or they can be established automatically by the software. In the first case, the user can move the acceptance and exclusion planes until they bound the desired volume of the stack. If the counting frame is set manually, it is important to avoid counting any segmented object that is truncated by the boundaries of the stack, since the measurements of their volume or surface will not be reliable. When the counting frame is set automatically, Espina will automatically create the largest counting frame that avoids counting truncated objects.

The counting frame can be *orthogonal* or *adaptive* (Fig. 4). An orthogonal counting frame creates a regular rectangular prism inside the stack, as described above, so it is appropriate, for example, to obtain the coordinates of synapses for spatial statistical analysis. However, the stack of serial sections is not usually a perfect rectangular prism. On the contrary, it commonly has an irregular surface, due to the small displacements that take place during image acquisition. An adaptive counting frame will adapt its shape to the contour of the stack in order to exploit most of the available volume while still avoiding truncated objects; this type of counting frame is more appropriate for the calculation of the density of synapses within the stack.

7 Measuring the Size of Synapses

Synapses are plastic structures that are continuously remodeled and replaced in response to functional changes [21–23]. Measuring the size of synapses is important for several functional reasons. On the presynaptic side, the size of the active zone is proportional to the number of docked synaptic vesicles and to the probability of release [24–27], while on the postsynaptic side, the size of the PSD is proportional to the number of postsynaptic receptors [28–32] and to the amplitude of postsynaptic potentials [33].

The simplest way to measure 3D objects is the Feret diameter [34]. The Feret diameter of a particle can be defined as the diameter of the smallest sphere that circumscribes the object. The main advantage of the Feret diameter is its simplicity, since it requires few computational resources and can be calculated quickly, immediately after finishing the segmentation of any given synaptic junction [17, 18]. Another advantage of the Feret diameter is that it is not very sensitive to the gray threshold used during the segmentation. For example, if a given synapse is segmented with two different thresholds, the volumes of the resulting 3D objects will be different, since the segmentations will be thicker or thinner depending on the threshold. However, the segmented synapse will still have approximately the same contour in spite of the different thresholds, so the Feret diameters will be similar. The main disadvantage of the Feret diameter is that it is a linear measure, so clearly some information about the shape of the object is lost, since very different objects can have the same Feret diameter.

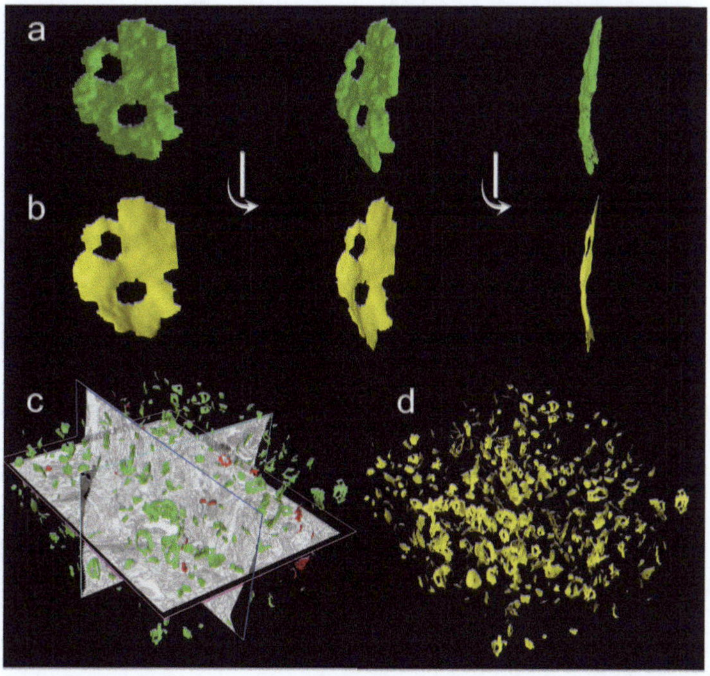

Fig. 5 Extraction of the synaptic apposition surface (SAS). (**a**) The segmentation of a synapse is a 3D object that comprises the darkest areas of the synaptic junction (mostly the PSD) (**b**). The SAS is a surface contained within the 3D segmentation of the synapse. The surface area of the SAS is biologically relevant since it is equivalent to the area of the active zone and to the area of the PSD. It also preserves the shape and curvature of the synaptic junction, as well as the indentations and perforations, if present. (**c**) For each synapse, it is possible to generate its own SAS, so a synaptic cloud has a corresponding SAS cloud (**d**)

Apart from the Feret diameter, Espina provides the volume and surface of any segmented object. These measurements are, however, sensitive to the thresholds used during the segmentation, as we have already mentioned. In search of a more reliable and biologically relevant measurement, we developed an algorithm for the extraction of the synaptic apposition surface (SAS), which is unique to Espina software (Fig. 5) [35]. The starting point of the algorithm is the 3D segmentation of a synaptic junction, which comprises the pre- and postsynaptic densities. Given that both densities are closely apposed and are approximately the same size [24, 36], they can be represented by a single surface, so the algorithm calculates a surface that lies in the middle of the 3D segmentation of the synaptic junction (Fig. 5b). The result is a new segmented object with a zero-volume sheet shape that represents the interface between the pre- and postsynaptic densities. The area of the SAS is relevant from a physiological perspective since it correlates both with the area of the presynaptic active zone and with the area of the PSD. In addition, the SAS preserves the contour and curvature of the synaptic junction, as well as the perforations and indentations, if present, thus providing information not only about the size of the synapse but also about its shape [37].

8 Identification of Presynaptic Elements and Postsynaptic Targets

Once all synapses that are present in a given stack have been segmented and classified, the pre- and postsynaptic elements that establish each synapse can be identified according to simple criteria. For example, axons can be subdivided into excitatory and inhibitory, depending on the type of synapses that they establish (asymmetric or symmetric, respectively). Postsynaptic elements such as dendritic spines, dendritic shafts, or cell bodies can also be identified, so any synapse can be tagged as axospinous, axodendritic, or axosomatic. In this way, we can calculate, for example, the proportions of excitatory and inhibitory synapses established on dendritic spines or dendritic shafts in different cortical layers [38].

9 The Tracing Tool

It is possible to use Espina to fully reconstruct selected structures, such as dendritic spines [39] or multivesicular bodies [40]. However, the software has not been designed to perform full or dense reconstructions of all the elements within the stack. Instead, we have implemented a tool that allows the user to trace individual axons and dendrites, providing a skeletonized 3D representation of these fibers (Fig. 6). For example, to trace an axon, the user must navigate the stack and click inside the axon with the mouse. A segment is created by the software between every two clicks, so the final tracing will be composed of several segments that will follow the course of the axon. This simplified representation of the axon, the skeleton, can be connected to the synapses that it establishes and saved in its own category (axon) and subcategory (excitatory or inhibitory), both created by the user, with a unique ID number. Dendrites are traced in a similar way, although they usually have dendritic spines (Fig. 6b). In this case, the dendritic spines will be traced as small branches arising from the shaft. The shaft and the spines will be tagged as such, and the entire skeleton will be connected to all the synapses that are established on the dendrite (Fig. 6c).

Using the skeletons, we can calculate the linear density of synapses established by excitatory and inhibitory axons (i.e., the number of synapses that they establish per unit length). In dendrites, we can calculate the linear density of spines, and the linear

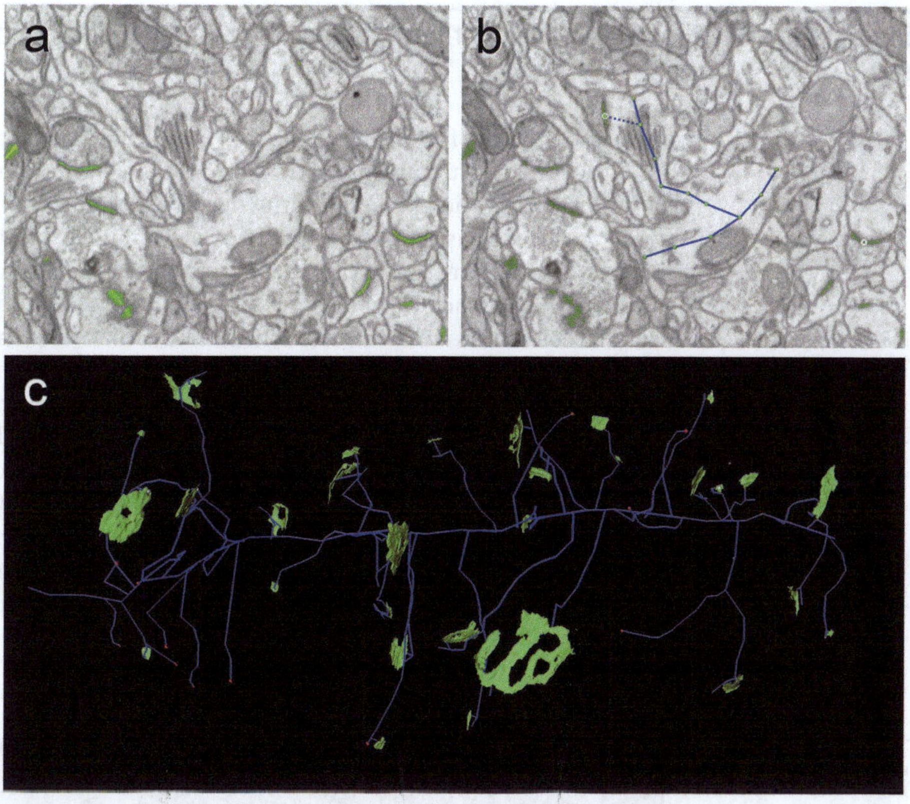

Fig. 6 Skeleton tracing tool of Espina. (**a**) Electron microscopy image from a stack of images of the cerebral cortex. Several segmented synapses are shown in green. (**b**) Some of the nodes and connections that make up the skeleton of one of the dendrites are shown. The dendritic shaft and a dendritic spine belonging to the same dendrite can be seen. An asymmetric synapse (green profile) is connected to the dendritic spine (**c**). Complete skeleton of a dendrite (blue) with its corresponding synapses (green). Each segment of the skeleton is associated with information about its type (shaft or dendritic spine), and the presence and location of synapses are also recorded

densities of synapses established on spines and on shafts. If we complete the tracing of all axons and dendrites within the stack, we will have a simplified representation of the connectivity within the stack, and we will be able to calculate the proportion of axons vs. dendrites and the total length of axons and dendrites, and, since each skeleton has its unique ID, we can determine exactly which fibers connect with which. Also, given that every element has been appropriately tagged as an excitatory or inhibitory synapse, as an excitatory or inhibitory axon, or as a dendritic shaft or spine, we can extract information about the connectivity of each category individually. For example, we can calculate the proportions of synapses that excitatory axons establish on dendritic spines and shafts and compare them with inhibitory axons.

10 Identification of Other Subcellular Structures and Their Relationship with Synapses

Espina can be used for the visualization and/or segmentation of other subcellular structures. Cortical circuits are not static structures and therefore need an energetic input to promote the flow of information as well as a system capable of maintaining and modifying the cellular structures involved. Therefore, it is important that, together with the study of synapses (Fig. 7a–d), we also study other subcellular elements such as mitochondria, which are the main source of energy [41, 42], and multivesicular bodies, which are responsible for a large proportion of the traffic of substances both inside and outside the cells, as a fundamental part of the endosomal system [43, 44]. Both mitochondria and multivesicular bodies can be easily identified in FIB-SEM serial sections (Fig. 7e–g). Quantification of these structures may be performed after 3D reconstruction with Espina, as in the case of multivesicular bodies [40] (Fig. 7e), or offline, with the help of stereological methods [45]. Other structures, such as astrocytic processes, may require specific labeling to facilitate their identification [46].

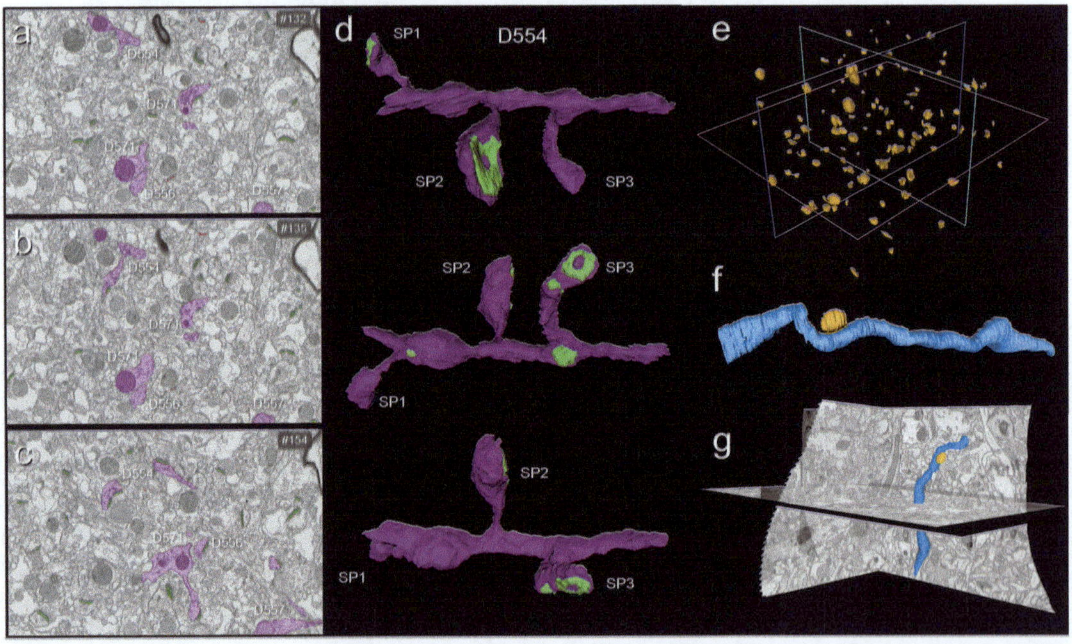

Fig. 7 Non-synaptic structures reconstructed with Espina. (**a–c**) Electron microscopy serial images showing dendritic segments (D554, D556, D557, and D571, purple) from a human brain sample. (**d**) 3D-reconstructed dendritic segment D554 showing three views after rotation of the major dendritic axis at different angles. Three dendritic spines (SP1 to SP3) are shown, establishing asymmetric synapses (green) with different morphologies. (**e**) Cloud of reconstructed multivesicular bodies (MVBs, orange) in a sample of mouse brain. (**f**) Three-dimensional reconstruction of a MVB (orange) docked to a mitochondrion (blue). (**g**) 3D view of the reconstructed MVB and mitochondrion in their original position within the tissue. (Images a to d have been taken from [39])

11 Raw Information Reports

As we have seen, Espina allows the segmentation of synapses and other structures. Each segmentation is classified within its own category and/or subcategory. Some predetermined objects have their own properties. For example, the SAS has perimeter and surface curvature but not volume, and the 3D segmentation of a synapse has volume and surface area but not perimeter. The software provides reports containing quantitative information of any structure that has been previously segmented. The user can select the category and subcategory from which to generate the report and then select the specific data that must be provided (Fig. 8a). For example, the user can generate a report containing only the Feret diameters of all synapses or another containing Feret diameter, 3D coordinates, the tags associated with the segmentations, and the IDs of postsynaptic elements of excitatory synapses only (Fig. 8b).

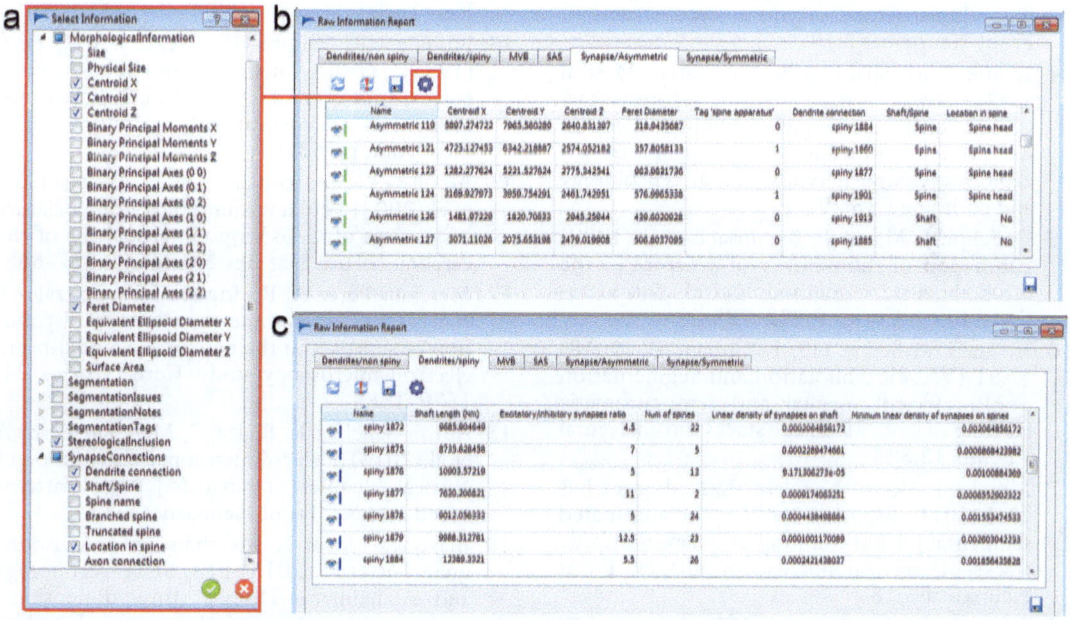

Fig. 8 Raw information reports in Espina software. (**a**) To generate a report from any category of segmented objects, the user selects the information to be displayed. (**b**) In this example, the information corresponding to several asymmetric synapses is displayed. The information shown in this particular case includes the spatial coordinates of each synapse, its Feret diameter, the presence/absence of spine apparatus in the target dendritic spine, the dendrite to which it is connected, the position on the dendritic shaft or spine, and the position in the head or neck of a dendritic spine. The table containing the desired information can be saved in Excel format. (**c**) Another example displaying the information corresponding to several dendrites with spines (spiny dendrites). In this case, the information includes the shaft length of the dendrite, the ratio between excitatory and inhibitory synapses, the number of spines, and the linear densities of synapses in the shaft and dendritic spines

Additionally, information about the SAS, such as its area and perimeter, can be obtained. Espina also provides information about the skeletons of dendrites, dendritic spines, and axons, such as their lengths, number of dendritic spines, or linear densities of synapses (Fig. 8c). The desired information is selected in the corresponding dialog boxes and is generated and saved as an Excel file (Fig. 8a).

References

1. Denk W, Horstmann H (2004) Serial block-face scanning electron microscopy to reconstruct three-dimensional tissue nanostructure. PLoS Biol 2:e329
2. Knott G, Marchman H, Wall D et al (2008) Serial section scanning electron microscopy of adult brain tissue using focused ion beam milling. J Neurosci 28:2959–2964
3. Merchán-Pérez A, Rodriguez J-R, Alonso-Nanclares L et al (2009) Counting synapses using FIB/SEM microscopy: a true revolution for ultrastructural volume reconstruction. Front Neuroanat 3:18
4. Schifferer M, Snaidero N, Djannatian M et al (2021) Niwaki instead of random forests: targeted serial sectioning scanning electron microscopy with reimaging capabilities for exploring central nervous system cell biology and pathology. 15:79
5. DeFelipe J, Marco P, Busturia I et al (1999) Estimation of the number of synapses in the cerebral cortex: methodological considerations. Cereb Cortex 9:722–732
6. Berger DR, Seung HS, Lichtman JW (2018) VAST (Volume Annotation and Segmentation Tool): efficient manual and semi-automatic labeling of large 3D image stacks. Front Neural Circuits 12:88
7. Morales J, Alonso-Nanclares L, Rodríguez J-R et al (2011) Espina: a tool for the automated segmentation and counting of synapses in large stacks of electron microscopy images. Front Neuroanat 5:18
8. Schneider CA, Rasband WS, Eliceiri KW (2012) NIH Image to ImageJ: 25 years of Image Analysis. Nat Methods 9:671–675
9. ImageJ, https://imagej.net/ij/index.html
10. Schindelin J, Arganda-Carreras I, Frise E et al (2012) Fiji: an open-source platform for biological-image analysis. Nat Methods 9: 676–682
11. Fiji: ImageJ, with "Batteries Included," https://fiji.sc/
12. Gray EG (1959) Axo-somatic and axo-dendritic synapses of the cerebral cortex: an electron microscope study. J Anat 93:420–433
13. Colonnier M (1968) Synaptic patterns on different cell types in the different laminae of the cat visual cortex. An electron microscope study. Brain Res 9:268–287
14. Peters A, Palay SL, Webster H (1991) The fine structure of the nervous system. Neurons and their supporting cells. Oxford University Press, New York
15. Houser CR, Vaughn JE, Hendry SHC et al (1984) GABA neurons in the cerebral cortex. In: Cerebral Cortex, Vol. 2, Functional properties of cortical cells. Plenum Press, New York, pp 63–89
16. Ascoli GA, Alonso-Nanclares L, Anderson SA et al (2008) Petilla terminology: nomenclature of features of GABAergic interneurons of the cerebral cortex. Nat Rev Neurosci 9:557–568
17. Merchán-Pérez A, Rodríguez J-R, González S et al (2014) Three-dimensional spatial distribution of synapses in the neocortex: a dual-beam electron microscopy study. Cereb Cortex 24: 1579–1588
18. Anton-Sanchez L, Bielza C, Merchan-Perez A et al (2014) Three-dimensional distribution of cortical synapses: a replicated point pattern-based analysis. Front Neuroanat 8:85
19. Blazquez-Llorca L, Merchán-Pérez A, Rodríguez J-R et al (2013) FIB/SEM Technology and Alzheimer's Disease: three-dimensional analysis of human cortical synapses. J Alzheimers Dis 34:995–1013
20. Howard CV, Reed MG (2005) Unbiased stereology: three-dimensional measurement in microscopy. Garland Science/BIOS Scientific Publishers, Oxon
21. Fauth M, Tetzlaff C (2016) Opposing effects of neuronal activity on structural plasticity. Front Neuroanat 10:75

22. Lisman J (2017) Glutamatergic synapses are structurally and biochemically complex because of multiple plasticity processes: long-term potentiation, long-term depression, short-term potentiation and scaling. Philos Trans R Soc Lond B Biol Sci 372:20160260
23. Popov VI, Davies HA, Rogachevsky VV et al (2004) Remodelling of synaptic morphology but unchanged synaptic density during late phase long-term potentiation (LTP): a serial section electron micrograph study in the dentate gyrus in the anaesthetised rat. Neuroscience 128:251–262
24. Schikorski T, Stevens CF (1997) Quantitative ultrastructural analysis of hippocampal excitatory synapses. J Neurosci 17:5858–5867
25. Branco T, Marra V, Staras K (2010) Examining size–strength relationships at hippocampal synapses using an ultrastructural measurement of synaptic release probability. J Struct Biol 172:203–210
26. Matz J, Gilyan A, Kolar A et al (2010) Rapid structural alterations of the active zone lead to sustained changes in neurotransmitter release. Proc Natl Acad Sci USA 107:8836–8841
27. Holderith N, Lorincz A, Katona G et al (2012) Release probability of hippocampal glutamatergic terminals scales with the size of the active zone. Nat Neurosci 15:988–997
28. Nusser Z, Lujan R, Laube G et al (1998) Cell type and pathway dependence of synaptic AMPA receptor number and variability in the hippocampus. Neuron 21:545–559
29. Ganeshina O, Berry RW, Petralia RS et al (2004) Differences in the expression of AMPA and NMDA receptors between axospinous perforated and nonperforated synapses are related to the configuration and size of postsynaptic densities. J Comp Neurol 468:86–95
30. Tarusawa E, Matsui K, Budisantoso T et al (2009) Input-specific intrasynaptic arrangements of ionotropic glutamate receptors and their impact on postsynaptic responses. J Neurosci 29:12896–12908
31. Kharazia VN, Weinberg RJ (1999) Immunogold localization of AMPA and NMDA receptors in somatic sensory cortex of albino rat. J Comp Neurol 412:292–302
32. Takumi Y, Ramírez-León V, Laake P et al (1999) Different modes of expression of AMPA and NMDA receptors in hippocampal synapses. Nat Neurosci 2:618–624
33. Holler S, Köstinger G, Martin KAC et al (2021) Structure and function of a neocortical synapse. Nature 591:111–116
34. Walton WH (1948) Feret's statistical diameter as a measure of particle size. Nature 162:329–330
35. Morales J, Rodríguez A, Rodríguez J-R et al (2013) Characterization and extraction of the synaptic apposition surface for synaptic geometry analysis. Front Neuroanat 7:20
36. Schikorski T, Stevens CF (1999) Quantitative fine-structural analysis of olfactory cortical synapses. Proc Natl Acad Sci USA 96:4107–4112
37. Santuy A, Rodriguez J-R, DeFelipe J et al (2018) Study of the size and shape of synapses in the juvenile rat somatosensory cortex with 3D electron microscopy. eNeuro 5: e0377–17.2017
38. Santuy A, Rodriguez J-R, DeFelipe J et al (2018) Volume electron microscopy of the distribution of synapses in the neuropil of the juvenile rat somatosensory cortex. Brain Struct Funct 223:77–90
39. Cano-Astorga N, DeFelipe J, Alonso-Nanclares L (2021) Three-dimensional synaptic organization of layer III of the human temporal neocortex. Cereb Cortex 31:4742–4764
40. Turegano-Lopez M, Santuy A, DeFelipe J et al (2020) Size, shape, and distribution of multivesicular bodies in the juvenile rat somatosensory cortex: a 3D electron microscopy study. Cereb Cortex 30:1887–1901
41. Rangaraju V, Lauterbach M, Schuman EM (2019) Spatially stable mitochondrial compartments fuel local translation during plasticity. Cell 176:73–84.e15
42. Rossoll W, Bassell GJ (2019) Crosstalk of local translation and mitochondria: powering plasticity in axons and dendrites. Neuron 101:204–206
43. Von Bartheld CS, Altick AL (2011) Multivesicular bodies in neurons: distribution, protein content, and trafficking functions. Prog Neurobiol 93:313–340
44. Shearer LJ, Petersen NO (2019) Distribution and Co-localization of endosome markers in cells. Heliyon 5:e02375
45. Santuy A, Turégano-López M, Rodríguez JR et al (2018) A quantitative study on the distribution of mitochondria in the neuropil of the juvenile rat somatosensory cortex. Cereb Cortex 28:3673–3684
46. Kikuchi T, Gonzalez-Soriano J, Kastanauskaite A et al (2020) Volume electron microscopy study of the relationship between synapses and astrocytes in the developing rat somatosensory cortex. Cereb Cortex 30:3800–3819

Chapter 8

Automated Imaging and Analysis of Synapses in Freeze-Fracture Replica Samples with Deep Learning

David Kleindienst, Tommaso Costanzo, and Ryuichi Shigemoto

Abstract

Sodium dodecyl sulfate-digested freeze-fracture replica labeling (SDS-FRL) is an electron microscope (EM) sample preparation technique which allows for high-resolution visualization of membrane proteins with high sensitivity. However, image acquisition of specific replica profiles such as synapses in a large field of EM view needs a valid experience and a long time for manual searching. Here, we describe how to utilize deep learning for automatizing image acquisition of specific profiles of interest in replica samples. This protocol facilitates the labor-intensive collection of EM images, in particular for rare profiles. We provide instructions for using SerialEM image acquisition software in conjunction with object detection by our newly developed deep learning software DarEM, to automatically acquire tilt series of all synapses in a selected region. We then show how to perform a mostly automated analysis of gold particle labeling in the acquired images by utilizing Darea software.

Key words Freeze-fracture replica, Immunogold labeling, Deep learning, Electron microscopy, Automation, Image acquisition

1 Introduction

Sodium dodecyl sulfate-digested freeze-fracture replica labeling (SDS-FRL) is a sample preparation technique for electron microscopy (EM). Epitopes are exposed on the surface of the replica, which allows for high accessibility to antibodies and high-sensitivity immunogold labeling [1, 2]. SDS-FRL has been successfully used to visualize and quantify key molecules involved in synapse function, including glutamate receptors, GABA receptors, and voltage-gated calcium channels [2–6]. However, the identification and precise demarcation of synaptic membrane specialization require expert experience, in particular for presynaptic sites because morphological clues are less clear than those in excitatory postsynaptic sites with densely packed intramembrane particles [2, 7]. Furthermore, even for well-experienced observers, the collection of a large

Joachim H. R. Lübke and Astrid Rollenhagen (eds.), *New Aspects in Analyzing the Synaptic Organization of the Brain*, Neuromethods, vol. 212, https://doi.org/10.1007/978-1-0716-4019-7_8,

number of synaptic profiles by manual searching is a time-consuming task.

In the last decade, deep learning has revolutionized computational image analysis, vastly outperforming traditional computer vision algorithms in object localization and semantic segmentation tasks and surpassing even human accuracy in image classification tasks [8, 9]. Here, we describe how to utilize the power of deep learning for automated EM acquisition and analysis of freeze-fracture replicas. We introduce DarEM (profile *d*etection for *a*utomated *r*eplica *e*lectron *m*icroscopy), a software we developed for automatic detection of profiles of interest in freeze-fracture replicas. DarEM is based on the deep learning framework TensorFlow [10] and the TensorFlow Objection Detection API [11]. We describe how to use DarEM in conjunction with the automated EM data acquisition software SerialEM [12, 13] for automatic imaging of profiles of interest from multiple tilting angles. We further describe how to efficiently analyze obtained images with Darea software [5] which utilizes deep learning for automated demarcation of the area of interest (such as the postsynaptic density (PSD)) and standard machine learning techniques for gold particle detection. We show how to use Darea to perform simulations to explore particle distributions, clustering patterns, and co-localization of two kinds of molecules. Finally, we show how to train deep neural networks (DNNs) for automated imaging and analysis of profiles of interest. No programming is necessary for any of the procedures described in this chapter.

Here, we focus on imaging and analysis of freeze-fracture replicas. For a detailed protocol on sample preparation and immunolabeling, we refer the reader to other book chapters [14, 15].

1.1 Deep Learning

Many excellent and comprehensive introductions to deep learning have been written [16–19], and providing another one is beyond the scope of this chapter. Instead, we will only briefly introduce terms and concepts necessary for training and judging training success of the DNNs used. One of the reasons for the great success of DNNs is that their basic architecture allows to perform many different functions (such as taking an input image and returning the coordinates of a particular feature). The exact function performed (i.e., whether the detected feature is a synapse or a cat) is determined by the *weights* of the DNN. Modern DNNs contain millions of such weights which are tuned to detect the feature of choice in a process called *training*. During training a DNN is first presented with one or more images (the number of simultaneously presented images is called the *batch size*) and allowed to make a prediction. The prediction is then compared to the ground truth (typically an expert annotation of the same image). In this comparison a *loss* value is computed by a loss function, which varies depending on the task. The loss describes how different the DNN's prediction is from the truth—the lower the loss, the better the prediction. The DNN then iteratively goes through the images of the training dataset and tunes its weights to minimize the loss. Such an iteration is called a training *step*, and performing training steps on each image of the dataset once is called an *epoch*. During training, the weights are saved periodically as *checkpoint*. During training we expect the loss to gradually decrease. If such a decrease is not observed, it points to a problem of either the training parameters (e.g., a too high *learn rate*—a value which describes how much the weights will be updated in each step) or of the training dataset, such as a problem of image preprocessing (e.g., during some conversion step, the image was accidentally made completely black). When training with a small training dataset, it can happen that rather than "understanding" the general feature (such as a PSD) it is supposed to detect, the DNN may instead learn the particular presented examples "by heart" in a phenomenon called *overfitting*. To detect overfitting, we usually divide the dataset into two parts—a training and a validation dataset. The loss of the validation dataset will be computed periodically during training. Loss of training and validation dataset should show a correlated decrease. Overfitting is indicated by an increase of validation loss, while the training loss keeps decreasing. Checkpoints after training and validation loss started to diverge should not be used. If a satisfying performance of the DNN is not achieved at the last checkpoint before the divergence, collecting additional training data is necessary.

1.2 Analysis

A major goal of our freeze-fracture replica analysis often is comparison of synaptic labeling density between different conditions, which requires accurate demarcation of the synaptic area and quantification of gold particles. Due to the size of primary and secondary antibodies, gold particles corresponding to an epitope near the border of the synaptic area may be localized outside of the area. To include such particles in the analysis, we also quantify particles in an outer rim of 20–30 nm surrounding the demarcated area, provided that the extrasynaptic target density is much lower than the density within synapses [5, 15].

Besides quantification of area and gold particle number and density, several other questions about the distribution of proteins of interest in synaptic membrane specializations can be addressed with freeze-fracture replica labeling: Is the protein distributed randomly or does it show clustering? Is a protein preferentially localized at the center or the periphery of the synapse? Are two proteins co-localized within the synapse? Monte Carlo simulations of gold particle distributions are available in Darea software to answer these questions. First, nearest neighbor distances (NNDs) between gold particles and distances between gold particles and the center and edge of the synaptic area can be quantified and compared between real and simulated particles to examine if real particle distribution is significantly different from random distribution, and if real particles are located significantly close to the center or edge of synapses [3, 5]. Random simulations will distribute an equal number of gold particles in the region of interest (ROI), with each pixel having the same probability of becoming the center of a simulated particle, while keeping a specifiable minimum distance between gold particles. This minimum distance results from steric hindrance between gold particle-conjugated antibodies and can be empirically determined. When particles in an outer rim surrounding the synapses are included in the analysis, two-step simulations, which first simulate the position of the antigenic epitope followed by simulation of the gold particle position within a defined range of deviation from the simulated epitope, can be used [5]. Second, gold particle clusters can be defined based on a minimum number of particles to form a cluster and a maximum distance between particles of the same cluster [3]. Then, properties of particle clusters, i.e., number of clusters per synapse, gold particles per cluster, cluster area, density of gold particles within clusters, NNDs between clusters, and distance of clusters from the center and edge of the synapse can be computed. For technical considerations on the suitability of a particular labeling for cluster analysis, *see* [20]. Third, to examine if there is significant co-localization or repulsion between two proteins, fitted simulations can additionally keep the NND distributions of simulated and real particles of one kind of protein similar by iteratively redistributing simulated particles (using the same rules as in random simulations) and keeping the result if the NND distribution is closer to that of real particles. This process is continued until a specifiable criterion of similarity (e.g., $p_{KS\text{-}test} > 0.1$) between real and simulated NNDs is reached. Finally, the distributions of distance from each particle of this kind to the nearest real particle of the second kind of protein are compared between the real and simulated particles to examine if distance between the two kinds of real particles is significantly shorter (co-localization) or longer (repulsion) than that between simulated and real ones [3].

2 Materials

2.1 Automated Imaging

1. A transmission electron microscope compatible with SerialEM (*see* https://bio3d.colorado.edu/SerialEM/ for a list; *see* **Note 1**).
2. A camera PC running Microsoft Windows.
3. A graphics card with at least 8 GB of dedicated graphics memory (*see* **Note 2**).

2.2 Image Analysis

1. A computer running Windows or Linux. A computer running macOS can be used for analysis, but not for training the DNN.
2. A graphics card with at least 8 GB of dedicated graphics memory (required for training the DNN, optional for analysis) (*see* **Notes 2** and **3**).
3. MATLAB (MathWorks) version 2018b or later.

3 Methods

3.1 Automated Imaging with SerialEM and DarEM

3.1.1 Software Installation and Setup

1. Install SerialEM [12] and configure it. SerialEM should be calibrated before use following the instruction at https://bio3d.colorado.edu/SerialEM/hlp/html/setting_up_serialem.htm. The most important calibrations for the workflow proposed here are pixel size, image shift, stage shift, and autofocus. These calibrations can be carried out by using standard cross grating grids with latex beads.
2. Download py-em [13] from https://git.embl.de/schorb/pyem, and install it according to the installation instructions provided there.
3. Download DarEM from https://github.com/DavidKleindienst/DarEM, and follow the installation instructions provided with the software.

3.1.2 Automated Imaging

A graphical overview of the automated imaging pipeline is shown in Fig. 1. Replicas have a three-dimensional structure of which a two-dimensional projection is seen in EM images, which can lead to errors when measuring an area of interest. To minimize this measurement error, the replica should be tilted until the area in the two-dimensional image is maximal when taking the image. For our automatic imaging pipeline, we therefore acquire a tilt series of the profile of interest, to allow for later selection of the best tilting angle:

1. Insert your sample into the SerialEM-configured microscope. Check beam alignment, set eucentric height normally, and move to the middle of the replica.
2. Start SerialEM, and select "New Montage" in the file menu to take a low-magnification overview image of the whole replica (*see* **Note 4**).
3. Open a new navigator file and add the overview as a new map.
4. Go to a medium magnification similar to what you would use when manually searching for profiles of interest (*see* **Note 5**). Adjust brightness and target defocus for the best image quality.
5. Align the low-mag overview map with the desired magnification for the medium resolution image acquisition: Place a point

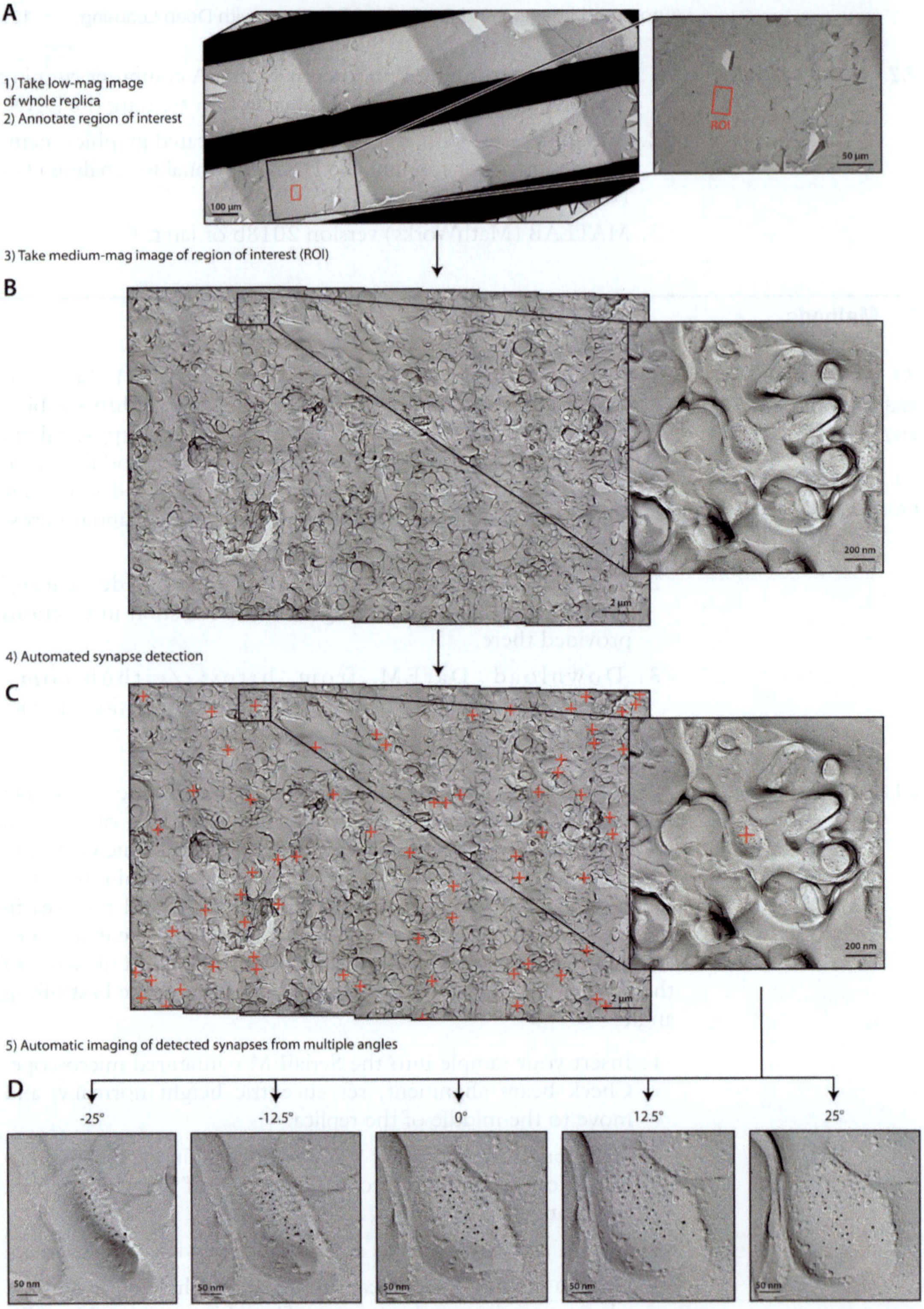

Fig. 1 Overview of the automated imaging pipeline. (**a**) Overview image of a replica from mouse hippocampus taken at 105 nm/pixel resolution with an annotated region of interest (red) in CA1 *stratum oriens*. (**b**) Medium magnification (0.83 nm/pixel resolution) tiled the image of the region of interest shown in red in (**a**). (**c**) Same as (**b**), but with red plusses indicating the position of algorithmically detected E-face PSDs. (**d**) High-resolution (0.26 nm/pixel) tilt series of an exemplary PSD

on the overview map, then find the same position with the medium magnification, and place a marker. Then select "shift to marker" in the navigator menu to align the two magnifications (*see* **Note 6**).

6. Use add polygon to select the region in which synapses should be detected.
7. Set up polygon montage to take medium-resolution images of the whole area of interest. Use a magnification similar to what you would use when manually searching for profiles of interest. Select "Series of TIFF files listed in an Autodoc file" as file format. Then load the resulting montage as a map.
8. Set the acquire flag on the new map, and then save the navigator file under a name of your choice in the folder containing the images.
9. Select "Find Features" in the tools menu, and select the desired DNN to perform the automated ROI detection, which will take a few minutes (*see* **Note 7**). Then select "Features to Points" in the tools menu to convert the detections into a new navigator file with the same name followed by the suffix "_with_points.nav."
10. Load the navigator file created in the previous step. You should now see a set of points displayed on the map, most of which should correspond to a profile of interest (*see* **Note 8**). At this step you may remove points which do not correspond to a profile of interest to avoid imaging of irrelevant profiles.
11. Open acquire at items, deselect everything in the "Initial actions" section in the navigator menu, and select run "DarEM_TiltSeries" script in the "primary task" to acquire a high-resolution tilt series for each detected synapse (*see* **Note 9**).

3.1.3 Training

1. Acquire a set of training images using the same parameters as in Subheading 3.1.2, **step 7** (*see* **Note 10**).
2. Annotate your profiles of interest with Darea, LabelImg [21], or any general-purpose image manipulation software such as Adobe Photoshop or GIMP (*see* **Note 11**). Be careful to annotate every profile of interest.
3. Open train.cmd (Windows) or train.sh (Linux) to launch the graphical user interface.
4. Click "Import Images for Training" to open the menu for converting your annotated dataset into the required .tfrecord file format.
5. Select the annotated datasets, name your feature of interest (e.g., "PSD"), select where to save the generated dataset, and click "Make" to start the conversion process.

6. Go back to the main menu, and click "Perform Training" to open the menu for setting the training parameters.
7. Select the .tfrecord dataset generated in **step 5**, a network to continue training from (*see* **Note 12**), the number of training steps, and the name of your new network. Click "Start" to run the training, which, depending on the number of steps and the computational power of the GPU, may take several days to complete.
8. Click "See Training result in TensorBoard" to get a link to see the training results in your browser. This may also be used to monitor training progress.
9. Verify training success in TensorBoard: In the scalars tab, and "total_loss" of both training ("train") and evaluation ("eval") should steadily decrease and eventually reach a plateau. Furthermore, precision ("DetectionBoxes_Precision") and recall ("DetectionBoxes_Recall") should steadily increase and eventually reach a plateau. Several issues may be discovered at this point: (1) Total loss of training increases. This can usually be solved by decreasing the learn rate. (2) Total loss of evaluation first decreases but then increases again. This is a sign of overfitting, and in this case, the last checkpoint before the increase in evaluation loss starts should be taken as a new DNN. If this checkpoint does not provide good-enough detection performance, more annotated images have to be added to the training dataset to avoid overfitting. (3) Total loss of both training and evaluation steadily decrease but do not reach a plateau. Similarly, precision and recall increase but do not reach a plateau. Not enough training steps were used, and the DNN should be fine-tuned with additional training steps. (4) Precision and recall remain −1 over the whole graph. This indicates that the network detected no profile of interest in the whole evaluation dataset. This can happen if your profile of interest is extremely rare and most images of the dataset do not contain a profile of interest. To combat this problem, the DNN should first be pre-trained on a dataset containing only images containing the profile of interest. Such a dataset can be created by ticking "Only include images containing objects" option when generating the .tfrecord dataset. The pre-trained DNN should then be added to DarEM (see below), and the training should then be performed on the full dataset, continuing training from the pre-trained network.
10. The checkpoints for the newly trained network are stored in the "models/<model_architecture>/<model_name>" folder and consist of an ".index" and a ".data-00000-of-00001" file. Copy the last checkpoint for which no overfitting was observed to the "models/<model_architecture>" folder, and rename it to the network name of your choice. The new model is now selectable in the predict menu.

3.2 Image Analysis with Darea

3.2.1 Installation

1. Download Darea software from https://github.com/DavidKleindienst/Darea.
2. Follow the installation instructions in the included handbook.

3.2.2 Create Project and Import Files

1. Organize your image files in the following manner: One project folder which contains a subfolder with images for each sample (e.g., each animal or replica). Samples immunolabeled for different molecules should be put into separate project folders, while samples from different groups (e.g., different brain regions or genotypes) should be put into the same project folder if a comparison between these groups is desired.
2. Start Darea, and click "Create Project" to open the image import dialog. Name the project, and select the project folder. For images acquired with SerialEM as described in Subheading 3.1.2, choose "Import SerialEM files with multiple angles," or select "Import .tif images" for conventionally taken tiff images. In case of SerialEM images, the resolution will be detected automatically. For conventional images, the resolution may be read from the filename (click "Magnification Options" for detailed options) and/or edited in the "Manage Images" menu after import.
3. Open "Project Settings," and specify the number of different gold particle sizes and their diameters. If you want to include gold particles in an outer rim around the feature of interest, choose the desired width now (*see* **Note 13**).

3.2.3 Annotation of Region of Interest and Gold Particles

1. Open the "Automated Prediction" menu, and run automated demarcation and/or automated particle annotation.
2. Open "Demarcate" menu to select the appropriate tilting angle (*see* **Note 14**), and revise the generated demarcations.
3. Open "Particles" menu to revise the detected particle annotations.

3.2.4 Analysis, Simulations, and Exporting Results

1. (Optional) Open the groups menu, and annotate the groups each image belongs to. You may specify multiple groupings (such as genotype, experimental condition, or brain region).
2. Open "Analysis" menu, and choose analysis parameters such as kind and number of simulations and particle cluster definition.
3. (Optional) Open "Visualize" menu to visually examine resulting simulations and clusters. This menu can also be used to export images such as those shown in Fig. 2.
4. Open "Figures" menu, and select which groupings should be distinguished. The data will be pooled over groupings not selected. This allows, for example, to easily choose whether

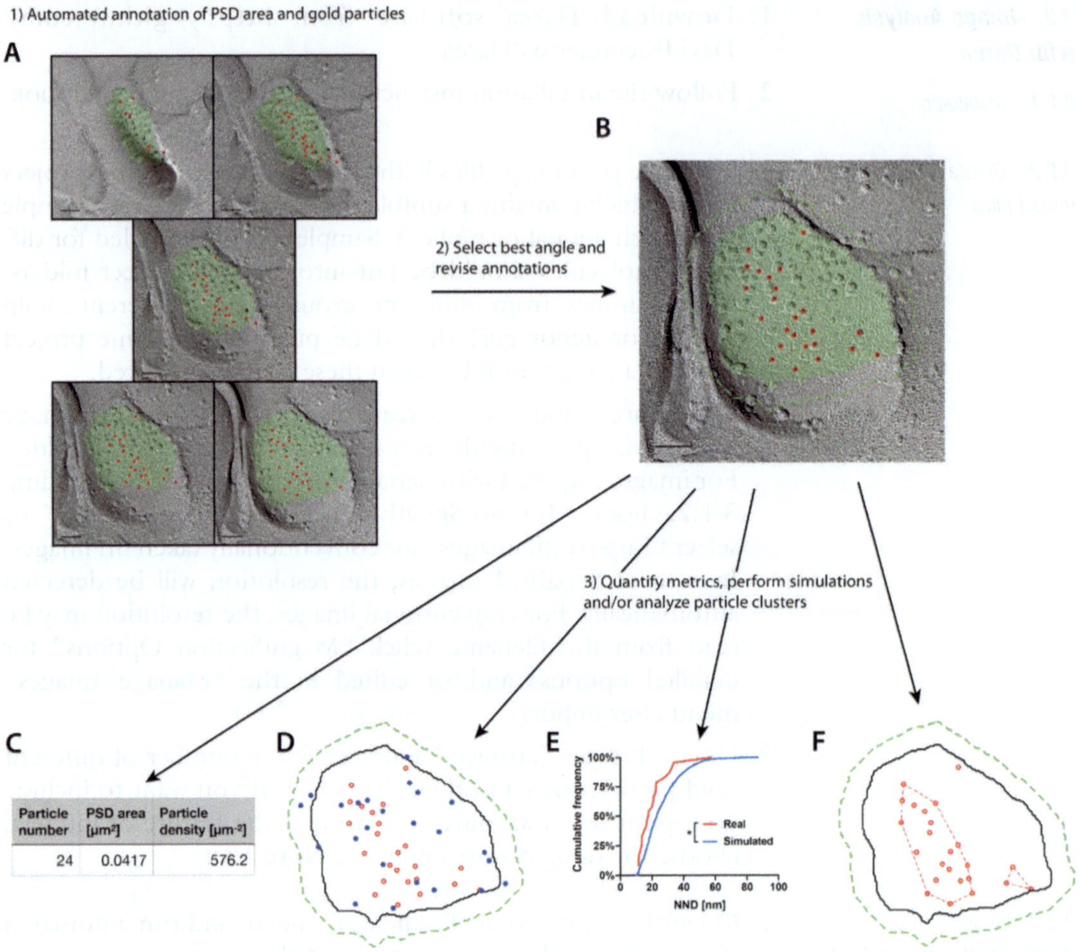

Fig. 2 Overview of the analysis pipeline. (**a**) Same tilt series shown in Fig. 1d with green overlay indicating detected PSD area and red circles indicating detected 5 nm gold particles. (**b**) Image of tilt series which shows the maximum PSD area with manually corrected PSD and gold particle annotations. (**c**–**f**) A few examples of analyses which can be performed with Darea: Quantification of PSD area, gold particle number and density (**c**), random simulation of the same number of gold particles (**d**), cumulative frequency distribution of real and simulated gold particles (**e**), and analysis of gold particles clusters (**f**). Scale bars indicate 50 nm

biological replicates should be pooled, by selecting the appropriate groupings. Then click "Make Figures" to export result tables.

3.2.5 Results

Three kinds of quantifications are distinguished by Darea—metrics, distances, and cluster parameters. Metrics are saved in the "Metrics.csv" file and encompass properties defined on the level of demarcated ROI, i.e., area and number and density of gold particles. Distances are defined on the level of gold particles. The following distances are measured and saved in the "Distances" folder: NNDs ("NND"), distances between all pairs of particles

("All_Distances"), distance from the center of gravity of the demarcation ("Distance_from_Center"), distance from the nearest edge of demarcation ("Distance_from_Edge"), and a normalized distance from the center ("normalized_Distance_from_Center"). This distance is calculated as $\frac{\text{Distance\_from\_Center}}{\text{Distance\_from\_Center}+\text{Distance\_from\_Edge}}$ and can be squared to obtain the center-periphery index. Quantifications for cluster analysis are saved in "ClusterParameters" folder and encompass the following parameters: number of clusters in an image ("NumberOfClusters"), cluster area defined as the convex hull around the particles of the cluster ("ClusterArea"), number of particles in a cluster ("ParticlesPerCluster"), density of gold particles within the cluster ("DensityWithinCluster"), NND to clusters of the same and clusters of different particle sizes, distance from edge and distance and normalized distance from the center, and overlap of cluster area between clusters of different sizes ("overlap"). In addition, information about the maximum distance between particles belonging to the same cluster ("maxDistance") and number of clusters that were excluded from the analysis ("excludedClusters") is available. Clusters are excluded from analysis when all the centers of their particles lie exactly on a line, which would cause the clusters to have no area. Typically, the number of excluded clusters should be at or very close to 0, and large numbers of excluded clusters may indicate a problem in the analysis. For each distance and cluster parameter, three kinds of statistical analyses are performed, and results are saved in files with different suffixes—"-descriptiveStats.csv," "1by1.csv," and "PopMeans.csv." "DescriptiveStats.csv" contains descriptive statistics (mean, median, standard deviation, standard error), "1by1.csv" contains information on the percentage of images where real and simulated distributions differ significantly from each other, and "PopMeans.csv" test if real and simulated distributions differ significantly from each other by performing a paired t-test (real and simulated distributions of the same image are paired). In addition to these analyses, quantifications for each synapse and each gold particle or cluster can be found in "individual_Metrics," "individual_Distances," and "individual_-ClusterParameters" folders.

3.2.6 Training a DNN for Synapse Demarcation

1. Gather and annotate a training dataset of at least 100 images. The higher the number of images, the better performance can be expected, and we usually use 1000 images and more for best results. For a robust network that will perform well even with difficult samples, training images should be taken from many different samples including samples with suboptimal structure.
2. Convert one or several Darea projects into a training dataset, or train directly from the projects (*see* **Note 15**).

3. Select the training parameters—image size, number of epochs, batch size, and learn rate (*see* **Note 16**)—and name your new DNN. You may also select an already trained network and continue training on it which may reduce necessary training time and improve network performance.
4. Start the training. The checkpoint with the best performance on the validation dataset will be automatically picked and added as selectable option in Darea.

3.2.7 Training of the Particle Detection Classifier

For particle detection we first use a circle finding algorithm. Detected circles are then classified by a trainable machine learning classifier—naïve Bayes or random forest—into gold particles and irrelevant detections [3, 5]:

1. Gather and annotate a training dataset. About 50 images usually suffice for decent results on images with similar brightness and contrast. All gold particles in either the image or the demarcated area have to be annotated.
2. Open the train particle detection submenu, select your algorithm of choice (*see* **Note 17**), detection parameters, and a name for your new classifier, and start the training.

4 Notes

1. We've used a JEM 2800 microscope (JEOL, Tokyo, Japan) with 200 kV acceleration voltage and a XF416ES (TVIPS GmbH, Gauting, Germany) camera or a Tecnai 12 (FEI, Hillsboro, OR, USA) with 120 kV acceleration voltage and a F216 (TVIPS GmbH) camera.
2. Our deep learning algorithms utilize the TensorFlow framework [10], which has been designed for running on NVIDIA graphics cards. The compatibility of TensorFlow with AMD graphics cards has been improved in recent years, and it should be possible to run this software on AMD graphics cards too, but we have not tested this option.
3. Automated demarcation of profiles of interest can be run without graphics card but may then take few hours to complete.
4. We acquire overview images with tiles of 4096 × 4096 pixels, using a resolution of 105 nm/pixel on JEM 2800, or with 2048 × 2048 pixels and a resolution of 126 nm/pixel on Tecnai 12.
5. We acquire images with autofocus every 4 × 4 tiles using 4096 × 4096 pixels and a resolution of 0.83 nm/pixel on JEM 2800 or 2048 × 2048 pixels and a resolution of 2.09 nm/pixels on Tecnai 12.

6. A video tutorial from SerialEM developers can be watched at https://www.youtube.com/watch?v=gdICBVMhmmE.
7. Detection of points of interest may also be performed in parallel to imaging on a different PC which is able to access the drive where the acquired images are being stored. To do this, save the navigator file before starting the imaging at medium magnification. On the PC you would like to use for analysis and open "predict.cmd" (Windows) or "predict.sh" (Linux), while the microscope is imaging. This will open the prediction user interface. Select the corresponding navigator file, choose a DNN, and pick "Run in parallel to imaging."
8. If most points do not correspond to profiles of interest and/or many profiles of interest are not marked by a point, retraining of the DNN should be performed (*see* Subheading 3.1.3).
9. We image at five equally spaced tilting angles from −25° to +25° and take 4096 × 4096 pixel images with a resolution of 0.26 nm/pixel on JEM 2800 or 2048 × 2048 pixel images with a resolution of 0.54 nm/pixel on Tecnai 12.
10. Training images should be representative for a wide range of possible conditions. By training on 967 images, we achieved a false-positive rate of 32% and a false-negative rate of 12% for detection of E-face PSDs.
11. When using a general-purpose software, make a corresponding image with white background and region of interests in black for each image, and save this image as <original_image_name>_mod.tif.
12. While DarEM supports different DNN architectures, we have so far used EfficientDet-D2 architecture [22] which provides a good balance of detection performance and computational demand. For usage of other DNN architectures, please refer to the DarEM handbook included in the software package.
13. If a comparison of outer rims with different widths is desired, we recommend starting with the largest one. This will allow you to first annotate all potentially included gold particles. Afterward, the outer rim size can be reduced without the need for reannotation, as annotated particles outside the outer rim can be excluded.
14. The angle may also be selected before performing the automated prediction to save computation time.
15. At this step, a validation dataset containing a specifiable percentage of images is randomly chosen. Thus, it is recommended to first convert your dataset to a training dataset to keep the selection of validation images the same when using it for multiple trainings.

16. The image is downscaled to the selected image size before training. Larger image size may improve network performance but require more graphics card memory. At a batch size of 1, the maximum image size is 448 × 448 if the graphics card is also used for the display or 512 × 512 if the graphics card is used solely for this computation. We recommend using 200–400 epochs for training, but fewer epochs can be useful for a quick initial test. Higher learn rates will reduce the number of epochs necessary, while smaller learn rates may improve the final network performance. Too high learn rates can lead to a failure of the network to learn the desired feature. It can be useful to first train the network with a higher learn rate and then retrain it with a smaller one on the same image set for fine-tuning.
17. In our experience, naïve Bayes works better for single labelings, while random forest works better for double labelings [5].

Acknowledgments

This research was supported by the European Research Council Advanced Grant 694539 to RS and by the Scientific Service Units of IST Austria through resources provided by the Electron Microscopy Facility.

References

1. Fujimoto K (1995) Freeze-fracture replica electron microscopy combined with SDS digestion for cytochemical labeling of integral membrane proteins. Application to the immunogold labeling of intercellular junctional complexes. J Cell Sci 108(11):3443–3449
2. Masugi-Tokita M, Shigemoto R (2007) High-resolution quantitative visualization of glutamate and GABA receptors at central synapses. Curr Opin Neurobiol 17(3):387–393. https://doi.org/10.1016/j.conb.2007.04.012
3. Luján R, Aguado C, Ciruela F, Cózar J, Kleindienst D, de la Ossa L, Bettler B, Wickman K, Watanabe M, Shigemoto R, Fukazawa Y (2018) Differential association of GABAB receptors with their effector ion channels in Purkinje cells. Brain Struct Funct 223(3):1565–1587. https://doi.org/10.1007/s00429-017-1568-y
4. Rebola N, Reva M, Kirizs T, Szoboszlay M, Lőrincz A, Moneron G, Nusser Z, DiGregorio DA (2019) Distinct nanoscale calcium channel and synaptic vesicle topographies contribute to the diversity of synaptic function. Neuron 104(4):693–710.e9. https://doi.org/10.1016/j.neuron.2019.08.014
5. Kleindienst D, Montanaro J, Bhandari P, Case MJ, Fukazawa Y, Shigemoto R (2020) Deep learning-assisted high-throughput analysis of freeze-fracture replica images applied to glutamate receptors and calcium channels at hippocampal synapses. IJMS 21(18):6737. https://doi.org/10.3390/ijms21186737
6. Bhandari P, Vandael D, Fernández-Fernández D, Fritzius T, Kleindienst D, Önal C, Montanaro J, Gassmann M, Jonas P, Kulik A, Bettler B, Shigemoto R, Koppensteiner P (2021) GABAB receptor auxiliary subunits modulate Cav2.3-mediated release from medial habenula terminals. elife 10:e68274. https://doi.org/10.7554/eLife.68274
7. Harris KM, Landis DMD (1986) Membrane Structure at Synaptic Junctions in Area CA1 of the Rat Hippocampus. Neuroscience 19(3):857–872. https://doi.org/10.1016/0306-4522(86)90304-0
8. He K, Zhang X, Ren S, Sun J (2015) Delving deep into rectifiers: surpassing human-level

performance on imagenet classification. In: Proceedings of the IEEE International Conference on Computer Vision (ICCV), pp 1026–1034

9. Serre T (2019) Deep learning: the good, the bad, and the ugly. Annu Rev Vis Sci 5(1): 399–426. https://doi.org/10.1146/annurev-vision-091718-014951

10. Abadi M, Agarwal A, Barham P, Brevdo E, Chen Z, Citro C, Corrado GS, Davis A, Dean J, Devin M, Ghemawat S, Goodfellow I, Harp A, Irving G, Isard M, Jia Y, Jozefowicz R, Kaiser L, Kudlur M, Levenberg J, Mane D, Monga R, Moore S, Murray D, Olah C, Schuster M, Shlens J, Steiner B, Sutskever I, Talwar K, Tucker P, Vanhoucke V, Vasudevan V, Viegas F, Vinyals O, Warden P, Wattenberg M, Wicke M, Yu Y, Zheng X (2016) TensorFlow: large-scale machine learning on heterogeneous distributed systems. arXiv:1603.04467 [cs]

11. Huang J, Rathod V, Sun C, Zhu M, Korattikara A, Fathi A, Fischer I, Wojna Z, Song Y, Guadarrama S, Murphy K (2017) Speed/accuracy trade-offs for modern convolutional object detectors. arXiv:1611.10012 [cs]

12. Mastronarde DN (2005) Automated electron microscope tomography using robust prediction of specimen movements. J Struct Biol 152(1):36–51. https://doi.org/10.1016/j.jsb.2005.07.007

13. Schorb M, Haberbosch I, Hagen WJH, Schwab Y, Mastronarde DN (2019) Software tools for automated transmission electron microscopy. Nat Methods 16(6):471–477. https://doi.org/10.1038/s41592-019-0396-9

14. Fukazawa Y, Masugi-Tokita M, Tarusawa E, Hagiwara A, Shigemoto R (2008) SDS-digested freeze-fracture replica labeling (SDS-FRL). In: Cavalier A, Spehner D, Humbel BM (eds) Handbook of cryo-preparation methods for electron microscopy. CRC Rress, Boca Raton, pp 559–576

15. Kaufmann WA, Kleindienst D, Harada H, Shigemoto R (2021) High-resolution localization and quantitation of membrane proteins by SDS-digested freeze-fracture replica labeling (SDS-FRL). In: Lujan R, Ciruela F (eds) Receptor and ion channel detection in the brain. Springer US, New York, pp 267–283. https://doi.org/10.1007/978-1-0716-1522-5_19

16. Maier A, Syben C, Lasser T, Riess C (2019) A gentle introduction to deep learning in medical image processing. Z Med Phys 29(2):86–101. https://doi.org/10.1016/j.zemedi.2018.12.003

17. Li Y, Huang C, Ding L, Li Z, Pan Y, Gao X (2019) Deep learning in bioinformatics: introduction, application, and perspective in the Big Data era. Methods 166:4–21. https://doi.org/10.1016/j.ymeth.2019.04.008

18. Géron A (2019) Hands-on machine learning with Scikit-Learn, Keras, and TensorFlow: concepts, tools, and techniques to build intelligent systems

19. Chollet F (2021) Deep learning with Python, Second Edition. Manning Publications Co. LLC, New York

20. Eguchi K, Montanaro-Punzengruber JC, Le Monnier E, Shigemoto R (2022) The number and distinct clustering patterns of voltage-gated calcium channels in nerve terminals. Front Neuroanat 16:846615. https://doi.org/10.3389/fnana.2022.846615

21. Tzutalin. LabelImg. https://github.com/tzutalin/labelImg (accessed 2021–11–28)

22. Tan M, Pang R, Le QV (2020) EfficientDet: scalable and efficient object detection. arXiv:1911.09070 [cs, eess]

Part II

Connectivity

Chapter 9

Multiplexed *d*STORM Imaging of Neuronal Tissue and Cells

Marina Ruth Hesse, Thomas Kuner, and Maja Klevanski

Abstract

The nervous system consists of neurons interconnected by densely packed synapses, which are composed of distinct sets of synaptic proteins mediating the highly demanding processes of synaptic transmission. Most classical single-molecule localization microscopy (SMLM) methods allow the visualization and identification of molecular nanostructures at approximately 20–30 nm lateral and 60 nm axial resolution but are often limited to simultaneous acquisition of only two labeled proteins. However, to visualize and understand the nano-architecture of complex protein machineries, such as the presynaptic active zone or postsynaptic density, many proteins need to be identified in the same sample at the highest possible resolution. Such highly multiplexed imaging experiments will help identify geometries of protein composition that may finally drive function. To achieve this, we developed maS^{3}TORM (multiplexed automated serial staining stochastic optical reconstruction microscopy), a microscopy approach allowing fully automated re-staining of the sample directly on the microscope by combining 3D dual-channel *d*STORM imaging with automated solution exchange using a pipetting robot. Using maS^{3}TORM, we can successfully identify more than 15 targets within the same presynaptic terminal, revealing protein distributions at a new level. Hence, this technique can be used to explore the nano-architecture of synapses or any other cellular protein machines and thereby help gain a better understanding of how complex macromolecular assemblies may drive function.

Key words Super-resolution imaging, Nanoscopy, Single-molecule localization microscopy, SMLM, *d*STORM, Multiplexed imaging, Multicolor, Synapses, Neurobiology, Tissue

1 Introduction

The investigation of the nanomolecular organization of molecular key players responsible for neurotransmission, information processing, and maintenance of synaptic homeostasis remains challenging. However, understanding of the molecular nano-architecture of these processes is required to unravel complex biological functions: for example, knowing the molecular composition of synaptic vesicle pools and their interplay with scaffolding proteins and the exocytosis machinery at the active zone (AZ) could open the door to a more comprehensive understanding of neurotransmission. Furthermore, it is crucial to understand the trans-synaptic nano-

Joachim H. R. Lübke and Astrid Rollenhagen (eds.), *New Aspects in Analyzing the Synaptic Organization of the Brain*, Neuromethods, vol. 212, https://doi.org/10.1007/978-1-0716-4019-7_9,

organization of pre- and postsynaptic sites, including the distribution of presynaptic vesicle fusion sites with respect to postsynaptic receptors, since even slight alterations in these structures can drive pathology [1, 2].

The resolution of conventional light microscopy is limited to approximately 200 nm, because of the diffraction limit of light. However, dendrites, axons, and in particular synaptic terminals are small, and the analysis of their composition requires a resolution below 200 nm. Most of the fundamental understanding of neurotransmission so far is based on (1) morphological examination of electron microscopic images, (2) in vitro biochemical approaches, (3) electrophysiology, and (4) conventional immunofluorescence microscopy providing information on (a) morphological features without knowledge on the molecular composition, (b) molecular composition without considering morphological features, and (c) electrical properties often combined with (d) diffraction-limited morphological features.

The Nobel Prize in Chemistry 2014 has been awarded to Eric Betzig, Stefan W. Hell, and William E. Moerner for their breakthrough discovery on how to surpass the limitation of light using single-molecule localization microscopy (SMLM) and stimulated emission depletion (STED) microscopy. Since then, various SMLM methods evolved that employed blinking fluorescent proteins, which sporadically emit light at positions far enough apart from each other so that their localizations can be precisely determined over an extended timeframe. Direct stochastic optical reconstruction microscopy, *d*STORM, uses photo-switchable fluorophores [3], while point accumulation for imaging in nanoscale topography (PAINT) [4] and DNA-PAINT [5] are based on reversibly binding fluorescent probes, and photoactivation localization microscopy (PALM) [6] uses genetically fused photoactivatable proteins [7]. Thus, SMLM opens new ways to resolve the molecular nano-organization of proteins and to answer fundamental biological questions leading to further developments in preclinical medicine and life sciences.

*d*STORM is a SMLM technique allowing super-resolution microscopy by acquiring image series of spatially and temporally separated fluorescent blinking events [8–11]. Photo-switching of fluorophores in *d*STORM is induced by thiols such as β-mercaptoethanol and high-power laser illumination, which allows imaging of non-overlapping emitters. The profile of individual emitters can be precisely analyzed and fitted with the point spread function or a Gaussian function, thereby defining the center of mass with nanometer precision [12, 13]. While many SMLM techniques enable simultaneous acquisition of two channels (dual-color imaging) utilizing dyes with similar but distinguishable excitation and emission spectra, the ability to image multiple targets in SMLM mode is limited due to physical properties of fluorophores and the accuracy at which different spectral channels can be distinguished. To achieve a more comprehensive insight into the molecular composition of synapses by detecting multiple targets within specific regions of interest, we developed a repetitive labeling and imaging approach. Multiplex imaging at a large scale is very time-consuming, and individual steps have to be performed under standardized conditions inevitably requiring an automated solution for re-staining and image acquisition. Therefore, we developed a hybrid system comprising a fully automated pipetting robot attached to a SMLM microscope—multiplexed automated serial staining STORM (maS^{3}TORM) [14]. For this approach (1) a custom-built 3D dual-color *d*STORM microscope, (2) a pipetting robot, and (3) a custom-written software for coordination of both systems are used. In principle, this solution is compatible with any setup that allows an integration of all three components. The maS^{3}TORM approach is based on a re-staining strategy realized by the exchange of antibodies, nanobodies, or any other binder labeled with a fluorophore. Our system design allows two strategies of signal removal: chemical elution and photo-bleaching. We provide a custom-written "Experiment Editor" software where the exact re-staining strategy for individual experimental designs can be specified. To ascertain precise registration of serially acquired images, we implemented a fiducial marker-based strategy. We adapted our system for multiplexed super-resolution imaging of brain tissue and neuronal cells and demonstrated the effective application of maS^{3}TORM.

2 Materials

2.1 Microscope, Pipetting Robot, and Software

Every microscope that is suitable for *d*STORM can be applied if *d*STORM-compatible labeling of the biological material is used. However, successful multiplexing at nanoscale resolution requires several conditions:

1. Physical focus stabilization, e.g., via correction of the z-position by a piezo stage realized by feedback tracking of the reflection angle of an infra-red laser beam.
2. High-precision XY microscope stage for choosing multiple regions of interest (ROIs).
3. Laser for *d*STORM acquisition: Depending on the biophysical properties of the fluorophores, it might be advantageous to choose high-power lasers (e.g., 100–200 mW for 661 nm laser) to ensure constant and reliable blinking in the whole field of view.
4. A fast CCD camera to enable acquisition of tens of thousands of frames with ≤30 ms/frame, allowing to capture fluorophores featuring fast blinking dynamics.

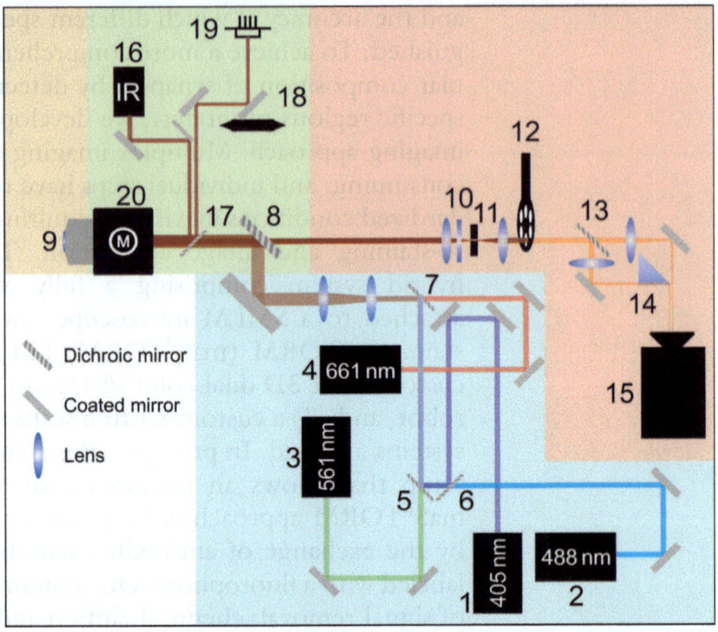

Fig. 1 Schematic overview of major microscope components of the custom-built maS^{3}TORM setup. The blue-shaded area corresponds to the excitation pathway with four lasers—405 nm #1, 488 nm #2, 561 nm #3, and 661 nm #4—followed by dichroic mirrors #5–#7 used to merge the four laser beams into a shared light path. The beams are widened and parallelized by two lenses, reflected by dichroic mirror #8, and collected by a 100× objective #9. The red-shaded area corresponds to the emission pathway containing a tube lens to focus the beam, the cylindrical lens #10 for 3D imaging, a slit to limit the size of the beam, and lens #11 that parallelizes the beam. The beams pass filter wheel #12 with emission filters selected with respect to the used laser line. Finally, the beam passes dichroic mirror #13 that splits the emission light into two channels and prism #14 that is used to direct the two channels onto the chip of high-speed camera #15. The green-shaded area corresponds to the focus stabilization system consisting of: infra-red diode #16, dichroic mirror #17, movable mirror #18, quadrant diode #19, and piezo-driven stage #20

5. Multiple laser lines for independent acquisition of position markers in epifluorescence mode (e.g., fiducials or labeled reference structures).
6. Possibility for automation, both technically, e.g., by ensuring physical space for a pipetting robot, and on the software side, e.g., by providing interfaces for external control software.

Figure 1 illustrates the design of the *d*STORM microscope, and the relevant components are listed in Table 1. A critical component for automated liquid exchange is the pipetting robot (Table 2). We use the pipetting robot PAL3 RTC (Axel Semrau, Germany); however, any similar robot can be used. Figure 2 provides an overview of the software components required to control and coordinate the robot and the microscope. In brief we recommend:

Table 1
Most relevant components for the custom-built microscope

Components microscope
Objective: 100×, 1.49 numerical aperture, 0.1 mm working distance
XY stage: with ca. 100 mm travel distance in x and y
Controller for XY stage and piezo mirror positioner
Focus lock components
Laser diode: 785 nm, ca. 5–20 mW
Controller and power supply for 785 nm diode
Piezo mirror positioner: ca. 100 μm range
Quadrant silicon photodiode: red enhanced
Piezo objective positioner: 100 μm range and controller for piezo objective positioner
Circuit board processing output from quadrant diode, feeding piezo objective positioner
Excitation pathway
Lasers: 661 nm 100 mW–200 mW (a laser in this spectral range is obligatory for *d*STORM imaging of AF647 and CF680) 405 nm 100 mW (optional to enhance transition of fluorophore from the dark to the ground state) 488 nm (recommended for markers acquired in a wide-field mode) 561 nm (recommended for acquisition of fiducial markers)
Clean-up filters for lasers, e.g., ZET 660/20 for the 661 nm laser
Emission pathway
Lenses: in our case tube lens with focal length 180 mm and two lenses with focal length 300 mm
Cylindrical lens for astigmatism-based 3D imaging: in our case 1000 mm focal length
Notch filter for 661 nm excitation
Slit to narrow channels projected to a camera chip: in our case 3.4 mm width
Emission filters: 525/45 (488 nm laser), 605/50 (561 nm laser), 700/75 (661 nm laser); all filters should block laser light in the infra-red range emitted from the 785 nm diode
Dichroic mirror 690 DCXR to split 661 nm laser-induced emission into two channels allowing subsequent spectral demixing
Prism: right angle, 40 mm
Camera: iXon Ultra 897

1. Microscope control software, e.g., our custom-written "Microscope Control".
2. Robot control software that allows to directly or indirectly (via a hand-shake principle) communicate with the microscope control software, e.g., "Chronos" (the software for robot

Table 2
Components of the pipetting robot

Pipetting robot
Robotic xyz autosampler (platform with pipetting arm), e.g., PAL3 RTC
Liquid tools (electronic pipettes) with 1000 μL and 100 μL syringes
Washing stations for syringes; we recommend two stations: one for liquids used to wash the syringe (e.g., H_2O) and one that enables loading of syringes with classical buffers (e.g., PBS)
Tray holders with trays for vials; for typical multiplex experiments, we recommend ≥54 vials

operation provided by Axel Semrau) and our custom-written "Chronos" plugin that enables an automated loading of external control commands.

3. Software allowing to set up user-defined experimental designs and to coordinate the execution of microscope- and robot-related commands, e.g., our custom-written "Experiment Editor."

More information about the components of the home-built microscope and download links to our software solutions can be found in our previous publication [14].

2.2 Coating of Glass-bottom Dishes with TetraSpeck Microsphere Fiducials

We recommend the usage of fiducials (e.g., TetraSpeck microspheres) for optimal registration:

1. Glass-bottom dishes with a thickness that is compatible with the microscope objective, for example, MatTek or Ibidi (e.g., dish diameter of 35 cm, glass with 14 cm diameter, and No. 0 or 1 thickness).
2. TetraSpeck microspheres with fluorophores in four spectra, 100 nm, Thermo Fisher Scientific.
3. Poly-L-lysine (PLL), 0.1% (w/v) in H_2O.
4. Phosphate buffered saline (PBS), pH 7.4.

2.3 Sample Preparation for dSTORM Imaging

2.3.1 Chemical Fixation and Vibratome Sectioning of Brain Tissue

1. Mice or rats; tissue from other species possible but can require protocol optimization.
2. Anesthesia and procedures according to the current national animal care guidelines.
3. PBS, pH 7.4.
4. 4% paraformaldehyde (PFA) in PBS, pH 7.4 stock solution.
5. Winged infusion set with two-channel tubing.
6. Falcons for 4% PFA in PBS and PBS.
7. Dissection tools: forceps, scissors, clamps, needles.
8. Falcons for brain tissue storage.

9. Superglue.
10. Vibratome.
11. Razor blades.
12. Forceps.
13. Fine brush for transfer of the tissue.
14. Six-well plate.

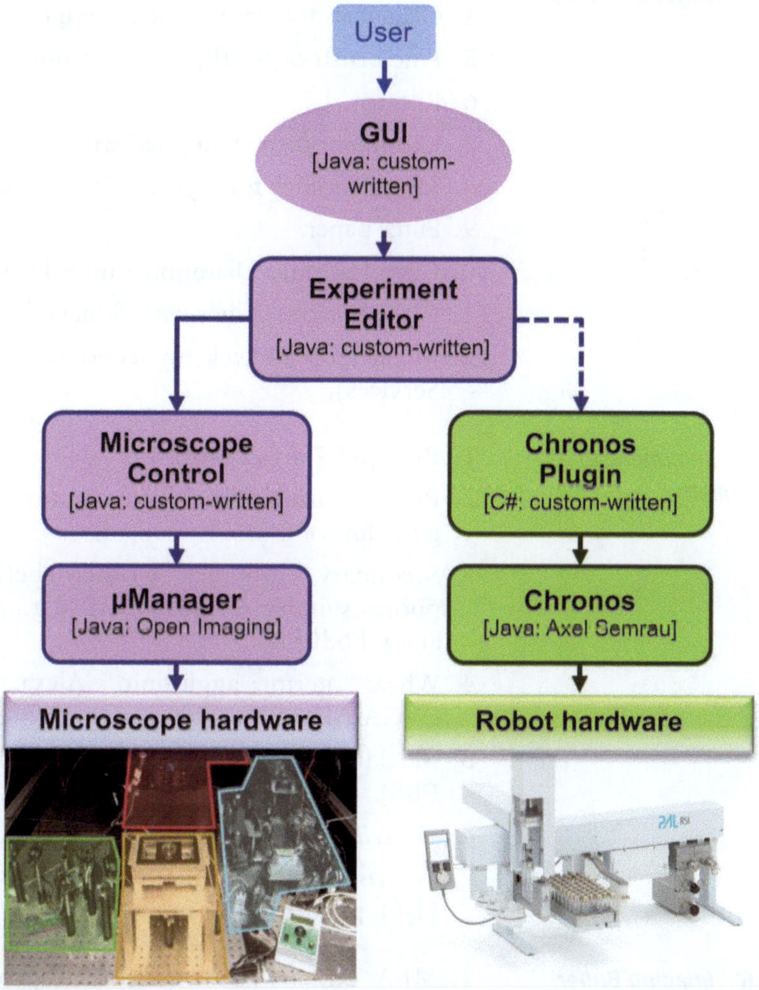

Fig. 2 Software components required to control and coordinate the robot and the microscope. Coordination is realized via the custom-written "Experiment Editor" software where users can design their individual experiments via the respective graphical user interface (GUI). "Experiment Editor" directly communicates with the "Microscope Control" software that controls all microscope components via μManager. Communication of the "Experiment Editor" software with the robot is realized via an exchange folder (indirect commination indicated by dashed line) that is monitored and modified by both, the "Experiment Editor" and a custom-written plugin of "Chronos", and the software for robot operation provided by Axel Semrau

2.3.2 Chemical Fixation of Hippocampal Cell Cultures

1. Prewarmed 4% PFA, pH 7.4 in cytoskeletal buffer (10 mM Pipes, 100 mM NaCl, 300 mM Sucrose, 1 mM EGTA, 1 mM $MgCl_2$, 1 mM DTT).
2. PBS, pH 7.4.

2.4 Semi-thin Cryo-sectioning of Brain Tissue According to the Tokuyasu Method

1. Razor blade.
2. 2.1 M and 2.3 M sucrose stored at 4 °C.
3. 2% methylcellulose stored at 4 °C.
4. Glass vials for tissue block storage.
5. Fine brush or toothpick for transfer of the tissue.
6. Rotator.
7. Ultramicrotome equipped with a cryo-chamber.
8. Sample holder for cryo-ultramicrotomy.
9. Filter paper.
10. Cryo Immuno Diamond Knife (Diatome) or glass knife.
11. Glass-bottom dishes with fiducials.
12. Metal loop to pick up sections, e.g., Perfect Loop (Science Services).

2.5 Immunohisto-chemistry

1. PBS, pH 7.4 (*see* **Note 1**).
2. Primary antibodies (concentrations according to company guidelines).
3. Secondary antibodies or other labels conjugated with fluorophores suitable for *d*STORM (e.g., Alexa Fluor 647 (AF647) and CF680).
4. Wheat germ agglutinin Alexa Fluor 488 conjugate (WGA-AF488), Invitrogen (*see* **Note 2**).
5. Blocking solution (e.g., 5% fetal calf serum, 100 mM Glycine in PBS).
6. *For cells:* Detergent, e.g., 0.1% Triton X-100.
7. *For tissue (optional):* 0.1% sodium dodecyl sulfate (SDS) in H_2O, pH 13 for 15 min.

2.6 Imaging Buffer

1. MEA buffer: 100 mM β-mercaptoethanol (MEA) in PBS, pH 8.0 (*see* **Note 3**).

2.7 Chemical Elution

1. 0.1% SDS in H_2O, pH 13 for 15 min.
2. PBS for washing, 3 × 5 min.

2.8 Photo-bleaching

1. PBS, pH 7.4.
2. 661 nm laser at 2 kW/cm^2 and 405 nm laser at 0.7 kW/cm^2 for 4 min.

3 Methods

3.1 dSTORM Imaging

We combine a fully automated *d*STORM setup with a pipetting robot for serial staining, which allows the identification of the nano-architecture of synapses in a multiplexed manner. Our *d*STORM setup is equipped with a 100× objective (1.49 numerical aperture) and allows astigmatism-based acquisition of biological structures in 3D [15], which is especially beneficial for protein machineries with complex molecular organization and allows to precisely evaluate the orientation of synaptic specializations, e.g., of the synaptic cleft. Dual-color imaging is realized using dyes with similar but distinguishable excitation and emission spectra (Alexa Fluor 647 and CF680), whose emission beams are split by a dichroic mirror. For *d*STORM acquisition, we recommend the usage of a 100 kW 661 nm laser at 100% power, which results in a sample illumination with 2 kW/cm^2 at our setup. For the best blinking dynamics of both individual fluorophores, we use an exposure time of 30 ms and collect 20,000 frames. Directly after the *d*STORM acquisition, the fiducials are imaged using the 561 nm laser at 3–5% power (0.08–0.12 kW/cm^2 at our setup), while the emission is collected via the far-red filter that is used for *d*STORM. Usage of a common filter for *d*STORM and fiducial imaging prevents chromatic aberration between the two imaging steps (*see* **Note 4**). Although the fiducials are constantly in the on-state and hence are not blinking, their sparse distribution (ca. 5–10 fiducials per field of view) allows direct application of point spread function or Gaussian fitting, resulting in an SMLM image. Therefore, to determine their position, we acquire images in a "pseudo STORM" mode (i.e., more like in conventional wide-field epifluorescence imaging) using only a few frames (in our case, 50) and an exposure time of 200 ms. After every single SMLM imaging round, the acquired image stacks (in tagged image file format (TIFF)) are automatically post-processed with RapidStorm [16]; the resulting output is further processed as described in Subheading 3.7 "Post-Processing of SMLM Acquisitions." For more details, *see* Fig. 1 and Table 1.

3.2 Coating of Glass-bottom Dishes and Fiducial Application

Application of fiducials to glass-bottom dishes is mandatory to enable drift correction during one imaging round and for registration of images acquired in different imaging rounds. The fiducials should be placed on top of the cover-slip of the glass-bottom dish, below the cells/tissue to ensure immobilization of fiducials and best conditions for the acquisition of fiducial signal:

1. *Cells:* Coat glass bottom with poly-L-lysine (PLL) or fibronectin for 30 min. The choice depends on the used cell line. Work under sterile conditions.

   Tissue: No coating is necessary.

2. Apply TetraSpeck beads with a dilution of ca. 1:500 for 10 min.
3. Wash 2× with 300 μL ddH_2O, and let the dishes dry completely.
4. *Tissue:* Let glass-bottom dish dry thoroughly. Coated glass-bottom dishes can be stored at 4 °C for several months. Continue with Subheading 3.3 "Perfusion and Vibratome Sectioning."

For cells continue with **steps 5–9***:*

5. Add 0.1% PLL or fibronectin for 10–20 min.
6. Remove PLL or fibronectin and wash 3× with 300 μL ddH_2O.
7. Dry thoroughly for 30–60 min under the cell culture hood.
8. Seed and culture the desired cell line (e.g., hippocampal neurons) [17].
9. Chemically fix cells by application of 4% PFA in cytoskeletal buffer.

3.3 Perfusion and Vibratome Sectioning

1. Anesthetize animals as recommended by current national animal care guidelines (e.g., intraperitoneal application of 100 mg/kg ketamine, 5 mg/kg xylazine for rats).
2. After reflex responses subside, fixation of the brain is achieved by transcardiac perfusion [18].
3. The brain is carefully removed and post-fixed in the fixative at 4 °C for 24 h.
4. Sections with a thickness of 200–300 μm are cut using a vibratome.
5. Sections should be stored in PBS at 4 °C and processed as soon as possible.

3.4 Preparation of Semi-thin Tissue Sections

1. Semi-thin tissue sections of approximately 400 nm thickness are generated following the Tokuyasu method developed for electron microscopy [19]. From vibratome sections, a block of the tissue with the dimensions of approximately 1 mm^3 is dissected with sharp fresh razor blades and transferred to a glass vial containing 2.1 M sucrose for cryo-protection. Use a rotator for optimal infiltration.
2. After at least 8 h of incubation at 4 °C, the tissue is completely infiltrated. Blocks can be stored for approximately 6 months at 4 °C.
3. Sample is mounted on a sample holder for cryo-ultramicrotomy, and excess sucrose is removed with sharp edges of two rectangularly cut filter papers until most excess sucrose is absorbed (*see* **Note 5**). Immediately after this step,

the sample is plunge frozen in liquid nitrogen. The tissue is cut with an ultramicrotome equipped with a cryo-stage at −70 to −90 °C in 400-nm-thick sections (*see* **Note 6**).

4. Pick up multiple sections (usually 3 to 7) with a freshly prepared 1:1 solution of 2.3 M sucrose and 2% methylcellulose (final concentration 1.15 M sucrose and 1% methylcellulose) (*see* **Note 7**) with a Perfect Loop or a custom-made platinum loop.
5. Place sections in the center of a glass-bottom dish.
6. Sections can be stored for up to 6 months at 4 °C.

3.5 Staining

For our multiplex strategy, we employ immunostaining using primary and secondary antibodies (e.g., F(ab')-fragments), directly labeled antibodies, nanobodies, and other types of labels like lectins or phalloidin. In the following, a typical protocol for primary and secondary antibodies is described. However, probe-specific adaptations might be necessary:

1. *For tissue:* To remove the sucrose, apply 300 μL PBS to the sucrose-mounted section for at least three times for a total of 15 to 20 min. Make sure that sections are completely rehydrated. If the multiplex experiment includes elution steps, a pre-incubation of the tissue with 0.1% SDS, pH 13 for 10 min, might be necessary, followed by 3 × 5 min wash with PBS.

   For cells: Wash cells 3 × 5 min with 300 μL PBS. Permeabilize cells with detergent, e.g., 0.1% Triton X-100 for 10 min.
2. Examine the tissue/cell quality with a light microscope.
3. Incubate the tissue/cells for 10 min with WGA-488 followed by 3 × 5 min wash with PBS (*see* **Note 2**).
4. Add 300 μL blocking solution for 10 min followed by 3 × 5 min wash with PBS.
5. Incubate the tissue/cells for 1 h with primary antibody diluted according to the company guidelines and experimental tests followed by 3 × 5 min washing with PBS.
6. Incubate the tissue/cells for 1 h with secondary antibody followed by 3 × 5 min washing with PBS.
7. Fix the glass-bottom dish on the microscope stage, and define multiple ROIs. It is recommended to maintain a minimal distance between the selected ROIs to avoid bleaching of adjacent ROIs.
8. Prepare vials with a defined amount of MEA powder for the MEA stock solution, a vial with 1 M KOH for calibration of the correct pH, and empty vials for the final 0.1% MEA solution in PBS pH 8.0 (filled by the robot). Prefill further vials with blocking solution, antibody/label solution(s),

and elution buffer (optional). In the case of the PAL3 RTC, PBS washing buffer and water are drawn directly from the syringe washing station. If a buffer/water reservoir is not available, provide vials prefilled with the required liquids.

9. Add freshly prepared imaging buffer to the tissue/cells. Make sure that the whole acquisition procedure (multiple ROIs) is not exceeding 2 h (*see* **Note 3**).

3.6 Experiment Design According to the maS^{3}TORM Method

Here, we want to outline general steps that need to be considered to design a multiplex experiment. While the design can vary depending on the application, we provide a general scheme for a multiplexed experiment in Fig. 3 (*see* **Note 8**). The experiment is set up by preparing vials with antibodies or other labels and buffers and the initial staining of the sample (*see* Subheading 3.5). In addition to the labeling of targets for the initial STORM acquisition, we recommend using a supplementary morphological marker such as Alexa Fluor 488-conjugated WGA (*see* **Note 2**), allowing to select optimal ROIs (in our case, between 4 and 8) and to control tissue integrity throughout the experiment. In the next step, the imaging buffer is freshly prepared by the robot. The robot automatically pipettes the required amount of PBS into the vial containing MEA powder and adjusts the pH by addition of a precalculated volume of 1 M KOH. The fresh imaging buffer is pipetted onto the sample, and sequential imaging of selected ROIs begins. For each ROI, three steps of imaging are performed: acquisition of (1) the tissue marker in an epifluorescence mode using the 488 nm laser, (2) targets that were stained in the current round in the *d*STORM mode using the 661 nm laser, and (3) fiducials using the 561 nm laser; more details can be found in Subheading 3.1 "*d*STORM Imaging."

Before the next pair of targets can be stained, it is important to remove the residual signal from the previous round. For this purpose, depending on the requirements for the next staining round, we use two main signal removal strategies. If cross-reaction is expected, e.g., when using primary antibodies from the same species in consecutive rounds, we employ chemical elution to physically remove all labels (*see* Subheading 2.7). To ensure maximal signal reduction, elution is always followed by photo-bleaching (*see* Subheading 2.8). If no cross-reaction between previous and upcoming labels is expected, it is sufficient to eliminate remaining signal via photo-bleaching alone. Subsequently, a second staining round is performed according to Subheading 3.5 "Staining" (**steps 5–7**). If a fluorophore-conjugated primary antibody is used, a single incubation step is required. From our experience, this procedure can be repeated and maintained for approximately seven to ten imaging rounds.

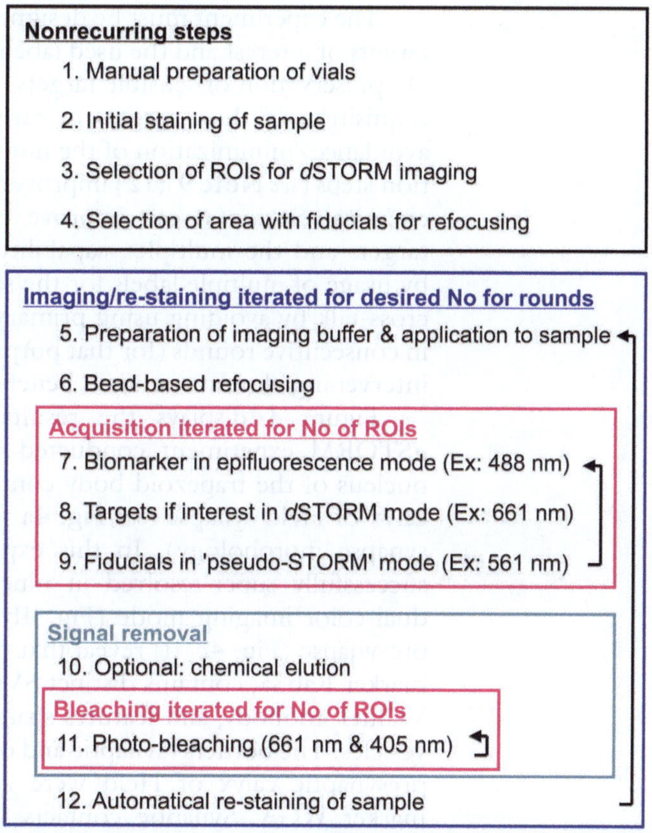

Fig. 3 General scheme for a multiplex experiment. Black box contains steps that must be performed once in the very beginning of a multiplex experiment. Blue box contains all steps of one imaging and re-staining round that are performed automatically and can be repeated for a desired number of rounds. Magenta boxes denote steps repeated for multiple regions of interest (ROIs) that are acquired within one individual staining round. Cyan box comprises steps required for signal removal. **Steps 1–4** are typically performed manually, albeit **step 2** could be principally done automatically by the pipetting robot. First, the vials containing all required compounds and solutions are prepared manually. Second, the initial staining is done according to Subheading 3.5 "Staining." Afterward, multiple ROIs (typically 4–8) and area with fiducials for refocusing are selected. **Steps 5–12** are performed fully automatically and can be repeated for a desired number of iterations. In **step 5**, the MEA buffer is freshly prepared and applied to the sample before the acquisition series starts. In **step 6**, bead-based refocusing is performed using fiducials selected in **step 4** as a reference. **Steps 7–9** refer to all imaging steps that are performed (for more detail, *see* Subheading 3.6). **Steps 10–11** indicate measures for signal removal, where chemical elution is optional and depends on the labeling strategy. **Step 12** indicates automated re-staining of the sample that is performed in analogy to **step 2**

The experiment must be designed in strong dependence on the targets of interest and the used labeling techniques. Major goals are (1) preservation of sensible targets, which can be reached, e.g., by acquisition of these targets in early imaging steps and often by avoidance/minimization of the number of preceding chemical elution steps (*see* **Note 9**); (2) improvement of labeling efficiency, e.g., chemical elution(s) can improve target accessibility for masked targets and the multiplex capability allows to enrich information by usage of multiple labels for the same target; and (3) preventing cross-talk by avoiding using primary antibodies from same species in consecutive rounds (for that purpose, maximizing the number of intervening elution rounds is beneficial).

Figure 4 displays the results of an exemplary multiplex *d*STORM experiment conducted in brain tissue of the medial nucleus of the trapezoid body containing the giant glutamatergic calyx of Held synapse (*see* Fig. 4a for schematic representation of synapse morphology). In this experiment, 15 targets could be successfully super-resolved in nine staining rounds acquired in dual-color imaging mode (Fig. 4b). Close up views of the calyx presynapse (Fig. 4c, d) reveal that it is densely filled with synaptic marker Rab3a, contains distinct SVs and SV clusters labeled using VGlut1 antibody, and features sparsely distributed clathrin-coated vesicles. The borders (synaptic and outer membrane, Fig. 4d) of the presynaptic calyx of Held were approximated using the lectin marker WGA. Synaptic contacts were identified using the AZ marker Bassoon that is located at the calyceal side of the synaptic membrane and the postsynaptic density (PSD) marker Homer that is juxtaposed to the AZ at the periphery of the postsynaptic principal cell. These results demonstrate that the maS^{3}TORM approach allows to super-resolve and correlate the ultrastructural distribution of SVs, AZ scaffolds, and other synaptic components.

3.7 Post-Processing of SMLM Acquisitions

We post-process our data in three steps. First, Gaussian fitting is realized using RapidStorm, resulting in files with x, y, and z coordinates and intensity values of localizations. Any equivalent fitting software like ThunderSTORM [20] can be used. Second, the transformation of coordinates of acquisitions from different rounds is realized based on fiducial information. Third, we apply our self-written post-processing software, which performs spectral demixing of the two fluorophores acquired in one round (AF647 and CF680), corrects the drift, and merges fluorophores that are in an on-state in consecutive frames to prevent overcounting of localizations. Finally, the software creates rendered 2D images and coordinate outputs that can be used for 3D visualization and analysis. However, a broad range of different post-processing and rendering software solutions are available, e.g., SMAP [21].

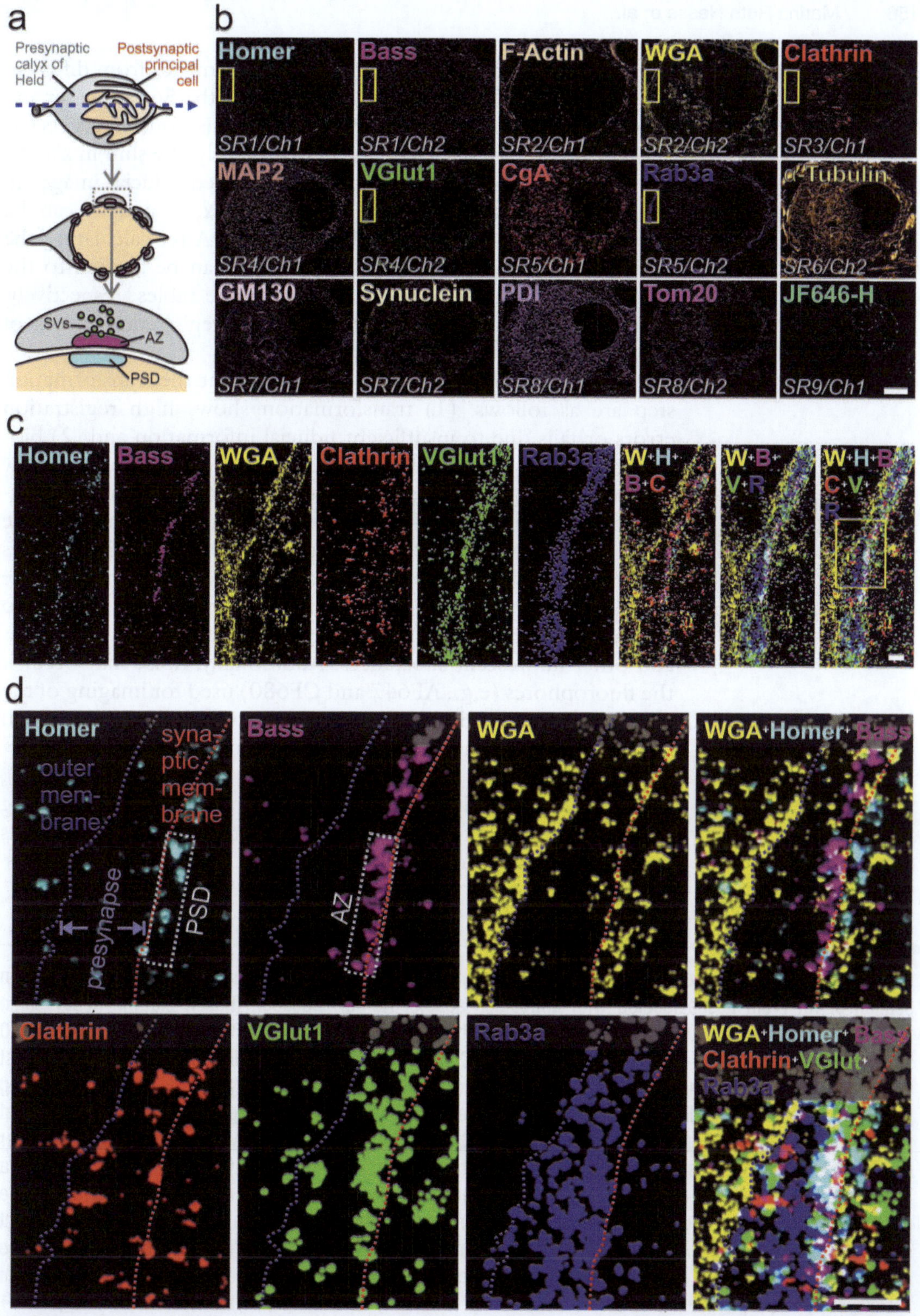

Fig. 4 Multi-target *d*STORM imaging of a single calyx of Held synaptic contact. (**a**) Graphical representation of a transverse section through a synaptic contact between the calyx of Held (gray) and the principal cell (beige). The presynaptic calyx harbors synaptic vesicles (SVs, green) and active zones (AZs, magenta), which are

3.8 Registration

Fiducials are used for the registration of the images from different imaging rounds. Fiducials are acquired using the 561 nm laser and post-processed similarly to the *d*STORM acquisitions of targets (*see* Subheading 3.1 for more details). Subsequently, the shift in x and y can be calculated from the overlay of rendered fiducial images in ImageJ, or the shift along all three axes (x, y, and z) can be calculated based on 3D fiducial coordinates. After calculating the shift, the resulting transformation matrices can be applied to the rendered 2D image or to the 3D coordinate tables, respectively. Using this approach, we typically achieve registration errors of down to 10 nm in subsequent rounds.

Two problems that can be encountered at the transformation step are as follows: (1) transformation shows high registration errors or fails due to insufficient fiducial information and (2) biomarkers are misaligned despite successful transformation with low registration errors. In case (1), evaluate whether fiducials are present and fully immobilized, fiducial signal is in an intensity and size range that is accepted by the fitting software, or fiducial signal is masked by noise signal. Noise signal can arise, for example, if the sample dried out or got contaminated (*see* **Note 1**) leading to increased autofluorescence and unspecific binding of fluorophores and other fluorescent substances. Additionally, noise may occur if the fluorophores (e.g., AF647 and CF680) used for imaging of the molecular targets did not sufficiently enter the dark state or bleach during the preceding *d*STORM acquisition and are still considerably excited by the 561 nm laser. When encountering case (2), check if sample quality, integrity, and immobilization are satisfying and remain constant throughout the experiment as tissue or cells could move independently of the fiducials.

Fig. 4 (continued) opposed by the postsynaptic densities (PSDs) of the postsynaptic principal cell. (**b**) Thin section of rat brain tissue labeled with antibodies against Homer, Bassoon (Bass), clathrin heavy chain 17 (CHC17), MAP2, VGlut1, chromogranin A (CgA), Rab3a, α-tubulin, GM130, α/β-synuclein, PDI, and Tom20 with phalloidin for F-actin staining, the lectin WGA-CF680 for membrane staining, and the JF646-Hoechst reversibly binding DNA. The 15 targets were acquired in nine staining rounds (SR) in a dual-color imaging mode with two spectrally discernable channels (Ch) per round. (**c**) Magnified images and overlays of different subsets of stainings: Homer (H), Bassoon (Bass/B), WGA (W), clathrin (C), VGlut1 (V), and Rab3a (R). (**d**) Further zoom-in of the region boxed in the last panel of (b) showing a fragment of a calyx of Held presynapse (area indicated by violet arrows) delineated by the outer membrane (blue-dashed line) and the synaptic membrane (red-dashed line) with the principal cell located to the right of the synaptic membrane. The synaptic contact comprises the presynaptic AZ stained using Bassoon antibody (dashed white box in the second panel) and the postsynaptic density (PSD) stained using Homer antibody (dashed white box in the first panel). The borders of the calyx were approximated based on WGA staining. The presynapse is filled with SVs stained using VGlut1 antibody and Rab3a that transiently binds to SVs. Clathrin-coated vesicles are visualized using CHC17. Scale bars: 5 μm in (a) and 500 nm in (b and c)

3.9 Conclusion

The maS^{3}TORM technique outlined in this chapter has allowed us to visualize the nano-architecture of synapses with unprecedented resolution and multiplexing capabilities. By combining 3D dual-channel *d*STORM imaging with automated solution exchange using a pipetting robot, we have successfully identified more than a dozen of targets within the same presynaptic terminal, generating a comprehensive view of the distribution of synaptic proteins in relation to one another at a nanometer scale. The application of this technique can be used to explore the nano-architecture of synapses or any other cellular protein machines and thereby help gain a better understanding of how complex macromolecular assemblies may drive function.

4 Notes

1. To prevent contaminations, it might be advantageous to include 0.02–0.05% of sodium azide in the PBS solution.
2. We use the Alexa Fluor 488 (AF488)-conjugated lectin wheat germ agglutinin (WGA) that stains specific types of glycosylation, typically resulting in a staining at the plasma membrane, parts of the Golgi, and nuclear pores. Images of WGA-AF488 can be acquired in the wide-field mode using the 488 nm laser before the actual STORM acquisition with the 661 nm laser. Especially the ability of WGA to stain the glycocalyx allows to evaluate tissue and cell quality.
3. The imaging buffer contains thiols (e.g., β-mercaptoethanol), which are reactive to oxygen. Therefore, it is mandatory to prepare the imaging buffer freshly before starting an acquisition series and before every single multiplex round. The basic imaging buffer used for *d*STORM can be further adjusted and optimized by addition of oxygen scavengers or salts. Furthermore, as properties of individual labels and microenvironment can influence the blinking of the fluorophores, we recommend testing all antibodies and probes separately and adjust buffers if necessary before setting up a multiplex experiment.
4. We use fluorescent fiducials for subsequent registration between images from different rounds. Here, it is important to mention that the fiducials ideally should be imaged after the *d*STORM acquisition as this will ensure that most fluorophores used for labeling will be in the dark state or bleached and will not mask fiducial signal.
5. It is necessary to prevent drying out of the sample and at the same time to ensure removal of excess sucrose. If too much sucrose remains at the sample, sucrose crystals could emerge during freezing, thereby damaging the sample.

6. The thickness of the sample can vary depending on the size of the biological region of interest and the used type of the tissue. In general, thicknesses between 100 and 1000 nm are possible. For thicker sections, a cryo-histo knife or a glass knife is recommended, while for thinner sections, a cryo-immuno knife is more favorable. The temperature during sectioning can vary from −70 to −90 °C and should be adjusted with respect to the thickness of the sections. For thicker sections, it is recommended to increase the temperature, while thinner sections can be cut under cooler conditions.
7. A pipette with a wide tip can be used for a bubble-free mixing. Since both solutions are highly viscous, the solutions should be mixed by pipetting up and down at least 20 times.
8. Figure 3 illustrates a multiplex experiment, showing the protocol for a classical immunohistochemistry staining using primary and secondary antibodies. Adjustments need to be done if fluorophore-labeled primary antibodies are used, as a second incubation step is not required in this case. Another helpful probe, which bypasses the usage of antibodies, is directly labeled chemical compounds, e.g., phalloidin specifically binding to polymerized actin or lectins, which stain specific types of glycosylation. Here, again, the second incubation step can be skipped. Please also note that not all antibodies can be efficiently removed with 0.1% SDS pH 13. Moreover, types of labels aside from antibodies can require alternative types of chemical elution.
9. To avoid chemical elution, a typical combination of primary antibody species could be, for example, mouse with rabbit in the first imaging round and chicken with guinea pig in the second imaging round.

Acknowledgments

We thank Mike Heilemann and Frank Herrmannsdörfer for discussions and advice. We gratefully acknowledge the data storage service SDS@hd supported by the Ministry of Science, Research, and the Arts Baden-Württemberg (MWK) through grant INST 35/1314–1 FUGG and INST 35/1503–1 FUGG and the Baden-Württemberg Stiftung MULT!nano (MET-ID13) and ISF2018-024 for their financial support.

References

1. Tang AH, Chen H, Li TP, Metzbower SR, MacGillavry HD, Blanpied TA (2016) A trans-synaptic nanocolumn aligns neurotransmitter release to receptors. Nature 536:210–214. https://doi.org/10.1038/nature19058
2. Margiotta A (2021) Role of snares in neurodegenerative diseases. Cells 10:1–15. https://doi.org/10.3390/cells10050991
3. Heilemann M, van de Linde S, Schüttpelz M, Kasper R, Seefeldt B, Mukherjee A,

Tinnefeld P, Sauer M (2008) Subdiffraction-resolution fluorescence imaging with conventional fluorescent probes. Angew Chem Int Ed Engl 47:6172–6176. https://doi.org/10.1002/anie.200802376

4. Sharonov A, Hochstrasser RM (2006) Wide-field subdiffraction imaging by accumulated binding of diffusing probes. Proc Natl Acad Sci USA 103:18911–18916. https://doi.org/10.1073/pnas.0609643104
5. Jungmann R, Steinhauer C, Scheible M, Kuzyk A, Tinnefeld P, Simmel FC (2010) Single-molecule kinetics and super-resolution microscopy by fluorescence imaging of transient binding on DNA origami. Nano Lett 10:4756–4761. https://doi.org/10.1021/nl103427w
6. Betzig E, Patterson GH, Sougrat R, Lindwasser OW, Olenych S, Bonifacino JS, Davidson MW, Lippincott-schwartz J, Hess HF (2006) Imaging intracellular fluorescent proteins at nanometer resolution. Science 313:1642–1646
7. Khater IM, Nabi IR, Hamarneh G (2020) A review of super-resolution single- molecule localization microscopy cluster analysis and quantification methods. Patterns 1:100038. https://doi.org/10.1016/j.patter.2020.100038
8. van de Linde S, Löschberger A, Klein T, Heidbreder M, Wolter S, Heilemann M, Sauer M (2011) Direct stochastic optical reconstruction microscopy with standard fluorescent probes. Nat Protoc 6:991–1009. https://doi.org/10.1038/nprot.2011.336
9. Möckl L, Moerner WE (2020) Super-resolution microscopy with single molecules in biology and beyond-essentials, current trends, and future challenges. J Am Chem Soc 142:17828–17844. https://doi.org/10.1021/jacs.0c08178
10. Lelek M, Gyparaki MT, Beliu G, Schueder F, Griffié J, Manley S, Jungmann R, Sauer M, Lakadamyali M, Zimmer C (2022) Single-molecule localization microscopy. Nat Rev Methods Primers 1:39. https://doi.org/10.1038/s43586-021-00038-x. PMID: 35663461
11. Sauer M, Heilemann M (2017) Single-molecule localization microscopy in eukaryotes. Chem Rev 117:7478–7509. https://doi.org/10.1021/acs.chemrev.6b00667
12. Endesfelder U, Malkusch S, Fricke F, Heilemann M (2014) A simple method to estimate the average localization precision of a single-molecule localization microscopy experiment. Histochem Cell Biol 141:629–638. https://doi.org/10.1007/s00418-014-1192-3
13. Vaughan JC, Jia S, Zhuang X (2012) Ultra-bright photoactivatable fluorophores created by reductive caging. Nat Methods 9:1181–1184. https://doi.org/10.1038/nmeth.2214
14. Klevanski M, Herrmannsdoerfer F, Sass S, Venkataramani V, Heilemann M, Kuner T (2020) Automated highly multiplexed super-resolution imaging of protein nano-architecture in cells and tissues. Nat Commun 11:1–11. https://doi.org/10.1038/s41467-020-15362-1
15. Huang B, Wang W, Bates M, Zhuang X (2008) Three-dimensional super-resolution reconstruction microscopy. Science (1979) 319: 810–813
16. Lsq S, Mle S, Lsq S, Mle S, Lsq S, Mle S, Mle S (2012) Correspondence rapidSTORM: accurate, fast open-source software for localization microscopy ORCAE: online resource for community annotation of eukaryotes. Nat Methods 9:1040–1042
17. Seibenhener ML, Wooten MW (2012) Isolation and culture of hippocampal neurons from prenatal mice. J Vis Exp 4–9. https://doi.org/10.3791/3634
18. Gage GJ, Kipke DR, Shain W (2012) Whole animal perfusion fixation for rodents. J Vis Exp 1–9. https://doi.org/10.3791/3564
19. Tokuyasu KT (1973) A technique for ultracryotomy of cell suspensions and tissues. J Cell Biol 57:551–565. https://doi.org/10.1083/jcb.57.2.551
20. Ovesný M, Křížek P, Borkovec J, Švindrych Z, Hagen GM (2014) ThunderSTORM: a comprehensive ImageJ plug-in for PALM and STORM data analysis and super-resolution imaging. Bioinformatics 30:2389–2390. https://doi.org/10.1093/bioinformatics/btu202
21. Ries J (2020) SMAP: a modular super-resolution microscopy analysis platform for SMLM data. Nat Methods 17:869–870. https://doi.org/10.1038/s41592-020-0938-1

Chapter 10

Detailed Phenotyping of the Physiology, Structure, and Connectivity of Individual Neocortical Neurons

Maria Medalla and Jennifer I. Luebke

Abstract

In this chapter we describe contemporary methods employed to characterize the detailed features of individual neocortical pyramidal neurons in the mammalian brain. First, a brief overview is provided of widely used approaches to study the intrinsic properties of neurons, including their morphology, physiology, and transcriptomic profile. Next, we describe in detail multi-modal methods that we and others have developed and employ to quantify both these intrinsic features and *extrinsic* features of identified pyramidal neurons such as presynaptic glutamatergic and GABAergic inputs and relationship to neuropil elements such as glial cells and vasculature. The use of this multi-modal approach enables a more comprehensive, nuanced, and detailed characterization of neuron types and the circuits to which they contribute.

Key words Pyramidal neuron, Morphology, Physiology, Transcriptomics, In vitro slice, Immunohistochemistry, Multiplexed immunohistochemistry, Synaptic inputs, Multi-modal

1 Introduction: Diversity of Cortical Pyramidal Neurons

Since the beginning of the modern neuroscience era, scientists have sought to understand the degree to which neurons can be categorized into distinctive groups. Classification of neurons was initially based exclusively on somato-dendritic morphological features; however neocortical pyramidal neurons differ only subtly at the somato-dendritic structural level, and this led to the common view that—unlike interneurons—pyramidal neurons were a relatively homogeneous population. With the advent of patch-clamp recordings in brain slices, detailed physiological properties were included as variables in classification schemes, opening the door to an enhanced understanding of neuronal diversity. Over the past decade, major methodological advances—particularly those incorporating molecular and genetic approaches—have become avail-

Joachim H. R. Lübke and Astrid Rollenhagen (eds.), *New Aspects in Analyzing the Synaptic Organization of the Brain*, Neuromethods, vol. 212, https://doi.org/10.1007/978-1-0716-4019-7_10,

able. These methods now enable researchers to demonstrate that cortical pyramidal neurons are by no means homogeneous but instead exist in many phenotypically distinct classes based on their diverse *intrinsic* features—including somato-dendritic *and* detailed axonal morphology, intrinsic physiology, and transcriptomic and epigenomic profile. While intrinsic features are clearly fundamental to a given neuron's phenotype, less commonly examined *extrinsic* features such as synaptic inputs and relationship to neuropil micro-environments may also define distinctive neuronal phenotypes and classes.

In this chapter, we describe multi-modal methods that we and others have developed and employ to quantify characteristic *extrinsic* features of identified pyramidal neurons such as presynaptic glutamatergic and GABAergic inputs and relationship to neuropil elements such as glial cells and vasculature. The use of this multi-modal approach reveals both intrinsic features *and* extrinsic features of neurons (Fig. 1), enabling a more comprehensive, nuanced, and detailed characterization of neuron types and the circuits to which they contribute.

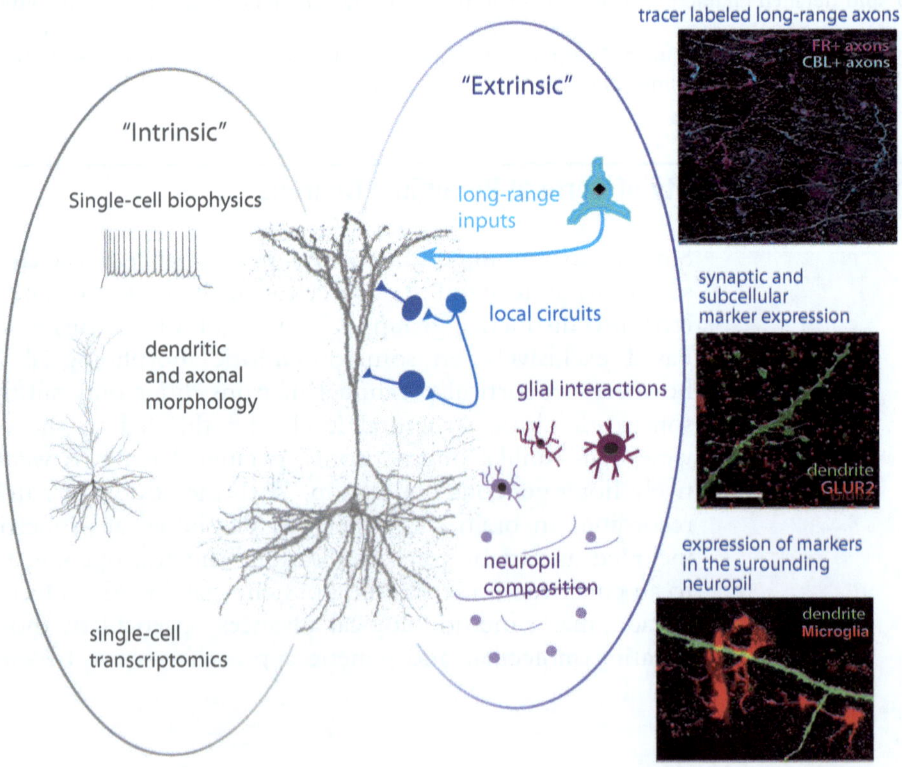

Fig. 1 Summary of the multi-modal intrinsic and extrinsic features that contribute to neuronal diversity

2 Methods to Characterize the Intrinsic and Extrinsic Features of Neurons

2.1 Somato-dendritic Morphology and Intrinsic Physiology

The existence of diverse classes of neurons has been recognized since the groundbreaking work of Ramon and Cajal, who employed the Golgi silver impregnation method to describe many neuron types in diverse brain regions and species based on their somato-dendritic morphology alone. Cajal documented the presence of pyramidal projection neurons in layers 1–6 of the neocortex that differed primarily in the size of their somata and dendritic arbors [1–4]. These and subsequent studies revealed that neocortical pyramidal neurons have a relatively homogeneous stereotypical shape and dendritic topology, particularly when compared to GABAergic interneurons. That said, pyramidal neurons do differ in size and density of dendritic spines depending on species, layer, and brain area (e.g., pyramidal neurons are smaller in L2 compared to L3; pyramidal "Betz" cells in L5 motor cortex are very large; and pyramidal cells in primary visual cortex are smaller and have less dense spines compared to those in the lateral prefrontal cortex) [1–3, 5–14]. While providing some hints about pyramidal neuron diversity, clearly the morphological features of the somato-dendritic compartments of pyramidal neurons alone are not distinctive enough to allow for comprehensive subclassification of these projection neurons. The development of patch-clamp recording combined with simultaneous cell filling in in vitro slices enabled researchers to incorporate distinct physiological features—such as rhythmic oscillating bursting versus regular action potential spiking behavior—into the classification of pyramidal neuron subtypes [e.g., 15–17]. Other studies have subcategorized pyramidal cells based on their morphological features—including local axonal arborization patterns—and their responses to sensory stimuli as in the mouse barrel cortex [e.g., 18]. These physiological methods, which elaborate on the initial descriptive structural studies by incorporating functional differences between pyramidal neurons, began to make it clear that pyramidal neurons are not generalizable across layers and regions but are in fact at least as diverse as inhibitory interneurons. Such structure–function studies are most commonly performed in rodents [18–25] but recently have been extended to non-human primate brain areas [e.g., 10, 12, 13, 26–28] and human brain tissue [29–34].

2.2 Axonal Morphology and Target Specificity

GABAergic interneurons, which possess relatively local axonal arbors, are categorized based not only on their neurochemical and somato-dendritic morphological features but also on their highly distinctive axonal ramification topologies and postsynaptic targets [35–41]. By contrast, pyramidal projection neurons in the neocortex possess extensive long-range axonal projections that until recently were not accessible to the same kind of characterization at scale because axons were typically severed in brain sections or not discernable over long distances. Thus, while it had been postulated that pyramidal neurons very likely are comprised of distinct subpopulations based on axonal targets and complete arbor topology [reviews 42, 43–45], until recently there was scant empirical evidence for this. That said, a number of excellent studies were able to demonstrate neuronal diversity based on differences in relatively local axonal anatomy in the mouse barrel field [18–25]. Only a few studies managed to reconstruct the complete morphology of individual neurons with the approach of cell filling, serial sectioning of the entire brain, and painstaking reconstruction of the entire neurons including their long-ranging axons across many serial sections [46–52]. The advent of sparse labeling of genetically defined neurons has opened a major new avenue for the characterization of neurons based on detailed axonal morphology. By injecting a low titer of AAV-Syn-iCre and a high-titer Cre-dependent reporter (eGFP or tdTomato), researchers are able to sparsely label neuronal populations in the mouse brain [53–59] for subsequent high-resolution imaging of the intact whole brain with two-photon imaging. In one such "MouseLight" experiment, Winubst et al. reconstructed 1000 projection neurons in the mouse motor cortex, thalamus, subiculum, and hypothalamus, revealing a great variety of target-specific neurons with distinctive axonal morphologies. But the question remains whether neurons with distinct axonal arbors and projecting to distinct target areas represent intrinsically different neuronal classes, per se. To address this question, parallel studies using retrograde tracing approaches have been used to label target-specific projection neurons and demonstrate diverse intrinsic and functional features based on target region [e.g. 28, 53, 54, 60].

2.3 Transcriptomic Analyses and Large-Scale Multi-modal Approaches

Over the past decade, the use of molecular approaches for single-cell RNA sequencing such as Drop-Seq, SMART-Seq, and Patch-Seq has opened new vistas for the elucidation of the diversity of neuron types based on their transcriptomic profile, often in combination with physiological and in some cases somato-dendritic morphological features. For example, Tasic and coworkers [60] employed single-cell RNA sequencing to reveal 19 distinct types of glutamatergic neurons in the adult mouse primary visual cortex. Using a similar approach, Zhu et al. found ten distinct clusters of excitatory neurons in the rhesus monkey PFC [61]. An open question remains whether and which transcriptomic distinctions map on to protein expression differences that may (or may not) be signatures of actual subclasses of pyramidal neurons.

Many research groups and institutes are currently using a combination of methods such as those described above in an effort to comprehensively characterize neuronal subtypes based on multimodal data on intrinsic properties. Perhaps the most large-scale effort currently underway is the US National Institutes of Health BRAIN Initiative Cell Census Network that began in October 2017 with the aim of cataloguing neuron types in human, monkey, and mouse brain. The recent publication of a comprehensive mapping of mouse primary motor cortex cell types used a multimodal assessment of single-cell transcriptomes, chromatin accessibility, DNA methylomes, morphological characteristics, and electrophysiological properties to create a framework of neuronal cell-type organization in this brain region (BICCN, 2021).

2.4 Inclusion of Information on Extrinsic Factors in Multi-modal Neuronal Profiling

The development of remarkable new molecular and anatomical tools has brought us to an exciting period in the study of cellular diversity of the brain that would have delighted Ramon and Cajal. What is emerging from this work is that there are many more cortical cell types discriminable based on their intrinsic features than previously appreciated. It is important to note however that the application of clustering algorithms on multi-modal datasets can generate many—indeed almost infinite—clustering schemes dependent on the variable set. New important questions have arisen as a consequence: where does one draw the line in defining a unique "cell type"? and which phenotypic details and cell classification schemes are relevant to the function of brain circuits? What has become clear from the current body of work is that it is the interactions of intrinsic and extrinsic factors which dictate *functional specialization* of known or proposed distinct cell types. For instance, the emergence of a specific axonal morphology of a given cell type is dictated by intrinsic genetic cues within the cell but also in response to extrinsic axon guidance cues that allow for connections and circuits to form.

Most studies seeking to unravel a realistic picture of neuronal diversity have focused exclusively on the intrinsic morphological, physiological, and transcriptomic features. However, superimposed upon a given neuron's intrinsic features are an array of extrinsic features which arguably provide an additional layer of information that should be taken into account in cell-type taxonomy schemes. Recent development studies have investigated this avenue and revealed that the process of functional phenotypic differentiation of a neuron is a product of both its intrinsic genetic "blue print" and how it is established within the extrinsic environment [62–65]. Furthermore, both morphological and physiological properties of neurons at any given point in time are heavily influenced if not determined by extrinsic synaptic inputs and even relationship to glial cells [66–70].

Relative to intrinsic features, far less information is available on the extrinsic features of neurons such as presynaptic glutamatergic and GABAergic inputs and relationship to neuropil elements such as glial cells and vasculature. In this chapter we describe multimodal methods that we and others have developed and employ to

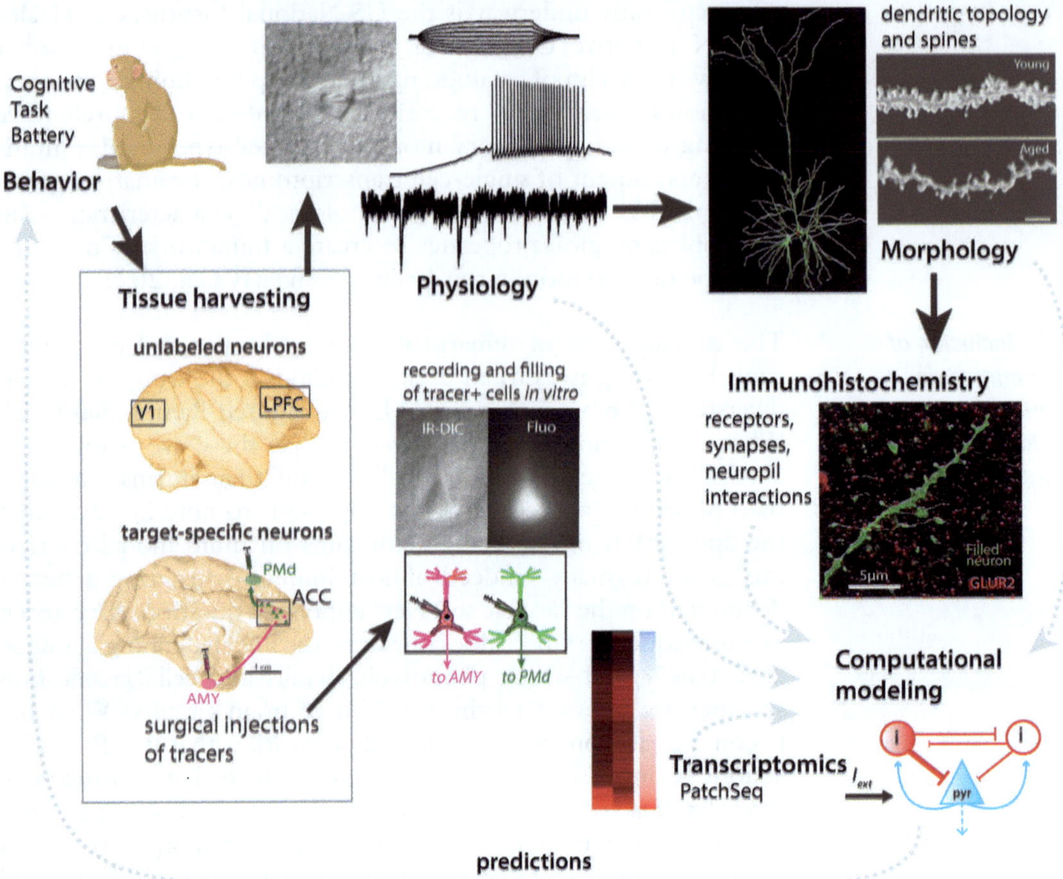

Fig. 2 Experimental design of characterization of single neuron intrinsic and extrinsic structural and functional properties

quantify characteristic *extrinsic* as well as intrinsic features of single-identified pyramidal neurons in distinct areas of the mammalian neocortex (Figs. 1 and 2).

3 Materials

3.1 Materials for In Vivo Labeling of Target, Developmental Lineage, or Genetically Specified Neurons

In many experiments, it is advantageous to label specific populations of neurons with fluorescent dyes in vivo in order to visualize them in subsequent in vitro recordings. Examples are as follows. In studies of target-specific neurons in rhesus monkeys, we have used retrograde tracers to label and then compare anterior cingulate pyramidal neurons that project to the amygdala (retrogradely labeled with FluoroRuby) versus those that project to the premotor cortex (retrogradely labeled with FluoroEmerald) [28] (*see* Figs. 2 and 3a–c). In studies of lineage-specific neurons in the mouse frontal [71] and somatosensory [72] cortices in the mouse, we

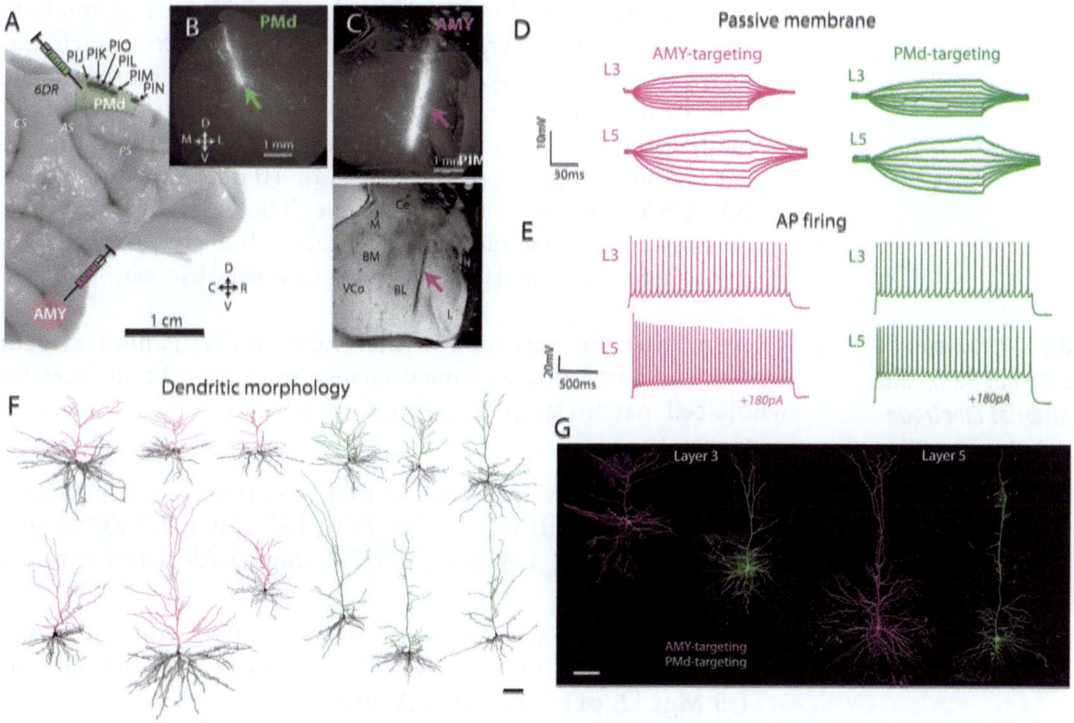

Fig. 3 Tract-tracing methods to assess single-cell properties of target-specific neurons. (**a**) Map of tracer injection sites in dorsal premotor cortex and the amygdala, shown in the lateral surface of the rhesus monkey brain. (**b**) Fluorescent image of coronal section through tracer injection site in PMd. (**c**) Fluorescent (top) and brightfield (bottom) image of coronal section through tracer injection site in basolateral AMY. Abbreviation of nuclei: Ce central, M medial, BM basomedial, BL basolateral, L lateral, VCo ventral cortical. (**d** and **e**) Example traces of biophysical responses of AMY- and PMd-targeting projection neurons in layer 3 (L3) and layer 5 (L5) of ACC: (**d**) Voltage responses to a series of 200 ms current pulse showing differences in Rn. (**e**) Action potential (AP) voltage firing responses to a 2 s + 180pA current step. (**f** and **g**) Morphological properties of AMY- and PMd-targeting projection neurons in L3 and L5 of ACC: (**f**) Representative 3D reconstructions of intracellularly target-specific projection neurons. (**b**) Z-maximum projections of confocal image stacks of representative pyramidal neurons. (Panels adapted and modified from [28])

have employed in utero genetic fate mapping using plasmids to compare neurons from Tbr2+ versus Tbr2- lineages. In studies of direct and indirect pathway medium spiny neurons and of parvalbumin positive neurons in the mouse striatum, we have recorded from neurons genetically tagged with GFP as D2 receptor containing or with tdTomato as parvalbumin containing [73, 74].

For the purposes of this chapter, we provide detailed methods for immunohistochemistry on biocytin-filled Alexa dye-labeled neurons with or without a tracer label in the rhesus monkey (Figs. 2 and 3). For details regarding assessment of neurons labeled with specific other methods in different brain areas and in the rodent, we refer the reader to our previous publications:

- *Bidirectional neural tracers* [28]: Used to identify afferents to and efferents from pyramidal neurons in diverse cortical circuits in the rhesus monkey. Dextran amine tracers, FluoroEmerald (FE, dextran fluorescein) and FluoroRuby (FR, dextran tetramethylrhodamine; Life Technologies), are injected into the target regions using a micro syringe (5 or 10 μL; Hamilton, Reno, NV, USA) mounted on a microdrive. These 3KDA isoforms of dextran amines are bidirectional tracers but are enhanced for retrograde transport ([75, 76] and personal observations).

3.2 Perfusate, External Bath, and Internal Electrode Solutions for Slice Preparation and Recordings

They are used for perfusion of the brains of rhesus monkeys, for fresh tissue harvesting and maintenance of in vitro brain slices for whole-cell patch-clamp recordings of labeled and/or unlabeled neurons:

- *Perfusate:* Krebs–Henseleit buffer solution; concentrations, in mM: 6.4 Na_2HPO_4, 1.4 Na_2PO_4, 137 NaCl, 2.7 KCl, 5 glucose, 0.3 $CaCl_2$, 1 $MgCl_2$; pH 7.4 (Sigma-Aldrich, St. Louis, MO, USA).
- *External recording solution:* Ringer's buffer, concentrations, in mM: 26 NaHCO3, 124 NaCl, 2 KCl, 3 KH2PO4, 10 glucose, 1.3 MgCl2, pH 7.4 (Sigma-Aldrich).
- *Patch pipette internal solution:* Potassium methane sulfonate-based solution (concentrations, in mM: 122 KCH_3SO_3, 2 $MgCl_2$, 5 EGTA, 10 NaHEPES, with 1% biocytin, pH 7.4; Sigma-Aldrich).

3.3 Materials for Visualizing the Somato-dendritic Structure of Neurons Filled with Biocytin During Recordings

- Following recordings slices are fixed 4% paraformaldehyde in 0.1 M phosphate-buffered saline (PBS; pH = 7.4). Subsequently they are incubated for 2 h in 1% Triton X-100 (in 0.1 M PBS), followed by 48 h in streptavidin-Alexa 405; streptavidin-Alexa 488, or streptavidin-Alexa 647 (diluted 1: 500 in 0.1 M PBS; Thermo Fisher).

3.4 Materials for Immuno-histochemistry on Identified Biocytin-Filled, Alexa-Labeled Neurons

To label inhibitory or excitatory terminals closely apposed to identified filled neurons, microwave-assisted immunohistochemistry is employed:

- *Antigen retrieval solution:* 50 mM glycine in 10 mM sodium citrate buffer.
- *Pre-block solution:* 5% normal goat serum (NGS) and 5% bovine serum albumin (BSA) with 0.1% Triton X-100.
- *Primary antibody solution:* In the example of GABA inputs, we stained for the vesicular GABA transporter VGAT (1:400, guinea pig polyclonal, catalog #131004, RRID:AB_887873; Synaptic Systems) and GluR2 (1:400, information included, Millipore) diluted in 1% NGS, 0.2% acetylated BSA (BSA-c, Aurion), and 0.1% Triton X-100.

- *Secondary antibody solution:* Anti-VGAT was labeled using a biotinylated goat anti-guinea pig F(ab) IgG secondary (1:200; Jackson ImmunoResearch Laboratories), and anti-PV was labeled using Alexa Fluor 488-conjugated goat anti-mouse F (ab)2 IgG secondary (1:200; Invitrogen).
- *Mounting medium:* Sections were mounted on glass slides and cover-slipped with ProLong anti-fade medium (Invitrogen).

4 Methods

The overall experimental approach is to assess relationships between neurochemical, physiological, connectional, and morphological features of neurons at the subcellular and single-cell levels of resolution in diverse cortical areas of behaviorally characterized rhesus monkeys across the adult lifespan (Fig. 2). Empirical structural and functional data are incorporated into computational models of networks, to predict the implications of age-related changes for network function and ultimately for cognition.

4.1 Injections of Neural Tracers to Label Distinct Projection Neurons in the Monkey Cerebral Cortex

The combination of neural tracer injections with subsequent recordings of identified target-specific neurons in vitro slices has been a powerful tool in understanding distinctive properties of diverse groups of projection neurons in mice [e.g., 55, 77, 78]. We have similarly used neural tracer injections in rhesus monkeys for the first time to investigate neurons in the anterior cingulate cortex that project to the amygdala versus premotor cortex [28] (Fig. 3). We employ survival surgical procedures to inject two different neural tracers in stereotaxic-based injection sites, as described previously [28 2009]. All procedures are reviewed and approved by the Institutional Animal Care and Use Committee at Boston University School of Medicine.

Rhesus monkeys undergo a structural T1-weighted MRI scan prior to surgery for calculation of stereotaxic coordinates of injection sites. Injection sites within the rostral dorsal premotor cortex (PMd) and basolateral amygdala (AMY) are calculated from MRI scans using the midline and betadine-filled ear bar tips as reference points. Dextran amine tracers, FluoroEmerald (FE, dextran fluorescein) and FluoroRuby (FR, dextran tetramethylrhodamine; Life Technologies), are injected into the premotor cortex and amygdala, respectively, with a micro syringe (5 or 10 μL; Hamilton, Reno, NV, USA) mounted on a microdrive. For cortical PMd injections, the dye is delivered in wo to three penetrations (spaced 0.5 mm apart) at a depth of 1.2–1.6 mm below the pial surface to deliver a total of 4–6 μL of tracer. For AMY injections, 3 μL of tracer is delivered into the basolateral complex (Fig. 3). At each injection site, the needle is left in place for 5 min to allow for diffusion and to prevent upward suction of the dye upon retraction of the needle. Accuracy of injection sites are verified histologically based on previous cytoarchitectonic maps, as described [28].

4.2 Perfusion and Preparation of Acute Brain Slices

Following a post-injection survival period of 18–21 days, harvesting of acute brain slices for in vitro recordings is performed using our standard two-stage Krebs–Paraformaldehyde prefusion method as described [10]. Subjects are sedated with ketamine hydrochloride (10 mg/kg), deeply anesthetized with sodium pentobarbital (to effect, 15 mg/kg, i.v.), and then perfused through the ascending aorta with ice-cold Krebs–Henseleit buffer. A 10-mm^3 block of tissue is taken from the area of interest (e.g., LPFC, ACC, PMd, or V1) and sectioned with a Leica vibrating microtome into 300-μm-thick coronal slices in ice-cold Ringer's solution. Slices are immediately placed into room temperature, oxygenated (95% O_2, 5% CO_2) Ringer's solution, and equilibrated for 1 h.

4.3 Whole-Cell Patch-Clamp Recording of Target-Specific and Unlabeled Neurons

Following the equilibration period, individual slices are placed into submersion-type recording chambers (Harvard Apparatus, Holliston, MA, USA) mounted on the stages of Nikon E600 infrared-differential interference contrast (IR-DIC) microscopes. Slices are continuously superfused with room temperature, oxygenated Ringer's solution at a rate of 2–2.5 mL/min, and physiological experiments are performed at room temperature (~26 °C) which improves the viability and duration of recordings from monkey. Indeed, rhesus monkey cortical slices remain viable for more than 24 h under these conditions. Standard tight-seal, whole-cell patch-clamp recordings with simultaneous biocytin filling are obtained from pyramidal cells as described in our publications [10, 79–81] (*see* Figs. 2 and 3b,c). Neurons are first visualized under epifluorescence illumination to identify either red (amygdala projecting) or green (premotor cortex projecting) retrogradely labeled neurons. Next, optics are switched to IR-DIC for recording with patch electrodes fabricated on a horizontal Flaming and Brown micropipette puller (Model P-87, Sutter Instruments, Novato, CA, USA). A potassium methanesulfonate-based internal solution is used as internal solution in electrodes with resistances of 3–6 megaOhm. Data are acquired using EPC-10 patch-clamp amplifiers (HEKA Elektronik, Lambrecht, Germany), using PatchMaster acquisition software (HEKA Elektronik). Series and access resistance are both monitored frequently during the 15-min recording period. Bessel filter frequency is 10 kHz, and sampling frequency is at 7 kHz for voltage clamp and 12 kHz for current clamp recordings. For cell inclusion in electrophysiological analyses, cells are required to have a resting membrane potential ≤ -55 mV, stable access resistance, an AP overshoot, and the ability to fire repetitive action potentials in response to prolonged depolarizing current steps [10, 80–82].

4.4 Visualization and Identification of Putative Inputs and Glial Appositions onto Filled Neurons

Following recordings slices containing biocytin-filled neurons are sandwiched between two pieces of filter paper and immersion fixed in 4% paraformaldehyde. To confirm retrograde tracer label, amplification of tracer label was first employed using immunohistochemical methods on the slices as described in our publications [26, 28] (Fig. 4).

Immunohistochemistry: For antigen retrieval slices are incubated at room temperature in glycine and then treated with 10 mM sodium citrate buffer. Next, slices are pre-blocked for 2 h in normal donkey serum and bovine serum albumin. Sections are incubated in the primary antibody (e.g., against neurotransmitter transporters or microglial cells) for over 7 days at 4 °C with mild agitation, with eight interspersed 10 min 150-watt microwave sessions at 30 °C (BioWave, Ted Pella Inc., Redding, CA) to assist in antibody penetration. This is followed by an incubation in Alexa-conjugated secondary antibody for 4 days (4 °C, with 4× 10 min 150 W, at 30 °C biowave sessions). *For tracer amplification*, slices are then incubated in the primary antibodies against FluoroEmerald (1:800 mouse monoclonal or goat polyclonal, Life Technologies) or FluorоRuby (1:800, rabbit polyclonal, Life Technologies Medalla et al., 2021). Slices are then incubated in a mixture of donkey anti-rabbit secondary F(ab')2 IgG conjugated to Alexa 568 (1:200, Life Technologies) to amplify FluoroRuby and donkey anti-mouse F(ab') 2 IgG conjugated to Alexa 488 (1:200, Jackson ImmunoResearch Laboratories) to amplify FluoroEmerald. Finally, slices are rinsed in PB and then mounted in glycerol with a temporary cover-slip for imaging of somata at the confocal to confirm tracer label. To assess receptor distributions within filled dendrites and spines, slices are incubated in primary antibody (such as a monoclonal primary antibody against the N-terminal extracellular domain of the GluR2 subunit 1:500; anti-GluR2 mouse monoclonal, catalog # MAB397, Millipore). This is followed by incubation in secondary antibody (such as donkey anti-guinea pig F(ab')2 IgG conjugated to Alexa 563 or 647 secondary antibody, Jackson ImmunoResearch Laboratories) [12]. To assess inhibitory terminals closely apposed to filled neurons, slices are incubated in the primary antibody against vesicular GABAergic transporter (VGAT; 1:400, guinea pig polyclonal, Synaptic Systems Cat# 131004; RRID: AB_887873) followed by secondary donkey anti-guinea pig F (ab')2 IgG conjugated to Alexa 647 (Jackson ImmunoResearch Laboratories) [12]. After labeling, slices are mounted on glass slides and cover-slipped with Prolong anti-fade medium (Life Technologies).

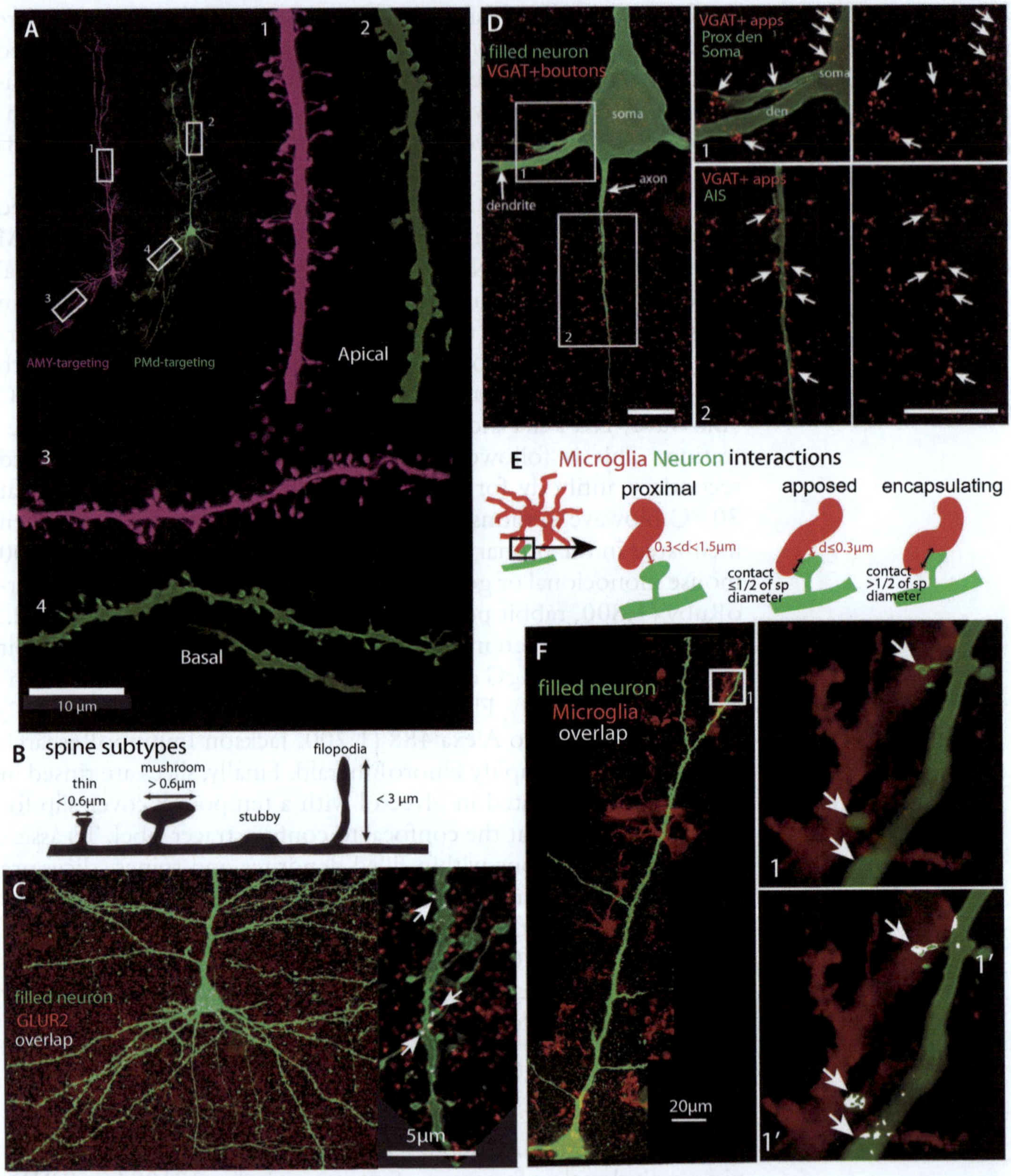

Fig. 4 High-resolution structural analyses of dendritic compartment-specific putative synaptic inputs and neuropil interactions. (**a**) High-resolution subsampling of apical and basal dendritic segments of filled target-specific (AMY targeting, magenta vs PMd-targeting, green) pyramidal neurons to study spine density and subclasses. (**b**) Schematic of criteria for spine subtype classification. (**c**) Examples of filled neuron co-labeled with GLUR2. High-mag inset showing GLUR2 + -labeled dendrites and spines (white arrows/overlay indicate overlapping signal). Panels adapted and modified from [12]. (**d**) Examples of filled neuron co-labeled with VGAT+ to label inhibitory boutons. High-mag inset showing VGAT+ boutons apposed (white arrows) to proximal dendrites and soma (panel 1) and axon initial segment (AIS, panel 2) of filled neurons to estimate putative inputs. Panels adapted and modified from [26]. (**e** and **f**) Filled neuron interactions with microglia labeled with P2RY12 + IBA1+: (**e**) Schematic of criteria for determining microglia–neuron interactions, based on the distance (d) and the size of contact between the microglia process and the filled neuronal spine/dendrite, as described in our previous work in mice [87]. (**f**) Montage scan of apical dendrite of filled neuron

4.5 Imaging and Analysis of Putative Inputs to Different Morphological Compartments of Pyramidal Neurons

Confocal image stacks of filled neurons labeled with Alexa 405 or 488-conjugated streptavidin are acquired using a using a Zeiss 710 confocal laser-scanning microscope (Zeiss, Jena, Germany). For experiments with both green and red tracers, neurons are visualized with streptavidin-Alexa 405, and each neuron is first classified as tracer labeled or unlabeled, using confocal imaging of the soma (Fig. 3d,e). Colocalization of green or red tracer labeling with biocytin-Alexa 405 is confirmed by multi-channel imaging using 405 nm diode, 488 nm argon, and 561 nm diode excitation laser lines. Image stacks through the entire z-extent of filled somata are acquired using a 40×/1.3 NA oil-immersion objective (210 μm working distance; Plan-Apochromat, Zeiss) at a resolution of 0.1 0.1 1.0 μm per voxel.

Neurons with complete somata and dendritic fills with no cut branches in the proximal third of the apical dendritic tree are selected for high-resolution imaging of dendrites. An image stack montage encompassing the whole biocytin-Alexa 405-filled cell is scanned at a resolution of 0.3 0.3 0.3 μm per voxel, using a 405 laser and a 40×/1.3 NA oil-immersion objective (210 μm working distance; Plan-Apochromat, Zeiss).

Individual image stacks are deconvolved using AutoQuant (Media Cybernetics, Bethesda, MD, USA), then stitched, and montaged using Volume Integration and Alignment System (VIAS) software [83] (Fig. 3d,e). Neurons are reconstructed in 3D using NeuronStudio [[83, 84]; RRID:SCR_013798] or Neurolucida software, and dendritic length, branching topology, and vertical and horizontal extents were quantified for apical and basal arbors, as described [26].

Neurons with high-quality filling are then assessed for filled spines and dendritic appositions with immunolabeled VGAT+ boutons (Fig. 4a–d). After whole-cell dendritic scanning, a second series of image stacks is acquired at a resolution of 0.04 × 0.04 × 0.3 μm per voxel via two-channel confocal imaging with 405 diode (for the filled neuron) and 630 Helium–Neon (for VGAT boutons) lasers and a 63×/1.4 NA oil-immersion objective (UPlanFL, Olympus or Plan-Apochromat, Zeiss). For each neuron, confocal stacks within 100 μm from the tissue surface are acquired to image one complete basilar dendritic branch, and the main apical trunk followed to the end of one complete distal apical dendritic branch, as described [26]. Image stacks are deconvolved (as above) and tiled and analyzed using Neurolucida (MBF Bioscience; RRID: nif-0000-10294).

Fig. 4 (continued) and P2RY12 + IBA1 + microglia in the surrounding neuropil of rhesus monkey PMd cortex (left). High-mag insets (right panels) showing microglia processes encapsulating/apposed to filled spines and dendrites (white arrows). White arrows indicate neuron–microglia appositions validated by red–green signal overlap in panel 1, marked as white overlay mask in panel 1′

Spines Spines along the imaged dendrite are counted and classified into subtypes as follows [10, 12, 26, 28]. Image stacks of dendritic segments obtained with a 63× or 100× 1.4 NA oil-immersion objectives are used to quantify spine subtypes and measure the size of spine heads (widest part of spine head perpendicular to the neck; Fig. 4a,b). Dendritic segments are reconstructed in 3D using the 64-bit version of NeuronStudio [83]. Dendritic spines are then manually marked on the reconstructed dendrites and automatically measured for the maximum spine head width perpendicular to the neck and maximum distance from the shaft, using the Rayburst-based algorithm [83]. Spines are manually classified into subtypes according to the following criteria [for review, *see* 10, 85]: spines with a head width of 0.6 μm are classified as either thin or filopodia, with thin spines having a neck length of 3 μm and filopodia having a neck length of >3 μm. Spines with a head width of >0.6 μm are classified as mushroom, and spines lacking a neck are classified as stubby (Fig. 4b). Densities of total spines and of each subtype are calculated as the number of spines per micron of dendritic length. The resulting measured spine head widths for each cell are then used to assess the size of spines, and K-means cluster analysis was used to separate the population of "large" versus "small" spines, as described [12]. Sholl analysis using concentric spheres placed at 20 μm increments and originating at the soma [86] is used to determine the number and mean maximum spine width of total, small, and large spines at different distances from the soma.

Receptor labeling in spines Filled dendrites are imaged under the multitrack mode, using an argon laser, and receptors [such as GluR2 receptors in [12]) are imaged with either a helium–neon 543 nm laser (red) or 633 nm laser (far-red). The channels are merged as an RGB stack in ImageJ (http://imagej.nih.gov/ij/; Rasband, WS, 1997–2014, National Institutes of Health, Bethesda, Maryland), and colocalization of immunolabel with spines is assessed using a manually set threshold for each channel [12] (Fig. 4c). Spines with colocalized points are marked and counted using either NeuronStudio or Neurolucida (MicroBrightField).

Neurotransmitter transporter-specific and glial appositions Appositions between the neurons and transporter-specific puncta (such as VGAT+ or Vglut1 or 2+) are classified as either apposed to dendritic shafts, spine head or neck of each subtype, or somata. An apposition is identified as a spatial overlap between the saturated red and green signals greater than >2 pixels in the x–y (~ >0.1 × 0.1 μm), >2 optical slices in the z (~ >0.4 um) for thin dendrites, or >4 optical slices in the z (~ >0.8 um) for highly saturated segments of thick dendrites and somata, as described [26] (*see* Fig. 4d, adapted from [26]). Densities of VGAT+ appositions on dendrites and spines are calculated as the number per micron dendritic length. Appositions on somata are quantified as number per surface area somata. Profiles of somata are automatically traced section by section, and surface area of somata is calculated by the Neurolucida software as a series of cylindrical sections capped by the end profiles (surface area per cylinder = perimeter of the profile × distance to the next profile). Sholl analyses with concentric spheres placed at 20 μm increments from the center of the soma [86] are employed to determine the distribution of the spine and VGAT+ appositions along subsampled apical and basal dendrites as a function of distance from the center of the soma. We have employed similar methods to assess neuronal interactions with microglia in our collaborative work in mice [87], which we are now applying to study neuronal interactions with glial subtypes and other non-neuronal elements of the neuropil in the rhesus monkeys (Fig. 4e,f). Appositions between the neurons and P2RY12 + IBA1+ microglia are classified as follows: encapsulation (within 0.3 μm; one z-plane, and microglia is surrounding at least ½ the spine head diameter), apposition (within 0.9 μm; three z-planes, and microglia is surrounding less than ½ the spine head diameter), and proximity (within 1.5 μm; five z-planes) to microglia (Fig. 4e). Distances are measured from the most saturated three pixels of the neuron and microglia within the same optical slice, which allows 500 nm resolution at the ×40 oil-immersion objective's NA 1.3.

5 Conclusions and Caveats

The use of multi-modal approaches such as those described in this chapter reveals characteristic features of individual neocortical pyramidal neurons at an unprecedented level of detail. As the key building blocks of neocortical circuits that endow distinct brain areas unique capabilities, it is vital that we gain as detailed an understanding of these neurons as possible. Certainly, every scientific approach has strengths and weaknesses that must be clearly recognized and incorporated into our thinking about the resulting data. The multi-modal approach that we employ enables elucidation of both intrinsic and extrinsic features of neurons in the non-human primate neocortex. It is important to emphasize that this approach is based on assessment of neuronal properties in an ex vivo preparation, namely, in vitro cortical slices. Obviously in such slices, which are typically 300 μm thick, long-range connections are severed, and shorter range connections are likely also truncated, as are the dendrites and axons of individual neurons within the slice. Furthermore, the cutting procedure itself renders in vitro slices damaged as a result of neuronal hyperexcitability and activation of inflammatory responses. Since both neurons and glial cells are highly plastic and rapidly responsive to insults such as anoxia and lesion, it is to be expected that data obtained from slices—even after a significant recovery period after cutting—is not identical to what exists in vivo. That said, the study of neurons in in vitro slices does provide an excellent approximation of many neuronal features that can in the future be validated with studies in the intact brain.

References

1. Cajal R (1894) Consideraciones Generales Sobre la Morfología de la Célula Nerviosa. Moya, Madrid
2. Cajal R (1904) Textura del Sistema Nervioso del Hombre y de los Vertebrados. Moya Madrid 1899
3. Cajal R (1933) Neuronismo o reticularismo Las pruebas objetivas de la unidad anatómica de las células nerviosas. Arch Neurobiol 13(217–91):579–646
4. DeFelipe JJ, Jones EG (1988) Cajal on the Cerebral Cortex. Oxford University Press, New York
5. Conel JL (1941) Postnatal development of the human cerebral cortex: the cortex of the one-month infant. Harvard University Press, Cambridge, MA
6. Conel JL (1967) Postnatal development of the human cerebral cortex: the cortex of the seventy-two-month infant. Harvard University Press, Cambridge, MA
7. DeFelipe J, Alonso-Nanclares L, Arellano JI (2002) Microstructure of the neocortex: comparative aspects. J Neurocytol 31(3–5):299–316. Epub 2003/06/20. https://doi.org/10.1023/a:1024130211265
8. Elston GN, DeFelipe J (2002) Spine distribution in cortical pyramidal cells: a common organizational principle across species. Prog Brain Res 136:109–133. Epub 2002/07/30. https://doi.org/10.1016/s0079-6123(02)36012-6
9. Elston GN (2003) Cortex, cognition and the cell: new insights into the pyramidal neuron and prefrontal function. Cereb Cortex 13(11):1124–1138. Epub 2003/10/25. https://doi.org/10.1093/cercor/bhg093
10. Amatrudo JM, Weaver CM, Crimins JL, Hof PR, Rosene DL, Luebke JI (2012) Influence of highly distinctive structural properties on the excitability of pyramidal neurons in monkey visual and prefrontal cortices. J Neurosci 32(40):13644–13660. Epub 2012/10/05. https://doi.org/10.1523/JNEUROSCI.2581-12.2012. PubMed PMID: 23035077; PubMed Central PMCID: PMCPMC3485081
11. Mohan H, Verhoog MB, Doreswamy KK, Eyal G, Aardse R, Lodder BN, et al (2015) Dendritic and axonal architecture of individual pyramidal neurons across layers of adult human neocortex. Cereb Cortex 25(12):4839–4853. Epub 2015/09/01. https://doi.org/10.1093/cercor/bhv188. PubMed PMID: 26318661; PubMed Central PMCID: PMCPMC4635923
12. Medalla M, Luebke JI (2015) Diversity of glutamatergic synaptic strength in lateral prefrontal versus primary visual cortices in the rhesus monkey. J Neurosci 35(1):112–127. Epub 2015/01/09. https://doi.org/10.1523/JNEUROSCI.3426-14.2015. PubMed PMID: 25568107; PubMed Central PMCID: PMCPMC4287137
13. Gilman JP, Medalla M, Luebke JI (2017) Area-specific features of pyramidal neurons-a comparative study in mouse and rhesus monkey. Cereb Cortex 27(3):2078–2094. Epub 2016/03/12. https://doi.org/10.1093/cercor/bhw062. PubMed PMID: 26965903; PubMed Central PMCID: PMCPMC6059164
14. Luebke JI (2017) Pyramidal neurons are not generalizable building blocks of cortical networks. Front Neuroanat 11:11. Epub 2017/03/23. https://doi.org/10.3389/fnana.2017.00011. PubMed PMID: 28326020; PubMed Central PMCID: PMCPMC5339252
15. McCormick DA, Connors BW, Lighthall JW, Prince DA (1985) Comparative electrophysiology of pyramidal and sparsely spiny stellate neurons of the neocortex. J Neurophysiol 54(4):782–806. Epub 1985/10/01. https://doi.org/10.1152/jn.1985.54.4.782

16. Connors BW, Gutnick MJ (1990) Intrinsic firing patterns of diverse neocortical neurons. Trends Neurosci 13(3):99–104. Epub 1990/03/01. https://doi.org/10.1016/0166-2236(90)90185-d
17. Yang CR, Seamans JK, Gorelova N (1996) Electrophysiological and morphological properties of layers V-VI principal pyramidal cells in rat prefrontal cortex in vitro. J Neurosci 16(5):1904–1921. Epub 1996/03/01. PubMed PMID: 8774458; PubMed Central PMCID: PMCPMC6578693
18. Oberlaender M, Ramirez A, Bruno RM (2012) Sensory experience restructures thalamocortical axons during adulthood. Neuron. 74(4):648–655. Epub 2012/05/29. https://doi.org/10.1016/j.neuron.2012.03.022. PubMed PMID: 22632723; PubMed Central PMCID: PMCPMC3564553
19. Schubert D, Staiger JF, Cho N, Kotter R, Zilles K, Luhmann HJ (2001) Layer-specific intracolumnar and transcolumnar functional connectivity of layer V pyramidal cells in rat barrel cortex. J Neurosci 21(10):3580–3592. Epub 2001/05/23. PubMed PMID: 11331387; PubMed Central PMCID: PMCPMC6762473
20. Schubert D, Kotter R, Luhmann HJ, Staiger JF (2006) Morphology, electrophysiology and functional input connectivity of pyramidal neurons characterizes a genuine layer va in the primary somatosensory cortex. Cereb Cortex 16(2):223–236. Epub 2005/05/06. https://doi.org/10.1093/cercor/bhi100
21. Feldmeyer D, Lubke J, Sakmann B (2006) Efficacy and connectivity of intracolumnar pairs of layer 2/3 pyramidal cells in the barrel cortex of juvenile rats. J Physiol 575 (Pt 2):583–602. Epub 2006/06/24. https://doi.org/10.1113/jphysiol.2006.105106. PubMed PMID: 16793907; PubMed Central PMCID: PMCPMC1819447
22. Oberlaender M, Boudewijns ZS, Kleele T, Mansvelder HD, Sakmann B, de Kock CP (2011) Three-dimensional axon morphologies of individual layer 5 neurons indicate cell type-specific intracortical pathways for whisker motion and touch. Proc Natl Acad Sci U S A 108(10):4188–4193. Epub 2011/03/04. https://doi.org/10.1073/pnas.1100647108. PubMed PMID: 21368112; PubMed Central PMCID: PMCPMC3053980
23. Feldmeyer D (2012) Excitatory neuronal connectivity in the barrel cortex. Front Neuroanat 6:24. Epub 2012/07/17. https://doi.org/10.3389/fnana.2012.00024. PubMed PMID: 22798946; PubMed Central PMCID: PMCPMC3394394
24. Staiger JF, Bojak I, Miceli S, Schubert D (2015) A gradual depth-dependent change in connectivity features of supragranular pyramidal cells in rat barrel cortex. Brain Struct Funct 220(3):1317–1337. Epub 2014/02/27. https://doi.org/10.1007/s00429-014-0726-8. PubMed PMID: 24569853; PubMed Central PMCID: PMCPMC4409644
25. Rojas-Piloni G, Guest JM, Egger R, Johnson AS, Sakmann B, Oberlaender M (2017) Relationships between structure, in vivo function and long-range axonal target of cortical pyramidal tract neurons. Nat Commun 8(1):870. Epub 2017/10/13. https://doi.org/10.1038/s41467-017-00971-0. PubMed PMID: 29021587; PubMed Central PMCID: PMCPMC5636900
26. Medalla M, Gilman JP, Wang JY, Luebke JI (2017) Strength and diversity of inhibitory signaling differentiates primate anterior cingulate from lateral prefrontal cortex. J Neurosci 37(18):4717–4734. Epub 2017/04/07. https://doi.org/10.1523/JNEUROSCI.3757-16.2017. PubMed PMID: 28381592; PubMed Central PMCID: PMCPMC5426565
27. Medalla M, Chang W, Calderazzo SM, Go V, Tsolias A, Goodliffe JW, et al (2020) Treatment with mesenchymal-derived extracellular vesicles reduces injury-related pathology in pyramidal neurons of monkey perilesional ventral premotor cortex. J Neurosci 40(17):3385–3407. Epub 2020/04/04. https://doi.org/10.1523/JNEUROSCI.2226-19.2020. PubMed PMID: 32241837; PubMed Central PMCID: PMCPMC7178914
28. Medalla M, Chang W, Ibanez S, Guillamon-Vivancos T, Nittmann M, Kapitonava A et al (2021) Layer-specific pyramidal neuron properties underlie diverse anterior cingulate cortical motor and limbic networks. Cereb Cortex 32:2170. Epub 2021/10/07. https://doi.org/10.1093/cercor/bhab347
29. Testa-Silva G, Verhoog MB, Linaro D, de Kock CP, Baayen JC, Meredith RM, et al (2014) High bandwidth synaptic communication and frequency tracking in human neocortex. PLoS Biol 12(11):e1002007. Epub 2014/11/26. https://doi.org/10.1371/journal.pbio.1002007. PubMed PMID: 25422947; PubMed Central PMCID: PMCPMC4244038
30. Eyal G, Verhoog MB, Testa-Silva G, Deitcher Y, Lodder JC, Benavides-Piccione R, et al (2016) Unique membrane properties and enhanced signal processing in human neocortical neurons. Elife 5. Epub 2016/10/07.

https://doi.org/10.7554/eLife.16553. PubMed PMID: 27710767; PubMed Central PMCID: PMCPMC5100995

31. Deitcher Y, Eyal G, Kanari L, Verhoog MB, Atenekeng Kahou GA, Mansvelder HD, et al (2017) Comprehensive morpho-electrotonic analysis shows 2 distinct classes of L2 and L3 pyramidal neurons in human temporal cortex. Cereb Cortex 27(11):5398–5414. Epub 2017/10/03. https://doi.org/10.1093/cercor/bhx226. PubMed PMID: 28968789; PubMed Central PMCID: PMCPMC5939232

32. Kalmbach BE, Buchin A, Long B, Close J, Nandi A, Miller JA, et al (2018) h-Channels contribute to divergent intrinsic membrane properties of supragranular pyramidal neurons in human versus mouse cerebral cortex. Neuron 100(5):1194–1208 e5. Epub 2018/11/06. https://doi.org/10.1016/j.neuron.2018.10.012. PubMed PMID: 30392798; PubMed Central PMCID: PMCPMC6447369

33. Mihaljevic B, Larranaga P, Bielza C (2021) Comparing the electrophysiology and morphology of human and mouse layer 2/3 pyramidal neurons with Bayesian networks. Front Neuroinform 15:580873. Epub 2021/03/09. https://doi.org/10.3389/fninf.2021.580873. PubMed PMID: 33679362; PubMed Central PMCID: PMCPMC7930221

34. Moradi Chameh H, Rich S, Wang L, Chen FD, Zhang L, Carlen PL, et al (2021) Diversity amongst human cortical pyramidal neurons revealed via their sag currents and frequency preferences. Nat Commun 12(1):2497. Epub 2021/05/05. https://doi.org/10.1038/s41467-021-22741-9. PubMed PMID: 33941783; PubMed Central PMCID: PMCPMC8093195

35. Kawaguchi Y, Kubota Y (1998) Neurochemical features and synaptic connections of large physiologically-identified GABAergic cells in the rat frontal cortex. Neuroscience 85(3):677–701. Epub 1998/06/25. https://doi.org/10.1016/s0306-4522(97)00685-4

36. Gupta A, Wang Y, Markram H (2000) Organizing principles for a diversity of GABAergic interneurons and synapses in the neocortex. Science 287(5451):273–278. Epub 2000/01/15. https://doi.org/10.1126/science.287.5451.273

37. Markram H, Toledo-Rodriguez M, Wang Y, Gupta A, Silberberg G, Wu C (2004) Interneurons of the neocortical inhibitory system. Nat Rev Neurosci 5(10):793–807. Epub 2004/09/21. https://doi.org/10.1038/nrn1519

38. DeFelipe J, Lopez-Cruz PL, Benavides-Piccione R, Bielza C, Larranaga P, Anderson S, et al (2013) New insights into the classification and nomenclature of cortical GABAergic interneurons. Nat Rev Neurosci 14(3):202–216. Epub 2013/02/07. https://doi.org/10.1038/nrn3444. PubMed PMID: 23385869; PubMed Central PMCID: PMCPMC3619199

39. Kepecs A, Fishell G (2014) Interneuron cell types are fit to function. Nature 505(7483):318–326. Epub 2014/01/17. https://doi.org/10.1038/nature12983. PubMed PMID: 24429630; PubMed Central PMCID: PMCPMC4349583

40. Kubota Y, Karube F, Nomura M, Kawaguchi Y (2016) The diversity of cortical inhibitory synapses. Front Neural Circuits 10:27. Epub 2016/05/21. https://doi.org/10.3389/fncir.2016.00027. PubMed PMID: 27199670; PubMed Central PMCID: PMCPMC4842771

41. Krienen FM, Goldman M, Zhang Q, R CHDR, Florio M, Machold R, et al (2020) Innovations present in the primate interneuron repertoire. Nature 586(7828):262–269. Epub 2020/10/02. https://doi.org/10.1038/s41586-020-2781-z. PubMed PMID: 32999462; PubMed Central PMCID: PMCPMC7957574

42. Svoboda K (2011) The past, present, and future of single neuron reconstruction. Neuroinformatics 9(2–3):97–98. Epub 2011/02/01. https://doi.org/10.1007/s12021-011-9097-y

43. Ascoli GA, Wheeler DW (2016) In search of a periodic table of the neurons: axonal-dendritic circuitry as the organizing principle: patterns of axons and dendrites within distinct anatomical parcels provide the blueprint for circuit-based neuronal classification. Bioessays 38(10):969–976. Epub 2016/08/16. https://doi.org/10.1002/bies.201600067. PubMed PMID: 27516119; PubMed Central PMCID: PMCPMC5148124

44. Rockland KS (2010) Five points on columns. Front Neuroanat 4:22. Epub 2010/07/01. https://doi.org/10.3389/fnana.2010.00022. PubMed PMID: 20589097; PubMed Central PMCID: PMCPMC2893004

45. Rockland KS (2020) What we can learn from the complex architecture of single axons. Brain Struct Funct 225(4):1327–1347. Epub 2020/01/12. https://doi.org/10.1007/s00429-019-02023-3

46. Blasdel GG, Lund JS (1983) Termination of afferent axons in macaque striate cortex. J

Neurosci 3(7):1389–1413. Epub 1983/07/01. PubMed PMID: 6864254; PubMed Central PMCID: PMCPMC6564440

47. Cowan RL, Wilson CJ (1994) Spontaneous firing patterns and axonal projections of single corticostriatal neurons in the rat medial agranular cortex. J Neurophysiol 71(1):17–32. Epub 1994/01/01. https://doi.org/10.1152/jn.1994.71.1.17
48. Kita T, Kita H (2012) The subthalamic nucleus is one of multiple innervation sites for long-range corticofugal axons: a single-axon tracing study in the rat. J Neurosci 32(17):5990–5999. Epub 2012/04/28. https://doi.org/10.1523/JNEUROSCI.5717-11.2012. PubMed PMID: 22539859; PubMed Central PMCID: PMCPMC3479642
49. Kuramoto E, Furuta T, Nakamura KC, Unzai T, Hioki H, Kaneko T (2009) Two types of thalamocortical projections from the motor thalamic nuclei of the rat: a single neuron-tracing study using viral vectors. Cereb Cortex 19(9):2065–2077. Epub 2009/01/29. https://doi.org/10.1093/cercor/bhn231
50. Ropireddy D, Scorcioni R, Lasher B, Buzsaki G, Ascoli GA (2011) Axonal morphometry of hippocampal pyramidal neurons semi-automatically reconstructed after in vivo labeling in different CA3 locations. Brain Struct Funct 216(1):1–15. Epub 2010/12/04. https://doi.org/10.1007/s00429-010-0291-8. PubMed PMID: 21128083; PubMed Central PMCID: PMCPMC3577356
51. Wittner L, Henze DA, Zaborszky L, Buzsaki G (2007) Three-dimensional reconstruction of the axon arbor of a CA3 pyramidal cell recorded and filled in vivo. Brain Struct Funct 212(1):75–83. Epub 2007/08/25. https://doi.org/10.1007/s00429-007-0148-y. PubMed PMID: 17717699; PubMed Central PMCID: PMCPMC2662726
52. Wu H, Williams J, Nathans J (2014) Complete morphologies of basal forebrain cholinergic neurons in the mouse. Elife 3:e02444. Epub 2014/06/05. https://doi.org/10.7554/eLife.02444. PubMed PMID: 24894464; PubMed Central PMCID: PMCPMC4038840
53. Yamashita T, Pala A, Pedrido L, Kremer Y, Welker E, Petersen CC (2013) Membrane potential dynamics of neocortical projection neurons driving target-specific signals. Neuron 80(6):1477–1490. Epub 2013/12/24. https://doi.org/10.1016/j.neuron.2013.10.059
54. Yamashita T, Vavladeli A, Pala A, Galan K, Crochet S, Petersen SSA, et al (2018) Diverse long-range axonal projections of excitatory layer 2/3 neurons in mouse barrel cortex. Front Neuroanat 12:33. Epub 2018/05/17. https://doi.org/10.3389/fnana.2018.00033. PubMed PMID: 29765308; PubMed Central PMCID: PMCPMC5938399
55. Economo MN, Clack NG, Lavis LD, Gerfen CR, Svoboda K, Myers EW, et al (2016) A platform for brain-wide imaging and reconstruction of individual neurons. Elife 5:e10566. Epub 2016/01/23. https://doi.org/10.7554/eLife.10566. PubMed PMID: 26796534; PubMed Central PMCID: PMCPMC4739768
56. Han Y, Kebschull JM, Campbell RAA, Cowan D, Imhof F, Zador AM, et al (2018) The logic of single-cell projections from visual cortex. Nature 556(7699):51–56. Epub 2018/03/29. https://doi.org/10.1038/nature26159. PubMed PMID: 29590093; PubMed Central PMCID: PMCPMC6585423
57. Winnubst J, Bas E, Ferreira TA, Wu Z, Economo MN, Edson P, et al (2019) Reconstruction of 1,000 projection neurons reveals new cell types and organization of long-range connectivity in the mouse brain. Cell. 179(1):268–281 e13. Epub 2019/09/10. https://doi.org/10.1016/j.cell.2019.07.042. PubMed PMID: 31495573; PubMed Central PMCID: PMCPMC6754285
58. Peng H, Xie P, Liu L, Kuang X, Wang Y, Qu L, et al (2021) Morphological diversity of single neurons in molecularly defined cell types. Nature 598(7879):174–81. Epub 2021/10/08. https://doi.org/10.1038/s41586-021-03941-1. PubMed PMID: 34616072; PubMed Central PMCID: PMCPMC8494643
59. Wang M, Liu K, Pan J, Li J, Sun P, Zhang Y, et al (2022) Brain-wide projection reconstruction of single functionally defined neurons. Nat Commun 13(1):1531. Epub 2022/03/24. https://doi.org/10.1038/s41467-022-29229-0. PubMed PMID: 35318336; PubMed Central PMCID: PMCPMC8940919
60. Tasic B, Yao Z, Graybuck LT, Smith KA, Nguyen TN, Bertagnolli D, et al (2018) Shared and distinct transcriptomic cell types across neocortical areas. Nature 563(7729):72–8. Epub 2018/11/02. https://doi.org/10.1038/s41586-018-0654-5. PubMed PMID: 30382198; PubMed Central PMCID: PMCPMC6456269

61. Zhu Y, Sousa AMM, Gao T, Skarica M, Li M, Santpere G, et al (2018) Spatiotemporal transcriptomic divergence across human and macaque brain development. Science 362(6420). Epub 2018/12/14. https://doi.org/10.1126/science.aat8077. PubMed PMID: 30545855; PubMed Central PMCID: PMCPMC6900982

62. Tuncdemir SN, Wamsley B, Stam FJ, Osakada F, Goulding M, Callaway EM, et al (2016) Early somatostatin interneuron connectivity mediates the maturation of deep layer cortical circuits. Neuron 89(3):521–535. Epub 2016/02/06. https://doi.org/10.1016/j.neuron.2015.11.020. PubMed PMID: 26844832; PubMed Central PMCID: PMCPMC4861073

63. Cocas LA, Fernandez G, Barch M, Doll J, Zamora Diaz I, Pleasure SJ (2016) Cell type-specific circuit mapping reveals the presynaptic connectivity of developing cortical circuits. J Neurosci 36(11):3378–3390. Epub 2016/03/18. https://doi.org/10.1523/JNEUROSCI.0375-15.2016. PubMed PMID: 26985044; PubMed Central PMCID: PMCPMC4792945

64. Mayer C, Hafemeister C, Bandler RC, Machold R, Batista Brito R, Jaglin X, et al (2018) Developmental diversification of cortical inhibitory interneurons. Nature 555(7697):457–462. Epub 2018/03/08. https://doi.org/10.1038/nature25999. PubMed PMID: 29513653; PubMed Central PMCID: PMCPMC6052457

65. Bandler RC, Vitali I, Delgado RN, Ho MC, Dvoretskova E, Ibarra Molinas JS, et al (2022) Single-cell delineation of lineage and genetic identity in the mouse brain. Nature 601(7893):404–409. Epub 2021/12/17. https://doi.org/10.1038/s41586-021-04237-0. PubMed PMID: 34912118; PubMed Central PMCID: PMCPMC8770128

66. Silbereis JC, Pochareddy S, Zhu Y, Li M, Sestan N (2016) The cellular and molecular landscapes of the developing human central nervous system. Neuron 89(2):248–268. Epub 2016/01/23. https://doi.org/10.1016/j.neuron.2015.12.008. PubMed PMID: 26796689; PubMed Central PMCID: PMCPMC4959909

67. Guerout N, Li X, Barnabe-Heider F (2014) Cell fate control in the developing central nervous system. Exp Cell Res 321(1):77–83. Epub 2013/10/22. https://doi.org/10.1016/j.yexcr.2013.10.003

68. Chaboub LS, Deneen B (2013) Astrocyte form and function in the developing central nervous system. Semin Pediatr Neurol 20(4):230–235. Epub 2013/12/25. https://doi.org/10.1016/j.spen.2013.10.003. PubMed PMID: 24365570; PubMed Central PMCID: PMCPMC3874818

69. Pont-Lezica L, Bechade C, Belarif-Cantaut Y, Pascual O, Bessis A (2011) Physiological roles of microglia during development. J Neurochem 119(5):901–908. Epub 2011/09/29. https://doi.org/10.1111/j.1471-4159.2011.07504.x

70. Li Z, Tyler WA, Zeldich E, Santpere Baro G, Okamoto M, Gao T, et al (2020) Transcriptional priming as a conserved mechanism of lineage diversification in the developing mouse and human neocortex. Sci Adv 6(45). Epub 2020/11/08. https://doi.org/10.1126/sciadv.abd2068. PubMed PMID: 33158872; PubMed Central PMCID: PMCPMC7673705

71. Tyler WA, Medalla M, Guillamon-Vivancos T, Luebke JI, Haydar TF (2015) Neural precursor lineages specify distinct neocortical pyramidal neuron types. J Neurosci 35(15):6142–6152. Epub 2015/04/17. https://doi.org/10.1523/JNEUROSCI.0335-15.2015. PubMed PMID: 25878286; PubMed Central PMCID: PMCPMC4397608

72. Guillamon-Vivancos T, Tyler WA, Medalla M, Chang WW, Okamoto M, Haydar TF, et al (2019) Distinct neocortical progenitor lineages fine-tune neuronal diversity in a layer-specific manner. Cereb Cortex 29(3):1121–38. Epub 2018/02/08. https://doi.org/10.1093/cercor/bhy019. PubMed PMID: 29415216; PubMed Central PMCID: PMCPMC6373699

73. Goodliffe JW, Song H, Rubakovic A, Chang W, Medalla M, Weaver CM, et al (2018) Differential changes to D1 and D2 medium spiny neurons in the 12-month-old Q175+/− mouse model of Huntington's disease. PLoS One 13(8):e0200626. Epub 2018/08/18. https://doi.org/10.1371/journal.pone.0200626. PubMed PMID: 30118496; PubMed Central PMCID: PMCPMC6097649

74. Goodliffe J, Rubakovic A, Chang W, Pathak D, Luebke J (2020) Structural and functional features of medium spiny neurons in the BACHDDeltaN17 mouse model of Huntington's Disease. PLoS One 15(6):e0234394. Epub 2020/06/24. https://doi.org/10.1371/journal.pone.0234394. PubMed PMID: 32574176; PubMed Central PMCID: PMCPMC7310706

75. Veenman CL, Reiner A, Honig MG (1992) Biotinylated dextran amine as an anterograde

tracer for single- and double-labeling studies. J Neurosci Methods 41(3):239–254. Epub 1992/03/01. https://doi.org/10.1016/0165-0270(92)90089-v

76. Reiner A, Veenman CL, Medina L, Jiao Y, Del Mar N, Honig MG (2000) Pathway tracing using biotinylated dextran amines. J Neurosci Methods 103(1):23–37. Epub 2000/11/14. https://doi.org/10.1016/s0165-0270(00)00293-4
77. Little JP, Carter AG (2013) Synaptic mechanisms underlying strong reciprocal connectivity between the medial prefrontal cortex and basolateral amygdala. J Neurosci 33(39):15333–15342. https://doi.org/10.1523/JNEUROSCI.2385-13.2013. PubMed PMID: 24068800; PubMed Central PMCID: PMCPMC3782616
78. Yamawaki N, Suter BA, Wickersham IR, Shepherd GM (2016) Combining optogenetics and electrophysiology to analyze projection neuron circuits. Cold Spring Harb Protoc 2016(10). Epub 2016/10/05. https://doi.org/10.1101/pdb.prot090084. PubMed PMID: 27698240; PubMed Central PMCID: PMCPMC5476926
79. Dickstein DL, Weaver CM, Luebke JI, Hof PR (2013) Dendritic spine changes associated with normal aging. Neuroscience 251:21–32. Epub 2012/10/17. https://doi.org/10.1016/j.neuroscience.2012.09.077. PubMed PMID: 23069756; PubMed Central PMCID: PMCPMC3654095
80. Luebke JI, Amatrudo JM (2012) Age-related increase of sI(AHP) in prefrontal pyramidal cells of monkeys: relationship to cognition. Neurobiol Aging 33(6):1085–1095. Epub 2010/08/24. https://doi.org/10.1016/j.neurobiolaging.2010.07.002. PubMed PMID: 20727620; PubMed Central PMCID: PMCPMC2992607
81. Chang YM, Rosene DL, Killiany RJ, Mangiamele LA, Luebke JI (2005) Increased action potential firing rates of layer 2/3 pyramidal cells in the prefrontal cortex are significantly related to cognitive performance in aged monkeys. Cereb Cortex 15(4):409–418. Epub 2005/03/08. https://doi.org/10.1093/cercor/bhh144
82. Luebke JI, Medalla M, Amatrudo JM, Weaver CM, Crimins JL, Hunt B, et al (2015) Age-related changes to layer 3 pyramidal cells in the rhesus monkey visual cortex. Cereb Cortex 25(6):1454–1468. Epub 2013/12/11. https://doi.org/10.1093/cercor/bht336. PubMed PMID: 24323499; PubMed Central PMCID: PMCPMC4428297
83. Rodriguez A, Ehlenberger DB, Hof PR, Wearne SL (2009) Three-dimensional neuron tracing by voxel scooping. J Neurosci Methods 184(1):169–175. Epub 2009/07/28. https://doi.org/10.1016/j.jneumeth.2009.07.021. PubMed PMID: 19632273; PubMed Central PMCID: PMCPMC2753723
84. Rodriguez A, Ehlenberger DB, Dickstein DL, Hof PR, Wearne SL (2008) Automated three-dimensional detection and shape classification of dendritic spines from fluorescence microscopy images. PLoS One 3(4):e1997. Epub 2008/04/24. https://doi.org/10.1371/journal.pone.0001997. PubMed PMID: 18431482; PubMed Central PMCID: PMCPMC2292261
85. Luebke J, Barbas H, Peters A (2010) Effects of normal aging on prefrontal area 46 in the rhesus monkey. Brain Res Rev 62(2):212–232. Epub 2009/12/17. https://doi.org/10.1016/j.brainresrev.2009.12.002. PubMed PMID: 20005254; PubMed Central PMCID: PMCPMC2822043
86. Sholl DA (1953) Dendritic organization in the neurons of the visual and motor cortices of the cat. J Anat 87(4):387–406. Epub 1953/10/01. PubMed PMID: 13117757; PubMed Central PMCID: PMCPMC1244622
87. Ikezu S, Yeh H, Delpech JC, Woodbury ME, Van Enoo AA, Ruan Z, et al (2021) Inhibition of colony stimulating factor 1 receptor corrects maternal inflammation-induced microglial and synaptic dysfunction and behavioral abnormalities. Mol Psychiatry 26(6):1808–1831. Epub 2020/02/20. https://doi.org/10.1038/s41380-020-0671-2. PubMed PMID: 32071385; PubMed Central PMCID: PMCPMC7431382

Chapter 11

3D Quantitative Analysis of Synaptic Ultrastructure in Identified Long-Range Projection Axons

Pablo J. Martin-Correa, Javier Rodriguez-Moreno, Astrid Rollenhagen, Joachim H. R. Lübke, and Francisco Clascá

Abstract

Thalamocortical synapses are key cellular links in sensory, motor, and cognitive information processing. Quantitative data on the subcellular structure of these synapses is important to model interactions between the thalamus and cortex.

Specifically, we chose the ventral posteromedial (VPM) and posterior (Po) thalamic nucleus axons that both converge in different layer patterns in the primary and secondary somatosensory cortex. Here, we set out to measure and compare the ultrastructure of these synapses and their postsynaptic targets.

Serial image samples of selected areas were obtained with serial sectioning electron microscopy, to further 3D reconstruct structures of interest. Our goal is to provide the reader with guidelines on how to apply techniques to selectively label long-range axonal projections and analyze its ultrastructure parameters as well as to highlight advantages and limitations of such approach.

Key words Electron microscopy, 3D reconstruction, Serial sectioning, TEM, Synapse, Thalamocortical, Axon, Somatosensory cortex, Dendritic spine

1 Introduction: Thalamocortical Projections as a Model for Synapse Studies

As for excitatory synapses in the mammalian cerebral cortex, glutamate is the major neurotransmitter released, while in inhibitory synapses, it is mainly GABA [1–3]. Despite the high morphological heterogeneity of synapses, they are usually grouped into discrete categories [4]. In this context, thalamocortical synapses are excitatory, as they release glutamate. Such synapses occur mainly between presynaptic axon enlargements called *boutons* and their postsynaptic cell targets found in specific cortical layers and areas, following distinct patterns.

Thalamocortical synapses represent the main conduit through which information from the periphery flows into the cortex and coordinates its activity despite comprising about a tenth of all

Joachim H. R. Lübke and Astrid Rollenhagen (eds.), *New Aspects in Analyzing the Synaptic Organization of the Brain*, Neuromethods, vol. 212, https://doi.org/10.1007/978-1-0716-4019-7_11,

cortical synapses [5]. It is then a widely used study model so as to establish structure–function relationships given its functional importance [3].

Regarding the postsynaptic element, the main recipient of the thalamocortical synapse is the dendritic spine. These are outgrowths of the dendritic stem, whose distal end remains free and usually has a thickening, called the head of the spine, where the neurotransmitter receptors are concentrated, in an area called postsynaptic density (PSD) [6]. That way, the dendritic spine is able to isolate and compartmentalize the postsynaptic response.

A matter of relevance in brain micro-anatomy is the different patterns of lamination in the cerebral cortex that can be found across different long-range projecting axons. As for thalamocortical neurons, all derive from the same forebrain segment and have similar gene expression profiles [7, 8]. All these neurons are glutamatergic in transmission, with a similar ion-channel profile and somatodendritic morphology [9]. Despite that, thalamocortical projections can be further classified by the target structures to which they project and the layer pattern of their terminal axonal arborizations [10, 11].

Studies at the level of individual neurons imply that different morphotypes of thalamocortical neurons connect with specific neurons and dendritic domains within its cortical microcircuit, and therefore, there may be structural differences at the level of the synapse (pre- and/or postsynaptic element), as well as different synaptic transduction mechanisms [12–14]. Cellular connectomics is the field devoted to study those differences as well as to define the synaptic microcircuitry underlying such projections, but not restricted to other projection sets like the spinal cord, hippocampal, or cerebellar circuits [15–17].

A greater synaptic efficacy has been attributed to thalamocortical synapses [3, 5]. The question then is whether synaptic efficacy is due to intrinsic characteristics derived from ultrastructure of the elements that form the synapse [5]. These would include synaptic vesicle pool, PSD area, mitochondrial volume, or dendritic spine volume [5, 14, 18, 19]. Or, on the contrary, it is due to the space–time convergence of the synapses, meaning the thalamic input on the same cortical neuron increasing its neurotransmitter release probability [20, 21]. Similar proposals can be further considered to study other regions such as hippocampal or prefrontal cortex synapse efficacy [22–25].

Three-dimensional volume electron microscopy techniques seem to be the perfect option to approach all these questions. Thalamocortical synapse studies have been carried out with anterograde tracers, but mainly in two dimensions, to provide data on maximum projection areas or distribution of terminals [26, 27]. More recent publications have attempted three-dimensional studies, but either using nonspecific labeling like vGLUT2 [28] or making assumptions on the thalamic origin of the projection based only on morphology [29].

1.1 Methods to Study Long-Range Projection Axons and Synapses Ultrastructure

Selective labeling of long-range projection axons avoids this unspecificity. Therefore, it poses as the most appropriate and anatomic-wise technique to provide reliable volumetric and morphometric data of thalamocortical projections, among others long projection axons [14, 19].

Since the introduction of electron microscopy and other techniques with subcellular resolution, knowledge about projection axons and synapse ultrastructure has increased exponentially, thus giving rise to numerous publications [19, 30–35]. Despite the increasing morphology and quantitative data available, one major drawback is the poor definitions employed and its inconsistency amongst publications.

Above all, long-range projection axons study has been approached with different methodologies, which complicates the comparison of existing data throughout the literature as there is no standardized methodology for obtaining quantitative data on the ultrastructure of the pre- and postsynaptic elements nor of the subcellular components associated.

For a proper morphological study of the elements that conform the synaptic bouton, electron microscopy is the main subcellular resolution technique (nanometric scale) that allows a reliable delineation and measurement of the ultrastructure of neuronal elements [36]. Serial sectioning transmission electron microscopy (ssTEM) imaging allowed for the first time the reconstruction of three-dimensional volumes of subcellular structures [37, 38]. In any case, ssTEM still constitutes to this day the ***gold standard*** in terms of resolution as it is the only technique that allows to assess parameters like the synaptic vesicle pool [14, 19].

Besides ssTEM, other serial sectioning techniques have gone under development in the last decade. For example, in the automated tape-collecting ultramicrotome (ATUM), sections are cut in a traditional way using an ultramicrotome and diamond knife with a water raft. The successive sections are picked up on moving tape and then imaged with a scanning electron microscope (SEM) [39].

FIB/SEM improved the section collection procedure by basically erasing it. FIB/SEM combines a focused gallium ion beam (FIB) with an SEM. The result is an automatic serial imaging of the sample, with similar appearance of transmission electron microscopy. After taking the photograph, the ion beam removes a volume of specific thickness. The result is a block of serial and pre-aligned images of the desired thickness, the only drawback being some resolution loss [40, 41]. On the other hand, the field of view of the image is more stable and can be increased in size, thus allowing to follow the same structure through microns. This results in some parameters that cannot be obtained otherwise, like non-varicose synapse occurrence and distribution, which are commonly disregarded, despite its frequency [42].

Further progress has also been made on increasing the field of view by coupling several SEM electron beams (up to 91) to obtain a bigger coverage of the samples [43], also resulting in high-throughput connectomic data [44, 45]. More recently, similar throughput of data was obtained also using several un-coupled TEM microscopes but state-of-the-art image stitching algorithms so as to increase the coverage of the field of view while maintaining maximum resolution, but still no selective labeling involved [46].

2 Materials

2.1 Test Animals and Buffers

As for electron microscopy experiments, we use adult wild-type mice strain C57BL/6. All procedures involving live animals were conducted under protocols approved by the University ethics committee, in accordance with the European Community Council Directive 2010/63/UE.

Throughout all the experimental procedures, both phosphate buffer (PB) and tris-acetate buffer (TAS) were used. For the PB, first prepare a 0.2 M stock solution. Dissolve 5.3 g of sodium dihydrogen phosphate monohydrate ($NaH_2PO_4 \cdot H_2O$, Sigma-Aldrich) and 28 g of potassium phosphate dibasic anhydrous (K_2HPO_4, Sigma-Aldrich) in 1 liter of distilled water. For the latter, 16.4 g of sodium acetate trihydrate ($CH_3COONa \cdot 3H_2O$, Sigma-Aldrich) needs to be dissolved in 1 L of distilled water. pH can be adjusted later on to 7.4 with a 1 N solution of sodium hydroxide.

2.2 Anesthesia, Surgery, and Perfusion

Mice were first anesthetized with an initial intraperitoneal injection of 0.075 mg/g body weight ketamine (Ketolar®, Parke-Davis) and 0.02 mg/g body weight xylazine (Rompun®, Bayer) using a 25-gauge needle. Anesthesia was maintained with oxygenated isoflurane (0.5–2%) (Isoflutek®, Alivira) throughout the surgical procedure with its overflow being collected with an activated carbon container with a vacuum pump (Fluovac® system, Harvard Apparatus). This procedure was performed in a stereotaxic frame (Kopf®, David Kopf Instruments), while the temperature of the animal was maintained using a circulating water pad at 37 °C (T-Pump®, Gaymar Industries).

As to proceed with the injection of the biotinylated dextran neuronal tracer (BDA 1000-lysine fixable, Molecular Probes, diluted at 2.5% in PB 0.01 M), the craniotomy was done with a trephine drill (Dremel® 255, Dremel-Europe). Additionally, a subcutaneous injection of 5% glucose solution was administred so as to mantain adequate levels of hydratation on the experimental animal, close to the end of the surgery procedure.

By the end of the surgery, the isoflurane flow was interrupted and a bolus of buprenorphine hydrochloride was injected subcutaneously (0.075 mg/kg body weight; Buprex®) for post-surgery analgesia. Ibuprofen (0.012 mg/mL) was also added to the drinking water to ensure analgesia during the survival period. At the time of sacrifice, animals were overdosed with sodium pentobarbital intraperitoneal injection (0.09 mg/g body weight). A peristaltic pump (MasterFlex, Cole-Palmer®) was used for perfusion. First saline (9 g of sodium chloride/L of distilled water) was used to clean all the vascular system.

The fixative of choice consisted of paraformaldehyde (PFA) at 4.0% and glutaraldehyde at 0.5% in 0.1 M PB at pH 7.4 and was used both for perfusion and post-fixation (1 h). Post-fixation should last the minimum time so the fixative damages the least possible the tissue ultrastructure while obtaining enough stiffness so as to proceed with the vibratome sectioning.

2.3 Tissue Permeabilization and Histochemistry

The post-fixed brain tissue was glued with cyanoacrylate adhesive (SuperGlue3, Loctite®) to the vibratome platen (Leica VT1200 S, Leica Microsystems GmbH) so as to obtain the 50 μm sections collected in 12-well plates (Costar®, Corning). Sections selected for electron microscopy were additionally cryo-protected with a 30% sucrose solution in 0.1 M PB to later be permeated with liquid nitrogen in a Dewar flask (Sigma-Aldrich). Once treated with H_2O_2 and washed with 0.1 M PB, brain tissue was incubated with avidin–biotin kit (ABC Elite®, Vector Laboratories; dilution 1: 100 + TX100, Sigma, at 2.0%; in 0.1 M PB).

In order to visualize the labeling (*see* Fig. 1a), we adapted the glucose oxidase reaction (0.2% D-glucose, 0.003% glucose oxidase, Sigma-Aldrich) with DAB (3,3′-diaminobenzidine tetrahydrochloride, Sigma, 0.02%) and nickel ammonium sulfate intensification (1.2% nickel sulfate, 0.02% ammonium chloride, Sigma) [47]. The reaction was carried out in 0.1 M TAS. The adapted protocol for electron microscopy sections is further explained in the following section.

2.4 Osmication, Dehydration, and Uranyl Acetate Staining

Selected sections were transferred to a disposable multi-well plate and treated with osmium tetroxide diluted at 1% in 0.1 M PB (4% aqueous solution, EMS Diasum) and later on washed with 0.1 M PB. Dehydration is started with 50° ethanol (Sigma-Aldrich). In between, sections are further treated with uranyl acetate (in powder, EMS Diasum) at 1% in 70° ethanol, which has been freshly prepared and filtered with a syringe filter (Millipore, Merck, 0.2 μm). Subsequently, dehydration is carried out with increasing graded ethanol solutions (70°–80°–90°–100°) and then transferred to tinfoil rafts (medium size, EMS Diasum) where the slices were treated with acetone (Sigma-Aldrich).

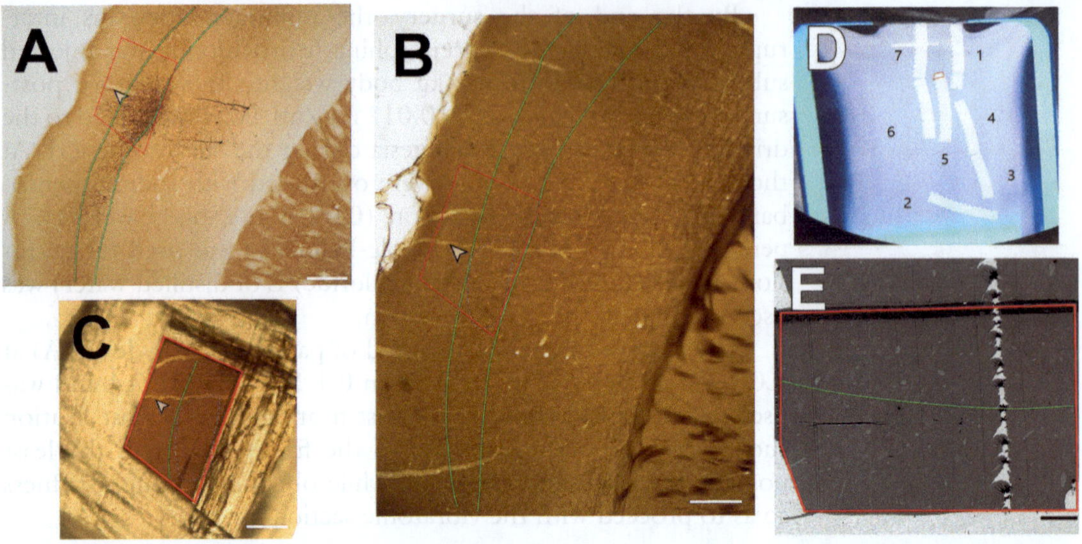

Fig. 1 Selective labeling of VPM axons in the mouse layer 4 somatosensory cortex. (**a**) BDA-labeled cVPM axons arborizing profusely in S2L4 cortex in light microscopy series. (**b**) EM embedded tissue of adjacent slice in the series. (**c**) Trimmed EM tissue sample showing region of interest. (**d**) Photograph of a continuous series of ultrathin sections made with a diamond knife. (**e**) Low-magnification TEM photography so as to locate the region of interest. Layer 4 is depicted in green lines and region of interest in red. The arrow pinpoints the same blood vessel across different panels. Scale bars: a = 200 μm; b = 250 μm; c = 300 μm e = 50 μm

2.5 Embedding in Epoxy Resin

Freshly made epoxy resin is needed for embedding. Durcupan® (EMS Diasum) was the resin of choice and was prepared carefully following this recipe: 10 g compound A + 10 g compound B + 0.3 g compound C + 0.3 g compound D. Later on, the components need to be properly mixed in agitation and degasified. A first mix of 1:1 acetone and Durcupan is used to start the slices embedding.

Once the acetone is fully evaporated, pure Durcupan is added to let the resin inclusion. Polymerization is carried out in a stable heater (Digitheat-TFT Premier, JP Selecta®) at 60° for 48 h. For doing that, Durcupan-embedded slices are transferred to Aclar® sheets (EMS Diasum) in a sandwich fashion (*see* Fig. 1b).

2.6 Ultrathin Sectioning

Previous to obtaining any ultrathin, cortical regions of interest are trimmed with a scalpel and glued with cyanoacrylate to a Durcupan block (*see* Fig. 1c). A razor blade is used to adjust the borders of the sample to the desired shape. An ultramicrotome (Leica Ultracut UC6, Leica Microsystems GmbH) is then used with the help of a diamond knife (ultra 45°, Diatome®) to obtain 50 nm slices, which are collected in single-slot copper grids (2 mm long × 1 mm wide, EMS Diasum) coated with Pioloform at 1% (Formvar®, EMS Diasum, 10 g/L of chloroform) (*see* Fig. 1d).

2.7 Electron Microscopy Imaging

Ultrathin section series suitable for imaging can be further contrasted using the Reynolds method [48]. Electron-dense labeled neural structures of interest were photographed with a TEM at a primary magnification of 8000× (Zeiss Libra® 120, Germany) using a Proscan 2 K digital camera and the panorama function of the SIS Analysis (Olympus Soft Imaging System, Germany) or ImageSP (Fa. Tröndle, Germany) software. Additionally, interesting structures or regions can be photographed at lower magnifications for further documentation, to establish fiduciary marks and cortical layer delimitation (*see* Fig. 1e). Further processing can easily be endured on the same embedded tissue on the epoxy resin block so as to repurpose it for other technologies like FIB/SEM imaging.

3 Methods

3.1 Anterograde Neural Tracing Technique

Following an intraperitoneal injection of a freshly prepared mixture of ketamine/xylazine (0.005 mL/g body weight) and checking the absence of patellar reflex, the mice scalp is carefully shaved to avoid damaging the whiskers in the process. Whisker trimming needs to be extra careful. It could heavily alter the results, as thalamocortical axon length shortening has been reported [49].

The animal is then placed on the stereotaxic frame to perform the surgery. To do so, the skin of the head was opened with a scalpel along the midline and then retracted, exposing the skull. Next, the position of the bregma and lambda reference points was checked, to then establish the anteroposterior and mediolateral coordinates of our region of interest [50]. Subsequently, a bilateral craniotomy was performed in the region of interest using a trephine drill to expose the meninges.

The 10 kDa BDA anterograde neuronal tracer used in the experiments completely fills the neuron's soma thanks to the iontophoresis technique and is transported anterogradely down the axon to its terminals in the cortex during the post-surgery time. The borosilicate micropipettes used for tracer injection (TW100F-4, World Precision Instruments) were adapted by means of a vertical stretcher (Model 700C, David Kopf Instruments) so that, in the first 5 mm, the internal diameter of the pipette was less than 100 μm, thus reducing the damage to the cerebral cortex. The end of the pipette was cut under magnification using a scalpel, and the pore was adjusted to a diameter of 6–10 μm.

Once the pipette was suffciently filled by the tracer (by simple capillarity it was attached to the stereotaxic arm, adjusted to the coordinate of interest, and slowly introduced into the brain until the selected coordinates were reached axis [50]. Injection by iontophoresis was performed by applying a positive current between 600 and 800 nA for 30–40 min in cycles of 1 s active/inactive through electrodes connected to a direct current source (Microiontophoresis Dual Current 260 source, World Precision Instruments).

The micropipette was finally removed after a few minutes to avoid any tracer reflux, and the muscle and skin were sutured. Only when patellar reflex is present again, mice are placed back on separate cages at the animal facility until the time of perfusion.

3.2 Perfusion, Post-fixation, and Vibratome Sectioning

After a survival period of approximately 5–7 days post-surgery, animals can be sacrificed by an intraperitoneal overdose of pentobarbital sodium to obtain the brain tissue. At the beginning, once the animal was deeply anesthetized and the areflexia of the animal was verified, it was fixed to the perfusion table with tape so as to expose the abdominal cavity. The abdominal cavity was then opened with an incision, and both the diaphragm and the rib cage were displaced to fully expose the heart.

A 25 G blunt-edge hypodermic needle attached to a peristaltic pump was then inserted into the left ventricle, a cut in the right atrium was made and then opened the valve to let saline flow. The perfusion of the animal began with the substitution of the blood with isotonic saline buffer. When all the blood is flushed out, the valve is closed and changed to the fixative solution. Approximately 80–100 mL of fixative solution is more than enough to acquire an optimal fixation of neural tissue. Besides that, the peristaltic pump fixative flow should be adjusted so as it takes around 20–25 min to pump all the fixative.

If the animal is correctly fixated, stiffness from head to tail should be present. Finally, the animal was decapitated, and its encephalon was carefully removed from the skull for post-fixation by immersion in the same fixative solution for a maximum of 1 h at 4 °C.

The post-fixed tissue block (after removing the olfactory bulbs and brain stem) was serially cut in the coronal plane into 50 μm sections using a vibratome. The sections were collected in multi-well boxes with 0.1 M PB in an orderly manner and in two alternate series, one for light microscopy (*see* Fig. 1a) and the other for electronic microscopy (*see* Fig. 1b).

3.3 Tissue Permeabilization and Histochemistry

Sections for electron microscopy were cryo-protected by incubation in a sucrose solution overnight at 4 °C under agitation. When sections were fully sunk in the solution, they were placed in suitable plastic beakers and subjected to three to five freeze–thaw cycles by approaching the beakers to liquid nitrogen. Finally, the tissue section permeated were washed and stored in PB until further use.

This process generates micro-tearings in the tissue that allow better uptake of the tracer. This is an alternative to using detergents (like Triton™ X-100), which can pierce and damage cell membrane integrity. This way, the ultrastructure of the tissue is altered to a lesser extent.

In the first series of sections (for light microscopy), peroxidase activity was blocked by incubation in PB-buffered H_2O_2 for 15 min, washed three times in 0.1 M PB (10 min each), and finally incubated in an avidin–biotin–peroxidase kit diluted in PB for 2 h at room temperature.

After washing in PB and TAS buffer (twice per buffer, 5 min each), labeling was visualized using the glucose oxidase DAB nickel sulfate-enhanced method [47]. The product of the reaction creates a brownish-black precipitate that labels the long-range projection axons of interest due to the iontophoresis BDA uptake during the surgery procedure (*see* Fig. 1a).

This reaction can be supervised with the naked eye and stopped at the desired level of precipitate by placing the sections back in TAS buffer (approximately 5–8 min would suffice). The chemical reaction is heavily affected on the lipid content of each test animal encephalon as well as the room temperature where the reaction takes place, so variations can be expected throughout the year and need to be adjusted accordingly.

The reverse order of TAS and PB buffer is then needed to properly wash the sections. Once the section with labeling is back in PB buffer, the counterstaining with the cytochrome oxidase reaction can be done [51]. Alternatively sections can also be counterstained with thionine. Finally, sections were mounted orderly onto gelatin-coated glass slides, dehydrated in graded ethanol, defatted in xylene, and coverslipped with DePeX® (*see* Fig. 1a).

These sections were used to perform a general cytoarchitectural delimitation with a light microscope, as well as the precise location of the tracer injection in the thalamus and the anterogradely labeled axons in the cerebral cortex (*see* Fig. 1a). If the location of the injection and its arborizations in cerebral cortex (in this case) were dense enough, then processing of the alternate series for electron microscopy would proceed (*see* Fig. 1b).

The second series of sections (electron microscopy) followed a protocol identical to that described above, except for the omission of oxygen–peroxide blocking step, which could severely damage the tissue ultrastructure. And, same way, the nickel sulfate-enhancement in the glucose oxidase-DAB reaction was omitted, as it can difficult the analysis of subcellular structures like synaptic vesicles under the transmission electron microscope.

3.4 Osmication, Dehydration, and Uranyl Acetate

Sections showing a dense labeling in the cerebral cortex region and layer of interest and in accordance with light microscopy series were selected for osmication (*see* Fig. 1b). Such sections were transferred to a new multi-well plate and were incubated in an osmium solution for 45–60 min in a fume hood in darkness. Osmium tetroxide is a strong oxidizing agent. Therefore, sections become really brittle after this step. Manipulation should be done carefully and using some sort of plain spoon to avoid further damage. So as to wash the osmium, a Pasteur pipette is used to remove the solution and replace it with PB buffer in three steps of 10 min each. Waste disposal and osmium tetroxide inactivation can be done in a beaker with milk powder diluted in tap water.

Afterwards, dehydration can be started anytime in the next 24 h. The first step involved incubating the section in 50° ethanol during 10 min and, subsequently, 40 min in a 1% uranyl acetate solution in 70° ethanol. Uranyl acetate is also photosensitive, so this step should also be carried out in dark. After this incubation step, sections were completely dehydrated in an ascending series of ethanol: 10 min in 70°, 10 min in 80°,10 min in 90°, and 2 × 10 min in 100°. Once finished, sections were carefully transferred to aluminum tinfoil rafts where dehydration continued with two incubation steps with acetone, 10 min each.

3.5 Flat Embedding in Durcupan

The epoxy resin of choice for tissue embedding was Durcupan, although other options such as Araldite are equally valid [52]. Fresh liquid Durcupan was previously prepared and degasified right before osmication. A 1:1 mixture of acetone and fresh resin was also prepared. Acetone was then removed from the dehydrated sections, the 1:1 mixture was added, and let it rest for 1 h until all acetone was evaporated. An extra step consisting of 2:1 of the same mixture can be done to enhance the embedding.

Once evaporated, the acetone–resin mixture was replaced with a pipette, and pure resin was added to proceed with the inclusion of the sections overnight (~12 h). Next day, several microscope slides were prepared with an Aclar® sheet that had been cropped to the same size as the slide. The sections with some resin remains were placed on top of the Aclar® sheet and covered with another Aclar® sheet and another slide. To flatten the section, you can put some small weights (e.g., microscope slides or bolts) onto the upper microscope slide. Air bubbles should be carefully removed applying slight pressure not directly over the included tissue section.

3.6 Ultrathin Serial Sectioning and TEM Imaging

After light microscopic inspection, samples containing dense labeling in the region of interest were cut out and glued onto prepolymerized resin blocks for serial section (the glue should be left to completely dry for 24 h) (*see* Fig. 1c).

The resulting blocks can be mounted on the ultramicrotome for sectioning. In the first place, with the help of a glass knife, the remaining epoxy resin needs to be removed to expose the embedded tissue. Next, the block has to be trimmed into a trapezoid shape containing only the region of interest (*see* Fig. 1c).

Additionally, to check the presence of tissue on the surface of the block, a toluidine blue staining on semithins can be made (0.5–2 μm thick). These semithins were collected on a microscope

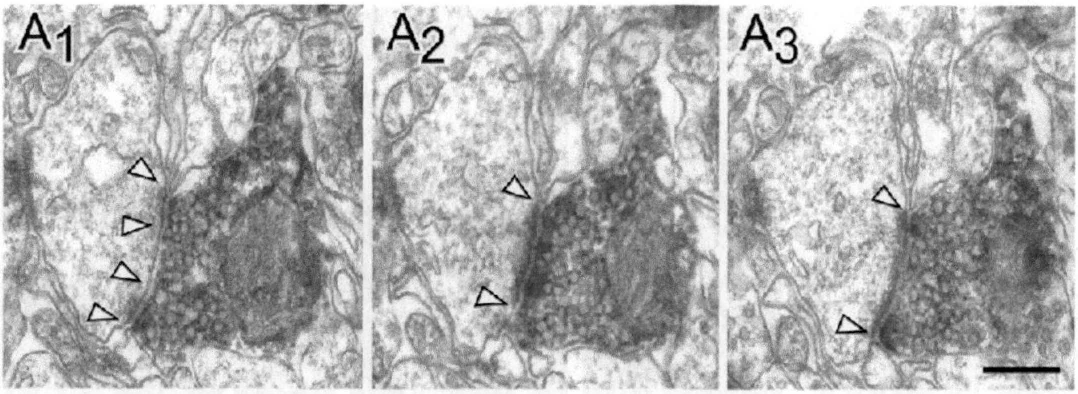

Fig. 2 Serial sectioning TEM for 3D reconstruction of a synaptic bouton in Po. (A_{1-3}). Three consecutive ultrathin section micrographs showing a labeled profile (black precipitate) of a synaptic bouton from a Po thalamocortical projection. Synaptic vesicles, shown in green, are distributed throughout the whole terminal. The perforated PSD is indicated by white arrowheads. Scale bar: $A_{1-3} = 250$ nm

slide. A syringe with a filter was used to pour a drop of toluidine blue solution onto each semithin section. Place the microscope slide on a hot plate at 60 °C, and then rinse with tap water. Under a light microscope, the presence of cell bodies and axons can be observed and assessed if a profitable ultrathin section series can be made.

Serial sectioning needs the use of a diamond knife with a water raft to collect the ultrathins in an orderly manner (*see* Fig. 1d). A proper diamond knife without scratches can reliably cut in continuous series of 50 nm ultrathin sections and ~90 sections/series. Ultrathins were collected on Pioloform coated single-slot copper grids (*see* Fig. 1d). Thereafter, copper grids can be treated both with UranyLess (1 min) and lead citrate (30 s–1 min) for further contrasting of neural tissue [48].

Once contrasted, ultrathins were examined with a transmission electron microscope (TEM) equipped with a bottom-mounted digital camera and the SIS Multi Image Acquisition software package (*see* Fig. 1e). At the EM level, BDA-labeled axons were easily identified due to the DAB reaction product being electron-dense. Serial digital images were taken at a magnification of 8000× and in a tile mosaic of 6 × 6 single images, which were utterly stored in a database until further use (*see* Fig. 2).

3.7 3D Reconstruction of Interest Profiles

Recognition of cellular and subcellular elements in electron microscopy images was based on descriptions by Peters' textbook [1]. Long-range projection axons in the neuropil are characterized by a smooth outline and usually maintain a constant diameter. These present protrusions called axonal varicosities that can present mitochondria of variable size and synaptic vesicles inside (*see* Fig. 2). Such varicosities generally establish synaptic contacts with

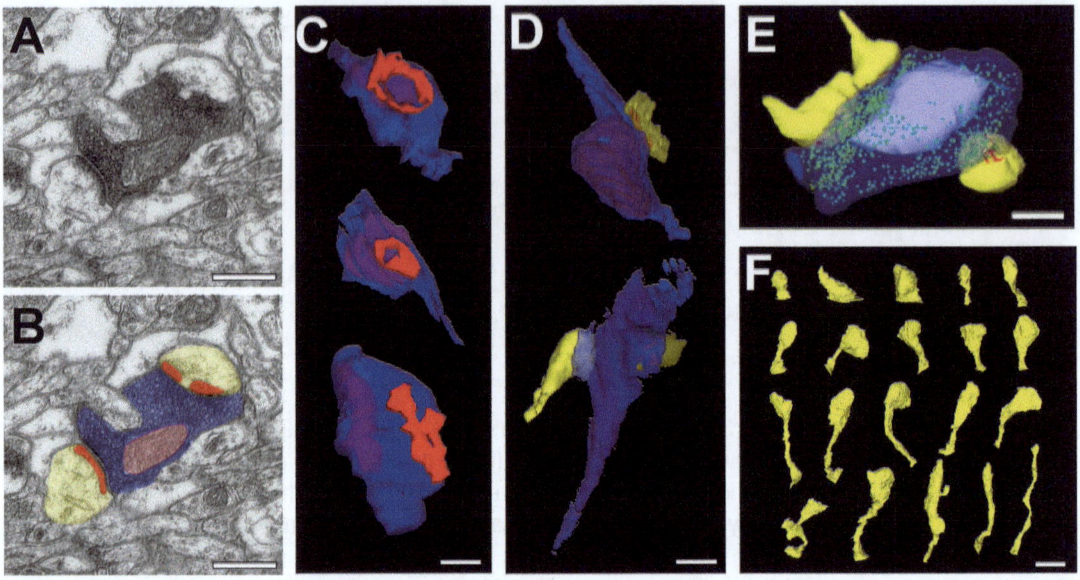

Fig. 3 3D reconstruction of selectively labeled VPM/Po axons in somatosensory cortex with ssTEM. (**a**) Micrograph of a BDA-labeled synaptic bouton. (**b**) The same synaptic bouton segmented. Bouton is depicted in blue, mitochondria in orange, dendritic spines in yellow, and PSDs in red. (**c**) Sample of synaptic boutons 3D reconstructed exposing the PSD. (**d**) Sample of mono- and multi-synaptic boutons showing the postsynaptic element. (**e**) 3D-reconstructed bouton with OpenCAR showing the synaptic vesicles in green. Mitochondria is represented in white so as to easily identify the vesicles within the bouton. (**f**) Sample of dendritic spines postsynaptic to thalamic boutons. Scale bars: a,b = 400 nm; c,d = 200; e = 250 nm; f = 500 nm

another profile/s. Therefore, they are also called synaptic boutons (*see* Figs. 2 and 3a, b).

So as to 3D reconstruct profiles of interest, random ultrathins throughout the whole series of grids were chosen, and BDA-labeled profiles were searched within the ultrathin (*see* Fig. 2). Regions with a higher density of labeled axons and presence of synaptic boutons were selected for imaging (*see* Fig. 1e). Once selected, tile mosaic digital photographs were taken throughout the entire extension of the series of grids at 8000× using the microscope-associated image acquisition program.

Only "complete" buttons were included in our study, that is, those where the synaptic bouton could be followed from beginning to end in consecutive sections (approximately 20–50 images), as well as the postsynaptic element if possible.

The resolution of the digital micrograph mosaics the TEM could obtain was good enough (~2 nm/px) to clearly identify most of subcellular structures of interest associated with boutons, which include the mitochondria, postsynaptic densities (PSDs), synaptic vesicles, and postsynaptic profiles that made contact with the marked presynaptic bouton (*see* Figs. 2 and 3).

Those postsynaptic profiles are dendrites and can be identified by their irregular outline, as well as their homogeneous microtubule content and the presence of mitochondria. They also do not have synaptic vesicles inside. Dendrites can present membrane specializations that consist of a dendritic protuberance called a spine composed of a small-caliber outline that ends in an oval enlargement, called the neck and head of the spine, respectively [53]. However, dendritic spines come in a wide variety of size and morphologies (*see* Fig. 3f) [54, 55].

These profiles usually contain within them different specializations of packed endoplasmic reticulum called spinal apparatus [53] or multivesicular bodies [56]. Besides that, boutons can establish synaptic contacts both with dendrite shafts and dendritic spines [57]. Also, dendritic spines do not contain mitochondria.

Regarding synapses, the presynaptic element is generally an axon that presents a dilation (the synaptic bouton), which contains synaptic vesicles that accumulate and release the neurotransmitter to the synaptic cleft. They have a variable diameter (40–50 nm) (*see* Figs. 2 and 3e).

Asymmetric synapses present a thickening of the postsynaptic density (PSD), while symmetrical synapses present pre- and postsynaptic densities of similar size [53]. Asymmetric synapses are either called type I or excitatory, while symmetrical synapses are called type II or inhibitory. On the other hand, synapses can also be classified into discrete categories by their morphology as macular, perforated, horseshoe, or fragmented (*see* Fig. 3c) [4]. Furthermore, due to the BDA-labeling reaction precipitate being electron-dense, we were only able to analyze the PSD, while the presynaptic active zone cannot be properly visualized (*see* Fig. 2) [19].

In the same way, BDA-labeling precipitate can obstruct proper synaptic vesicle counting and measurement (or any other presynaptic subcellular structure). As synaptic vesicles don't exceed 50 nm in diameter and ultrathin sectioning is 50 nm wide, vesicles are counted only once across sections (*see* Fig. 2). Regarding the labeling, the critical step is the glucose oxidase reaction. The aim is to unambiguously identify vesicles while still being able to find the labeled profile. To do so, the reaction can be shortened down so there is less precipitate (~4 min). Minor contrast changes can also be obtained with time adjustments in the uranyl-acetate step as well as the lead citrate step so that the synaptic bouton elements can be more easily identified.

3.8 Image Alignment and Analysis

All series of tile mosaic images that included at least one complete synaptic bouton were subsequently analyzed with the three-dimensional digital reconstruction software *OpenCAR* [37], although other options are also available and are equally useful for the same purpose, like Fiji plugin *TrakEM2* [58, 59], IMOD [60], or *Reconstruct* [61]. In any case, all images on the series were manually aligned with rigid transformations (translation and/or rotation) so as to create a block of images where all the structures of interest were well defined and included within (*see* Fig. 2).

Subsequently, the profiles of interest were segmented. Segmentation involves a series of tools with the purpose of creating a contour, or mask, in each image (*see* Fig. 3b). Once all the contours/masks series are made a three-dimensional volumetric reconstruction can be extrapolated through different algorithms (depending on the software of election). Surface and volume measurements can be obtained from these reconstructions, among others (*see* Fig. 3c–f).

Quantitative information of any structure that has been segmented within the stack of sections can be extracted. This information includes the number, position, and size parameters of each segmented object, as well as the category to which it belongs, so morphological classification analysis can be also made.

For the quantitative ultrastructural analysis of labeled long-range projection axons, each profile was classified as axon, bouton, mitochondria, synaptic vesicles, PSD, dendritic spine, and dendritic shaft. The surface measurements and volume of the dendritic spine were obtained only from the area identified as the spine head, although spine neck measurements are also possible [14]. With the quantitative data of the size of these profiles, it was calculated the mean average, standard deviation (SD), median, range of variation, and coefficient of variation (CV), for each group of data.

3.9 Correlative Light–Electron Microscopy

Probably the most critical step in the whole process is the light–electron microscopy correlation, as it's the only way to ensure objects segmented within the same stack, the same region, area, and/or layer of the brain are being properly sampled (*see* Fig. 1).

Such correlations are difficult to establish in long-range projection axons because different specific systems are organized in a convergent fashion in the cerebral cortex. Therefore, a great level of resolution is required to distinguish different subsets of populations. That way, electron microscopy inter-area/inter-layer studies are highly valuable as they are technically challenging.

As previously stated, two alternate series are collected after vibratome sectioning. This means each EM slice will be separated by a distance of 50 microns from the closest light microscopy counterpart. An stereotaxic brain atlas can be used to assess the coronal levels [50], and fiduciary marks made to correlate both images (*see* Fig. 1a–b). Once images are properly scaled, using rigid transformations with software like *Fiji*, light and electron microscopy slices can be compared and regions of interest selected. Other bioimage editing tools like the *Icy* software suite [62] include specific algorithms like *ec-CLEM* further enhance these correlations and allow better sampling [63].

4 Conclusions

None of the ultrastructural parameters here described has had any correction factor applied for tissue shrinkage due to its processing (aldehyde fixation and dehydration). Comparing aldehyde fixation with tissue cryofixation processes, several studies in the mouse cerebral cortex have described a marked difference in the volume of the extracellular space and glial cells but no differences in the length or volume size of axons or dendrites and its spines [64, 65].

In addition, other studies have shown that there are no significant differences in the values of structural parameters corresponding to the synapses of samples prepared by freezing at high pressure or cryosubstitution compared with samples prepared with a conventional processing of embedding tissue in resin, such as the one described in this chapter [66].

Although the most time-consuming of all, ssTEM still constitutes the gold standard in three-dimensional reconstruction and provides the best imaging resolution (*see* Fig. 3a, b). This allows clear identification of all subcellular structures mentioned above and its segmentation.

In this sense, it is noteworthy that each subset of long-range projection axons can have its own ultrastructural specializations (in the cerebral cortex in this case), such as protrusions of the dendritic spine within the synaptic bouton [14]. Taking this into account is crucial for not overestimating synaptic bouton volumes, in this case.

Although selective labeling can be easily adapted to other electron microscopy technologies like FIB/SEM, these details cannot be properly estimated as in ssTEM. It is also true that other parameters like synapses in non-varicose segments and extra-synaptic mitochondria can only be analyzed with FIB/SEM as these structures can be followed only through a larger stack of images.

Similarly, the density of labeled profiles in the cerebral cortex may also depend on the type of long-range projetion axon system of interest and its cortical layer target. Some thalamocortical projections into cortical layer 1 for example might be pretty scarce [14]. Technical procedure should take precedence and enhance sampling of the region of interest (being more restrictive on the arborizations selected in light–electron microscopy correlations) and adjusting the size of the iontophoretic injection to increase the number of labeled terminals.

In conclusion, selective labeling of long-range projection axons coupled with electron microscopy techniques like ssTEM (or FIB/SEM) still offers a reliable and unbiased option to a wide range of microanatomy research inquiries and poses far away from being obsolete.

Acknowledgments

This research was supported by the Spanish Ministry of Universities Predoctoral Fellowship Program (FPU) and the Human Brain Project-SGA3, Workpackage 2 (Partner 64 UAM).

References

1. Peters A, Palay SL, Webster H (1991) The fine structure of the nervous system: neurons and their supporting cells. Oxford University Press
2. Kharazia VN, Weinberg RJ (1994) Glutamate in thalamic fibers terminating in layer IV of primary sensory cortex. J Neurosci 14(10):6021
3. Castro-Alamancos MA, Connors BW (1997) Thalamocortical synapses. Prog Neurobiol 51(6):581–606
4. Santuy A, Rodríguez JR, DeFelipe J, Merchán-Pérez A (2018) Study of the size and shape of synapses in the juvenile rat somatosensory cortex with 3D electron microscopy. eNeuro 5(1):1–14
5. Gil Z, Connors BW, Amitai Y (1999) Efficacy of thalamocortical and intracortical synaptic connections: quanta, innervation, and reliability. Neuron 23(2):385–397
6. Yuste R (2010) Dendritic spines. MIT Press
7. Puelles L, Martinez-de-la-Torre M, Ferran JL, Watson C (2012) Diencephalon. In: The mouse nervous system. Elsevier Inc., pp 313–336
8. The Cerebral Cortex and Thalamus Oxford University PressNew York
9. Nakagawa Y, Shimogori T (2012) Diversity of thalamic progenitor cells and postmitotic neurons. Eur J Neurosci 35(10):1554–1562
10. Clascá F, Porrero C, Galazo MJ, Rubio-Garrido P, Evangelio M (2016) Anatomy and development of multispecific thalamocortical axons. In: Axons and brain architecture, pp 69–92
11. The Thalamus Thalamic Output Pathways Cambridge University Press 45-70
12. Viaene AN, Petrof I, Sherman SM (2011) Properties of the thalamic projection from the posterior medial nucleus to primary and secondary somatosensory cortices in the mouse. Proc Natl Acad Sci USA 108(44): 18156–18161
13. Mo C, Petrof I, Viaene AN, Sherman SM (2017) Synaptic properties of the lemniscal and paralemniscal pathways to the mouse somatosensory thalamus. Proc Natl Acad Sci USA 114(30):E6212–E6221
14. Rodriguez-Moreno J, Porrero C, Rollenhagen A, Rubio-Teves M, Casas-Torremocha D, Alonso-Nanclares L et al (2020) Area-specific synapse structure in branched posterior nucleus axons reveals a new level of complexity in thalamocortical networks. J Neurosci 40:2886–2819
15. Svara FN, Kornfeld J, Denk W, Bollmann JH (2018) Volume EM reconstruction of spinal cord reveals wiring specificity in speed-related motor circuits. Cell Rep 23(10):2942–2954
16. Wilke SA, Antonios JK, Bushong EA, Badkoobehi A, Malek E, Hwang M et al (2013) Deconstructing complexity: serial block-face electron microscopic analysis of the hippocampal mossy fiber synapse. J Neurosci 33(2):507–522
17. Osorno T, Rudolph S, Nguyen T, Kozareva V, Nadaf NM, Norton A et al (2022) Candelabrum cells are ubiquitous cerebellar cortex interneurons with specialized circuit properties. Nature Neuroscience 25(6):702–713
18. Casas-Torremocha D, Porrero C, Rodriguez-Moreno J, García-Amado M, Lübke JHR, Núñez Á et al (2019) Posterior thalamic nucleus axon terminals have different structure and functional impact in the motor and somatosensory vibrissal cortices. Brain Struct Funct 224(4):1627–1645
19. Rodriguez-Moreno J, Rollenhagen A, Arlandis J, Santuy A, Merchan-Pérez A, DeFelipe J et al (2018) Quantitative 3D ultrastructure of thalamocortical synapses from the "Lemniscal" ventral posteromedial nucleus in mouse barrel cortex. Cereb Cortex 28(9):3159–3175
20. Bruno RM, Sakmann B (2006) Cortex is driven by weak but synchronously active thalamocortical synapses. Science (1979) 312(5780):1622–1627
21. Schoonover CE, Tapia JC, Schilling VC, Wimmer V, Blazeski R, Zhang W et al (2014) Comparative strength and dendritic organization of thalamocortical and corticocortical synapses onto excitatory layer 4 neurons. J Neurosci 34(20):6746–6758
22. Menon V, Musial TF, Liu A, Katz Y, Kath WL, Spruston N et al (2013) Balanced synaptic

impact via distance-dependent synapse distribution and complementary expression of AMPARs and NMDARs in hippocampal dendrites. Neuron 80(6):1451–1463

23. de Rossi P, Nomura T, Andrew RJ, Masse NY, Sampathkumar V, Musial TF et al (2020) Neuronal BIN1 regulates presynaptic neurotransmitter release and memory consolidation. Cell Rep 30(10):3520–3535.e7
24. Buss EW, Corbett NJ, Roberts JG, Ybarra N, Musial TF, Simkin D et al (2021) Cognitive aging is associated with redistribution of synaptic weights in the hippocampus. Proc Natl Acad Sci USA 118(8):e1921481118
25. Nava N, Chen F, Wegener G, Popoli M, Nyengaard JR (2014) A new efficient method for synaptic vesicle quantification reveals differences between medial prefrontal cortex perforated and nonperforated synapses. J Comp Neurol 522(2):284–297
26. Lu SM, Lin RCS (1993) Thalamic afferents of the rat barrel cortex: a light-and electron-microscopic study using phaseolus vulgaris leucoagglutinin as an anterograde tracer. Somatosens Mot Res 10(1):1–16
27. White EL, Weinfeld E, Lev DL (2004) Quantitative analysis of synaptic distribution along thalamocortical axons in adult mouse barrels. J Comp Neurol 479(1):56–69
28. Bopp R, Holler-Rickauer S, Martin KAC, Schuhknecht GFP (2017) An ultrastructural study of the thalamic input to layer 4 of primary motor and primary somatosensory cortex in the mouse. J Neurosci 37(9):2435–2448
29. Hua Y, Loomba S, Pawlak V, Voit KM, Laserstein P, Boergens KM, et al (2022) Connectomic analysis of thalamus-driven disinhibition in cortical layer 4. Cell Rep [Internet] 2022 Oct 11 [cited 2023 Jan 12];41(2)
30. Arellano JI, Benavides-Piccione R, DeFelipe J, Yuste R (2007) Ultrastructure of dendritic spines: correlation between synaptic and spine morphologies. Front Neurosci 1(1):131–143
31. da Costa NM, Martin KAC (2011) How thalamus connects to spiny stellate cells in the cat's visual cortex. J Neurosci 31(8):2925–2937
32. da Costa NM, Martin KAC (2013) Sparse reconstruction of brain circuits: or, how to survive without a microscopic connectome. Neuroimage [Internet] 80:27–36
33. Rollenhagen A, Klook K, Sätzler K, Qi G, Anstötz M, Feldmeyer D et al (2015) Structural determinants underlying the high efficacy of synaptic transmission and plasticity at synaptic boutons in layer 4 of the adult rat 'barrel cortex'. Brain Struct Funct 220(6):3185–3209
34. Dufour A, Rollenhagen A, Sätzler K, Lübke JHR (2016) Development of synaptic boutons in layer 4 of the barrel field of the rat somatosensory cortex: a quantitative analysis. Cereb Cortex 26(2):838–854
35. Sun Y, Smirnov M, Kamasawa N, Yasuda R (2021) Rapid ultrastructural changes in the PSD and surrounding membrane after induction of structural LTP in single dendritic spines. J Neurosci 41(33):7003–7014
36. Knott G, Genoud C (2013) Is EM dead? J Cell Sci 126:4545–4552
37. Sätzler K, Söhl LF, Bollmann JH, Borst JGG, Frotscher M, Sakmann B et al (2002) Three-dimensional reconstruction of a calyx of Held and its postsynaptic principal neuron in the medial nucleus of the trapezoid body. J Neurosci 22(24):10567–10579
38. Petrak LJ, Harris KM, Kirov SA (2005) Synaptogenesis on mature hippocampal dendrites occurs via filopodia and immature spines during blocked synaptic transmission. J Comp Neurol 2005 Apr 4 [cited 2022 Jun 15];484 (2):183–190
39. Baena V, Schalek RL, Lichtman JW, Terasaki M (2019) Serial-section electron microscopy using Automated Tape-Collecting Ultramicrotome (ATUM). Methods Cell Biol 152:41
40. Knott G, Marchman H, Wall D, Lich B (2008) Serial section scanning electron microscopy of adult brain tissue using focused ion beam milling. J Neurosci 28(12):2959–2964
41. Merchán-Pérez A, Rodriguez JR, Alonso-Nanclares L, Schertel A, DeFelipe J (2009) Counting synapses using FIB/SEM microscopy: a true revolution for ultrastructural volume reconstruction. Front Neuroanat 3:944
42. Gordon MG, Kristen S, Harris M (1998) Three-dimensional structure and composition of CA3→CA1 axons in rat hippocampal slices: implications for presynaptic connectivity and compartmentalization. J Neurosci Res 18 (20):8300–8310. https://doi.org/10.1523/JNEUROSCI.18-20-08300.1998
43. Eberle AL, Mikula S, Schalek R, Lichtman J, Knothe Tate ML, Zeidler D (2015) High-resolution, high-throughput imaging with a multibeam scanning electron microscope. J Microsc 259(2):114
44. Motta A, Berning M, Boergens KM, Staffler B, Beining M, Loomba S et al (2019) Dense connectomic reconstruction in layer 4 of the somatosensory cortex. Science 366:eaay3134
45. Shapson-Coe A, Januszewski J, Berger DR, Pope A, Wu Y, Blakely T et al (2024) A petavoxel fragment of human cerebral cortex reconstructed at nanoscale resolution. Science

384(6696). https://doi.org/10.1126/science.adk4858

46. Turner NL, Macrina T, Bae JA, Yang R, Wilson AM, Schneider-Mizell C et al (2022) Reconstruction of neocortex: Organelles compartments cells circuits and activity. Cell 185 (6):1082–1100.e24. https://doi.org/10.1016/j.cell.2022.01.023
47. Shu S, Ju G, Fan L (1988) The glucose oxidase-DAB-nickel method in peroxidase histochemistry of the nervous system. Neurosci Lett 85(2):169–171
48. Reynolds ES (1963) The use of lead citrate at high pH as an electron-opaque stain in electron microscopy. J Cell Biol 17(1):208–212
49. Oberlaender M, Ramirez A, Bruno RM (2012) Sensory experience restructures thalamocortical axons during adulthood. Neuron 74(4):648–655
50. Franklin K, Paxinos G (eds) (2019) The mouse brain in stereotaxic coordinates, 5th edn
51. Margaret TT, Michael W-R, Merzenich M, Leake PA (1978) Changes in endogenous enzymatic reactivity to DAB induced by neuronal inactivity. Brain Res 141(1):185–192. https://doi.org/10.1016/0006-8993(78)90629-7
52. Cossart R, Dinocourt C, Hirsch JC, Merchan-Perez A, de Felipe J, Ben-Ari Y et al (2001) Dendritic but not somatic GABAergic inhibition is decreased in experimental epilepsy. Nat Neurosci 4(1):52–62
53. Gray EG (1959) Electron microscopy of synaptic contacts on dendrite spines of the cerebral cortex. Nature 183(4675):1592–1593
54. Hering H, Sheng M (2001) Dendritic spines: structure, dynamics and regulation. Nat Rev Neurosci 2(12):880–888
55. Ofer N, Berger DR, Kasthuri N, Lichtman JW, Yuste R (2021) Ultrastructural analysis of dendritic spine necks reveals a continuum of spine morphologies. Dev Neurobiol 81(5):746–757
56. Cooney JR, Hurlburt JL, Selig DK, Harris KM, Fiala JC (2002) Endosomal compartments serve multiple hippocampal dendritic spines from a widespread rather than a local store of recycling membrane. J Neurosci 22(6):2215–2224
57. Gray EG (1959) Axo-somatic and axo-dendritic synapses of the cerebral cortex: an electron microscope study. J Anat 93:420–433
58. Schindelin J, Arganda-Carreras I, Frise E, Kaynig V, Longair M, Pietzsch T et al (2012) Fiji: an open-source platform for biological-image analysis. Nat Methods 9:676–682
59. Cardona A, Saalfeld S, Schindelin J, Arganda-Carreras I, Preibisch S, Longair M et al (2012) TrakEM2 software for neural circuit reconstruction. PLoS One 7(6):e38011
60. Kremer JR, Mastronarde DN, McIntosh JR (1996) Computer visualization of three-dimensional image data using IMOD. J Struct Biol 116(1):71–76
61. Fiala JC (2005) Reconstruct: a free editor for serial section microscopy. J Microsc 218 (Pt 1):52–61
62. de Chaumont F, Dallongeville S, Chenouard N, Hervé N, Pop S, Provoost T et al (2012) Icy: an open bioimage informatics platform for extended reproducible research. Nat Methods 9(7):690–696
63. Heiligenstein X, Paul-Gilloteaux P, Belle M, Raposo G, Salamero J (2017) EC-Clem: flexible multidimensional registration software for correlative microscopies with refined accuracy mapping. Microsc Microanal 23(S1):360–361
64. Korogod N, Petersen CCH, Knott GW (2015) Ultrastructural analysis of adult mouse neocortex comparing aldehyde perfusion with cryo fixation. elife 4:e05793
65. Tamada H, Blanc J, Korogod N, Petersen CC, Knott GW (2020) Ultrastructural comparison of dendritic spine morphology preserved with cryo and chemical fixation. elife 9:1–15
66. Zhao S, Studer D, Chai X, Graber W, Brose N, Nestel S et al (2012) Structural plasticity of hippocampal mossy fiber synapses as revealed by high-pressure freezing. J Comp Neurol 520(11):2340–2351

Chapter 12

Quantitative Analysis of the Synaptic Organization of the Human Temporal Lobe Neocortex

Astrid Rollenhagen, Kurt Sätzler, and Joachim H. R. Lübke

Abstract

Synapses are key structural elements that govern the computational properties of any given network in the brain. Although the term synapse was introduced more than 100 years ago, our knowledge about these structures was considerably improved only recently by applying modern techniques including high-end fine-scale transmission and focused ion beam electron microscopy. One of the most important parameters in synaptic neuroscience is the size and shape of the active zone (AZ, equivalent to a functional neurotransmitter release site) and the size and organization of synaptic vesicle pools. Both structural parameters critically determine synaptic efficacy and strength but also the modulation of synaptic plasticity. Meanwhile, there are numerous quantitative studies about the synaptic organization in various mammalian synapses. However, whether findings in experimental animals can be transferred to the human brain one to one is still rather unknown and is thus discussed controversially. Here, we provide a protocol suitable to describe in great detail the synaptic organization in the human brain, as exemplified by the human temporal lobe neocortex (TLN) taken from biopsy samples during epilepsy surgery.

Key words Human brain, 3D volume reconstructions, Quantitative 3D synaptic models, Transmission electron microscopy for 3D reconstructions

1 Introduction

For centuries, the belief that the structure of the brain and its elements reflecting their function had been the "driving force" for investigations and the source of major discoveries in neuroscience. Meanwhile, a wide spectrum of methods including high-end fine-scale transmission electron microscopy (TEM), block-face and focused ion beam scanning electron microscopy (FIB-SEM), and EM tomography have extended these investigations to offer a detailed description of the synapse, the elementary building block of any given neural network of the brain. Since then, a wealth of information was obtained that deepened our understanding of synaptic structures in both health and disease, which were mainly carried out in experimental animals.

Joachim H. R. Lübke and Astrid Rollenhagen (eds.), *New Aspects in Analyzing the Synaptic Organization of the Brain*, Neuromethods, vol. 212, https://doi.org/10.1007/978-1-0716-4019-7_12,

Numerous quantitative studies in various animal species and brain regions have investigated synaptic boutons by describing different synaptic parameters in great detail [1–18].

A further development and important step was to work on synaptic structures in the human brain, including the human neocortex [19–31].

Two approaches have been developed to work on synaptic complexes in the human brain: Firstly, the so-called postmortem brains are taken in a limited time window up to 4 h between the decease of the donor and the removal of the brain, e.g., [21, 29, 30].

Due to a comparatively "bad" ultrastructural preservation of "postmortem" tissue samples, it turned out that such brain tissue is only conditionally suitable for high-end fine-scale EM investigations of synaptic structures, e.g., to describe the organization of synaptic vesicle pools.

An alternative method of choice is the use of biopsy material taken during epilepsy- or tumor neurosurgery, now a worldwide established method for structural and functional investigation on the human brain. It has been demonstrated that such neocortical access tissue samples quite distant from the epileptic focus or tumor can be considered non-affected as demonstrated by structural and functional studies using this experimental approach; see, for example, [19, 25, 26, 31–43]. Hence, it turned out that biopsy tissue is well suited for investigations of the neuronal and synaptic structures in the human brain at the high-resolution EM level due to an excellent preservation of the ultrastructure.

The human TLN, located at the basolateral side of the cerebral hemispheres, is regarded as a highly specialized associative, homotypic granular, six-layered neocortex [44, 45]. Besides various other functions, it is involved in auditory, visual, vestibular, linguistic, and olfactory information processing. Hence, the human TLN represents a higher-order multimodal but not primary neocortex [46]. Being the area of origin and onset of temporal lobe epilepsy, the temporal lobe has become an important field in neuroscience.

Using the human TLN as a model, we provide an experimental protocol describing the synaptic organization in different layers of the human neocortex.

2 Materials and Methods

2.1 Tissue Removal and Fixation

During epilepsy surgery, tissue samples from the human TLN are transected to access the hippocampus via a transcortical, transventricular pathway (Fig. 1a). After their removal, the extracted neocortical tissue samples are immediately immersion-fixed in fresh ice-cold 4% paraformaldehyde and 2.5% glutaraldehyde diluted in 0.1 M phosphate buffer (PB, pH 7.4, 300 mOsm) and then left

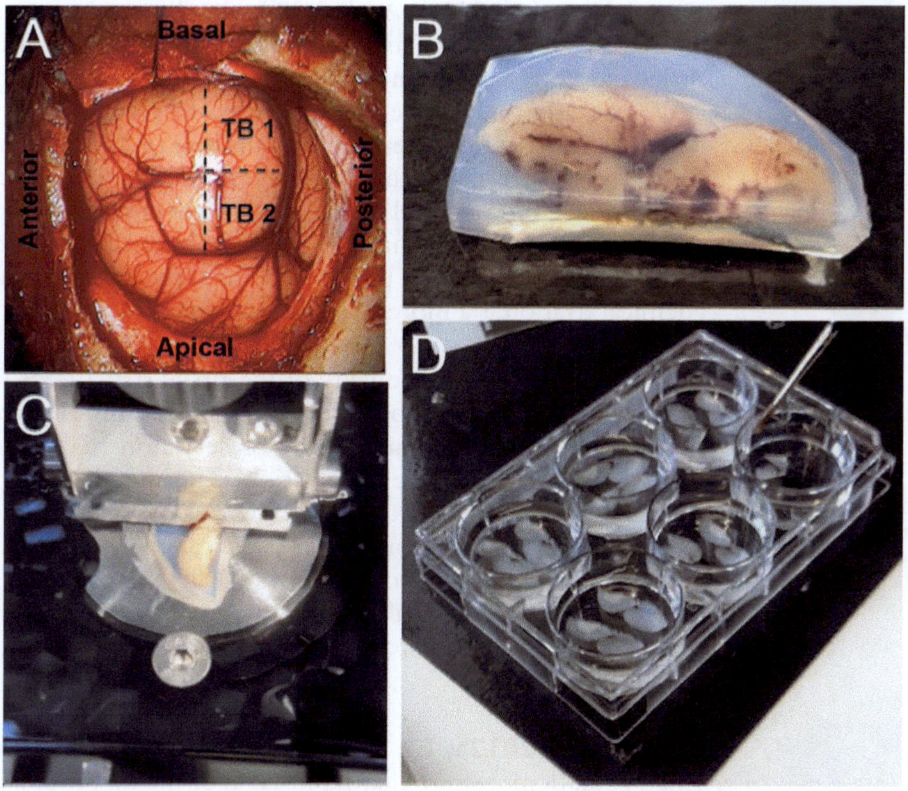

Fig. 1 Removal, embedding, vibratome sectioning, and collection of the human tissue samples. (**a**) View through the operation microscope onto part of the temporal lobe neocortex (gyrus inferior and medialis) during epilepsy surgery. TB1 and TB2 are the neocortical areas that were removed and sampled for electron microscopic embedding and analysis. (**b**) Prior to vibratome sectioning, neocortical tissue samples were embedded in 5% agar-agar to guarantee stability of the tissue block during slicing. (**c**) Vibratome sectioning of the neocortical tissue block in the frontal plane. (**d**) Collecting the vibratome sections in 6-well plates in PB

(in place) for 24–72 h at 4 °C. The glutaraldehyde is added to the paraformaldehyde fixative shortly before the tissue removal from the brain. After 4 h, the fixative is replaced by the same but fresh solution, and tissue samples are stored at 4 °C until further use.

2.1.1 Alternatives in Fixation of Human Tissue Samples for EM Investigations

Cano-Astorga et al. (2021, 2023): Human brain tissues from autopsies with a postmortem delay of less than 4 h were post-fixed for 24 h in a solution containing 2% paraformaldehyde, 2.5% glutaraldehyde, and 0.003% $CaCl_2$ in 0.1 M sodium cacodylate buffer [29, 30].

Domínguez-Álvaro et al. (2019, 2021): The human brain tissue was taken from autopsies with short postmortem delays of less than 3.5 h after the decease of the donor. On removal, brain tissue was fixed in cold 4% paraformaldehyde in 0.1 M sodium phosphate buffer (pH 7.4, for 24–48 h) [27, 28].

Korogod et al. (2015): Analysis of brain ultrastructure using EM typically uses chemical fixation that is thought to cause tissue shrinkage. Alternatively, in this study, rapid, high-pressure freezing (HPF) on adult mouse neocortex is used to compare both methods. As a result, HPF reveals larger numbers of "docked" synaptic vesicles, a smaller glial volume, and a less intimate glial coverage of synapses and blood vessels compared to chemical fixation. However, it was pointed out that with HPF, only small areas within the tissue block are optimally preserved. Thus, the authors stated that chemical fixation is a more than adequate alternative [47].

Sele et al. (2019): The quality of the tissue structure is strongly influenced by the postmortem interval (PMI), the time between the decease of the patient and the preservation (fixation) of the tissue samples. Here, the preparation procedure for the human frontal lobe was optimized to preserve the ultrastructure despite a long PMI up to 16 h. Combining chemical pre- and post-fixation with cryo-fixation and cryo-substitution ("hybrid freezing"), it was possible to improve the ultrastructural preservation of the human brain samples compared to the "standard" epoxy resin embedding method [48].

Zhao et al. (2012a, b): In contrast to Korogod et al. (2015), Zhao et al. demonstrate that rapid HPF of the hippocampal tissue preserves the structural details of mossy fiber synapses and shows no major differences to tissue fixed by aldehyde fixation [49, 50].

2.1.2 Vibratome Sectioning and Embedding

Prior to vibratome sectioning, human TLN tissue samples are thoroughly rinsed in cold 0.1 M PB (pH 7.2–7.4) and embedded afterward in 5% agar-agar (Sigma, Munich, Germany) diluted in 0.1 M PB (Fig. 1b) to guarantee stability of the tissue blocks during slicing. To obtain the best results, it is recommended that slicing should be carried out at high frequency but slow speed. Tissue blocks are cut in the coronal plane through the human TLN with a Vibratome VT1000S (Leica Microsystems GmbH, Wetzlar, Germany) into 150–200-μm-thick sections (Fig. 1c) that are collected in 6-well plates containing 0.1 M PB (Fig. 1d). Afterward, they are washed thoroughly 3× in 0.1 M PB and transferred to 0.5–1% PB-sucrose buffered OsO_4 (300 mOsm, pH 7.4) for 60–90 min. After visual inspection to check the degree of osmification (light-brown appearance), sections are washed several times in 0.1 M PB and left overnight at 4 °C. The next day, they are dehydrated and weighed down by round glass coverslips to prevent the wrinkling of the tissue, through an ascending series of 20%, 30%, 50%, 60%, 70%, 80%, and 90% ethanol (15 min for each step), followed by 95% ethanol (20 min), and finally in 100% ethanol (twice for 30 min).

After the absolute ethanol step, the glass coverslips are removed, and sections are transferred to glass vials and then shortly incubated in propylene oxide (twice for 2 min) followed by a mixture of propylene oxide and Durcupan™ (1:1 and 2:1, 60 min for each step, Fa. Fluka, Neu-Ulm, Germany). Sections are stored after a final change in pure Durcupan™ overnight at room temperature. The next day, sections are flat-embedded in fresh Durcupan™ between ACLAR™ foils on glass object slides and polymerized at 60 °C for 48 h. Section can then be stored in a preparation box until further use.

2.1.3 Alternatives in Embedding Human Tissue Samples

Domínguez-Álvaro et al. (2019): After vibratome sectioning (150 μm thickness), sections are washed in sodium cacodylate buffer (0.1 M) and treated with 1% OsO_4, 0.1% potassium ferrocyanide, and 0.003% $CaCl_2$ in 0.1 M sodium cacodylate buffer for 1 h at room temperature. After washing in 0.1 M PB, sections are stained with aqueous 2% uranyl acetate (2 h), dehydrated in an ascending series of ethanol, and finally flat-embedded in Araldite for 48 h at 60 °C [27].

2.1.4 En Bloc Staining for TEM

During the dehydration procedure with 20%, 30%, 50%, 60% (15 min for each step), in the 70% ethanol, an en block staining with 1% uranyl acetate in 70% ethanol for 1 h can be included. After that, the embedding procedure is identical with the standard protocol, followed by further dehydration in 80%, 90%, and 95% ethanol (20 min) and finally in 100% ethanol (twice for 30 min). As already described above, sections are then incubated in propylene oxide (twice for 2 min), thereafter left in mixtures of propylene oxide and Durcupan™ (1:1 and 2:1, 60 min for each step, Fa. Fluka, Neu-Ulm, Germany), and stored in pure Durcupan™ overnight at room temperature. Finally, sections are flat-embedded in fresh Durcupan™ between ACLAR™ foils and polymerized at 60 °C for 48 h.

2.1.5 En Bloc Staining for Large Tissue Samples Used for FIB-SEM

Hua et al. (2022) and Song et al. (2023): In these publications, a modified en bloc staining protocol for tissue samples ~2 mm in size, e.g., mouse hemispheres, whole brains, and human cortex samples, is reported. Tissue samples are fixated as described above for TEM preparations. In contrast to TEM preparations, vibratome sections were collected in multi-well plates containing 0.15 M cacodylate buffer (CB) + 2 mM $CaCl_2$ and washed several times in the same buffer solution (5× for 3 min). For the en bloc staining, sections are incubated in 1.5% potassium ferricyanide, 2% osmium tetroxide, and 2 mM $CaCl_2$ diluted in CB for 1 h on ice. This is followed by a treatment in 1% aqueous thiocarbohydrazide for 20 min at room temperature (RT) and another subsequent incubation for 30 min in 2% aqueous osmium tetroxide solution at RT. Then, sections are block contrasted with filtered aqueous 1% uranyl acetate overnight at 4 °C before contrasting them further with Walton's lead aspartate staining (aqueous solution of 20 mM lead nitrate in 30 mM aspartic acid) for 30 min at 60 °C. After thorough washing with deionized H_2O (5× for 3 min), sections are dehydrated through an ascending series of ethanol on ice and flat embedded in Durcupan™ as described above [51, 52].

Karlupia et al. (2023): With their protocol for immersion fixation of human and rodent biopsy samples, together with a modified heavy metal staining, their tissue processing method preserves an excellent ultrastructure throughout large samples of volumes up to 37 mm^3.

In contrast to other fixatives used in standard protocols performed at 4 °C, a modified cacodylate buffer (0.15 mM, 295 mOsm, pH 7.4) containing 2% PA and 2.5% GA plus 2 mM $CaCl_2$ is applied, and 3% mannitol (445 mOsm) is added. The add-on of mannitol particularly maintains the extracellular space.

To achieve a uniform and strong heavy metal staining in large tissue samples, several crucial steps are modified when compared to standard protocols. The OsO_4 (2% in cacodylate buffer +2 mM $CaCl_2$) incubation is divided into two steps: first, a 3 h incubation at room temperature for the fixation of the lipids and, second, a 12 h incubation at 4 °C in fresh fixative for staining and improvement of the contrast. A decreased duration (~3 h at room temperature) of potassium ferrocyanide plus 2 mM $CaCl_2$, shorter than the first osmium step, is applied to mitigate the expansion and the brittleness of the tissue. To increase the membrane contrast and reduce breaks, 2 mM $CaCl_2$ is added to all solutions prior to an incubation in thiocarbohydrazide. Furthermore, the incubation with cold (4 °C) uranyl acetate is suggested to reduce cytoplasmic darkening. Between all incubation steps, thorough and multiple washes in cacodylate buffer plus 2 mM $CaCl_2$ or distilled water either at room temperature or at 4 °C are required.

Finally, slices are dehydrated in an ascending series of acetonitrile in distilled water and embedded in LX-112 resin [53].

2.1.6 Semi- and Ultrathin Sectioning

Embedded and osmicated tissue samples (Fig. 2a) of the human TLN are inspected light microscopically to determine a region of interest (ROI), which is then excised and glued onto a pre-polymerized Durcupan™ block. To look for the quality (preservation of the tissue) and to define the final ROI, semithin sections (typically 500 nm) are cut using a histo-diamond knife (Fa. Diatome, Nidau, Switzerland) with a Leica UltracutS ultramicrotome (Leica Microsystems, Vienna, Austria) and stained with methylene blue (Fig. 2b). After inspection and definition of the final ROI, the block is further trimmed to its final size for serial ultrathin sectioning (Fig. 2c).

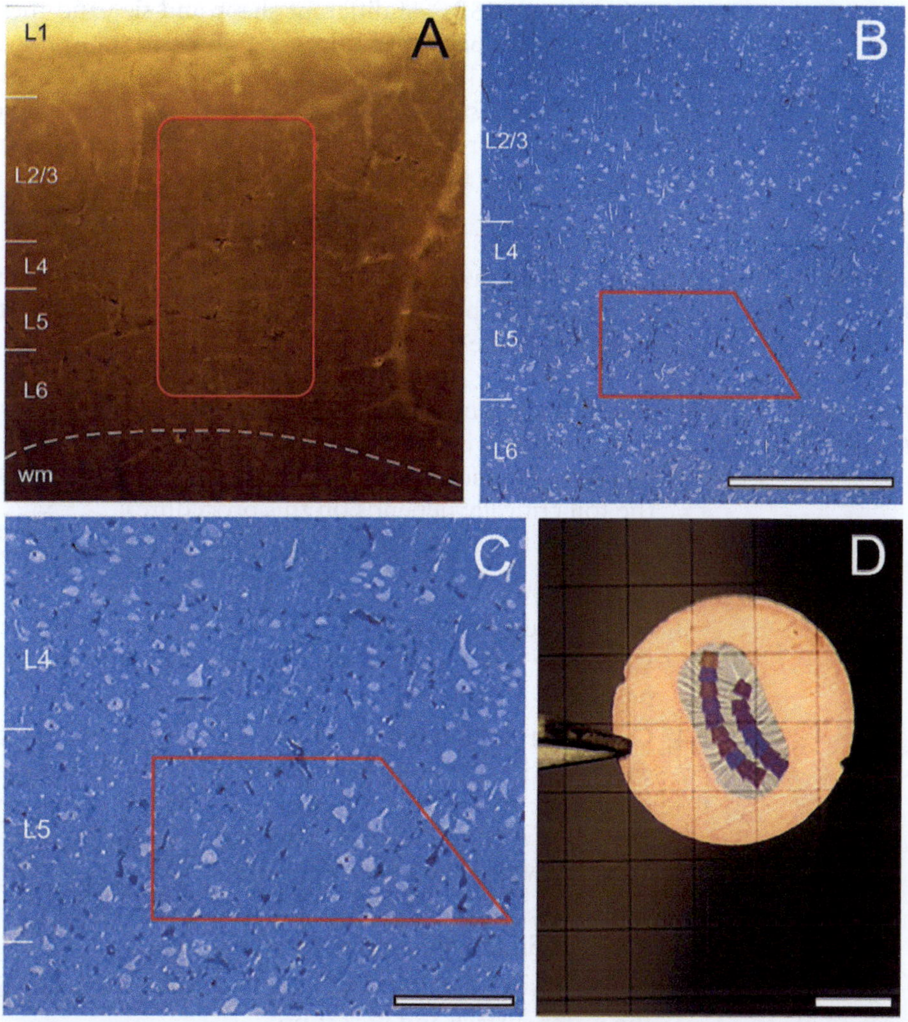

Fig. 2 Definition of a ROI in osmium-treated and Durcupan-embedded vibratome sections. (**a**) Osmium-treated and embedded section through the gyrus medialis of the human TLN showing the characteristic six-layered structure (L1–L6) and the underlying white matter (wm, dashed line). The framed area is shown in Fig. 2b at higher magnification. (**b**) Methylene blue-stained section through the cortical layers 2/3, 4, 5, and 6 as shown by the framed area in **a** (framed area). The red rectangle indicates the ROI in layer 5. Scale bar 2 mm. (**c**) High power of the framed area shown in **b**. In this case, synaptic structures in layer 5 of the human TLN were taken as the ROI, identified by the typical large somata of the layer 5 pyramidal neurons, some of which are highlighted in the framed area. Scale bar 250 μm. (**d**) Series of ultrathin section transferred to a Formvar-coated slot copper grid

Serial ultrathin sections (50 ± 5 nm in thickness, silver to gray interference contrast) are cut with the same Leica UltracutS using a diamond knife (Fa. Diatome, Nidau, Switzerland) and collected on Formvar-coated slot copper grids (see preparation protocol below; Fa. Plano, Wetzlar, Germany; Fig. 2d). An individual series comprises usually between 75 and 150 ultrathin sections to reconstruct synaptic boutons (SBs) of different shape and size and their target structures to meet both structural variability and statistical variance. Prior to EM examination and to increase the contrast, ultrathin sections are finally stained with 5% aqueous uranyl acetate or UranyLess (Science Services, Munich, Germany) for 15–20 min and lead citrate for 3–5 min.

2.1.7 Preparation of a Formvar Solution (0.25–0.3%)

Dissolve 0.25–0.35 g Formvar in 100 mL dichloroethane and/or allow at least 24 h for the Formvar to dissolve before use. The Formvar solution thus prepared is stable for years at room temperature but should be stored in the dark.

2.1.8 Production of Formvar Films

1. Use cleaned, dust-free glass slides. Pour the Formvar solution into a staining cuvette and dip the glass slide into the cuvette (approx. 2 s) and pull it out quickly and evenly (the thickness of the foil critically depends on this procedure).
2. Leave the glass slide to dry vertically and dust-free for about 10 min.
3. Fill a large beaker with double-distilled water up to the rim. Clean the water surface with a glass rod from dust or other particles.
4. Carefully run a scalpel along the lower edges of the glass slide covered with Formvar film. Breathe on the filmed glass slide and, while the surface is still moist, slide the narrow side into the water at an angle of approx. 25° until the Formvar film is completely detached from the slide.
5. Using fine tweezers, place slot grids, shiny side up, on top of Formvar film floating on the water surface.
6. Cover the foil with a slightly larger Parafilm while rolling it on the water surface and pick it up.
7. Place the strips of Parafilm covered with Formvar film and underlying slide meshes in Petri dishes and let them air-dry.
8. Pour the Formvar solution from the cuvette back into the storage bottle and store tightly closed in the dark.
9. Clean the device with dichloroethane.

2.1.9 Alternative Sampling of Ultrathin Sections on Wafers Combined with Scanning EM Imaging

Horstmann et al. (2012): In this approach, serial 35 nm ultrathin sections were collected on silicon wafer and imaged with a SEM at a lateral pixel resolution of 3.7 nm by recording the backscattered electrons with the in-lens detector of the SEM. The resulting electron micrographs are qualitatively comparable to those obtained by conventional TEM and can be used for 3D reconstructions of larger synaptic structures [54].

2.1.10 EM Data Acquisition

TEM

Ultrathin sections are examined with a Zeiss Libra 120 TEM (Fa. Zeiss, Oberkochen, Germany) equipped with a bottom-mounted ProScan 2 K digital camera using the ImageSP software (Fa. Tröndle, Moorenweis, Germany) at a magnification of 8000× with panorama frames of 6 × 6 images (10,444 × 11,129 final pixel size) for the 3D reconstruction of synaptic complexes. After the definition of one to three different ROIs in the ultrathin sections, the different ROIs are imaged in consecutive sections throughout the entire series of ultrathin sections. Afterward, images are stored in a database until further processing and quantitative analysis of synaptic parameters (see below).

2.1.11 Alternatives (See Also Chap. 6 of This Book)

FIB-SEM

An alternative to TEM is a two-beam EM that integrates a fine focus gallium (Ga) ion beam (FIB) into a scanning electron microscope (SEM) (Crossbeam 540 with a STEM detector and ATLAS 3D visualization software, Carl Zeiss AG, Oberkochen, Germany). The FIB system automatically polishes and fines down the surface of the block, whereas the electron optics is used to consecutively generate a series of block-face images from the stepwisely exposed sample planes. Besides a high throughput of experimental samples, the main advantage of this system is the creation of comparatively larger-scaled ROIs of up to 150 × 150 μm with a nearly unlimited z-dimension. In addition, the fine-down process can be reduced to 5 nm by an automatically adjusted fine focus Ga ion beam. The surface of the block is then scanned and imaged with the SEM system at any given magnification with high spatial resolution.

However, using FIB-SEM, the x–y resolution is significantly worse compared to TEM resolution. In previous FIB-SEM measurements, the contours of biological membranes, in particular, those of the active zones (AZs) and synaptic vesicles, appeared blurry at high magnifications. Thus, it is difficult to unequivocally identify these subcellular compartments in a thorough quantitative analysis [55].

2.1.12 Block-Face SEM Imaging

In contrast to conventional serial ultrathin sectioning, the diamond knife for cutting the surface of the block is integrated in the scanning EM microscope [56–59]. Like with FIB-SEM, the surface of the block is imaged, and then a section of defined thickness is cut from the block surface. Advantage: the sectioning and imaging is done automatically, but the resolution is limited by the largeness of the ROI and resolution like with FIB-SEM. For example, the resolution of structural subelements like AZs and synaptic vesicles cannot be accurately determined in x and y resolution when compared with TEM images at a comparable magnification, which is a mandatory factor for quantitative measurements.

2.1.13 ATUMtome-SEM (See Also Chap. 6 of This Book)

Automated tape-collecting ultramicrotome (ATUM)-SEM and ATUM-transmission electron microscopy (TEM) have several features developed for microcircuit investigations using three-dimensional reconstructions of neural elements in very large tissue volumes. This method enables the acquisition of data from 1 cubic millimeter or larger volumes of brain tissue. 3D reconstruction of neural elements can be done via AI-assisted automated segmentation allowing the analysis of synaptic connections in a given brain microcircuit [60].

2.1.14 Correlated Light and Electron Microscopy (CLEM)

CLEM is an approach that benefits from imaging the same object by at least two different techniques to understand structure–function relations in Life Sciences; for review, see, for example, [61–63] While, for example, electron microscopy with TEM or FIB-SEM is able to disclose the structural layout of cells and the macro molecular arrangement of proteins, it cannot directly follow, e.-g., dynamics in living cells. The latter can be achieved with fluorescence microscopy and the CLEM approach; for review, see, for example, [64–67].

Another approach to study structure–function relations with CLEM is to investigate functional aspects of a given network with optogenetic [68] or chemical (Chap. 15 of this book) stimulations as well as two-photon Ca^{2+} imaging [69] combined with EM tomography, TEM, or FIB-SEM.

Furthermore, patch-clamp experiments in living brain slices, e.g., [12, 13, 70, 71], or in organotypic cell cultures, e.g., [72], followed by cell injections of appropriate dyes (e.g., biocytin, dextran amines, or horseradish peroxidase), in combination with subsequent light microscopic and TEM or SEM-FIB analyses, are well-established CLEM techniques. Here, after functional physiological inspection of, for example, a single or paired cortical neuron (s), the overall morphology of the same neuron(s) can be traced and analyzed (see "Alternative to Golgi-Cox Impregnation of Biopsy Material") by means of, e.g., the Neurolucida® software (MicroBrightField Inc., Williston, VT 640, USA), and subsequently investigated by TEM or SEM-FIB imaging. Several protocols describing the histological procedures for this approach are available, at least for animals, e.g., [73, 74].

If functional studies in living human slices are not possible due to the non-availability of appropriate material, but an overall morphology of neurons is desired, an elaborate protocol is described in Chap. 3 of this book (see also [75] and "Alternative to Golgi–Cox Impregnation of Biopsy Material" in this chapter). Here, a method for the analysis of labeled structures at the combined light and electron microscopic level after single-neuron labeling with a mixture of Lucifer yellow and HRP in fixed tissue is given.

2.1.15 LIONESS (Live Information-Optimized Nanoscopy Enabling Saturated Segmentation)

With this elegant approach, the authors have developed an integrated optical/machine learning technology for studying the dynamic functional (nano-)architecture of living brain structures. By imaging the tissue multiple times, it is possible to observe and measure dynamic changes of cellular structures resulting in a 3D reconstruction. The advantages of LIONESS are above all the refined optics and the two layers of deep learning that form the core of the system: the first layer improves the image quality, while the second one identifies the different cellular structures in the dense neural environment. With LIONESS, one can examine living brain tissue, thereby generating 3D images without damaging the living tissue sample. By using Ca^{2+} imaging or patch-clamp recordings combined with intracellular dye fillings, for example, LIONESS offers the opportunity to investigate synaptic contacts between recorded neurons three-dimensionally as well as to study dynamic changes of neuronal structures, e.g., synaptic bouton size [76].

2.1.16 Quantitative Analysis and Three-Dimensional (3D) Volume Reconstructions of Synaptic Structures in the Human TLN

Serial EM images taken with either TEM or FIB-SEM can be imported into OpenCAR (Contour Alignment Reconstruction, [4]) running under Windows on a conventional PC or on a Macintosh computer under OSX64. For the alignment, two consecutive EM images are superimposed via rotation and translation, such that the corresponding structures like mitochondria and membrane specializations (PreAZs or PSDs) in the two sections lie on top of each other (Fig. 3a). This procedure is applied for the entire series of digital images.

After alignment, contouring of the membranes of interest can be done by defining separate contour markers for all synaptic structures of interest (e.g., presynaptic bouton, yellow contour; PreAZ and PSD, red contours; presynaptic mitochondria, white contours; synaptic vesicles, green lines; dense core vesicles, magenta lines; postsynaptic dendrite or spine, blue contours). A subsequent linear interpolation between these markers results in a polygonal membrane "contour" (Fig. 3b). The 3D geometry is then reconstructed from the stacks of all contoured images within a series. By filling these stacks with tetrahedra of known surface area (triangulation: [77–79]), volumes and surface areas of the presynaptic and postsynaptic structures of interest can be calculated.

For the quantitative analysis and 3D volume reconstructions of synaptic structures in the human TLN as described in detail by [25, 26, 31], the following structural parameters were analyzed quantitatively: (1) surface area and volume of synaptic boutons (SBs); (2) volume of presynaptic mitochondria; (3) surface area of the presynaptic active zones (PreAZs) and postsynaptic densities (PSDs); (4) number and diameter of clear synaptic and dense core vesicles (DCVs); and (5) total pool of SVs and the readily releasable (RRP), recycling (RP), and resting pool.

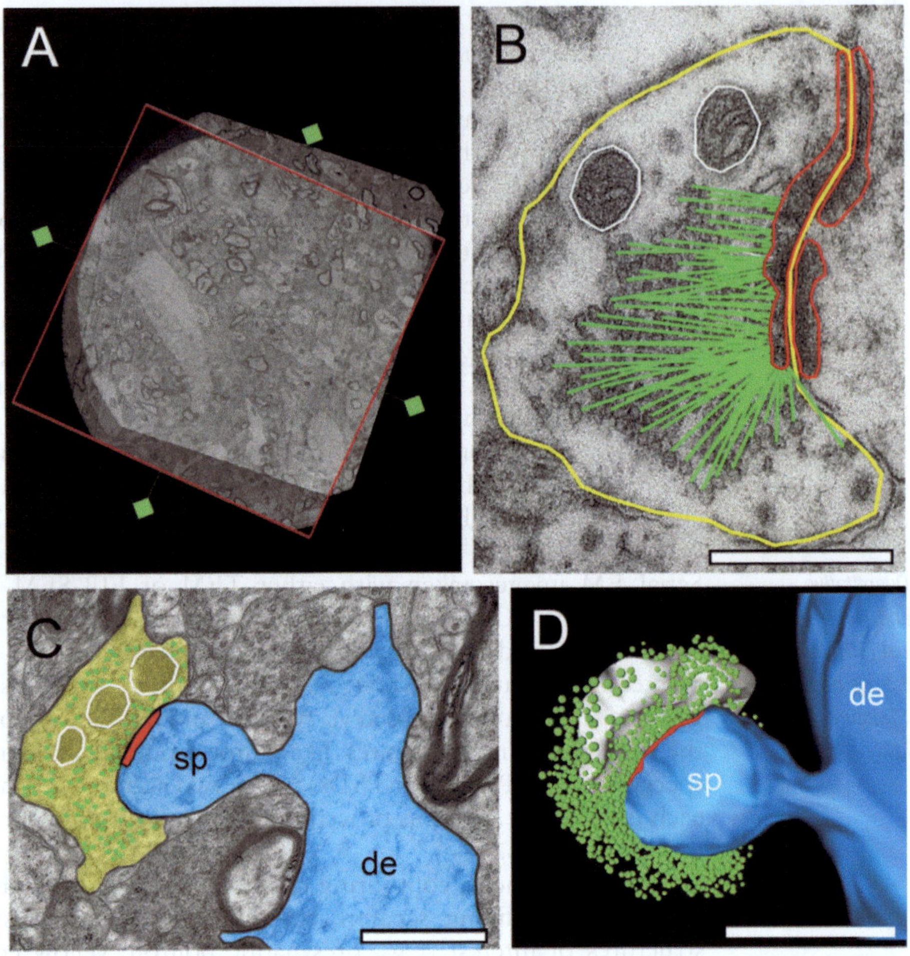

Fig. 3 Alignment, contouring, and 3D volume reconstructions of synaptic structures in OpenCAR. (**a**) Alignment process in OpenCAR. Two digital EM images were aligned and/or rotated until the corresponding structures in both images were superimposed. This alignment was carried out in all images within a series. (**b**) Contouring of the PreAZ, PSD (red contours), mitochondria (white contours), and pool of synaptic vesicles. By drawing an imaginary line, the distance of each synaptic vesicle to the PreAZ was measured in OpenCAR. (**c**) Summary of all contours of interest in the pre- and postsynaptic element. The contour of the presynaptic element is given in yellow, mitochondria as white contours, PreAZ and PSD as red contours, and synaptic vesicles as green lines. The postsynaptic element is always given as a blue contour. Scale bar 0.5 μm. (**d**) 3D volume reconstruction of the mushroom spine shown in **c**. The same color code as in **c**. Scale bar 0.5 μm

All synaptic structures of interest were outlined on the outer edge of their membranes throughout the series of EM images.

For contouring of SBs, a series was inspected for SBs that can be followed from the beginning to their end. A SB was considered completely captured when it was possible to follow the axon in both directions up to the occurrence (swelling) of a SB (*en passant* boutons) or the enlargement of the axon leading to an end terminal bouton. The start of a bouton was defined by the typical widening of the axon and the abrupt occurrence of SVs.

2.1.17 Measurements of PreAZ and PSD

Per definition, the PreAZ and PSD constituting the AZ are regions of electron-dense, dark-appearing material, condensed at the pre- and postsynaptic apposition zone. The surface areas of the PreAZ and PSD are computed separately by first generating a 3D surface model of the SB. Then, the PreAZ is measured by extracting this area from the reconstructed presynaptic bouton membrane covered by this membrane specialization (i.e., where the contour line coincides with <30 nm distance from the presynaptic membrane). Hence, the length (l) of the PreAZ (l PreAZ) and the surface area (SA) of the PreAZ (SA PreAZ) are already known. The size of the PSD opposing the PreAZ is estimated under the following assumptions: (1) both membrane specializations PreAZ and PSD run parallel to each other at the pre- and postsynaptic apposition zone (Figs. 3 and 4), and (2) for both membrane specializations, a contour line is drawn determining their actual length (l PreAZ and l PSD). The surface area of the PSD (SA PSD) is measured using the following equation:

$$\mathrm{SAPSD} = \mathrm{SAPre} * \mathrm{l\ PSD}/\mathrm{l\ PreAZ}$$

which is the perimeter ratio between the outlines of the PSD and that of the synaptic contact.

2.1.18 Synaptic Cleft Measurements

The synaptic cleft diameter is measured because of its importance for the transient increase of glutamate concentration, reversible binding of glutamate to appropriate glutamate receptors, and eventual uptake and diffusion of glutamate out of the cleft. To a large extent, these processes are governed by the geometry of the synaptic cleft and the shape and size of the PreAZs and PSDs. Synaptic cleft width measurement was performed only on synaptic contacts cut perpendicular to the AZ and showing the typical broadening of the synaptic cleft. The distance between the outer edge of the pre- and postsynaptic membranes at the center of the synaptic contact and at the two lateral edges was measured and averaged for each synaptic contact. The two values for the lateral edges were averaged, and a mean ± SD was calculated for each patient. Finally, a total mean ± SD over all patients was given.

2.1.19 Quantitative Analysis of the Three Pools of Synaptic Vesicles

Synaptic vesicles are, besides the shape and size of AZs, key structural elements in storing and releasing neurotransmitters and hence play a fundamental role in synaptic transmission as well as in modulating short- and long-term synaptic plasticity [27]. Their distribution in synaptic terminals and organization into three distinct functionally defined pools, namely, the readily releasable (RRP), the recycling (RP), and the resting pool, regulate synaptic efficacy and strength and determine the mode and probability of release (P_r; uni- vs. multi-vesicular; uni- vs. multi-quantal [80–89]).

To analyze the total pool of SVs, the RRP, RP, and resting pool, all SVs were marked throughout each SB, and their diameters (outer to outer membrane) were measured individually (Figs. 3b

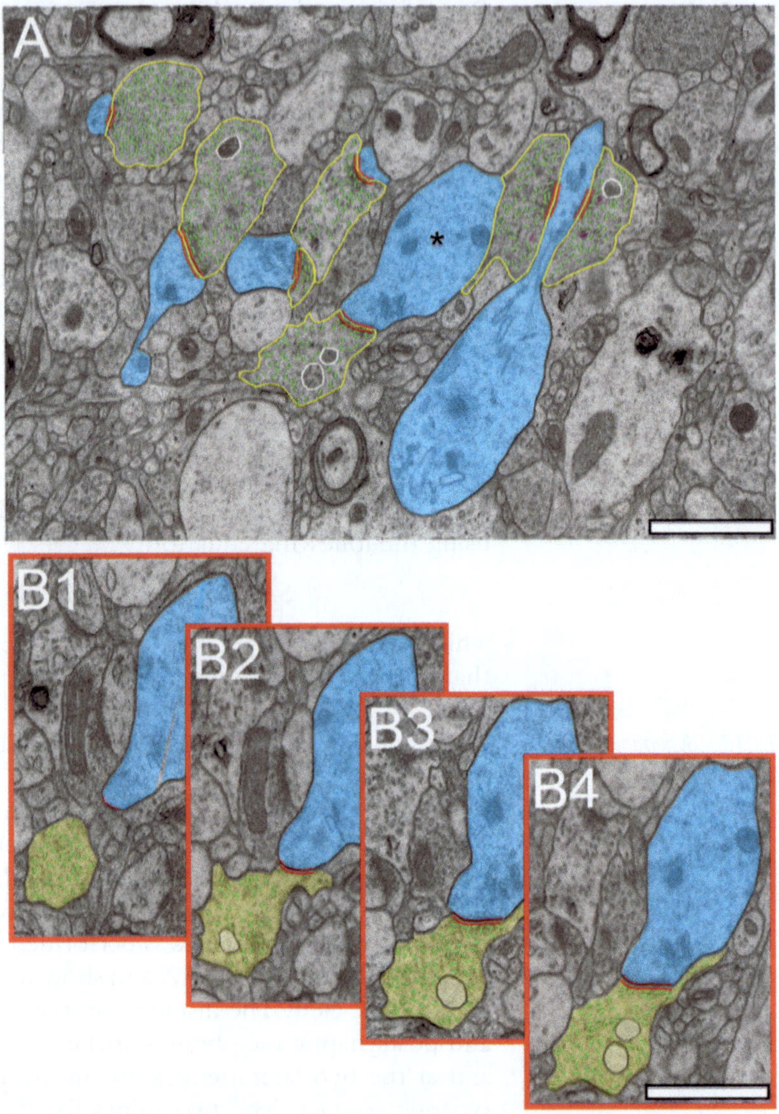

Fig. 4 Contouring structures of interest in consecutive EM micrographs. (**a**) Low-power EM micrograph of a section through layer 6 of the human TLN. Here, several synaptic complexes composed of either a dendritic shaft or spine are given as contours. The postsynaptic elements are always given in transparent blue, the presynaptic elements are given as yellow countors, PreAZs and PSDs as red lines, mitochondria as white contours, and synaptic vesicles as green lines. (**b**1–**b**4) Four consecutive EM images through the stubby spine marked by the black asterisk in **a**. The same color code as in **a**. Scale bars in **a** and **b**1–**b**4 0.5 μm

and 4). To determine the distribution profile of the SVs, the minimal distance between each SV membrane to the contour lines of the PreAZ (Fig. 4b) was calculated throughout the SB in every single image of the series. This led not only to the estimation of the total pool of SVs (Fig. 5b1–3) but also the calculation of the number of

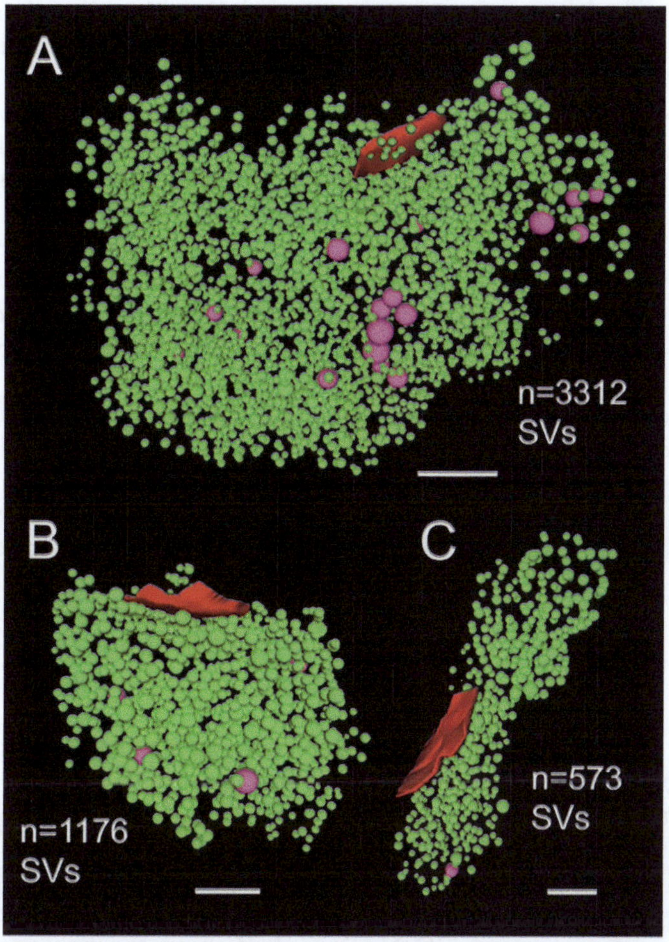

Fig. 5 3D volume reconstructions of the pool of SVs. (**a–c**) Three representative examples of the total pool of SVs (green dots) and DCVs (magenta dots) homogeneously distributed throughout the SB. The PreAZ is given in red. Scale bars: **a**, 0.25 μm, and **b** and **c**, 0.5 μm

SVs in the respective vesicle pools (RRP, RP, and resting pool). To this date, there is no morphological correlate to the three functionally defined pools of SVs. However, using a concentric ring analysis (perimeter analysis) with defined ring size of 10 and 20 nm from the PreAZ (RRP), which is less than a vesicle diameter (~35 nm), 60–100 nm (RP) and >200 nm (resting pool) made it possible to determine the three pools morphologically.

Large DCVs are only counted in the image where they appeared largest.

2.1.20 Alternative Software Packages for 3D Volume Reconstructions of Synaptic Structures

Reconstruct: Reconstruct is a software package developed in the lab of Prof. Kristen Harris (SynapseWeb: http://synapseweb.clm.utexas.edu). The program allows users to trace and annotate images, thereby creating 3D reconstructions as described here [90–92].

EspINA (https://cajalbbp.es/espina/#workflow, Chap. 7 of This Book): EspINA is a tool performing segmentation and analysis of elements in a reconstructed 3D volume of the brain. It facilitates and accelerates these processes allowing visualization and segmentation of large image stack datasets, both from electron microscopy (e.g., FIB/SEM) and confocal laser microscopy [93].

IMOD: If enough image features are similar from one section to the next, IMOD offers an automatic alignment. This software package also allows changes in size and stretching along an arbitrary axis as well as image translation and rotation [94].

2.1.21 Automated Counting of Synaptic Vesicles

Imbrosci et al. (2022): The manual contouring of synaptic vesicles is highly time-consuming and limits the fast throughput of large datasets. To overcome this problem, Imbrosci et al. provide an algorithm automatically detecting and localizing synaptic vesicles, particularly in TEM electron micrographs. The number, coordinates, areas, and nearest neighbor distances of synaptic vesicles can be measured. They also provide a graphical user interface to guide the users through the image analysis, visualization of the results, and proofreading [95].

2.1.22 Golgi–Cox Impregnation of Biopsy Material

If enough biopsy material is available, some of the tissue can be used for Golgi–Cox impregnation (Fig. 6) using the commercially available Hito Golgi-Cox OptimStain Kit™ (Hitobiotec Corp, Kingsport, TE, USA). This approach allows direct comparison of light with electron microscopy and the visualization of the neuronal organization of a brain area of interest, in our case the human TLN, since data about the neuronal organization of the human neocortex are still quite rare, but see, for example, [25, 26, 31, 37, 96]. After removal of the biopsy material, selected tissue samples are briefly rinsed twice in double-distilled water (dd H_2O) and transferred into the impregnation solution overnight at room temperature. The next day, samples are incubated in a fresh impregnation solution and stored for 14 days in the dark at room temperature. Samples are then transferred into solution 3 and kept in the dark at room temperature for 1 day. Thereafter, they are placed into fresh solution 3 in the dark at room temperature for 6 additional days. Then, solution 3 is exchanged, and samples are stored at 4° in the dark overnight. Tissue blocks are embedded in 5% agarose (Carl Roth, Karlsruhe, Germany) diluted in dd H_2O and sectioned with a vibratome in the coronal plane at 100–250 μm thickness and transferred to dd H_2O. After careful removal of the agarose, free-floating sections are incubated in solution 3 for 2–3 min in the dark at room temperature and then immediately placed into dd H_2O, washed several times, and stored overnight. The next day, sections are put into a mixture of solutions 4 and 5 for

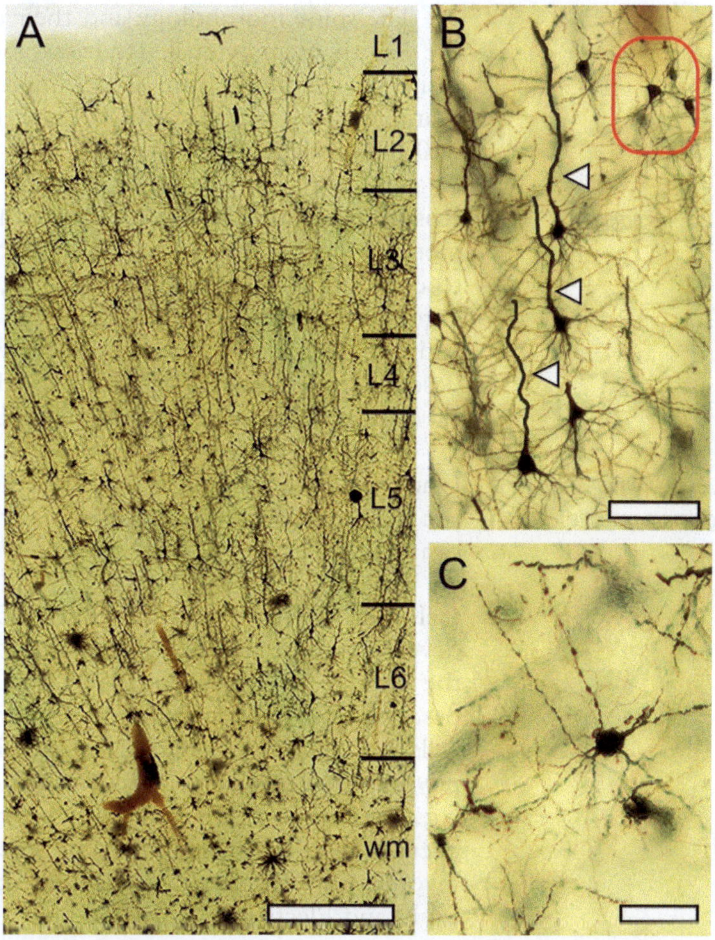

Fig. 6 Golgi impregnation of a part of the medial gyrus of the human TLN. (**a**) Lower-power micrograph of a cortical column (L1-wm) in the human TLN. Scale bar 2 mm. (**b**) Pyramidal neurons in layer 6A (asterisks) identifiable by their characteristic vertical orientation of the apical dendrites. Scale bar 100 μm. (**c**) Putative GABAergic interneuron with a radial orientation of its dendrites in layer 6B. Scale bar 50 μm. (Taken from Schmuhl-Giesen et al. 2022)

10 min at room temperature. Afterward, they are rinsed twice in dd H_2O for 4 min each; dehydrated in 50%, 70%, and 95% ethanol for 5 min each; transferred into absolute ethanol (3 × 5 min); degreased in xylene; and finally embedded in Eukitt™ (Sigma-Aldrich, Taufkirchen, Germany), coverslipped, and air-dried. Finally, sections are examined and imaged with an Olympus BX61 light microscope equipped with the CellSense software package (Olympus, Hamburg, Germany). For publication, selected images are further processed using Adobe Photoshop™ and Adobe Illustrator™ (e.g., [31]).

The spine morphology and their density can be defined, and the length of individual dendrites and number of branches can easily be measured from Golgi-impregnated material using, for example, Neurolucida (MicroBrightField Inc., Williston, VT 640, USA; see, e.g., [96]).

2.1.23 Alternative

Rosoklija et al. (2014): In this study, a modified Golgi–Cox method was applied to fresh human brain tissue, but in contrast to conventional Golgi–Cox protocols, a longer incubation in the chromation solution (at least 10 weeks vs. 2–3 weeks) was used. In addition, the use of Parlodion for a very gradual infiltration was described to eliminate background and to reduce the risk of physical disruption during cutting or handling of sections [97].

2.1.24 Alternatives to Golgi Impregnations

One principal limitation of the Golgi method is the inconsistency of the impregnation. To overcome this problem, single-cell injections can be applied.

Cell Filling with Biocytin by Patch-Clamping In Situ Since the development of the patch-clamp method by Sakmann and Neher in the late 1970s and early 1980s, this method has been used to describe both the physiological properties of single and/or paired cells and their morphological features simultaneously. After electrophysiological recordings, the patched cells were either filled with, e.g., biocytin or HRP and thereafter processed immunohistochemically to visualize and analyze the filled cells for LM and EM investigations (for example, [98–100]; see also above "Correlated Light and Electron Microscopy").

Turegano-Lopez et al. (2022) (See Also Chap. 3 of This Book): As already mentioned above, the availability of "living" human tissue is limited. Here, a protocol for filling cells with a mixture of Lucifer yellow and HRP in fixed tissue and subsequent FIB-SEM analysis followed by 3D reconstructions is described [75].

2.1.25 Measurement of the Synaptic Density

In order to examine and further analyze structural parameters that allow more detailed information in the structural and functional organization of the human brain [21, 27, 29, 30, 32, 101–105], a stereological unbiased estimation of the density of synaptic contacts can be performed using a physical dissector technique [90, 106]. For this analysis, the same series of ultrathin sections can be used as for the 3D reconstructions. To calculate the synaptic density for each patient and cortical layer, a stack of 20 successive frames (also called dissector) within each series was selected, aligned, and edited with Fiji ImageJ [107 https://fiji.sc]. A grid was layered over each image, and newly emerged synapses were marked and counted by using the previous section as a reference to avoid double counting. The results were then inserted in the following equation [90]:

$$Nv = \frac{\sum d\ Q_d}{\sum d\ VQ_d}$$

In this case, Nv is the density of synaptic contacts in a virtual volume calculated for the number of synaptic contacts per volume, Q_d is the number of synaptic contacts per dissector, and V_d is the volume of the dissector given by the number of dissectors x frame area x section thickness.

2.1.26 Alternatives

Merchán-Pérez et al. (2009): Most studies quantifying synaptic densities use stereological methods. Merchán-Pérez et al. [108] estimated the number of synapses by the size–frequency method. Synaptic junctions were counted as described for the dissector method comprising the pre- and postsynaptic membrane densities of each synaptic junction. In order to measure the cross-sectional lengths of the synaptic junctions, they were skeletonized and measured using the ImageJ program [107 https://fiji.sc]. The number of synapses per unit volume (NV) was estimated using the formula NV = NA/d, where NA is the number of synaptic junctions per unit area and d the mean cross-sectional length of the synaptic junctions [109].

Santuy et al. (2020): In this publication, the measurement of the number of synapses in the hippocampus and other brain regions is described by volume electron microscopy and genetic labeling. Genetic labeling of PSD95 and SAP102 was performed, and subsequently spinning disk confocal microscopy (SDM) was used to estimate the number of fluorescent puncta in the CA1 area of the hippocampus. FIB-SEM was the method of choice to measure the actual numbers of synapses. Thereafter, the ratio between the three-dimensional densities obtained with FIB-SEM (synapses/μm^3) and the bi-dimensional densities obtained with SDM (puncta/100 μm^2) was calculated [110].

2.1.27 Automated Detecting of Synapses

Kreshuk et al. (2011) (Ilastik, http://www.ilastik.org/synapse-detection): Ilastik is able to detect automatically and segment synapses in serial electron microscopic images taken with either FIB/SEM or TEM. The procedure is based on interactive machine learning and only requires a few labeled synapses for training [111].

Staffler et al. (2017) (SynEM): Here, a method for automated detection of synapses from conventionally en bloc stained 3D electron microscopy image stacks is reported. The approach is based on segmentation of image data and focuses on classifying borders between neuronal processes as synaptic or non-synaptic. SynEM yields 97% precision and recall in binary cortical connectomes with no user interaction. It scales to large volumes of cortical neuropil, plausibly even whole-brain datasets. SynEM replaces manual synapse annotation for large densely mapped connectomes [112].

2.1.28 Quantitative Analysis of the Astrocytic Coverage of Synaptic Complexes in the Human TLN

To quantify the astrocytic coverage of synaptic complexes that constitute the "tripartite" synapse in the human neocortex (Fig. 7), the interactive software ImageJ [107 https://fiji.sc] can be used. The first, the middle, and the last images of a digital series can be taken for a further quantitative volumetric analysis. In each section of the same series used for the 3D volume reconstructions, a grid (grid size $0.8 \times 0.8\ \mu m^2$) is placed over the EM image, and in each square, the abundance of fine astrocytic processes is documented throughout these images and averaged. Astrocytic profiles can easily be identified by their structural features in the digital images. Using the Cavalieri method [Unbiased Stereology: Three-Dimensional Measurement in Microscopy (Advanced Methods) Paperback—January 7, 2005, by Vyvyan Howard and Matthew Reed], the (absolute) volume contribution of astrocytic processes is determined according to the Cavalieri estimator:

$$V = a(p) \times \Sigma P \times t$$

where $a(p)$ is the size of one square ($0.8 \times 0.8\ \mu m^2$), P is the number of squares counted, and t is the thickness of the slice.

3 Discussion

High-end fine-scale transmission electron microscopy in combination with computer-assisted 3D volume reconstruction and on-demand pre- or post-embedding immunohistochemistry is a powerful tool to unravel the synaptic organization of the brain in experimental animals and humans. The advanced development of various fixation and embedding techniques led to relatively good preservation of the ultrastructure in both "postmortem" brain samples and biopsy material taken during neurosurgery.

3.1 Different Fixation Protocols

Various chemical fixation methods lead to an excellent preservation of the ultrastructure. Working with human brain tissue, one critical point is the time window from the decease of the donor to the removal of the brain tissue sample. Even in "postmortem" brain, a good ultrastructural preservation is still achieved between 3 and 6 h after removal of the brain tissue samples and immediate fixation in glutaraldehyde-containing fixatives [27–30].

This approach is particularly useful in "postmortem" brain material taken from donors suffering from neurological or

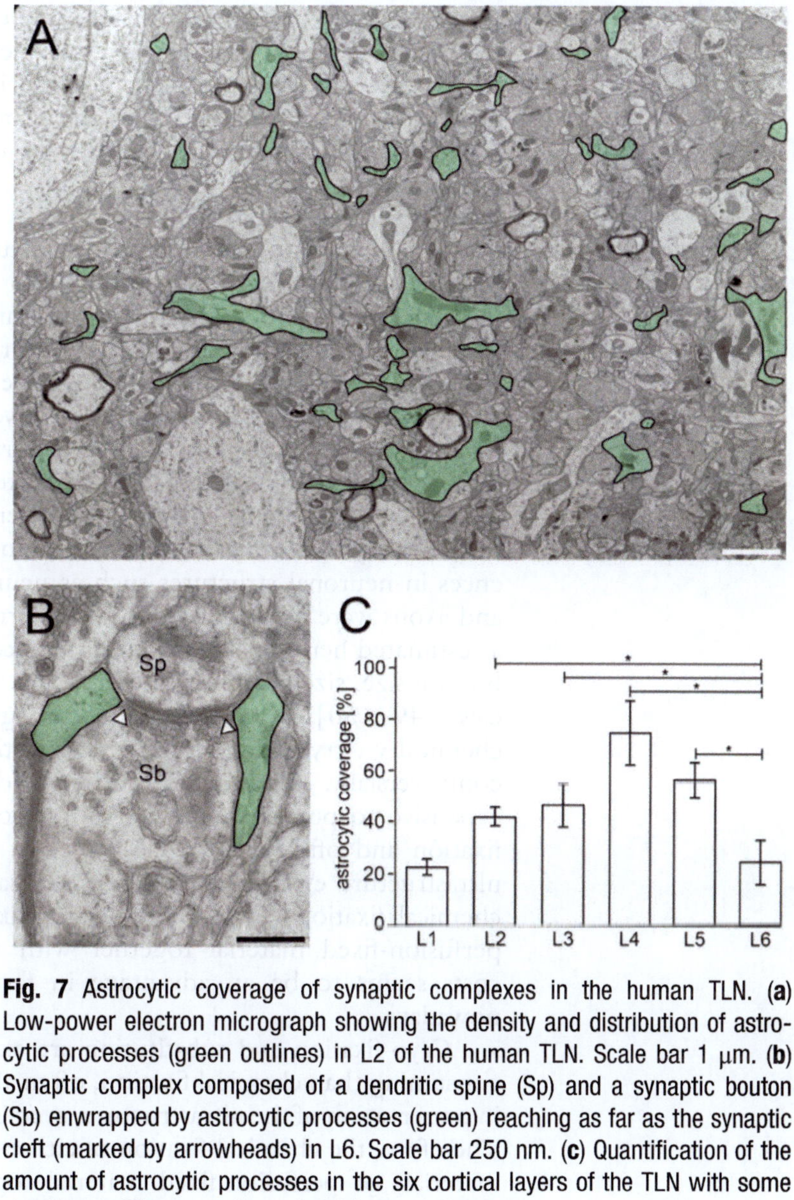

Fig. 7 Astrocytic coverage of synaptic complexes in the human TLN. (**a**) Low-power electron micrograph showing the density and distribution of astrocytic processes (green outlines) in L2 of the human TLN. Scale bar 1 μm. (**b**) Synaptic complex composed of a dendritic spine (Sp) and a synaptic bouton (Sb) enwrapped by astrocytic processes (green) reaching as far as the synaptic cleft (marked by arrowheads) in L6. Scale bar 250 nm. (**c**) Quantification of the amount of astrocytic processes in the six cortical layers of the TLN with some significant differences (*$p < 0.05$) between layers

neurodegenerative disorders such as Alzheimer's or Parkinson disease. On the other hand, biopsy material taken during epileptic or tumor surgery has the advantage of immediate fixation of the brain tissue samples and guarantees an excellent preservation of the ultrastructure in glutaraldehyde-containing fixatives in most cases [25, 26, 31]. Furthermore, biopsy samples taken immediately after their removal during neurosurgery can be used for functional electrophysiological studies of various aspects of brain microcircuits; see, for example, [37, 38, 42, 43].

In our approach, brain tissue samples were fixed after their removal, over a period of 24–72 h to achieve an excellent preservation of the ultrastructure for fine-scale high-resolution electron microscopic investigations. Here, the duration of the fixation does not seem to influence the quality of the ultrastructure (own observations).

3.2 Tissue Shrinkage

In this and other studies on the ultrastructure of the brain, aldehyde fixation was used that is still thought to induce tissue shrinkage, thereby biasing the structural quantification of various synaptic parameters (but see [47, 113]). However, a direct comparison of structural parameters obtained from either aldehyde or cryo-fixed and substituted brain tissue samples showed differences in cortical thickness (~16% larger in cryo-fixed material) and volume of extracellular space (~ sixfold larger in cryo-fixed material) and a slight increase in glial volume and the overall density of synaptic contacts (~14% in cryo-fixed material) [47]. Strikingly, no significant differences in neuronal structures such as neuronal somata, dendrites, and axons were found. Moreover, concerning synaptic parameters as estimated here, no significant difference was found for synaptic bouton size, size of mitochondria, active zones, and synaptic vesicles [49, 50]. Hence, the advantage or disadvantage of chemical vs. cryo-fixation and cryo-substitutions is still discussed controversially. Since cryo-fixation and cryo-substitution are expensive to purchase and more time-consuming than chemical fixation and often led to differences in the preservation of the ultrastructure even in the same tissue sample, the advantage of chemical fixation of either immersion-fixed (biopsy material) or perfusion-fixed material together with their comparably lower costs seems to be an advantage in favor of chemical fixation nowadays.

Cryo-fixation and substitution are expensive and more time-consuming than chemical fixation, often also leading to differences in preservation of the ultrastructure, even in the same tissue sample. Therefore, the chemical fixation of either immersion-fixed (biopsy material) or perfusion-fixed material together with comparably lower costs is an advantage mostly used today.

In summary, no correction for tissue shrinkage seems to be required to quantify various structural parameters that represent morphological correlates for synaptic transmission and plasticity as estimated here after chemical fixation and conventional embedding for electron microscopy. Together with others, we are thus convinced that the generation of quantitative 3D models of synaptic structures is accurate, mirrors the in vivo situation, and can be directly used for detailed computational models of synaptic complexes. Large-scale preservation for ultrastructural analysis will therefore continue to rely on chemical fixation approaches, due to the limited preservation of the ultrastructure in cryo-fixed material as stated in [47].

Serial ultrathin sectioning is very labor- and time-consuming. Thus, serial ultrathin sectioning and imaging of a region of interest (ROI) in stacks of digital electron microscopic images can easily take several weeks. By using a focused ion beam scanning electron microscope (FIB-SEM), a relatively new electron microscopic technique, this time can be significantly reduced. The FIB gun automatically polishes and fines down the surface of the block prepared for electron microscopy, while the electron optics consecutively generates a series of digital images from the stepwisely exposed sample planes. The main advantage of this system is, besides a high throughput of experimental samples, the creation of large-scale ROIs of up to 150 × 150 μm with a nearly unlimited z-dimension. The fine-down process can be adjusted down to 5 nm by an automatically adjusted fine focus gallium ion beam. The surface of the block (block-face imaging) is scanned and imaged at any given magnification with a relatively high spatial resolution with the SEM system.

In comparison, the advantage of the FIB-SEM approach over serial ultrathin sectioning in combination with transmission electron microscopy (TEM) is that:

1. FIB-SEM analyses are less labor- and time-consuming. Employing a FIB-SEM, this time can be reduced to 1–2 days per sample block.
2. The fragile Pioloform or Formvar film on slot grids can be distorted or even burst during imaging, leading to a gap in following up the sections, which, in the worst case, cannot be compensated.
3. Due to several technical hard- and software limitations, the maximum ROI area for a manual serial sectioning approach using a TEM is 25 × 25 μm. One can define a second or third ROI in the series of ultrathin sections with the risks of bursting of the film. It is nearly impossible to create a stack of ultrathin sections beyond 100–150 μm in z-dimension due to sectioning and handling of ultrathin sections.

Unfortunately, the x–y resolution with FIB-SEM is significantly worse compared to a TEM analysis. In FIB-SEM samples at high resolution, the contours of biological membranes, in particular, those of the AZs and synaptic vesicles, appeared blurry, which made it difficult to unequivocally identify these compartments in a thorough quantitative analysis [55]. By modifying the sample preparation process as well as relevant image acquisition parameters, the overall image quality of FIB-SEM technology for the analysis of synaptic parameter can be further improved. That way, both EM techniques can be used side by side in the future as a comprehensive toolbox to adequately address questions about the synaptic organization of the brain.

3.3 Automated vs. by Eye and Hand Reconstruction of Synaptic Structures

Computer-assisted 3D reconstructions of synaptic structures by eye and manual tracing are time-consuming and labor-intensive. Therefore, recent attempts were made for semi- or even fully automated reconstructions of synaptic structures from digital electron microscopic images. One major problem was that nerve tissue contains a high density of synaptic structures, for example, about 3–5 per μm^3 in the mammalian cerebral cortex. Even for small blocks of nerve tissue, dense connectomic mapping requires the identification of millions to billions of synaptic structures with a comparably "bad" signal-to-noise ratio, i.e., the discrimination between a given "real" structural synaptic element and background noise. This is much easier to handle for reconstructing labeled neurons and their processes, which are filled intracellularly with marker substances such as biocytin, a commonly used intracellular tracer. Here, individual neurons and their processes are easily distinguishable from a clear background; see, for example, [114, 115]. While at the beginning the focus of connectomic data analysis has been on dendrites and axon reconstruction, e.g., [115–119], synapse detection became limiting when datasets grew in size and dense mapping was required. However, attempts were made by several groups; see, for example, [94, 111, 112, 120], that developed different software tools for the automated detection of synaptic structures, including the pool of synaptic vesicles from conventionally en bloc stained 3D electron microscopy image stacks or electron microscopic tomography (*see* Chap. 13 by Walkenfort et al. of this book). All approaches are based on the segmentation of the image data and focus on classifying borders between neuronal processes as synaptic or non-synaptic and the shape of synaptic vesicles. Thus, such software tools may help to replace the extremely time-consuming manual synapse annotation for large densely mapped connectomes.

Computer-aided quantitative 3D reconstruction from a series of ultrathin sections through the brain area of interest to obtain quantitative 3D models allows a detailed description of the synaptic organization of the brain, as exemplified by our investigations of the cortical column in the human temporal lobe neocortex in both health and disease. In addition, the comparison with other types of CNS synapses in experimental animals leads to an improved understanding of synaptic complexes embedded in different neuronal networks. Furthermore, the detailed quantitative 3D models and the quantification of morphological parameters of these synapses provide the basis for realistic numeric or Monte Carlo simulations of various parameters of synaptic transmission and plasticity that are still only partially available from electrophysiological or optical measurements, at least in humans.

References

1. Harris KM, Sultan P (1995) Variation in the number, location and size of synaptic vesicles provides an anatomical basis for the nonuniform probability of release at hippocampal CA1 synapses. Neuropharmacology 34: 1387–1395
2. Spirou GA, Rowland KC, Berrebi AS (1998) Ultrastructure of neurons and large synaptic terminals in the lateral nucleus of the trapezoid body of the cat. J Comp Neurol 398: 257–272
3. Xu-Friedman MA, Harris KM, Regehr WG (2001) Three-dimensional comparison of ultrastructural characteristics at depressing and facilitating synapses onto cerebellar Purkinje cells. J Neurosci 21:6666–6672
4. Sätzler K, Söhl LF, Bollmann JH, Borst JG, Frotscher M, Sakmann B, Lübke JH (2002) Three-dimensional reconstruction of a calyx of Held and its postsynaptic principal neuron in the medial nucleus of the trapezoid body. J Neurosci 22:10567–10579
5. Xu-Friedman MA, Regehr WG (2003) Ultrastructural contributions to desensitization at cerebellar mossy fiber to granule cell synapses. J Neurosci 23:2182–2192
6. Marrone DF, LeBoutillier JC, Petit TL (2005) Ultrastructural correlates of vesicular docking in the rat dentate gyrus. Neurosci Lett 378:92–97
7. Freese JL, Amaral DG (2006) Synaptic organization of projections from the amygdala to visual cortical areas TE and V1 in the macaque monkey. J Comp Neurol 496:655–667
8. Rollenhagen A, Sätzler K, Rodriguez EP, Jonas P, Frotscher M, Lübke JH (2007) Structural determinants of transmission at large hippocampal mossy fiber synapses. J Neurosci 27:10434–10444
9. Popov VI, Stewart MG (2009) Complexity of contacts between synaptic boutons and dendritic spines in adult rat hippocampus: three-dimensional reconstructions from serial ultrathin sections in vivo. Synapse 63:369–377
10. Wilke SA, Antonios JK, Bushong EA, Badkoobehi A, Malek E, Hwang M, Terada M, Ellisman MH, Ghosh A (2013) Deconstructing complexity: serial block-face electron microscopic analysis of the hippocampal mossy fiber synapse. J Neurosci 33: 507–522
11. Elston GN, Manger P (2014) Pyramidal cells in V1 of African rodents are bigger, more branched and more spiny than those in primates. Front Neuroanat 8:4
12. Rollenhagen A, Klook K, Sätzler K, Qi G, Anstötz M, Feldmeyer D, Lübke JH (2015) Structural determinants underlying the high efficacy of synaptic transmission and plasticity at synaptic boutons in layer 4 of the adult rat 'barrel cortex'. Brain Struct Funct 220:3185–3209
13. Rollenhagen A, Ohana O, Sätzler K, Hilgetag CC, Kuhl D, Lübke JHR (2018) Structural properties of synaptic transmission and temporal dynamics at excitatory layer 5B synapses in the adult rat somatosensory cortex. Front Synaptic Neurosci 10:24
14. Bopp R, Holler-Rickauer S, Martin KA, Schuhknecht GF (2017) An ultrastructural study of the thalamic input to layer 4 of primary motor and primary somatosensory cortex in the mouse. J Neurosci 37:2435–2448
15. Hsu A, Luebke JI, Medalla M (2017) Comparative ultrastructural features of excitatory synapses in the visual and frontal cortices of the adult mouse and monkey. J Comp Neurol 525:2175–2191
16. Rodriguez-Moreno J, Rollenhagen A, Arlandis J, Santuy A, Merchan-Perez A, DeFelipe J, Lübke JHR, Clasca F (2018) Quantitative 3D ultrastructure of thalamocortical synapses from the "Lemniscal" ventral posteromedial nucleus in mouse barrel cortex. Cereb Cortex 28:3159–3175
17. Rodriguez-Moreno J, Porrero C, Rollenhagen A, Rubio Teves M, Casas-Torremocha D, Alonso-Nanclares L, Yakoubi R, Santuy A, Merchan-Pérez A, DeFelipe J, Lübke JHR, Clasca F (2020) Area-specific synapse structure in branched posterior nucleus axons reveals a new level of complexity in thalamocortical networks. J Neurosci 40:2663–2679. https://doi.org/10.1523/JNEUROSCI.2886-19.2020
18. Rollenhagen A, Lübke JH (2006) The morphology of excitatory central synapses: from structure to function. Cell Tissue Res 326: 221–237
19. Alonso-Nanclares L, Gonzalez-Soriano J, Rodriguez JR, DeFelipe J (2008) Gender differences in human cortical synaptic density. Proc Natl Acad Sci USA 105:14615–14619
20. DeFelipe J (2011) The evolution of the brain, the human nature of cortical circuits, and intellectual creativity. Front Neuroanat 5:29
21. Blazquez-Llorca L, Merchan-Perez A, Rodriguez JR, Gascon J, DeFelipe J (2013) FIB/-SEM technology and Alzheimer's disease:

three-dimensional analysis of human cortical synapses. J Alzheimers Dis 34:995–1013

22. Kay KR, Smith C, Wright AK, Serrano-Pozo A, Pooler AM, Koffie R, Bastin ME, Bak TH, Abrahams S, Kopeikina KJ, McGuone D, Frosch MP, Gillingwater TH, Hyman BT, Spires-Jones TL (2013) Studying synapses in human brain with array tomography and electron microscopy. Nat Protoc 8: 1366–1380
23. Bernhardt BC, Hong S, Bernasconi A, Bernasconi N (2013) Imaging structural and functional brain networks in temporal lobe epilepsy. Front Hum Neurosci 7:624
24. Liu XB, Schumann CM (2014) Optimization of electron microscopy for human brains with long-term fixation and fixed-frozen sections. Acta Neuropathol Commun 2:42
25. Yakoubi R, Rollenhagen A, von Lehe M, Miller D, Walkenfort B, Hasenberg M, Sätzler K, Lübke JH (2019) Ultrastructural heterogeneity of layer 4 excitatory synaptic boutons in the adult human temporal lobe neocortex. Elife 8
26. Yakoubi R, Rollenhagen A, von Lehe M, Shao Y, Sätzler K, Lübke JHR (2019) Quantitative three-dimensional reconstructions of excitatory synaptic boutons in layer 5 of the adult human temporal lobe neocortex: a fine-scale electron microscopic analysis. Cereb Cortex 29:2797–2814
27. Dominguez-Alvaro M, Montero-Crespo M, Blazquez-Llorca L, DeFelipe J, Alonso-Nanclares L (2019) 3D electron microscopy study of synaptic organization of the normal human transentorhinal cortex and its possible alterations in Alzheimer's disease, vol 6, p eNeuro
28. Dominguez-Alvaro M, Montero-Crespo M, Blazquez-Llorca L, DeFelipe J, Alonso-Nanclares L (2021) 3D ultrastructural study of synapses in the human entorhinal cortex. Cereb Cortex 31:410–425
29. Cano-Astorga N, DeFelipe J, Alonso-Nanclares L (2021) Three-dimensional synaptic organization of layer III of the human temporal neocortex. Cereb Cortex 31:4742–4764
30. Cano-Astorga N, Plaza-Alonso S, DeFelipe J, Alonso-Nanclares L (2023) 3D synaptic organization of layer III of the human anterior cingulate and temporopolar cortex. Cereb Cortex 15:bhad232. https://doi.org/10.1093/cercor/bhad232
31. Schmuhl-Giesen S, Rollenhagen A, Walkenfort B, Yakoubi R, Sätzler K, Miller D, von Lehe M, Hasenberg M, Lübke JHR (2021) Sublamina-specific dynamics and ultrastructural heterogeneity of layer 6 excitatory synaptic boutons in the adult human temporal lobe neocortex. Cereb Cortex
32. Alonso-Nanclares L, Kastanauskaite A, Rodriguez JR, Gonzalez-Soriano J, Defelipe J (2011) A stereological study of synapse number in the epileptic human hippocampus. Front Neuroanat 5:8
33. Testa-Silva G, Verhoog MB, Goriounova NA, Loebel A, Hjorth J, Baayen JC, de Kock CP, Mansvelder HD (2010) Human synapses show a wide temporal window for spike-timing-dependent plasticity. Front Synaptic Neurosci 2:12. https://doi.org/10.3389/fnsyn.2010.00012
34. Testa-Silva G, Verhoog MB, Linaro D, de Kock CP, Baayen JC, Meredith RM, De Zeeuw CI, Giugliano M, Mansvelder HD (2014) High bandwidth synaptic communication and frequency tracking in human neocortex. PLoS Biol 12:e1002007. https://doi.org/10.1371/journal.pbio.1002007
35. Testa-Silva G, Rosier M, Honnuraiah S, Guzulaitis R, Megias AM, French C, King J, Drummond K, Palmer LM, Stuart GJ (2022) High synaptic threshold for dendritic NMDA spike generation in human layer 2/3 pyramidal neurons. Cell Rep 41:111787
36. Navarrete M, Perea G, Maglio L, Pastor J, Garcia de Sola R, Araque A (2013) Astrocyte calcium signal and gliotransmission in human brain tissue. Cereb Cortex 23:1240–1246
37. Mohan H, Verhoog MB, Doreswamy KK, Eyal G, Aardse R, Lodder BN, Goriounova NA, Asamoah B, AB BB, Groot C, van der Sluis S, Testa-Silva G, Obermayer J, Boudewijns ZS, Narayanan RT, Baayen JC, Segev I, Mansvelder HD, de Kock CP (2015) Dendritic and axonal architecture of individual pyramidal neurons across layers of adult human neocortex. Cereb Cortex 25:4839–4853
38. Molnar G, Rozsa M, Baka J, Holderith N, Barzo P, Nusser Z, Tamas G (2016) Human pyramidal to interneuron synapses are mediated by multi-vesicular release and multiple docked vesicles. Elife 5
39. Obermayer J, Verhoog MB, Luchicchi A, Mansvelder HD (2017) Cholinergic modulation of cortical microcircuits is layer-specific: evidence from rodent, monkey and human brain. Front Neural Circuits 11:100
40. Goriounova NA, Heyer DB, Wilbers R, Verhoog MB, Giugliano M, Verbist C, Obermayer J, Kerkhofs A, Smeding H, Verberne M, Idema S, Baayen JC, Pieneman AW, de Kock CP, Klein M, Mansvelder HD

(2018) Large and fast human pyramidal neurons associate with intelligence. Elife 7

41. Kramvis I, Mansvelder HD, Meredith RM (2018) Neuronal life after death: electrophysiologic recordings from neurons in adult human brain tissue obtained through surgical resection or postmortem. Handb Clin Neurol 150:319–333
42. Seeman SC, Campagnola L, Davoudian PA, Hoggarth A, Hage TA, Bosma-Moody A, Baker CA, Lee JH, Mihalas S, Teeter C, Ko AL, Ojemann JG, Gwinn RP, Silbergeld DL, Cobbs C, Phillips J, Lein E, Murphy G, Koch C, Zeng H, Jarsky T (2018) Sparse recurrent excitatory connectivity in the microcircuit of the adult mouse and human cortex. Elife 7
43. Mansvelder HD, Verhoog MB, Goriounova NA (2019) Synaptic plasticity in human cortical circuits: cellular mechanisms of learning and memory in the human brain? Curr Opin Neurobiol 54:186–193
44. Zilles K, Bacha-Trams M, Palomero-Gallagher N, Amunts K, Friederici AD (2015) Common molecular basis of the sentence comprehension network revealed by neurotransmitter receptor fingerprints. Cortex 63:79–89
45. Zilles K, Palomero-Gallagher N (2017) Multiple transmitter receptors in regions and layers of the human cerebral cortex. Front Neuroanat 11:78
46. Insausti R (2013) Comparative neuroanatomical parcellation of the human and nonhuman primate temporal pole. J Comp Neurol 521: 4163–4176
47. Korogod N, Petersen CC, Knott GW (2015) Ultrastructural analysis of adult mouse neocortex comparing aldehyde perfusion with cryo fixation. Elife 4
48. Sele M, Wernitznig S, Lipovsek S, Radulovic S, Haybaeck J, Birkl-Toeglhofer AM, Wodlej C, Kleinegger F, Sygulla S, Leoni M, Ropele S, Leitinger G (2019) Optimization of ultrastructural preservation of human brain for transmission electron microscopy after long post-mortem intervals. Acta Neuropathol Commun 7:144
49. Zhao S, Studer D, Chai X, Graber W, Brose N, Nestel S, Young C, Rodriguez EP, Saetzler K, Frotscher M (2012a) Structural plasticity of hippocampal mossy fiber synapses as revealed by high-pressure freezing. J Comp Neurol 520:2340–2351. https://doi.org/10.1002/cne
50. Zhao S, Studer D, Graber W, Nestel S, Frotscher M (2012b) Fine structure of hippocampal mossy fiber synapses following rapid high-pressure freezing. Epilepsia 53 (Suppl 1):4–8. https://doi.org/10.1111/j.1528-1167.2012.03469.x
51. Hua Y, Loomba S, Pawlak V, Voit KM, Laserstein P, Boergens KM, Wallace DJ, Kerr JND, Helmstaedter M (2022) Connectomic analysis of thalamus-driven disinhibition in cortical layer 4. Cell Rep 41:111476. https://doi.org/10.1016/j.celrep.2022.111476
52. Song K, Feng Z, Helmstaedter M (2023) High-contrast en bloc staining of mouse whole-brain and human brain samples for EM-based connectomics. Nat Methods 20: 836–840. https://doi.org/10.1038/s41592-023-01866-3
53. Karlupia N, Schalek RL, Wu Y, Meirovitch Y, Wei D, Charney AW, Kopell BH, Lichtman JW (2023) Immersion fixation and staining of multicubic millimeter volumes for electron microscopy-based connectomics of human brain biopsies. Biol Psychiatry 94:352–360. https://doi.org/10.1016/j.biopsych.2023.01.025
54. Horstmann H, Korber C, Sätzler K, Aydin D, Kuner T (2012) Serial section scanning electron microscopy (S3EM) on silicon wafers for ultra-structural volume imaging of cells and tissues. PLoS One 7:e35172
55. Rollenhagen A, Walkenfort B, Yakoubi R, Klauke SA, Schmuhl-Giesen SF, Heinen-Weiler J, Voortmann S, Marshallsay B, Palaz T, Holz U, Hasenberg M, Lübke JHR (2020) Synaptic organization of the human temporal lobe neocortex as revealed by high-resolution transmission, focused ion beam scanning, and electron microscopic tomography. Int J Mol Sci 21
56. Denk W, Horstmann H (2004) Serial block-face scanning electron microscopy to reconstruct three-dimensional tissue nanostructure. PLoS Biol 2:e329. https://doi.org/10.1371/journal.pbio.0020329
57. Karimi A, Odenthal J, Drawitsch F, Boergens KM, Helmstaedter M (2020) Cell-type specific innervation of cortical pyramidal cells at their apical dendrites. Elife 9:e46876. https://doi.org/10.7554/eLife.46876
58. Denk W, Briggman KL, Helmstaedter M (2012) Structural neurobiology: missing link to a mechanistic understanding of neural computation. Nat Rev Neurosci 13:351–358. https://doi.org/10.1038/nrn3169
59. Helmstaedter M (2013) Cellular-resolution connectomics: challenges of dense neural circuit reconstruction. Nat Methods:501–507. https://doi.org/10.1038/nmeth.2476

60. Kubota Y, Sohn J, Kawaguchi Y (2018) Large volume electron microscopy and neural microcircuit analysis. Front Neural Circuits 12:98
61. Begemann I, Galic M (2016) Correlative light electron microscopy: connecting synaptic structure and function. Front Synaptic Neurosci 8:28
62. Hauser M, Wojcik M, Kim D, Mahmoudi M, Li W, Xu K (2017) Correlative super-resolution microscopy: new dimensions and new opportunities. Chem Rev 117:7428–7456
63. Mironov AA, Beznoussenko GV (2023) Algorithm for modern electron microscopic examination of the Golgi complex. Methods Mol Biol 2557:161–209
64. Möbius W, Posthuma G (2019) Sugar and ice: immunoelectron microscopy using cryosections according to the Tokuyasu method. Tissue Cell 57:90–102. https://doi.org/10.1016/j.tice.2018.08.010
65. Iwasaki H, Ichinose S, Tajika Y, Murakami T (2022) Recent technological advances in correlative light and electron microscopy for the comprehensive analysis of neural circuits. Front Neuroanat 16:1061078. https://doi.org/10.3389/fnana.2022.1061078
66. Hayashi S, Ohno N, Knott G, Molnár Z (2023) Correlative light and volume electron microscopy to study brain development. Microscopy 72:279–286. https://doi.org/10.1093/jmicro/dfad002
67. Tanner H, Sherwin O, Verkade P (2023) Labelling strategies for correlative light electron microscopy. Microsc Res Tech 86:901–910. https://doi.org/10.1002/jemt.24304
68. Chakrabarti R, Jaime Tobon LM, Slitin L, Redondo Canales M, Hoch G, Slashcheva M, Fritsch E, Bodensiek K, Ozcete OD, Gultas M, Michanski S, Opazo F, Neef J, Pangrsic T, Moser T, Wichmann C (2022) Optogenetics and electron tomography for structure-function analysis of cochlear ribbon synapses. Elife 11
69. Eltes T, Kirizs T, Nusser Z, Holderith N (2017) Target cell type-dependent differences in Ca(2+) channel function underlie distinct release probabilities at hippocampal glutamatergic terminals. J Neurosci 37:1910–1924
70. Anstötz M, Huang H, Marchionni I, Haumann I, Maccaferri G, Lübke JH (2016) Developmental profile, morphology, and synaptic connectivity of Cajal-Retzius cells in the postnatal mouse hippocampus. Cereb Cortex 26:855–872
71. Gonzalez-Burgos G, Miyamae T, Krimer Y, Gulchina Y, Pafundo DE, Krimer O, Bazmi H, Arion D, Enwright JF, Fish KN, Lewis DA (2019) Distinct properties of layer 3 pyramidal neurons from prefrontal and parietal areas of the monkey neocortex. J Neurosci 39:7277–7290. Harris KM, Sultan P (1995) Variation in the number, location and size of synaptic vesicles provides an anatomical basis for the nonuniform probability of release at hippocampal CA1 synapses. Neuropharmacology 34:1387–1395
72. Sun Y, Smirnov M, Kamasawa N, Yasuda R (2021) Rapid ultrastructural changes in the PSD and surrounding membrane after induction of structural LTP in single dendritic spines. J Neurosci 41:7003–7014
73. Marx M, Günter RH, Hucko W, Radnikow G, Feldmeyer D (2012) Improved biocytin labeling and neuronal 3D reconstruction. Nat Protoc 7:394–407. https://doi.org/10.1038/nprot.2011
74. Perkins GA, Jackson DR, Spirou GA (2015) Resolving presynaptic structure by electron tomography. Synapse 69:268–282
75. Turegano-Lopez M, Santuy A, Kastanauskaite A, Rodriguez JR, DeFelipe J, Merchan-Perez A (2022) Single-neuron labeling in fixed tissue and targeted volume electron microscopy. Front Neuroanat 16: 852057. Pyramidal Neurons Cell Rep 41: 111787. https://doi.org/10.1016/j.celrep.2022.111787
76. Velicky P, Miguel E, Michalska JM, Lyudchik J, Wei D, Lin Z, Watson JF, Troidl J, Beyer J, Ben-Simon Y, Sommer C, Jahr W, Cenameri A, Broichhagen J, Grant SGN, Jonas P, Novarino G, Pfister H, Bickel B, Danzl JG (2023) Dense 4D nanoscale reconstruction of living brain tissue. Nat Methods 20:1256–1265
77. Boissonnat JD (1988) Shape reconstruction from planar cross sections. Comput Vis Graph Image Process 44:1–29
78. Boissonnat JD, Geiger B (1992) Three-dimensional reconstruction of complex shapes based on the Delaunay triangulation. INRIA Tech Rep:1697
79. Eils R, Sätzler K (1999) Shape reconstruction from volumetric data. In: Jähne B, Haussecker H, Geissler P (eds) Handbook of computer vision and applications. Academic, San Diego, pp 791–815
80. Südhof TC (2012) The presynaptic active zone. Neuron 75:11–25
81. Silver RA, Lübke J, Sakmann B, Feldmeyer D (2003) High-probability uniquantal

transmission at excitatory synapses in barrel cortex. Science 302:1981–1984

82. Saviane C, Silver RA (2006) Fast vesicle reloading and a large pool sustain high bandwidth transmission at a central synapse. Nature 439:983–987
83. Watanabe S, Trimbuch T, Camacho-Perez M, Rost BR, Brokowski B, Sohl-Kielczynski B, Felies A, Davis MW, Rosenmund C, Jorgensen EM (2014) Clathrin regenerates synaptic vesicles from endosomes. Nature 515:228–233
84. Schikorski T (2014) Readily releasable vesicles recycle at the active zone of hippocampal synapses. Proc Natl Acad Sci USA 111: 5415–5420
85. Vaden JH, Banumurthy G, Gusarevich ES, Overstreet-Wadiche L, Wadiche JI (2019) The readily-releasable pool dynamically regulates multivesicular release. Elife 8
86. Schneggenburger R, Sakaba T, Neher E (2002) Vesicle pools and short-term synaptic depression: lessons from a large synapse. Trends Neurosci 25:206–212
87. Rizzoli SO, Betz WJ (2005) Synaptic vesicle pools. Nat Rev Neurosci 6:57–69
88. Neher E (2015) Merits and limitations of vesicle pool models in view of heterogeneous populations of synaptic vesicles. Neuron 87: 1131–1142
89. Chamberland S, Toth K (2016) Functionally heterogeneous synaptic vesicle pools support diverse synaptic signalling. J Physiol 594:825–835
90. Fiala JC, Harris KM (2001) Extending unbiased stereology of brain ultrastructure to three-dimensional volumes. J Am Med Inform Assoc 8:1–16
91. Fiala JC (2005) Reconstruct: a free editor for serial section microscopy. J Microsc 218:52–61
92. Fiala JC, Sorra K, Harris KM (2009) Reconstruct user manual
93. Morales J, Alonso-Nanclares L, Rodriguez JR, Defelipe J, Rodriguez A, Merchan-Perez A (2011) Espina: a tool for the automated segmentation and counting of synapses in large stacks of electron microscopy images. Front Neuroanat 5:18
94. Kremer JR, Mastronarde DN, McIntosh JR (1996) Computer visualization of three-dimensional image data using IMOD. J Struct Biol 116:71–76
95. Imbrosci B, Schmitz D, Orlando M (2022) Automated detection and localization of synaptic vesicles in electron microscopy images. eNeuro:9
96. Wahle P, Sobierajski E, Gasterstädt I, Lehmann N, Weber S, Lübke JHR, Engelhardt M, Distler C, Meyer G (2022) Neocortical pyramidal neurons with axons emerging from dendrites are frequent in non-primates, but rare in monkey and human. Elife 20(11):e76101. https://doi.org/10.7554/eLife.76101
97. Rosoklija GB, Petrushevski VM, Stankov A, Dika A, Jakovski Z, Pavlovski G, Davcheva N, Lipkin R, Schnieder T, Scobie K, Duma A, Dwork AJ (2014) Reliable and durable Golgi staining of brain tissue from human autopsies and experimental animals. J Neurosci Methods 230:20–29
98. Markram H, Lübke J, Frotscher M, Roth A, Sakmann B (1997) Physiology and anatomy of synaptic connections between thick tufted pyramidal neurones in the developing rat neocortex. J Physiol 500(Pt 2):409–440
99. Feldmeyer D, Lübke J, Sakmann B (2006) Efficacy and connectivity of intracolumnar pairs of layer 2/3 pyramidal cells in the barrel cortex of juvenile rats. J Physiol 575:583–602
100. Kubota Y, Kondo S, Nomura M, Hatada S, Yamaguchi N, Mohamed AA, Karube F, Lubke J, Kawaguchi Y (2015) Functional effects of distinct innervation styles of pyramidal cells by fast spiking cortical interneurons. Elife 4
101. Scheff SW, Price DA (1993) Synapse loss in the temporal lobe in Alzheimer's disease. Ann Neurol 33:190–199
102. Marco P, DeFelipe J (1997) Altered synaptic circuitry in the human temporal neocortex removed from epileptic patients. Exp Brain Res 114:1–10
103. DeFelipe J, Marco P, Busturia I, Merchan-Perez A (1999) Estimation of the number of synapses in the cerebral cortex: methodological considerations. Cereb Cortex 9:722–732
104. Anton-Sanchez L, Bielza C, Merchan-Perez-A, Rodriguez JR, DeFelipe J, Larranaga P (2014) Three-dimensional distribution of cortical synapses: a replicated point pattern-based analysis. Front Neuroanat 8:85
105. Finnema SJ, Nabulsi NB, Eid T, Detyniecki K, Lin SF, Chen MK, Dhaher R, Matuskey D, Baum E, Holden D, Spencer DD, Mercier J, Hannestad J, Huang Y, Carson RE (2016) Imaging synaptic density in the living human brain. Sci Transl Med 8: 348ra396
106. Mayhew TM (1996) How to count synapses unbiasedly and efficiently at the ultrastructural level: proposal for a standard sampling

and counting protocol. J Neurocytol 25:793–804

107. Schindelin J, Arganda-Carreras I, Frise E, Kaynig V, Longair M, Pietzsch T, Preibisch S, Rueden C, Saalfeld S, Schmid B, Tinevez JY, White DJ, Hartenstein V, Eliceiri K, Tomancak P, Cardona A (2012) Fiji: an open-source platform for biological-image analysis. Nat Methods 9: 676–682

108. Merchán-Pérez A, Rodriguez JR, Alonso-Nanclares L, Schertel A, Defelipe J, Counting Synapses Using FIB/SEM Microscopy (2009) A true revolution for ultrastructural volume reconstruction. Front Neuroanat 5(3):18. https://doi.org/10.3389/neuro.05.018.2009

109. Colonnier M, Beaulieu C (1985) An empirical assessment of stereological formulae applied to the counting of synaptic disks in the cerebral cortex. J Comp Neurol 231:175–179

110. Santuy A, Tomás-Roca L, Rodríguez JR, González-Soriano J, Zhu F, Qiu Z, Grant SGN, DeFelipe J, Merchan-Perez A (2020) Estimation of the number of synapses in the hippocampus and brain-wide by volume electron microscopy and genetic labeling. Sci Rep 10:14014. https://doi.org/10.1038/s41598-020-70859-5

111. Kreshuk A, Straehle CN, Sommer C, Koethe U, Cantoni M, Knott G, Hamprecht FA (2011) Automated detection and segmentation of synaptic contacts in nearly isotropic serial electron microscopy images. PLoS One 6:e24899

112. Staffler B, Berning M, Boergens KM, Gour A, Smagt PV, Helmstaedter M (2017) SynEM, automated synapse detection for connectomics. Elife 6

113. Eyre MD, Freund TF, Gulyas AI (2007) Quantitative ultrastructural differences between local and medial septal GABAergic axon terminals in the rat hippocampus. Neuroscience 149:537–548

114. Feldmeyer D, Egger V, Lübke J, Sakmann B (1999) Reliable synaptic connections between pairs of excitatory layer 4 neurones within a single 'barrel' of developing rat somatosensory cortex. J Physiol 521(Pt 1): 169–190

115. Feldmeyer D, Lübke J, Silver RA, Sakmann B (2002) Synaptic connections between layer 4 spiny neurone-layer 2/3 pyramidal cell pairs in juvenile rat barrel cortex: physiology and anatomy of interlaminar signalling within a cortical column. J Physiol 538:803–822

116. Kasthuri N, Hayworth KJ, Berger DR, Schalek RL, Conchello JA, Knowles-Barley S, Lee D, Vazquez-Reina A, Kaynig V, Jones TR, Roberts M, Morgan JL, Tapia JC, Seung HS, Roncal WG, Vogelstein JT, Burns R, Sussman DL, Priebe CE, Pfister H, Lichtman JW (2015) Saturated Reconstruction of a Volume of Neocortex. Cell 162: 648–661

117. Januszewski M, Kornfeld J, Li PH, Pope A, Blakely T, Lindsey L, Maitin-Shepard J, Tyka M, Denk W, Jain V (2018) High-precision automated reconstruction of neurons with flood-filling networks. Nat Methods 15:605–610

118. Klinger E, Motta A, Marr C, Theis FJ, Helmstaedter M (2021) Cellular connectomes as arbiters of local circuit models in the cerebral cortex. Nat Commun 12:2785

119. Loomba S, Straehle J, Gangadharan V, Heike N, Khalifa A, Motta A, Ju N, Sievers M, Gempt J, Meyer HS, Helmstaedter M (2022) Connectomic comparison of mouse and human cortex. Science 377: eabo0924

120. Kaltdorf M, Dandekar T, Naseem M (2017) Reconstruction of an immune dynamic model to simulate the contrasting role of auxin and cytokinin in plant immunity. Methods Mol Biol 1569:83–92

Part III

Synaptic Vesicles

Chapter 13

Counting Synaptic Vesicles: As Simple as That?

Bernd Walkenfort, Astrid Rollenhagen, Mike Hasenberg, and Joachim H. R. Lübke

Abstract

One critical parameter determining synaptic transmission and the modulation of long- and short-term synaptic plasticity is the size of the three functionally defined pools of synaptic vesicles (SVs) in particular that of the so-called readily releasable pool of SVs (RRP). However, due to the random distribution of SVs within a given synaptic bouton, it is impossible to describe a structural correlate to the functionally defined pools. Thus, the so-called "docked" SVs fused with the presynaptic density (PreAZ) or omega-shaped bodies already opened that released a quantum of neurotransmitter were regarded to represent the RRP. However, whether only "docked" or omega-shaped bodies represent the structural correlate of the RRP still remains controversially discussed.

Here, we represent and compare two methods that can be used to describe the size of the RRP, a perimeter analysis where SVs are classified according to their distance to the PreAZ and high-resolution EM tomography by which the number of "docked" or omega-shaped SVs along the active zone are counted in individual tilt series along the PreAZ.

Key words Human brain, Electron microscopic tomography, Quantitative analysis of docked synaptic vesicles, Readily releasable pool

1 Introduction

Synaptic vesicles (SVs) are, besides the shape and size of active zones (AZ), the structural equivalent for a functional neurotransmitter release site, another key structural element that contains, stores, and releases neurotransmitters, and hence play a fundamental role in synaptic transmission and in modulating short- and long-term synaptic plasticity [1–4]. Their distribution in synaptic terminals and organization into three distinct functionally defined pools, namely, the readily releasable pool (RRP), the recycling pool (RP), and the resting pool, regulate synaptic efficacy and strength and determine the mode and probability of release (P_r; uni- vs. multi-vesicular; uni- vs. multi-quantal [5–14]). In particular, the size of the RRP and RP play a pivotal role in the functional

Joachim H. R. Lübke and Astrid Rollenhagen (eds.), *New Aspects in Analyzing the Synaptic Organization of the Brain*, Neuromethods, vol. 212, https://doi.org/10.1007/978-1-0716-4019-7_13,

"behavior" of a given synaptic terminal by controlling the induction, maintenance, and termination of synaptic transmission and modulating synaptic plasticity.

The endeavor and description of the three functionally defined pools of SVs was an important step to explain the various signal cascades underlying the functionality of a synapse [10, 12]. By using various tagging strategies in fluorescence microscopy, capacitance measurements, and highly resolving stimulated emission depletion (STED) microscopy, it became possible to visualize individual SVs and their behavior during low- and high-frequency stimulation conditions [15–17]. This finally led to the definition of the three functional pools of SVs. This concept and division of separate pools of SVs serving different functions was later confirmed by studies using the visualization of vesicle exo- and endocytosis by pHluorins [18–20].

It is now well established that the RRP is constituted by those SVs that are immediately available for release into the synaptic cleft and are already "docked" to the presynaptic AZ (PreAZ). Thus, the RRP represents the most important vesicle pool for neurotransmitter release and includes the number of the so-called "docked" SVs fused with the presynaptic active zone or omega-shaped vesicles, which have already opened and released a quantum of neurotransmitter. However, the RRP contains only a tiny fraction of the total pool of SVs, for example, the calyx of Held-principal neuron synapse in the auditory brain stem contains ~70,000 SVs distributed over 550 AZs, but the RRP contains only 1100 SVs, which represents 1.5% of the total pool of SVs [21]. Interestingly, the RRP seems to be a fraction of the RP of SVs, defined as all vesicles that can be labeled with FM1–43 when synapses are stimulated extensively [22–24] but are not always "docked" with the presynaptic active zone but close to it, approx. Two vesicle diameters apart [25–27]. After extensive high-frequency stimulation, the RRP can be depleted, and thus additional SVs are recruited for exocytosis from the so-called RP and resting pool to the active zone. It is thought that the RRP and the resting pool together constitute the RP of SVs. Under normal physiological conditions, the resting pool is not used and is regarded a pool of non-active SVs.

Interestingly, in contrast to the total pool of SVs, the size of the three functionally defined pools of SVs can vary substantially between synaptic boutons in the central nervous system (CNS) and is not always correlated with the size of the end terminal and PreAZ; see, for example, [11, 21, 25–32]. These marked differences in the three pools of SVs strongly suggest also substantial differences in the functional "behavior" of synaptic boutons embedded in different networks and microcircuits of the brain. Thus, it is important to describe these pools, in particular, the RRP, using morphologically based high-resolution methods, for example, STED imaging or transmission electron microscopy

(TEM), focused ion beam scanning (FIB-SEM), and electron microscopic (EM) tomography.

One critical parameter in characterizing synaptic transmission and modulation of long- and short-term synaptic plasticity is the size of the readily releasable pool of SVs (RRP) [10, 12–14]. However, it is impossible to describe a structural correlate to the functionally defined pools due to the random distribution of SVs within a given synaptic bouton. Thus, the so-called "docked" SVs fused with the PreAZ or omega-shaped bodies already opened that released a quantum of neurotransmitter were regarded to represent the RRP [12, 33–36]. However, whether only "docked" or omega-shaped bodies represent the structural correlate of the RRP still remains controversially discussed. In contrast to the functionally described pools, it is impossible to define a morphological correlated pool of SVs, since SVs are randomly distributed over the entire synaptic terminal. However, at PreAZs, "docked" SVs and vesicles close to it can be observed that may represent the RRP and the RP of SVs. Despite "docked" SVs undoubtedly representing the RRP, it is still controversially discussed whether other SVs, within a perimeter of less than a vesicle diameter, assuming an average diameter of SVs of ~35 nm, may also contribute to the RRP.

Here, we represent and compare two methods that can be used to describe the size of the RRP, a perimeter analysis (Fig. 1) where SVs are classified according to their distance to the PreAZ and high-resolution EM tomography (Fig. 4) by which the number of "docked" or omega-shaped SVs along the active zone can be easily determined in a virtual, three-dimensional (3D) scenario in tilt series along the AZ.

2 Methods

2.1 Fixation of Tissue Samples for High-Resolution TEM and EM Tomography

Brain tissue samples can be obtained from small experimental animals (preferentially rodents such as rats or mice) as well as from human donors.

For analyzing animal brain tissue in this work, animals were overdosed with sodium pentobarbital (0.09 mg/g, i.p.) and immediately perfused for 2–3 min through the heart with phosphate-buffered saline (PBS, 0.1 M) followed by 4% paraformaldehyde and 2.5% glutaraldehyde in 0.1 M phosphate buffer (PB) for 30 min. Subsequently, brains were carefully removed from the skull and submersed in the same but fresh fixative solution.

For targeting human material, access tissue taken during epilepsy or tumor surgery is well suited. After careful evaluation, this access tissue can be regarded as "healthy" tissue [37] if it is taken with significant distance from the focus of the epilepsy or tumor. When obtaining human biopsies, perfusion fixation obviously is not

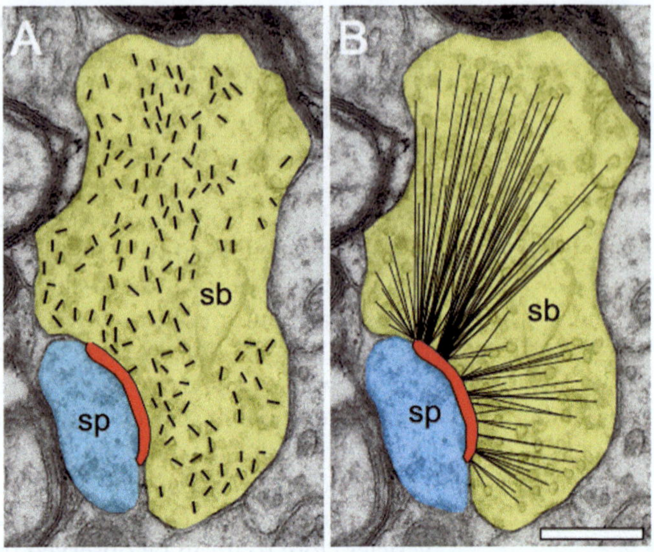

Fig. 1 Measuring the total number and diameter of the SVs at synapses in the human temporal lobe neocortex. (**a**) Here, the presynaptic terminal (sb) is given in transparent yellow, the postsynaptic spine (sp) is given in blue, and the AZ is highlighted in red. The diameter of SVs was determined by line contours (black) from the outer to the outer edge of all SVs within a serial ultrathin section. (**b**) By placing a contour line to individual SVs, also their distance to the PreAZ was measured by an algorithm automatically, here visualized as black solid lines between individual SVs and the PreAZ by their shortest distance. The same color code as in **a**. Scale bar 0.25 μm

possible. In the respective analyses of which the results are presented here, brain access tissue material from an ongoing surgery was roughly trimmed to blocks with a maximum diameter of 2–5 cm, right after explantation, still in the operation room, and transferred into the same fixative as used for animal tissue processing.

In both cases, the brain biopsies were then incubated for 24–72 h at 4 °C in the same fixative. Four hours after the onset of incubation, the fixative was replaced by the same but fresh solution and left in place for 24–72 h at 4 °C. Next, the brain samples were processed by vibratome sectioning to improve their accessibility for different solutions, in the following steps. Tissue blocks were thoroughly rinsed in cold PB and afterward embedded in 5% agar-agar (Sigma, Munich, Germany) dissolved in PB. Using a vibratome (VT1000 S, Leica Microsystems GmbH, Wetzlar, Germany), brains or blocks of tissue were cut in the coronal plane into 150–200-μm-thick sections, which were collected in PB in individual chambers of six-well plates. Afterward, they were washed again in PB and transferred to 0.5–1% PB-sucrose buffered osmium tetroxide (OsO_4; 300 mOsm, pH 7.4; Sigma, Munich, Germany) for 60–90 min. After visual inspection, sections were thoroughly washed several times in PB and left overnight at 4 °C. The next day, sections were dehydrated with an ascending series of ethanol starting at 20%, 30%, 50%, 60%, 70%, 80%, and 90% (15 min for each step), followed by 95% ethanol (20 min) and finally in 100% ethanol (twice for 30 min). Subsequently, sections were transferred briefly into propylene oxide (twice for 2 min) and then incubated in a mixture of propylene oxide and Durcupan™ (2:1 and 1:1, 60 min for each step; Fa. Fluka, Neu-Ulm, Germany) and stored in pure Durcupan™ overnight (ON) at room temperature (RT). Finally, sections were flat-embedded in fresh Durcupan™ between two Aclar® foils and polymerized at 60 °C for 48 h.

2.2 Ultrathin Sectioning for TEM

Embedded vibratome sections were inspected light-microscopically to determine the region of interest (ROI), which was then trimmed out (approximately 1 × 1 cm) with a new razor blade and glued onto a pre-polymerized resin block using a drop of fresh, unpolymerized Durcupan™. Before proceeding, the resin drop was cured at 60 °C ON. To evaluate the sample quality and to define the final ROI, semithin sections were cut using a histo-diamond knife (Fa. Diatome, Switzerland) with a Leica UltracutS ultramicrotome (Leica Microsystems, Vienna, Austria) and stained with methylene blue. After inspection and definition of the final ROI, the block surface area was further trimmed to its final size of 3 × 2 mm for serial ultrathin sectioning.

Serial ultrathin sections (50 ± 5 nm in thickness; silver to gray interference contrast) were cut with the same Leica UltracutS and collected on Pioloform-coated slot copper grids (Fa. Plano, Wetzlar, Germany). An individual series usually comprised between 75 and 150 ultrathin sections to reconstruct SBs of different shape and size to meet both structural variability and statistical variance. Prior to EM examination and to increase the contrast, ultrathin sections were finally stained with 5% aqueous uranyl acetate for 15–20 min and lead citrate for 3–5 min. Ultrathin sections were examined with a Zeiss Libra 120 TEM (Fa. Zeiss, Oberkochen, Germany) equipped with a bottom-mounted ProScan 2 K digital camera using the ImageSP software (Fa. Tröndle, Moorenweis, Germany) at various magnifications ranging from 8000 to 25,000× with frames of 6 × 6 images (10,444 × 11,129 final pixel size). All images were stored in a database until further processing.

2.2.1 Production of Formvar Films

1. Use cleaned, dust-free glass slides. Pour the Formvar solution into a staining cuvette and dip the glass slide into the cuvette (approx. 2 s) and pull it out quickly and evenly (the thickness of the foil critically depends on this procedure).
2. Leave the glass slide to dry vertically and dust-free for about 10 min.

3. Fill a large beaker with double-distilled water up to the rim. If required, fill and clean the water surface with a glass rod from dust or other particles.
4. Carefully run a scalpel along the lower edges of the glass slide covered with Formvar film. Breathe on the filmed glass slide, and, while the surface is still wet, place the narrow side on the water surface at an angle of approx. 25° and wait until the Formvar film has completely detached from the slide.
5. Using a fine tweezer, place slot grids, shiny side (bottom) up, on top of Formvar film floating on the water surface.
6. Cover the foil with a slightly larger Parafilm® while rolling it on the water surface and pick it up.
7. Place the strips of Parafilm® covered with Formvar film and underlying slide meshes in Petri dishes and let them air-dry.
8. Pour the Formvar solution from the cuvette back into the storage bottle and store tightly closed in the dark.
9. Clean the device with dichloroethane.

2.3 Ultrathin Sectioning for EM Tomography

For high-resolution EM tomography, a ROI was defined on selected vibratome sections according to the procedure for standard TEM (see above). Ideally, this ROI could be identified on consecutive sections obtained from the blocks used for serial ultrathin sectioning or alternatively from new specimens, which were prepared as described above. Also, identical to the protocol listed above, the vibratome sections were trimmed to the ROIs using sharp razor blades and glued to resin blocks. In contrast to the procedure for serial ultrathin sectioning, the thickness of sections for EM tomography ranged between 200 and 300 nm. After cutting, sections were transferred to Formvar-coated copper slot grids (for details, see above) and post-stained with uranyl acetate and lead citrate as described above.

2.4 TEM of Serial Ultrathin Sections and Analysis of SV Pools Using a Perimeter Analysis

After the definition of one to three different ROIs, individual SBs and their target structures were photographed throughout consecutive sections in the entire series of ultrathin sections.

Besides other structural parameters, the presynaptic active zone (PreAZ) and postsynaptic density (PSD) constituting the AZ are regions of densely packed, electron-dense, dark-appearing material, condensed at the pre- and postsynaptic apposition zones (Fig. 1). The surface areas of the PreAZ were computed in OpenCAR separately by first generating a 3D surface model of the SB. The PreAZ was then measured by extracting this area from the reconstructed presynaptic bouton membrane that was covered by this membrane specialization (i.e., where the contour line coincided with less than 30 nm distance from the presynaptic membrane). Hence, the length (l) of the PreAZ (l PreAZ) and the surface area

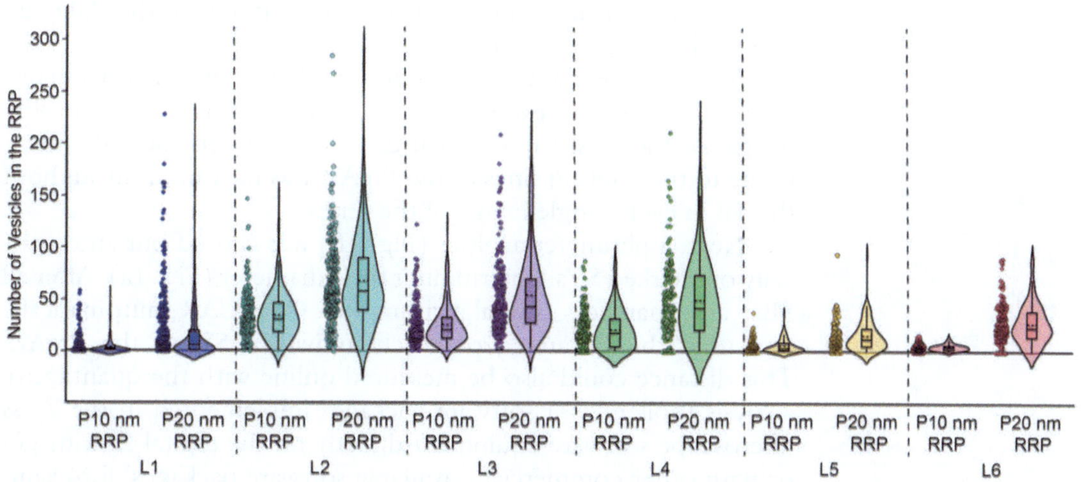

Fig. 2 Violin and dot plots of SVs in the p10 nm and p20 nm RRP for the cortical layers (L1–L6) in the human temporal lobe neocortex as revealed by the perimeter analysis

Table 1
Comparison of the RRP as estimated by the perimeter analysis and by EM tomography

Cortical layer	Perimeter analysis: p10 nm	Tomography: number of "docked" SVs	Perimeter analysis: p20 nm
L1	3.32 ± 9.03	3.56 ± 1.36	17.32 ± 31.47
L2	33.89 ± 22.68	–	66.80 ± 43.63
L3	26.60 ± 8.57	–	56.50 ± 18.81
L4	20.20 ± 18.58	7.32 ± 3.96	48.59 ± 39.02
L5	5.42 ± 4.09	–	15.21 ± 9.02
L6	12.70 ± 4.32	4.71 ± 3.18	28.70 ± 9.84

Note the layer-specific differences in the RRP using the perimeter and EM tomography analysis. All values are significantly different ($p < 0.01$ and $p < 0.001$, respectively)

(SA) of the PreAZ (SA PreAZ) are already known. The size of the PSD opposing the PreAZ is estimated under the following assumptions: (1) both membrane specializations, PreAZ and PSD, run parallel to each other at the pre- and postsynaptic apposition zone (Figs. 3 and 4) and (2) for both membrane specializations, a contour line is drawn determining their actual length (l PreAZ and l PSD). The surface area of the PSD (SA PSD) is measured using the following equation:

$$\text{SAPSD} = \text{SAPre} * \text{l PSD}/\text{l PreAZ}$$

which is the perimeter ratio between the outlines of the PSD and that of the synaptic contact.

All SVs were marked throughout each SB, and their diameters were individually measured (Fig. 1a). To determine the distribution profile of the SVs, the minimal distance between each SV membrane to the contour lines of the PreAZ was measured throughout the SB in every single image of the series.

Next, a perimeter analysis (Fig. 1b) was carried out automatically on marked SVs, determining their diameter (Fig. 1a). After all SBs are marked, the algorithm of OpenCAR automatically measured the distance from each individual SV to the PreAZ. This distance could also be measured online with the quantitative analysis tools of EM software packages implemented in the Zeiss microscope software equipment directly on the digital EM images or with other commercially available software packages, for example, Amira, Dreams3D, or ESPINA (see Chap. 7 of this book).

According to the established criteria [10], a RRP, RP, and resting pool can be defined using the perimeter analysis. Those SVs within a perimeter of 10 nm or 20 nm apart from the PreAZ that is less than the average vesicle diameter of 30 to 35 nm were regarded to belong to the RRP. The RP is constituted by SVs within a distance between p60 and p200 nm from the PreAZ, and the remaining more than 200 nm away from the PreAZ represent the resting pool. Using this criterion, it became possible to define a morphological correlate to the three functionally defined pools of SVs in serial ultrathin sections; see, for example, [25–27]. From this analysis, comparative tables (Table 1) and violin plots (Fig. 2) were generated to, for example, identify layer-specific differences in the composition of the three pools of SVs in the human temporal lobe neocortex.

For the definition of the RRP, a layer-specific difference in the p10 and p20 nm RRP in the human temporal lobe neocortex was found (Fig. 2). It should be pointed out that in all layers, the p20 nm RRP increased by ~2–3-fold (Fig. 2). Thus, to our belief, the RRP should not only be defined by the number of "docked" SVs.

2.5 EM Tomography of "Docked" SVs

The ultrastructural 3D analysis of virtual reconstructions based on serial ultrathin sections harbors the big problem of a relatively poor z-resolution as a natural result of technical limitations during ultramicrotomy (UM). Using this technique, it is impossible to obtain a much better resolution than ~50 nm in z-direction. With this restriction, the analysis error of fine cellular ultrastructure very often is of high significance. For example, regarding the description of SBs, it is impossible to precisely locate synaptic vesicles and to individually judge on their distance from the PreAZ, which is a prerequisite to unequivocally sort them into a certain vesicle sub-pool. Given a size of smaller than 50 nm, an entire synaptic

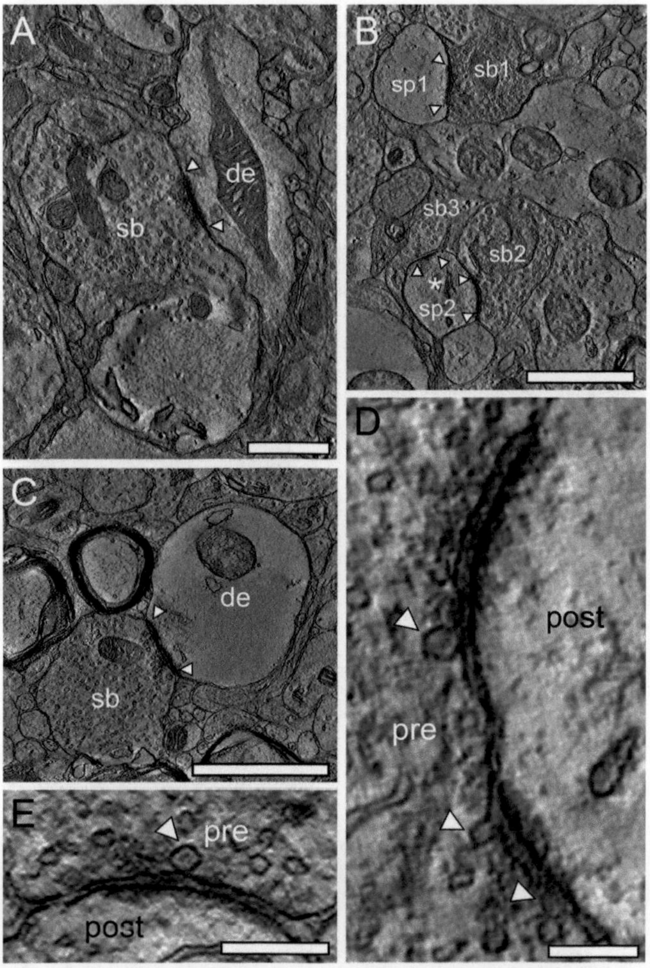

Fig. 3 EM tomography of synaptic complexes in the human temporal lobe neocortex. (**a**) Low-power EM tomographic image showing a dendrite (de) with a terminating large synaptic bouton (sb) in cortical layer 6. Scale bar 0.5 μm. (**b**) Low-power EM tomographic image of two spines (sp1, sp2) in cortical layer 4. Note that the upper spine (sp1) received a single synaptic bouton (sb1), whereas on the other spine (sp2), two synaptic boutons (sb2, sb3) were established. Scale bar 1 μm. (**c**) Large synaptic bouton (sb) synapsing on a dendritic segment (de) in cortical layer 1. Scale bar 1 μm. In **a**–**c**, active zones are marked by arrowheads. (**d**) High power of an active zone in cortical layer 4 showing three "docked" synaptic vesicles (arrowheads) fused with the presynaptic density (pre) at the pre- and postsynaptic (post) apposition zone. Scale bar 0.1 μm. (**e**) Another example of an active zone, composed of the pre- (pre) and postsynaptic density (post), with a single but large "docked" synaptic vesicle (arrowhead). Scale bar 0.1 μm

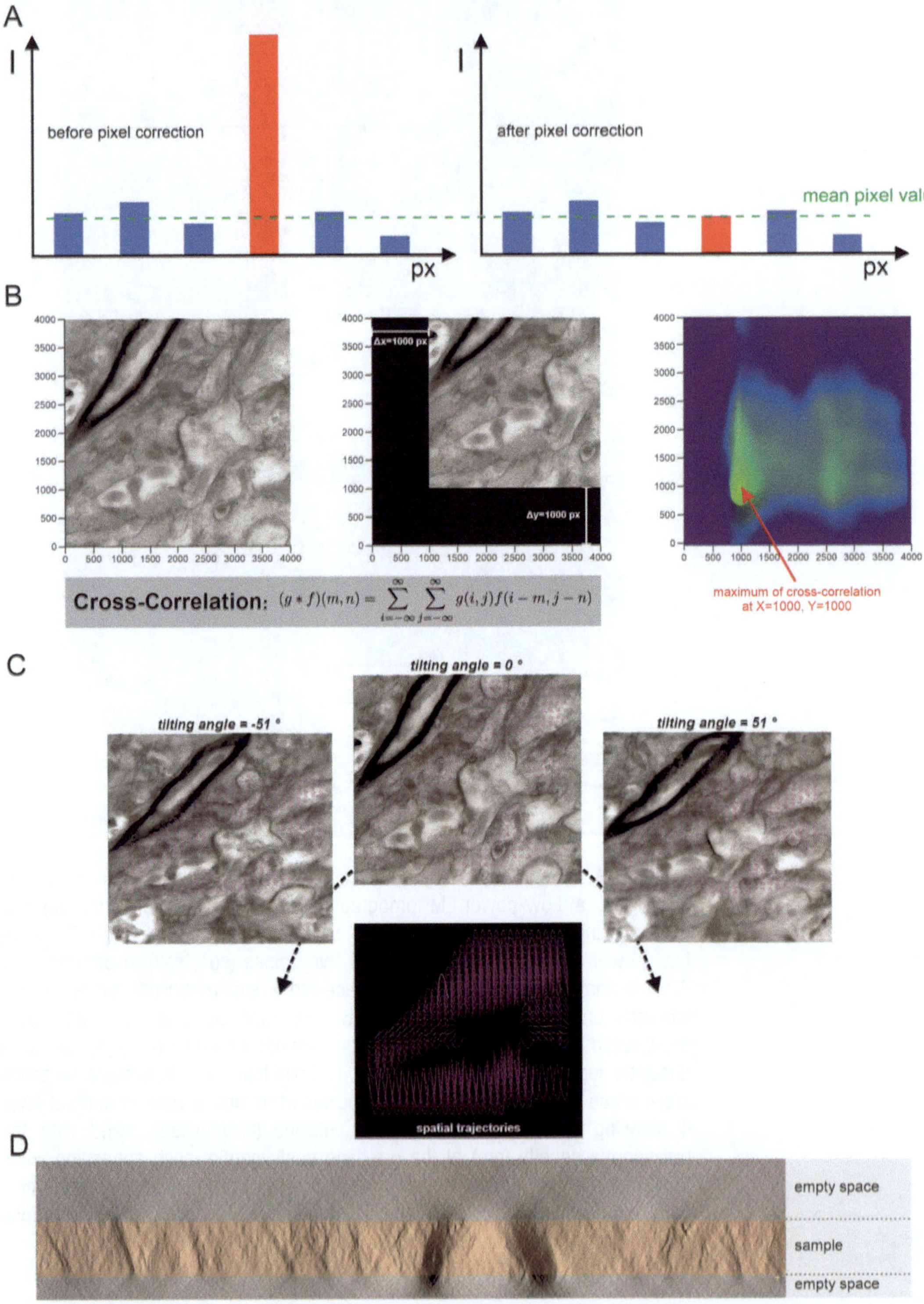

Fig. 4 Generation of a one-axis tilt series using the software package IMOD with the graphical user interface Etomo. (**a**) The workflow starts with correction of overdriven pixels. These are detected by comparing the intensity values of neighboring pixels in a predefined radius, and if a pixel is identified significantly exceeding mean intensity values (red), its intensity value is replaced by the average value of the intensities of neighboring pixels. In (**b**), the basic principle of cross-correlation is visualized. Cross-correlation is a mathematical

vesicle might be even neglected in a quantification approach if one section of an ultrathin section series unwantedly gets lost. Additionally, numerous SVs will not be cut in the middle plane of the sphere so that from a TEM analysis of an ultrathin section, it is not possible to objectively assess their full 3D dimension and consequently to speculate on their localization relative to other nanostructures, but see [38].

To overcome this problem, EM tomography is a method of choice [39–43]. During this, the entire volume of an UM section much thicker than 50 nm is analyzed with sub-nanometer resolution in z-direction so that the 3D of several entire vesicle spheres and their exact cellular localization can be determined within this sample volume (Fig. 3). The background of this methodology is subsequent, computer-aided 3D reconstruction from the image stack. The drawbacks of this approach are the limited x–y–z dimensions and the long acquisition time for each tilt series. Additionally, a significant amount of time and effort must be invested in the 3D reconstruction process.

For the examples shown here, EM tomography was carried out on 200–300-nm-thick sections from human temporal lobe neocortex biopsies (Fig. 3). These sections were examined with a JEOL

Fig. 4 (continued) operation frequently used in image analysis to measure the similarity or alignment between two images. The cross-correlation process begins with two input images, the target image and the kernel image. The kernel image is initially placed at a certain position (usually the top-left corner) of the target image. Then, at each position, the corresponding pixel values of the kernel and the target image are multiplied before all resulting products from the multiplication step are summed together to obtain a single value for this image comparison. Next, the kernel image is shifted by a small number of pixels in x- or y-direction to a new position in the target image. The multiplication and summation steps are then repeated for the new position. After repeating the kernel image shifting across a predefined area and the subsequent multiplication and summation steps, a so-called correlation map can be generated, which visualizes the image similarities in relation to kernel image movement. By this analysis, it is possible to coarsely align all individual images of a tilt series as their image content does not differ significantly among small-angle tilting steps. The coarse image alignment was followed by fine image alignment using trajectories (**c**). Image fine alignment using trajectories represents a computer-aided image post-processing method during which a sequence of images is aligned by estimating and tracking the trajectories of certain landmarks across the images. This process aims at minimizing the localization between corresponding points in the images, effectively aligning them spatially. In the initial step of the alignment process, distinctive fiducial markers are defined in each and subsequently detected and matched across all image(s) of the tilt series. Then, the trajectories of all fiducial markers are estimated by connecting them over the entire image stack. The estimated trajectories are analyzed to determine any inconsistencies or outliers. By applying geometric transformations, the estimated trajectories are then aligned, resulting in a fine alignment of the individual images. This method can effectively correct for spatial variations, rotations, translations, and other deformations present in the sequence of images and therefore is very helpful in the eradication of shrinkage artifacts, which are inevitable in the process of TEM sample preparation. Before ultimate 3D image reconstruction, it is meaningful to define the reconstruction space (**d**). The investigated ultrathin resin section does not fully match the reconstruction space in z-direction. To optimize computation time and efficiency, the section is precisely defined before 3D reconstruction. I = intensity, px = pixel(s)

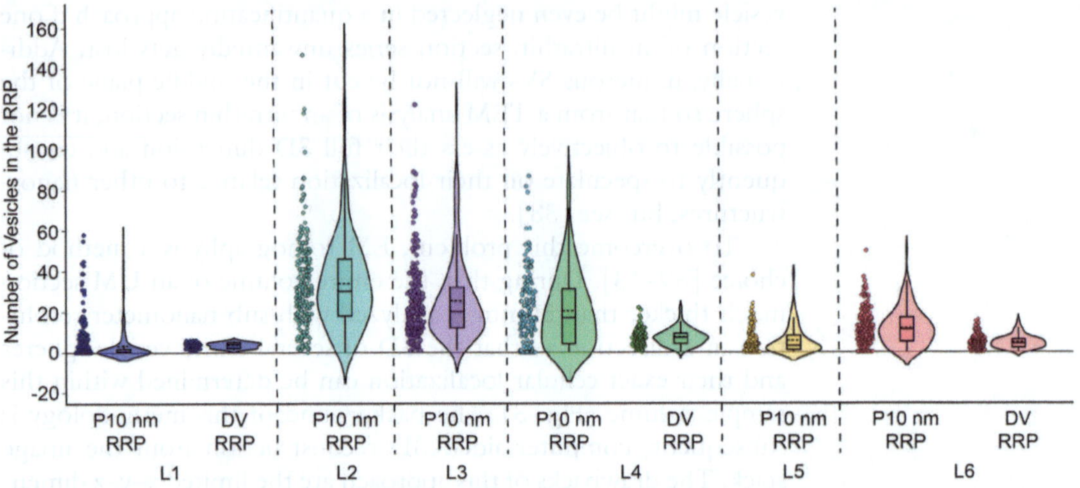

Fig. 5 Violin and dot plots showing the distribution of SVs in the p10 nm RRP by the perimeter analysis and "docked" SVs estimated by EM tomography. Data distributions for each patient are indicated by the medians (horizontal bars), IQRs (framed areas), and minimum and maximum (vertical lines). Abbreviations: L, cortical layers; RRP, readily releasable pool of SVs; DV, "docked" SVs

JEM 1400Plus TEM, operating at 120 kV and equipped with a 4096 × 4096 pixel CMOS camera (TemCam-F416, TVIPS, Gauting, Germany) and a precise goniometer to allow for the acquisition of tilt series. At an overview magnification of 5000×, resulting in a pixel size of 2.408 nm, numerous synaptic complexes (~30 per section) were selected, before, in the next step, tilt series were acquired automatically at these ROIs over an angular range of −60° to +60° at 2-degree increments using SerialEM (ver. 3.58; [44]). For the acquisition of these tilt series, a magnification of 10,000×, resulting in a pixel size of 1216 nm, was used.

The total acquisition time for an individual tilt series was about 20 min. Thus, one can easily run the entire acquisition of data overnight or during the weekend. The resulting tilt series individually covered a maximum sample volume of 5 μm × 5 μm × 300 nm.

After image acquisition, the raw image data of a tilt series contained several artifacts, which had to be irradiated to provide suitable data for the 3D reconstruction algorithm. Typical image artifacts included:

1. *Individual "spurious pixels" with high intensities. These were caused either by faulty pixels generated by the CMOS camera or by the impact of high-energy particles from cosmic rays.*
2. *Different mean values of the histogram distribution (brightness) and signal to noise ratios (SNR) due to fluctuating beam intensities during image acquisition.*
3. *Lateral offset of the image, caused by mechanical inaccuracies of the goniometer during positioning or sample tilting.*

4. *Shrinkage of the plastic section due to exposure to the electron beam.*

Manual correction of the image errors would be time inefficient and, in some cases, not accurate enough. Therefore, different software packages, which fulfill this purpose and implement powerful 3D reconstruction algorithms, have been applied.

For EM analyses, a frequently used software package is represented by IMOD [44, 45], which contains over 180 single programs to map the entire workflow from image stack quality improvement to final 3D reconstruction. IMOD can be operated mostly over the command line but also offers graphical user interfaces (3dmod, Midas, Etomo), which make all operations more comfortable and through which the results of the individual steps can be viewed directly in a preview image. IMOD, in versions 4.9.7 to 4.9.12, was used to visualize the EM tomograms included in this book chapter.

In the first step, individual overdriven "spurious pixels" were corrected using the program Etomo of the IMOD software suite (Fig. 4a). To detect these pixels, the intensity values in the neighborhood (with a given radius) are compared in each case. If the value of this pixel deviates too much, the intensity value is replaced by the average value of the intensities of the neighboring pixels. The above listed image histogram adjustments (brightness and SNR) are then performed passively following the routine Etomo workflow. Next, a coarse image alignment by means of cross-correlation was carried out. Since two images with small tilt angle difference should still look similar to each other, one can superimpose them for a coarse alignment and shift them until the closest possible match is present, which is the general idea of cross-correlation (Fig. 4b). In the output image, an integral quantity is specified at each point of the image that measures the correspondence between the two images.

The pixel position of the maximum of this function indicates the relative offset of the two images to each other. The coarse alignment of the images by means of cross-correlation corrects only a lateral offset of the image. Local changes such as shrinkage of the plastic section are not considered here. Therefore, after coarse image alignment at all tilting angles by cross-correlation, a fine alignment using a system of reference points, which can be tracked over the entire tilt series, was conducted (Fig. 4c). In the examples shown here, unchanging sample-inherent nanostructures were used as such fiducial markers. It would have also been possible to add additional particles, such as gold nanoparticles, to the sample in a resin-submersed form, which, after resin polymerization, could serve as reference points during this step. For fine alignment, the full image at each tilting angle is split into smaller sub-images of approximately 300 px × 300 px.

For each angle, the partial image is corrected for the projection-induced compression and superimposed on the original image using cross-correlation. Hence, the trajectory of the partial image can be followed over the entire tilt. If the centers of the partial images move closer together, one is dealing with a local shrinkage of the specimen.

As the last step of data preparation, right before 3D reconstruction, the reconstruction space is precisely defined (Fig. 4d). Depending on the pixel size (magnification) selected during image acquisition, an ultrathin resin section does not necessarily occupy the entire reconstruction space, but only a small fraction of it. In this next step, three sample tomograms are therefore calculated, which can subsequently be used to determine the spatial location of the plastic slice. All voxels that are not located in the slice do not need to be calculated either, which significantly reduces memory and computation time.

For 3D reconstruction, the filtered back projection method was used, which is also implemented in IMOD. Filtered back projection is a mathematical technique frequently used in image post processing to reconstruct the 3D model of an object from a series of two-dimensional projections such as the TEM tilt series. It starts with a filter step to remove high-frequency noise and artifacts from the projections. The filtered projections are then back-projected onto a virtual 3D grid. This means that the pixel values from each filtered projection are assigned to their corresponding locations in the 3D grid, for every tilt angle. After back projection, the pixel values in the 3D grid are summed or averaged across all the projections.

This step helps accumulate the contributions from all angles and improves the final image quality. After 3D reconstruction, the resulting images of each z-stack plane were slightly noisy. To improve their appearance, they were filtered using a median filter with a window size of 2 pixels.

The central parts of the ultimate data analysis were the classification and the quantification of "docked" or omega-shaped SVs using EM tomography (Fig. 3). In this process, "docked" vesicles were counted in reconstructed 3D models separated for spine and shaft SBs in individual tilt series. As one result from these analyses, the so-called violin and box plots were generated with open-access customized or self-programmed software to evaluate layer-specific differences in the organization of the RRP (Fig. 5). When compared, the values for the p10 nm RRP as measured with the perimeter analyzed exceed in most layers that of the values estimated by EM tomography. However, it should be noted that only "docked" or omega-shaped SVs were measured by EM tomography, whereas in the perimeter analysis, all SVs were counted within 10 nm from the PreAZ regardless of whether they are "docked" or not.

3 Discussion

Here two methods are described that may represent a morphological correlate of the functionally defined RRP by estimating the number of either SVs within a distance of p10 nm or p20 nm away from the PreAZ or counting the number of "docked" SVs or omega-shaped bodies that already released a quantum of neurotransmitter using high-resolution EM tomography.

For a long time, it was thought and is still controversially discussed whether only the number of "docked" SVs represent the RRP. However, using the perimeter analysis, this view may be revisited since it was shown by this study and two other publications [25–27] that are using the p10 nm criterion that the number of SVs representing the RRP is similar or even higher when compared with "docked" vesicles at least at SVs in the human temporal lobe neocortex (Table 1 and Figs. 2 and 5). In addition, the number of SVs in the p20 nm, which is still less than the average diameter of an SV, can exceed that of the p10 nm RRP by two- to fourfolds (see [25–27]) and that of "docked" SVs. The results of this study using both the perimeter analysis and EM tomography about the size of the RRP are comparable with findings as measured by FM1–43 staining/destaining (RRP 5–9 SVs [46]). In cultures of CA1 neurons, the number of SVs that are "docked" to the AZ determined by EM was also found to be approximately five to ten SVs per active zone [24, 47]. Furthermore, the RRP estimated by capacitance measurements at the mossy fiber bouton [48] was also in that range suggesting that the size of the RRP seemed to be similar at some but not all CNS synapses embedded in different microcircuits.

However, the perimeter analysis on serial digital images strongly suggests that both the p10 nm and p20 nm population of SVs may be easily recruited, for example, by either repetitive low- or high-frequency stimulation and hence contribute to the population of easily recruitable SVs suggesting a re-evaluation of the role of "docked" SVs as the only ones constituting the RRP.

Hence, both methods, the perimeter analysis and EM tomography, are well-suited morphological tools to define the RRP, one of the key synaptic parameters to explain and even predict the "functional behavior" of synaptic complexes in any given network of the brain exemplified here by estimating the RRP in different layers in the human temporal lobe neocortex. The results of such an analysis are severalfold: firstly, multi-vesicular release from the RRP seems to be the predominant mechanism at most CNS synapses as, for example, shown by the quantitative analysis of the vesicle pools with serial ultrathin sectioning and subsequent TEM analysis; see, for example, [21, 25–30]. However, the number of SVs constituting the RRP varies substantially between CNS SBs also suggesting a strong modulation in the response properties of a given SB.

Secondly, at the majority of CNS synapse, no or only a weak but rarely a good correlation was found between the size of PreAZ and the RRP suggesting an independent regulation of SVs constituting the RRP at least at cortical SBs in the human temporal lobe neocortex, although layer-specific differences were found [25–27, 29, 30].

Finally, it must be pointed out that both the perimeter analysis basal on serial TEM images and the subsequent analysis of structural and synaptic parameters and EM tomography of SVs at the PreAZ for the identification of "docked" SVs are technically elaborate labor intensive und hence time-consuming, and the entire analysis could last several weeks.

3.1 Perspectives

To overcome this problem, automated methods for the detection of SVs are therefore needed to increase the analytical throughput, to reduce manual labor, and to improve standardization. Several studies in the field of connectomics have successfully employed convolutional neural networks to compute large-scale 3D reconstructions of neuronal circuits; see, for example, [49–51]. A first successful attempt to identify SVs in presynaptic terminals required tomographic 3D reconstructions [52] and EM tomograms while providing detailed 3D information on single vesicles, which are nevertheless lengthy to acquire. However, just recently, Imbrosci et al. [53] demonstrated an automated sampling of SVs at the hippocampal mossy fiber boutons that came to similar results than a quantitative 3D model of the mossy fiber boutons achieved by 3D volume reconstructions of serial digital EM images (Rollenhagen et al. 2010). Nevertheless, Imbrosci et al. [53] automatically calculated the total pool of SVs but did not differentiate between the three functionally defined pools of SVs. Hence, more further developments are needed with respect to the discrimination of the three pools of SVs.

Another important analysis is the differentiation of SBs with different origins in the brain, e.g. intracortical vs. thalamocortical boutons. To classify SBs according to their origin, anterograde labeling of axons by tracer injection at their site of origin is the only reliable method. For example, by placing a tracer depot in the thalamic ventral posterior nucleus, it is possible to trace the axonal fibers up to layer 4 of the primary sensory cortex [54, 55]. Unfortunately, the labeling entirely masked the interior, in particular, the PreAZ, PSD, and SVs (Fig. 6a). In a first attempt, such labeled SBs were examined with EM tomography. As a result, it became possible to clearly identify structural sub-elements including the PreAZ, PSD, and "docked" vesicles (Fig. 6b). This approach of identifying "docked" SVs in labeled SBs would open a new window to address

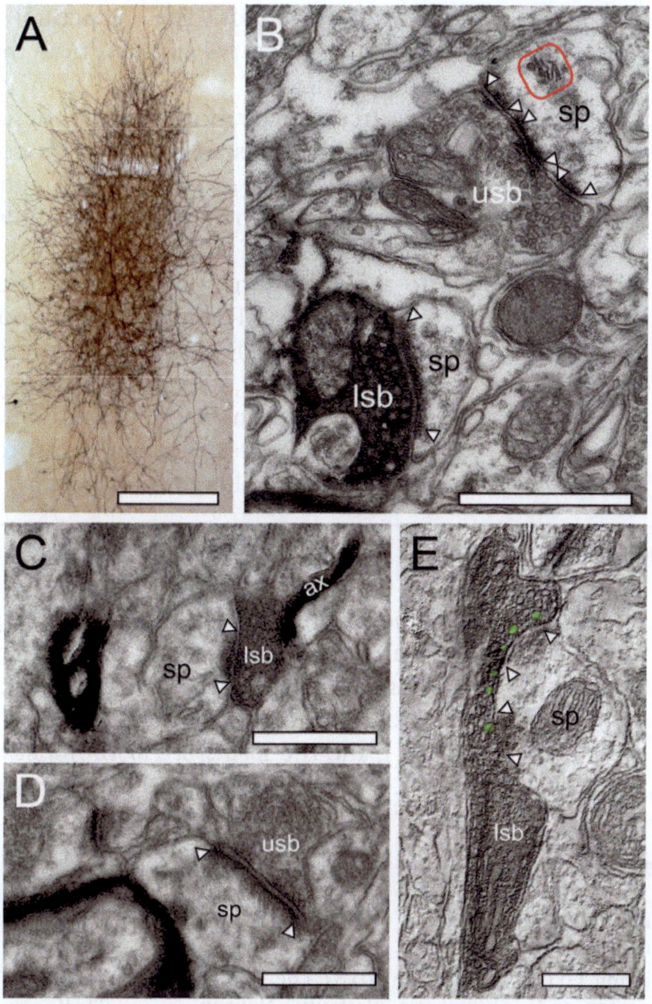

Fig. 6 EM tomography of labeled vs. unlabeled SBs in the rat neocortex. (**a**) Low-power light micrograph of the dense projection of thalamic fibers in layer 4 of the neocortex after the injection of a tracer in the ventral posterior nucleus of the thalamus. Scale bar 5 mm. (**b**) High-power TEM micrograph of a labeled SB (lsb) terminating on a dendritic spine (sp), identifiable by the content of a dark reaction product, and an unlabeled SB (usb) directly above. Active zones are marked by arrowheads. Scale bar 0.5 μm. (**c** and **d**) Selected labeled (lsb in **c**) and unlabeled (usb in **d**) SBs terminating both on a spine (sp) before taking a tilt series. Active zones are marked by arrowheads. Scale bar 0.5 μm. (**e**) Single image from a tilt series through a labeled bouton (lsb) with two active zones (arrowheads). Note the identification of "docked" synaptic vesicles (transparent green) at both active zones. At this bouton, a total of 12 "docked" synaptic vesicles were identified. Scale bar 0.25 μm

questions about the structural and functional impact of thalamocortical vs. intracortical inputs in synaptic transmission and plasticity.

In summary, both methods described here are well-suited morphological tools to identify SVs constituting the RRP of SVs at different CNS synapses. Although technically elaborate, labor-intensive, and hence time-consuming, both methods seem to be the methods of choice to morphologically identify SVs belonging to the three functionally defined pools of SVs, those constituting the RRP, the RP, and the resting pool. It is demonstrated that the size of the RRP is a crucial criterion describing and comparing synaptic efficacy, strength, but also synaptic modulation in different CNS synapses in experimental animals, in animal models of neurological and neurodegenerative disorders, but also in the normal and pathologically altered human brain.

Acknowledgments

The authors would like to thank Brigitte Marshallsay, Sylvia Voortmann, and Tayfun Palaz for excellent and constant technical support and the Research Centre Jülich GmbH (J.L and A.R.), the Medical Faculty of the University of Duisburg-Essen (B.W. and M. H.), and the *Deutsche Forschungsgemeinschaft* (J.L., A.R., B.W. and M.H.) for long-lasting financial support.

References

1. Südhof TC (2012) The presynaptic active zone. Neuron 75:11–25
2. Diering GH, Huganir RL (2018) The AMPA receptor code of synaptic plasticity. Neuron 100:314–329
3. Magee JC, Grienberger C (2020) Synaptic plasticity forms and functions. Annu Rev Neurosci 43:95–117
4. Fuchsberger T, Paulsen O (2022) Modulation of hippocampal plasticity in learning and memory. Curr Opin Neurobiol 75:102558
5. Silver RA, Lübke J, Sakmann B, Feldmeyer D (2003) High-probability uniquantal transmission at excitatory synapses in barrel cortex. Science 302:1981–1984
6. Saviane C, Silver RA (2006) Fast vesicle reloading and a large pool sustain high bandwidth transmission at a central synapse. Nature 439: 983–987
7. Watanabe S, Rost BR, Camacho-Perez M, Davis MW, Sohl-Kielczynski B, Rosenmund C, Jorgensen EM (2013) Ultrafast endocytosis at mouse hippocampal synapses. Nature 504:242–247
8. Schikorski T (2014) Readily releasable vesicles recycle at the active zone of hippocampal synapses. Proc Natl Acad Sci USA 111:5415–5420
9. Vaden JH, Banumurthy G, Gusarevich ES, Overstreet-Wadiche L, Wadiche JI (2019) The readily-releasable pool dynamically regulates multivesicular release. Elife 8
10. Rizzoli SO, Betz WJ (2004) The structural organization of the readily releasable pool of synaptic vesicles. Science 303:2037–2039
11. Rizzoli SO, Betz WJ (2005) Synaptic vesicle pools. Nat Rev Neurosci 6:57–69
12. Denker A, Rizzoli SO (2010) Synaptic vesicle pools: an update. Front Synaptic Neurosci 2: 135
13. Neher E (2015) Merits and limitations of vesicle pool models in view of heterogeneous populations of synaptic vesicles. Neuron 87: 1131–1142
14. Chamberland S, Toth K (2016) Functionally heterogeneous synaptic vesicle pools support diverse synaptic signalling. J Physiol 594:825–835

15. Wilhelm BG, Groemer TW, Rizzoli SO (2010) The same synaptic vesicles drive active and spontaneous release. Nat Neurosci 13:1454–1456
16. Alvelid J, Damenti M, Sgattoni C, Testa I (2022) Event-triggered STED imaging. Nat Methods 19:1268–1275
17. An SJ, Stagi M, Gould TJ, Wu Y, Mlodzianoski M, Rivera-Molina F, Toomre D, Strittmatter SM et al (2022) Multimodal imaging of synaptic vesicles with a single probe. Cell Rep Methods 2:100199
18. Sankaranarayanan S, De Angelis D, Rothman JE, Ryan TA (2000) The use of pHluorins for optical measurements of presynaptic activity. Biophys J 79:2199–2208
19. Fernandez-Alfonso T, Ryan TA (2004) The kinetics of synaptic vesicle pool depletion at CNS synaptic terminals. Neuron 41:943–953
20. Fredj NB, Burrone J (2009) A resting pool of vesicles is responsible for spontaneous vesicle fusion at the synapse. Nat Neurosci 12:751–758
21. Sätzler K, Söhl LF, Bollmann JH, Borst JG, Frotscher M, Sakmann B, Lübke JHR (2002) Three-dimensional reconstruction of a calyx of Held and its postsynaptic principal neuron in the medial nucleus of the trapezoid body. J Neurosci 22:10567–10579
22. Henkel AW, Lübke J, Betz WJ (1996) FM1-43 dye ultrastructural localization in and release from frog motor nerve terminals. Proc Natl Acad Sci USA 93:1918–1923
23. Südhof TC (2000) The synaptic vesicle cycle revisited. Neuron 28:317–320
24. Schikorski T, Stevens CF (2001) Morphological correlates of functionally defined synaptic vesicle populations. Nat Neurosci 4:391–395
25. Yakoubi R, Rollenhagen A, von Lehe M, Miller D, Walkenfort B, Hasenberg M, Sätzler K, Lübke JHR (2019a) Ultrastructural heterogeneity of layer 4 excitatory synaptic boutons in the adult human temporal lobe neocortex. Elife 8
26. Yakoubi R, Rollenhagen A, von Lehe M, Shao Y, Satzler K, Lübke JHR (2019b) Quantitative three-dimensional reconstructions of excitatory synaptic boutons in layer 5 of the adult human temporal lobe neocortex: a fine-scale electron microscopic analysis. Cereb Cortex 29:2797–2814
27. Schmuhl-Giesen S, Rollenhagen A, Walkenfort B, Yakoubi R, Sätzler K, Miller D, von Lehe M, Hasenberg M et al (2022) Sublamina-specific dynamics and ultrastructural heterogeneity of layer 6 excitatory synaptic boutons in the adult human temporal lobe neocortex. Cereb Cortex 32:1840–1865
28. Rollenhagen A, Sätzler K, Rodriguez EP, Jonas P, Frotscher M, Lübke JHR (2007) Structural determinants of transmission at large hippocampal mossy fiber synapses. J Neurosci 27:10434–10444
29. Rollenhagen A, Klook K, Sätzler K, Qi G, Anstötz M, Feldmeyer D, Lübke JHR (2015) Structural determinants underlying the high efficacy of synaptic transmission and plasticity at synaptic boutons in layer 4 of the adult rat 'barrel cortex'. Brain Struct Funct 220:3185–3209
30. Rollenhagen A, Ohana O, Sätzler K, Hilgetag CC, Kuhl D, Lübke JHR (2018) Structural properties of synaptic transmission and temporal dynamics at excitatory layer 5B synapses in the adult rat somatosensory cortex. Front Synaptic Neurosci 10:24
31. Rollenhagen A, Lübke JHR (2006) The morphology of excitatory central synapses: from structure to function. Cell Tissue Res 326: 221–237
32. Rollenhagen A, Lübke JHR (2010) The mossy fiber bouton: the "common" or the "unique" synapse? Front Synaptic Neurosci 2:2
33. Miller RF, Stenback K, Henderson D, Sikora M (2002) How voltage-gated ion channels alter the functional properties of ganglion and amacrine cell dendrites. Arch Ital Biol 140:347–359
34. Ruiz R, Cano R, Casanas JJ, Gaffield MA, Betz WJ, Tabares L (2011) Active zones and the readily releasable pool of synaptic vesicles at the neuromuscular junction of the mouse. J Neurosci 31:2000–2008
35. Trigo FF, Sakaba T, Ogden D, Marty A (2012) Readily releasable pool of synaptic vesicles measured at single synaptic contacts. Proc Natl Acad Sci USA 109:18138–18143
36. Kaeser PS, Regehr WG (2017) The readily releasable pool of synaptic vesicles. Curr Opin Neurobiol 43:63–70
37. Mansvelder HD, Verhoog MB, Goriounova NA (2019) Synaptic plasticity in human cortical circuits: cellular mechanisms of learning and memory in the human brain? Curr Opin Neurobiol 54:186–193
38. Abercrombie M (1946) Estimation of nuclear population from microtome sections. Anat Rec 94:239–247. https://doi.org/10.1002/ar.1090940210
39. Nagwaney S, Harlow ML, Jung JH, Szule JA, Ress D, Xu J, Marshall RM, McMahan UJ (2009) Macromolecular connections of active zone material to docked synaptic vesicles and

presynaptic membrane at neuromuscular junctions of mouse. J Comp Neurol 513:457–468

40. Szule JA, Harlow ML, Jung JH, De-Miguel FF, Marshall RM, McMahan UJ (2012) Regulation of synaptic vesicle docking by different classes of macromolecules in active zone material. PLoS One 7:e33333
41. Perkins GA, Jackson DR, Spirou GA (2015) Resolving presynaptic structure by electron tomography. Synapse 69:268–282
42. Bruckner JJ, Zhan H, Gratz SJ, Rao M, Ukken F, Zilberg G, O'Connor-Giles KM (2017) Fife organizes synaptic vesicles and calcium channels for high-probability neurotransmitter release. J Cell Biol 216:231–246
43. Imig C, Cooper BH (2017) 3D analysis of synaptic ultrastructure in organotypic hippocampal slice culture by high-pressure freezing and electron tomography. Methods Mol Biol 1538:215–231
44. Mastronarde DN (2005) Automated electron microscope tomography using robust prediction of specimen movements. J Struct Biol 152:36–51
45. Kremer JR, Mastronarde DN, McIntosh JR (1996) Computer visualization of three-dimensional image data using IMOD. J Struct Biol 116:71–76
46. Murthy VN, Stevens CF (1999) Reversal of synaptic vesicle docking at central synapses. Nat Neurosci 2:503–507
47. Schikorski T, Stevens CF (1997) Quantitative ultrastructural analysis of hippocampal excitatory synapses. J Neurosci 17:5858–5867
48. Hallermann S, Pawlu C, Jonas P, Heckmann M (2003) A large pool of releasable vesicles in a cortical glutamatergic synapse. Proc Natl Acad Sci USA 100:8975–8980
49. Ciresan D, Meier U, Masci J, Schmidhuber J (2012) Multi-column deep neural network for traffic sign classification. Neural Netw 32:333–338
50. Boergens KM, Berning M, Bocklisch T, Braunlein D, Drawitsch F, Frohnhofen J, Herold T, Otto P et al (2017) webKnossos: efficient online 3D data annotation for connectomics. Nat Methods 14:691–694
51. Loomba S, Straehle J, Gangadharan V, Heike N, Khalifa A, Motta A, Ju N, Sievers M et al (2022) Connectomic comparison of mouse and human cortex. Science 377: eabo0924
52. Kaltdorf KV, Schulze K, Helmprobst F, Kollmannsberger P, Dandekar T, Stigloher C (2017) Fiji macro 3D ART VeSElecT: 3D automated reconstruction tool for vesicle structures of electron tomograms. PLoS Comput Biol 13:e1005317
53. Imbrosci B, Schmitz D, Orlando M (2022) Automated detection and localization of synaptic vesicles in electron microscopy images. eNeuro 9
54. Rodriguez-Moreno J, Porrero C, Rollenhagen A, Rubio-Teves M, Casas-Torremocha D, Alonso-Nanclares L, Yakoubi R, Santuy A et al (2020) Area-specific synapse structure in branched posterior nucleus axons reveals a new level of complexity in thalamocortical networks. J Neurosci 40: 2663–2679
55. Rodriguez-Moreno J, Rollenhagen A, Arlandis J, Santuy A, Merchan-Perez A, DeFelipe J, Lübke JHR, Clasca F (2018) Quantitative 3D ultrastructure of thalamocortical synapses from the "Lemniscal" ventral posteromedial nucleus in mouse barrel cortex. Cereb Cortex 28:3159–3175

Chapter 14

Readily Releasable Pool

Sina Zobel and Christian Rosenmund

Abstract

In order to understand the functioning of signal transmission at synapses, it is crucial to understand how fusion of synaptic vesicles takes place. The number and energetic states of the pool of synaptic vesicles as well as the coupling to the calcium trigger are crucial for the efficiency of neurotransmitter release. Having tools available that allow us to distinguish pool parameters and calcium secretion coupling parameters is important to understand how release may differ from synapse to synapse and how molecular alterations affect a specific component of release. In this chapter, we introduce methods that specifically define and analyze the pool size of the readily releasable pool (RRP) and discuss their advantages and limitations.

Key words Neurotransmitter release, Readily releasable pool, Electrophysiology, Imaging, Ultrastructural analysis, Hypertonic solution

1 Introduction

The transduction of presynaptic action potentials into neurotransmitter release requires the influx of Ca^{2+} that in turn induces the fusion of release-competent vesicles.

In order to understand the mechanisms of neurotransmitter release, it is important to dissect the three main parameters that determine the output strength of the presynaptic terminal: (1) the number of fusion-competent vesicles, termed the readily releasable pool (RRP), (2) the energetic state (fusogenicity) of the RRP, and (3) the coupling efficiency of incoming Ca^{2+} to the release apparatus. In addition, sustained release during ongoing activity depends on how quickly consumed synaptic vesicles are replenished by newly primed ones. Having analytical and experimental tools available that allow us to distinguish between different vesicle pool and Ca^{2+} secretion coupling parameters is important to understand how release may differ from synapse to synapse and how molecular alterations affect a specific release parameter at distinct synapses. Furthermore, they help us to understand how synaptic plasticity

Joachim H. R. Lübke and Astrid Rollenhagen (eds.), *New Aspects in Analyzing the Synaptic Organization of the Brain*, Neuromethods, vol. 212, https://doi.org/10.1007/978-1-0716-4019-7_14,

takes place or how neurological disorders may be caused by specific presynaptic pathophysiologies.

In this chapter, we will not examine the processes related to Ca^{2+} secretion coupling but focus on several methods that have probed the size and behavior of the RRP. The RRP is only a fraction of the total vesicles in the synapse, and other pools were defined in the course of time. The vesicles in the recycling pool maintain release on moderate stimulation and are continuously recycled. A third pool from which vesicles are drawn more slowly and only during intense stimulation is called the reserve pool [1–5]. Because the RRP size is particularly critical for the output of synapses, approaches to study recycling pool and reserve pool will not be discussed here.

Research on the definition and measurement of RRP behavior is based on different methods including electrophysiology, imaging, and ultrastructural analysis. Since multiple parameters determine the pool behavior, modeling and theoretical analysis has helped significantly to dissect the contribution of individual parameters particularly in dynamic processes such as short-term plasticity. Again, we will focus here solely on discussing experimental approaches that define and probe readily releasable vesicles.

The goal of protocols that measure RRP size is to count the number of synaptic vesicles that are fusion competent, ready to be released, either spontaneously or action potential triggered. Those vesicles have characteristics different to other synaptic vesicles, such as that they are in close apposition to the plasma membrane, and whose fusion apparatus is fully assembled.

2 Application of Hypertonic Solution

A precise and well-characterized method to measure the RRP is inducing vesicle exocytosis by application of hypertonic solution.

In 1995, Stevens and Tsujimoto defined the readily releasable pool by applying hypertonic solution to a small population of identified hippocampal synapses for several seconds and counting every quantum that was released. They hypothesized that the pool of readily releasable vesicles is thereby formed by the population of vesicles docked to the active zone [1, 6].

The following year, Rosenmund and Stevens determined the relationship of the functionally defined readily releasable pool to the one drawn upon by action potential-evoked release. They utilized single neurons grown on micro islands (*see* Fig. 1b) and pulse-applied solutions with defined hypertonicities. In this system, release induced by application of hypertonic solution and by action potentials could be directly compared and their interdependence probed. The results demonstrated that both release processes share a common pool of readily releasable quanta [7]. This shared readily

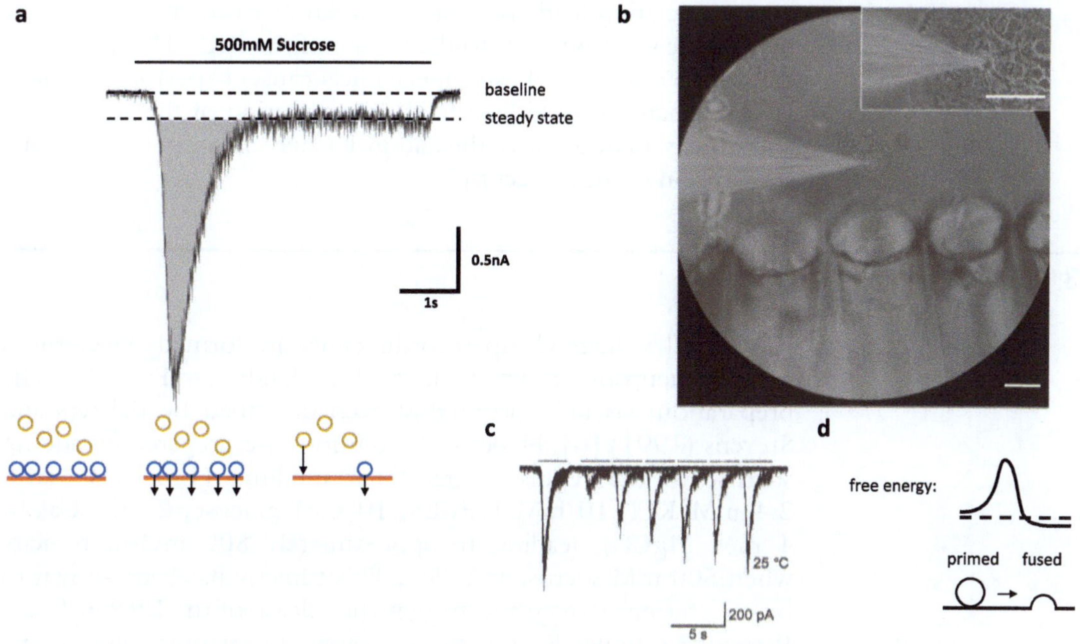

Fig. 1 Application of hypertonic solution to measure readily releasable pool (RRP) size. (**a**) Whole-cell recording of a response to hypertonic sucrose solution (500 mM) of a wild-type hippocampal autaptic neuron (*top*). The transient current component represents the release of all fusion-competent vesicles, and by integrating that component (gray area), the size of the RRP can be determined. The steady-state component represents the stage when pool refilling rate and release rate are at equilibrium. Sketches (*bottom*) represent the corresponding states of synaptic vesicles (SVs). *Left*: SVs are docked (blue circles) at the active zone (AZ; solid brown line) or close to the AZ (gold circles). *Middle*: Arrows indicate release of fusion-competent vesicles upon application of hypertonic sucrose solution. *Right*: Refilling of the RRP is in equilibrium with SV release. (**b**) Photo of microdot island with autaptic glutamatergic neuron and glass pipette directly on the neuron in whole-cell mode. An array of quartz flow pipes ensures full perfusion of the cell with solution. Scale bar: 200 μm. Insert: autaptic glutamatergic neuron and glass pipette in higher magnification. Scale bar: 50 μm. (**c**) Example trace of paired pulses with varying inter-pulse time intervals (1, 4, 7, 10, 13, and 16 s) at 25 °C. With increasing inter-pulse time intervals, the response of the second pulse increases. (Modified from [18]). (**d**) Schematic of free energy necessary for activation of SV fusion and corresponding vesicle model

releasable pool was subsequently correlated with the population of morphologically docked vesicles [8–10].

Another important finding of that study was that release activated by application of hypertonic solution is independent of intra- and extracellular Ca^{2+}. Neither the block of Ca^{2+} influx nor preventing elevation of intracellular Ca^{2+} by Ca^{2+} chelator perfusion impaired synaptic vesicle exocytosis produced by hypertonic solution. Thus, hypertonicity triggers release by bypassing the Ca^{2+}-triggering mechanism, allowing for direct analysis of exocytosis processes downstream of Ca^{2+} triggering and in addition allowing for the analysis of Ca^{2+}-dependent and Ca^{2+}-independent processes that affect the size and the properties of the RRP [4, 11]. While there is strong support for the usefulness in defining the RRP by hypertonic stimuli, there is still little known how mechanistically hypertonic solution leads to the release of the RRP. The increase in exocytosis rate might have a mechanical cause: hypertonic induced volume changes could lead to the deformation of the active zone and, thus, to a decrease of the energy barrier that must be overcome for a fusion event to occur [7].

3 Method

Whole-cell voltage-clamp recordings are performed on cultured autaptic neurons grown on microdot islands (*see* Fig. 1b). The preparation has first thoroughly been described by Bekkers and Stevens (1991) [6]. Hypertonic solutions are prepared by adding sucrose to the regular extracellular medium (140 mM NaCl, 2.4 mM KCl, 10 mM HEPES, 10 mM glucose, 2 mM $CaCl_2$, 4 mM $MgCl_2$), leading to approximately 800 mOsm tonicity when 500 mM sucrose is added. This tonicity has been shown to be of sufficient strength to trigger the release of the RRP in 1–2 s. Particular care needs to be taken to apply hypertonic solution in a pulsed fashion, utilizing a fast-flow system. For a fast-flow application of the solutions, an array of quartz flow pipes (400–500 μm diameters) is positioned within 100–200 μm of the neuron, and this ensures that the entire dendritic tree of the neuron is fully perfused (*see* Fig. 1b). The flow pipes are connected to gravity-fed valve-controlled reservoirs, and the flow rate is adjusted by the height of the reservoir to be 1–2 ml per minute and results in a flow speed of 13–26 mm/s sweeping across microdots with diameters of 200–400 μm. In a typical experiment, one of the flow pipes contains normal extracellular solution, while the neighboring flow pipe contains the hypertonic solution. Data acquisition software triggers the switch of the flow pipe position via step motors, and simultaneously switching the flowing solution from control to sucrose. After approximately 4–5 s, the solutions are switched back to control. This system ensures that all autapses are uniformly bathed with external solution or hypertonic solution and provides reliable and precise solution exchanges with time constants of approximately 20–30 ms [12].

The RRP of synaptic vesicles is estimated by application of 500 mM sucrose solution. Application of 500 mM sucrose using the fast-flow system leads to a kinetically well-defined two component response: first, a transient current component, representing the release of all fusion competent vesicles and, second, a steady-state component at the stage when the rate of pool refilling and the rate of release are at equilibrium (*see* Fig. 1a). Typically, 4–5-s-long applications are sufficient to distinguish between the transient and steady-state component. The charge carried by the transient component of the sucrose response curve is used to determine the RRP size [7]. The application of hypertonic solution triggers synaptic vesicle release in all synapses along the dendritic tree, and therefore, the method is particularly suitable for autaptic neurons since all synaptic inputs of a neuron consist of a defined presynaptic input. Using mass cultures, it is possible to measure the pool of all ingoing synapses on a neuron, but since the synaptic inputs arise from many different neurons, some parameters such as release probability are not readily accessible.

4 Application

One application of this method is the calculation of the total number of vesicles in the RRP, which is sampled as a total from all input synapses. This can be readily computed when the mean charge of spontaneous mEPSC events has been obtained from additional recordings of spontaneous release activity. The number of vesicles in the pool can then be computed by dividing RRP charge by the average charge of mEPSC events. Taking the mEPSC frequency into account, further parameters can be obtained. One of those parameters is the spontaneous vesicular release rate and can be computed by dividing the spontaneous mEPSC frequency by the number of RRP vesicles. This parameter reflects the average release behavior of individual vesicles in absence of Ca^{2+} triggering events and reflects the height of the energy barrier for fusion-competent vesicles to fuse spontaneously. Typically, the spontaneous release rate is in the order of 0.001 s^{-1} and 0.0025 s^{-1}, i.e., a fusion-competent vesicle resides in the ready to fuse state for 250–1,000 s before it fuses spontaneously [13–17]. Knowing that rate for a given cell type has the additional advantage that the RRP size then can be derived from the spontaneous mEPSC frequency, by dividing the mEPSC frequency by the vesicular mEPSC rate. For example, if the mEPSC frequency is 10 Hz and the spontaneous release rate of a glutamatergic neuron is 0.001 s^{-1}, this results in a pool size estimate of 10,000 vesicles. If, additionally, the number of synapses has been determined by morphological methods like immunolabeling with synaptic marker and subsequent counting of the total number of synaptic punctae using microscopy, one can calculate the number of readily releasable vesicles per synapse. For example, if optical counting of synaptic marker spots reveals 1000 release sites, and the RRP contains 10.000 vesicles, the mean number of fusion-competent vesicle per release site would be 10.

Another very important parameter for synaptic function is the vesicular release probability (P_{vr}), which is the mean likelihood that an individual fusion-competent vesicle fuses upon an action potential triggering event. This parameter can be readily obtained in autaptic neurons, where all synapses can be triggered to release by action potential. P_{vr} can simply be determined by dividing the charge of the EPSC by the charge of the RRP. As for example, a mean RRP charge of 150 pC and a mean EPSC charge of 25 pC results in a P_{vr} of 16%. These calculations help to identify mechanisms of synaptic short- and long-term plasticity and also provide insights in the behavior of distinct types of synapses in the nervous system, how efficient they transduce presynaptic action potentials into postsynaptic potentials, and what their behavior to trains of action potentials may look like.

Utilization of pulsed application of hypertonic solution to determine RRP behavior can also be used to study the dynamics of RRP replenishment, a process that is critical in sustained release activity when presynaptic neurons fire action potentials at high rates and vesicles are consumed at high rates. Here, a pair of two hypertonic pulse applications is utilized while the interval between the two pulses is varied (*see* Fig. 1c). Recovery of the RRP response over time occurs within seconds and can be described by one or two time constant functions [1, 7, 18, 19].

The autaptic preparation is also useful in examining whether action potential-triggered release additionally influences the dynamics and size of the RRP. In 1999, Stevens and Wesseling found that the size of the RRP is not increased by repetitive electrical stimulation but that the repeated action potential stimulation increases the vesicular release by increasing vesicle fusogenicity [20]. Fusogenicity is a biophysical measure of the height of the energy barrier that a primed vesicle must overcome before it can fuse with the membrane (*see* Fig. 1d) and, as mentioned above, defines spontaneous vesicular release rates and will define how efficient the action potential-induced Ca^{2+} influx triggers release and thus influences vesicular release probability [14, 17]. Fusogenicity of vesicles is best determined by measuring vesicle fusion rates independent of a Ca^{2+} trigger, e.g., by probing the sensitivity of release to hypertonic solution. Response kinetics and amplitudes to hypertonic solution depend both on the fusogenic state of vesicles and on the strength of hypertonicity. By creating dose–response functions or analyzing the magnitude, kinetics, and onset time of hypertonically induced synaptic vesicle fusion, sensitive assays are available to probe vesicle fusogenicity. A simple experiment to determine sensitivity of release to hypertonic solution is to perform a two-data point dose–response analysis. An intermediate and a high tonicity solution (e.g., 250 mM and 500 mM sucrose) are applied, and the relative response sizes induced by these solutions are set in relation [14, 17, 19, 21, 22]. Typical wild-type neurons show a fractional response of 35% for the ratio of 250 mM/500 mM responses. That fraction increases under phorbol ester treatment (which is known to cause an increase in P_{vr}) to 45%. Thus, the change in P_{vr} induced by phorbol esters is at least in part caused by an increase in vesicle fusogenicity. Besides varying the release behavior in response to varying the tonicity of the sucrose solution, changes in fusogenicity also are reflected in the sucrose response kinetics leading to alterations in onset time of the response or the peak amplitude [14]. Indeed, acceleration or deceleration of the sucrose-induced release has been detected in mutants of the vesicle release machinery proteins such as Syntaxin-1, complexin, and Munc13-1, and thus these proteins were identified as modifiers of vesicle fusogenicity and vesicular release probability [14, 19, 21, 22]. As in one example, Gerber et al. examined the role of a Syntaxin-1 mutation that affects its interaction with Munc18 and its transition between a "closed" and "open" form by keeping Syntaxin-1 constitutively "open." As a result, glutamatergic central synapses showed two changes to the RRP. First, the RRP pool was smaller, probably due to lower expression levels of Syntaxin-1 and Munc18. Second, synaptic vesicles within the pool had a drastically higher fusogenicity, as evident by the accelerated sucrose-mediated release kinetics, higher sensitivity to intermediate sucrose concentrations, and higher spontaneous vesicle release rates. The increased vesicle fusogenicity then caused a robust increase in vesicular release probability and pronounced increase in paired pulse depression [19]. A second example is related to the function of complexin, the best characterized SNARE complex binding protein. complexin-deficient mammalian neurons displayed a reduced release probability and reduced spontaneous release rate [11, 23]. Examining the RRP by probing with hypertonic solutions revealed that the RRP size was not altered but that the release kinetics were slowed and the sensitivity to hypertonicity was reduced, typically seen for a heightened energy barrier for vesicle fusion. These experiments therefore demonstrated that mammalian complexins facilitate vesicular release by enhancing the vesicle fusogenicity [21, 22, 24].

These findings illustrate that the characterization of neuron release behavior to hypertonic solutions can be very beneficiary to understand the molecular mechanisms of release itself. They particularly help to probe whether molecular or pharmacological manipulations or alterations that cause a change in vesicle release are due to effects on the fusion process itself or are linked to alternative contributors such as action potential wave form, Ca^{2+} channel function, or Ca^{2+} secretion coupling.

Another important parameter that varies with the size of the pool of readily releasable quanta is the release probability (P_r), which quantifies the efficiency or reliability of individual synapses to release at least one vesicle from a bouton upon arrival of an action potential. That efficiency is determined by the number of fusion-competent vesicles per synapse, and, thus, the size of the RRP helps determine P_r [7, 9, 25]. The strength of an evoked postsynaptic response (I) (or synaptic strength) depends on the release probability (P_r) from a pool of quanta (N) of defined quantal amplitude (Q):

$I = Q\,P_r\,N$ [26]. Rapid changes in synaptic strength as facilitation or depression of the postsynaptic response is called short-term plasticity and describes a highly abundant form of rapid, activity-dependent modulation of synaptic efficacy [27, 28]. Furthermore, it has been postulated that the pool size may change due to plasticity [29].

Moreover, associations between altered pool size and neurodevelopmental or neurodegenerative diseases show that the maintenance of proper pool sizes is crucial for normal synaptic communication [30]. The transcriptional regulator methyl-CpG binding protein 2 (MeCP2) is critical for normal brain function, and mutations that lead to loss or duplication of MeCP2 protein levels lead to severely disrupted neuronal network functions and neurological symptoms, indicating that the level of MeCP2 protein must be tightly controlled by the nervous system [31–33]. In mouse models that replicate the genetics causing the reduced or increased MeCP2 protein levels, RRP measurements from single neurons by hypertonic solutions showed severe reduction in RRP size within the glutamatergic neurons lacking MeCP2, while the neurons overexpressing MeCP2 showed a near doubling of RRP size, and both models showed no change in the vesicular release probability. Subsequent analysis showed that the alteration in RRP size was paralleled by alteration in the formation of synapse numbers [34]. Since the RRP size depends on the number of fusion-competent vesicles per synapse and the number of synapses per neuron, RRP size measurements can be utilized to test whether modifications in the formation and maintenance of synapse numbers are associated with a certain form of plasticity or pathophysiology.

5 Advantages

One of the major advantages of this very precise and well-characterized method to evoke fusion of synaptic vesicles is the Ca^{2+} independence and the resulting ability to separate Ca^{2+}-dependent from Ca^{2+}-independent processes. Moreover, in conjunction with the use of single neuron microdot island cultures, RRP measurements by application of hypertonic solution can be utilized together with measurements of evoked and spontaneous release to obtain a comprehensive picture of the overall release process.

6 Disadvantages/Drawbacks

One disadvantage of this method is that an average is formed globally across all synapses of one neuron, and it is not possible to measure individual synapses. Therefore, it is difficult to capture the functional heterogeneity of synapses. In addition, hypertonic sucrose application is only readily applicable in dissociated cell cultures since rapid changes in osmotic pressure are critical to properly quantify the transient component of the sucrose-induced response. In intact preparations, in acute or organotypic slice cultures, the convection/diffusion of the applied solution is not fast enough, leading to responses from nerve terminals that see gradients in hypertonic strength and where the delayed buildup of hypertonic strength leads to responses where RRP measurements are mixed up with release associated with pool refilling activity.

An additional limitation of the release of the RRP by hypertonic solution application is that the exact mechanism of how fusion-competent vesicles are released is not known.

7 Other Electrophysiological Methods

There are other electrophysiological methods that can be utilized to report the functional RRP. All these methods have in common that they rely on the use of Ca^{2+} to trigger the release of the RRP. When direct electrical access to the presynaptic terminal can be obtained, capacitance measurements combined with strong presynaptic depolarizations or Ca^{2+} uncaging can readily report functional RRP size. Furthermore, RRP estimates by application of action potential trains are useful in experimental settings where more direct methods are not readily available.

7.1 Capacitance Measurements

Synaptic vesicle membranes are added to the plasma membrane upon successful vesicle fusion, which in turn leads to a measurable increase in the membrane capacitance. Capacitance measurements require the application of a high-frequency sine-wave voltage stimulation and the measurement of the resulting sinusoidal current. That current is roughly proportional to the membrane capacitance, and the resulting errors can be dealt with by a phase-sensitive lock-in amplifier [35–38].

Most presynaptic capacitance measurements have been performed on large synapses, like squid giant synapses, bipolar neurons from goldfish retinae, or mammalian calyx of Held synapses, due to its experimental accessibility [4, 39–42]. High-resolution capacitance measurements from direct nerve terminal patch clamp recordings are difficult but possible and have been shown to be sensitive enough to detect single-vesicle fusion events from hippocampal glutamatergic mossy fiber terminals in hippocampal slices. Due to the small size of synaptic terminals, and considering that a synaptic vesicle contributes 70 aF of capacitance, the sensitivity of the technique operates near the absolute limits of resolution [43, 44]. RRP size measurements can generally be obtained by examining the step-wise increase in membrane capacitance induced by strong depolarizing voltage commands that induces Ca^{2+} influx that triggers the fusion of all fusion-competent vesicles [37, 39, 45–47]. This way, Hallermann and colleagues estimated the number of releasable vesicles in mossy fiber terminals to be 1400 vesicles, which is consistent with functional and ultrastructural assessments of RRP size [44].

If neurons are small and the presynaptic nerve terminal is close to the soma, e.g., in retinal bipolar neurons or All-amacrine cells, somatic patch clamp recordings are also sufficient to accurately report vesicle exocytosis from capacitance increases [46, 48–50]. While not directly related to the definition of functional pools, capacitance measurements allow for the precise and with the highest temporal resolution analysis of exocytosis-induced endocytosis and therefore allow to study how endocytosis temporally and functionally couples to release from the RRP. For example, Ritzau-Jost and colleagues have utilized this approach to functionally verify the existence of ultrafast endocytosis from central nerve terminals [38].

7.2 Ca^{2+} Jumps

As mentioned above, the definition of functional RRPs requires the use of a strong and rapid-acting stimulus to distinguish the release of the RRP from release caused by vesicles that are already reentering the RRP by replenishment actions. Typically, 30 ms lasting depolarizations of presynaptic terminals cause sufficient Ca^{2+} influx to trigger the complete release of all fusion-competent vesicles [51–53]. This, however, requires that the RRP size is detected by the activation of synaptic currents from the postsynaptic neuron, and issues of postsynaptic receptor saturation and desensitization need to be taken into account. In addition, the intracellular Ca^{2+} concentration can be increased by increasing the number of ions that enter the cells in response to an action potential by increasing the extracellular Ca^{2+} concentration. That way the Ca^{2+} sensors required for vesicle fusion are saturated, and all vesicles in the RRP are expected to fuse synchronously [54]. Another approach to increase intracellular Ca^{2+} is to use a voltage-gated potassium channel blocker like 4-aminopyridine (4AP) to extend the duration of the action potential and at the same time to extend the time that the Ca^{2+} channels open [55]. That approach could show better results in increasing the exocytosis in response to an action potential than increasing the extracellular Ca^{2+} concentration. Exocytosis as a function of the relative increase in intracellular Ca^{2+} concentration can be fitted by a generalized Hill model where the size of the RRP is indicated by a saturation value [54].

Alternatively to utilizing voltage-gated Ca^{2+} channels as a source of rapid elevation of intraterminal Ca^{2+}, presynaptic Ca^{2+} concentrations can be rapidly and uniformly increased by Ca^{2+} uncaging methods. Here sudden increases of intracellular Ca^{2+}

from nanomolar concentrations to tens of micromolar can be accomplished by UV light-induced photolysis of a caged Ca^{2+} chelator such as 1-(2-nitro-4,5-dimethoxyphenyl)-N,N,N′,N′--tetrakis[(oxycarbonyl)methyl]-1,2-ethanediamine (DM-nitrophen). To make sure that all vesicles are released, a second laser pulse can be applied with a 30 ms time interval [56–62]. While proper control and simultaneous measurement of the post flash-induced Ca^{2+} concentration is demanding, the approach has been highly successful in defining many release parameters besides the definition of the functional RRP. An additional advantage of the Ca^{2+} uncaging methods is that they allow for the discrete stimulation of individual synapses using spatially confined uncaging. In 2012, Trigo and colleagues were able to investigate the RRP of an individual synaptic connection by measuring quantal events evoked postsynaptically by presynaptic Ca^{2+} uncaging with localized laser light [63].

7.3 Action Potential Trains

Another method to quantify RRP sizes is the application of action potential trains at high frequencies. The idea behind this approach is that neurotransmitter output will have two phases: a large amplitude component that is transient and that represents the consumption of readily releasable vesicles and a chronic component that reflects the renewed formation of fusion-competent vesicles. In general, there are three different ways to use action potential trains to measure the RRP size.

In the first method, the train method, the cumulative EPSC amplitude is plotted as a function of stimulus number. With a frequency of 100 Hz, a steady-state region for times larger than 100 ms corresponds to vesicle replenishment, and a linear regression fitted to that steady-state region is extrapolated. The intercept with the *y*-axis marks the estimate of the cumulative EPSC amplitudes before steady-state depression, thus, the RRP size (number of vesicles in the RRP multiplied by the quantal size). Critical for the approach to work is that synaptic transmission takes place with high initial release probabilities, so that depletion of RRP vesicles dominates against rates of vesicle replenishment. Therefore, synapses that display rapid depression during action potential trains are more accurately reporting their RRP with this method [47, 64, 65].

The second method takes into account that for a depressing synapse with high release probability, depletion of the RRP dominates the amplitude of the synaptic responses in the beginning of a high-frequency train [65]. Based on a method developed for use at the frog neuromuscular junction (NMJ) by Elmqvist and Quastel in 1965 [66], the EPSC amplitudes can be plotted as a function of the cumulative EPSC, and the quick depression during the first few responses leads to a linear relationship between EPSC amplitude and cumulative EPSC representing the almost complete depletion of the RRP followed by a decaying single-exponential curve. The straight line formed by the first few stimuli can be extrapolated, and the intersection with the *x*-axis corresponds to the RRP size assuming no replenishment during the first few stimuli. Since it cannot be precluded that replenishment of the pool takes place during the first few stimuli, this method overestimates the size of the RRP.

Lastly, the third method of analysis assumes that depression of synaptic responses is a consequence of depletion and the decay of EPSC amplitudes during high-frequency stimulations can be used to directly measure the release probability. When plotting EPSC amplitudes during a high-frequency stimulation versus stimulus numbers, the EPSC amplitudes decay exponentially over time according to the equation $EPSC_n / EPSC_0 = e^{(-n/\lambda)}$, where n is the stimulus number and λ is the decay constant of a single-exponential fit to EPSC amplitudes throughout a train. This method assumes that if the decrement is due to depletion and there is no significant replenishment, p can be estimated by $p_{decay} = 1 - e^{(-1/\lambda)}$ [67].

8 Anatomical and Imaging Approaches

8.1 Electron Microscopy

Besides electrophysiological methods, several optical methods have been established to measure the RRP. At the level of electron microscopy, the proximity of SV to the plasma membrane across the postsynaptic density was utilized to define the morphological equivalent of functionally defined RRPs [9, 10, 68, 69]. In many studies, the size of the RRP has been defined by counting the number of docked vesicles. In some cases, the vesicles have been considered as docked vesicles when they were located opposite a postsynaptic density and appeared immediately adjacent to the active zone membrane and directly touched the presynaptic membrane without any visible space between the adjacent membranes [8, 10, 70–72]. Other analyses took the limited z resolution of electron micrographs from ultrathin sections into consideration and defined SVs within a distance up to 50 nm of the AZ as docked vesicles [73–75].

It turned out that the method of fixation of the tissue plays an important role for defining SV docking. Chemical fixation methods led to much less SVs in immediate proximity of the plasma membrane, which made it difficult to define when a SV is still considered to be docked. A breakthrough was the utilization of high-pressure freezing tissue preservation techniques. The prevention of ice crystal formation during rapid freezing under high pressure combined with the lack of fixatives during the conservation step led to much reduced structural changes during the fixation process.

In 2006, Weimer used high-pressure freezing to examine the nematode *Caenorhabditis elegans* (*C. elegans*) by electron microscopy. Treatment of the tissue by cryo-fixation, freeze substitution,

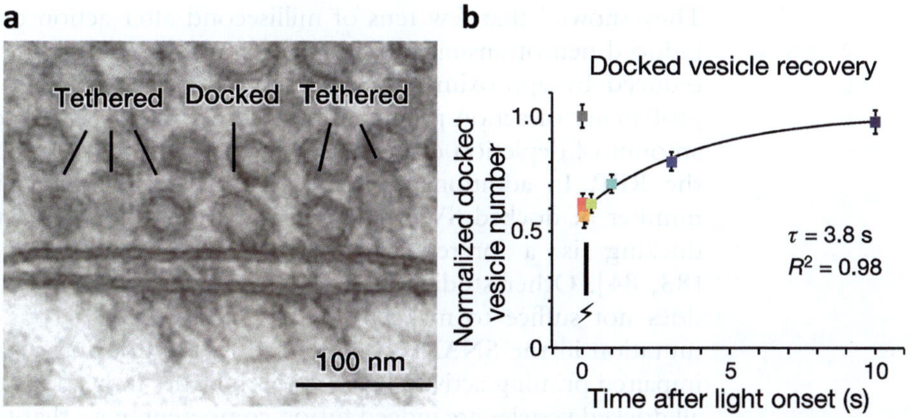

Fig. 2 Ultrastructural quantification of the readily releasable vesicle pool using flash and freeze electron microscopy. (Modified from [84]). (**a**) Micrograph of cultured mouse hippocampal neurons showing docked and tethered synaptic vesicles. (**b**) Recovery of docked vesicles normalized to non-stimulated control (0 ms). Flash indicates the time point of light onset when a brief single light stimulus was applied. The number of docked vesicles was reduced by approximately 40% and recovered with a time constant of 3.8 s

and subsequent embedding led to cellular architecture with exceptional spatial preservation [76]. In 2007, Siksou and colleagues performed high-pressure freezing experiments on rat hippocampal slices to analyze the three-dimensional architecture of the presynaptic cytomatrix in electron tomograms [77]. Due to the improved spatial resolution, synaptic vesicles are considered as docked vesicles within a distance of 0–2 nm from the presynaptic membrane [69, 78]. How SV docking relates to the functional status of readily releasable vesicles came then to discussion in several studies that utilized mutants of essential proteins of the release apparatus. In 2007, Hammarlund and colleagues used high-pressure freezing and EM analysis to outline a molecular mechanism for docking in *C. elegans.* By analyzing Syntaxin mutants and Unc13 mutants that were known to cause the loss of functional RRP, they found that these mutations also caused a loss of SV docking [79]. Subsequent studies in mammalian Munc13 mutants confirmed the strong correlation of docking and SV priming activity [80], and this study was important as previous studies using chemical fixation in Munc13-deficient synapses did not find a change in SV docking activity, which led to the initial wrong interpretation that Munc13 is an essential SV priming factor that acts after SV docking [73, 81, 82]. Finally, more systematic comparison of various presynaptic protein mutants in murine synapses strongly supported that SV docking and priming are highly correlated processes [69]. Beyond correlation, Watanabe et al. tested the relation of SV docking and priming directly using flash and freeze tissue conservation, where SV release was induced by optogenetic stimulation of an action potential followed by timed high-pressure freezing (*see* Fig. 2).

They showed that few tens of millisecond after action potential-induced neurotransmitter release, the number of docked SVs was reduced by approximately 40%, similar to the estimates of the probability of action potential-induced SV release. Therefore, the amount of depletion of docked SVs matched the consumption of the RRP. In addition, replenishment kinetics of RRP size and number of docked SVs were similar, further supporting that SV docking is a prerequisite for SV priming (*see* Fig. 2b) [83, 84]. Other studies, however, also showed that SV docking does not suffice to make SV fully fusion competent, as specific mutation in the SNARE complex led to full docking activity but impaired priming activity [85]. Also, there are indications that not all docked vesicles are indeed fusion competent, e.g., that there is a heterogeneity of release readiness among docked vesicles. Comparison of imaging experiments using pHluorin and EM has suggested that only a fraction of docked vesicles is released by action potentials [54, 47, 69, 86]. One possible explanation is that the molecular requirements for docking are lower than that for priming. Indeed, the impact of progressively reducing the concentrations of Munc13 in the synapse showed that SV priming is more sensitive to loss of Munc13-1 compared to SV docking [87].

Nevertheless, SV docking as measured by high-pressure fixed synapses is an excellent predictor for SV priming and therefore for the RRP size.

8.2 RRP Labeling with Dyes

Synaptic vesicles can also be labeled by externally applied fluorescent membrane dyes when they cycle through the exo- and endocytosis cycle. Since the presynapse responds with an equivalent amount of endocytosis to the amount of exocytosed membrane, the degree of fluorescence taken up by dyes can be utilized to measure the amount of exocytosis, and, thus given an appropriate stimulus, the RRP can be quantified. One of the most widely used membrane dyes for neurotransmitter release mechanisms is the styryl dye FM1-43. To quantify the dye uptake, the extracellular membrane is stained by adding the dye in the external media, while the synaptic exocytosis is stimulated. The resulting endocytosis traps the dye in the vesicle and can be readily quantified after the dye is washed from the external membrane [4, 88–90]. Key for this method is to use an appropriate stimulus that represents the RRP, e.g., by measuring the amount of dye uptake by pulses of hypertonic solution or by applying a short action potential train [5, 91, 92]. An additional advantage of FM dye labeling is that it can be also combined with electron microscopy. Via photoconversion, the induction of fluorescence of FM1-43 causes local free radical production, which in turn can be used for photoconversion of 3,3′-diaminobenzidine (DAB) to an electron-dense EM-compatible product that labels individual synaptic vesicles. This opens up an opportunity to identify the SV that participate in the exo-/

endocytosis cycle [4, 90, 93]. Schikorski and Stevens used this approach to correlate the number of docked vesicles to the number of stained vesicles in individual synapses and, thus, developed a method to quantify the population of the morphologically docked synaptic vesicles corresponding to the RRP [10].

Beyond defining RRP size, the method also reports about which fraction of SVs participated in SV cycling, e.g., it helps to define the recycling pool of vesicles. Harata et al. were able to visualize the recycling of synaptic vesicles in hippocampal neurons and found that only a minority of SVs were stained in most boutons after stimulation known to cause the functional vesicular pool to turn over completely [90], while the fraction of actively recycling vesicles is higher in slice cultures compared to primary cultures [94].

8.3 pHluorin

Synaptic vesicles are acidified by proton pumps to enable the uptake of neurotransmitters. These properties have been used in a clever approach to follow synaptic vesicles through their exo-/endocytosis cycle. By expressing a (pHluorin) variant of GFP in the lumen of synaptic vesicles (e.g., by tagging it onto the luminar part of the vesicular SNARE protein synaptobrevin), SV fusion can be monitored in live imaging experiments. The tag resides in an acidified environment when the SVs are filled with NTs, and thus is dimmed, but becomes fluorescent when the SV fuses and its environment becomes pH neutral [95, 96]. Compared to the use of FM styryl dyes, pHluorin-labeled synapses are even more versatile and can be used to define the RRP, as well as the recycling pool and the reserve pool of synaptic vesicles, and how SVs can switch between these pools [97–99].

8.4 Quantum Dots

A third method for optically probing synaptic vesicle trafficking and thus also to quantify RRP size is based on the use of photoluminescent quantum dots (QDs). Those semiconductor-based nanocrystals can be used as "fluid phase" SV uptake and release markers based on their small size. Since they are stable, bright, non-photobleaching light emitters with superior photo-properties, they allow for high signal to noise ratio imaging, which makes the quantification of exo- and endocytosis much easier [4, 100, 101]. One advantage of this method is that the QDs with an organic coating bearing carboxyl groups have a diameter of about 15 nm and cannot escape the vesicles during the nonclassical mode of vesicle recycling "kiss-and-run" where fusion by a transient fusion pore without intermixing of the membranes takes place. In that way, QDs are used to distinguish classical full collapse fusion from "kiss-and-run" modes of vesicle recycling [101–103]. One further application of this method would be the use of QDs of different colors to study the exchange between different SV pools within a nerve terminal [101, 104].

References

1. Stevens CF, Tsujimoto T (1995) Estimates for the pool size of releasable quanta at a single central synapse and for the time required to refill the pool. Proc Natl Acad Sci U S A 92(3):846–849. https://doi.org/10.1073/pnas.92.3.846. PMID: 7846064; PMCID: PMC42717
2. Rizzoli SO, Betz WJ (2005) Synaptic vesicle pools. Nat Rev Neurosci 6(1):57–69. https://doi.org/10.1038/nrn1583. PMID: 15611727
3. Denker A, Rizzoli SO (2010) Synaptic vesicle pools: an update. Front Synaptic Neurosci 2: 135. https://doi.org/10.3389/fnsyn.2010.00135. PMID: 21423521; PMCID: PMC3059705
4. Alabi AA, Tsien RW (2012) Synaptic vesicle pools and dynamics. Cold Spring Harb Perspect Biol 4:a013680. https://doi.org/10.1101/cshperspect.a013680
5. Gan Q, Watanabe S (2018) Synaptic vesicle endocytosis in different model systems. Front Cell Neurosci 12:171. https://doi.org/10.3389/fncel.2018.00171. PMID: 30002619; PMCID: PMC6031744
6. Bekkers JM, Stevens CF (1991) Excitatory and inhibitory autaptic currents in isolated hippocampal neurons maintained in cell culture. Proc Natl Acad Sci U S A 88(17):7834–7838. https://doi.org/10.1073/pnas.88.17.7834. PMID: 1679238; PMCID: PMC52398
7. Rosenmund C, Stevens CF (1996) Definition of the readily releasable pool of vesicles at hippocampal synapses. Neuron 16(6):1197–1207. https://doi.org/10.1016/s0896-6273(00)80146-4. PMID: 8663996
8. Schikorski T, Stevens CF (1997) Quantitative ultrastructural analysis of hippocampal excitatory synapses. J Neurosci 17(15):5858–5867. https://doi.org/10.1523/JNEUROSCI.17-15-05858.1997. PMID: 9221783; PMCID: PMC6573206
9. Dobrunz LE, Stevens CF (1997) Heterogeneity of release probability, facilitation, and depletion at central synapses. Neuron 18(6):995–1008. https://doi.org/10.1016/s0896-6273(00)80338-4. PMID: 9208866
10. Schikorski T, Stevens CF (2001) Morphological correlates of functionally defined synaptic vesicle populations. Nat Neurosci 4(4):391–395. https://doi.org/10.1038/86042. PMID: 11276229
11. Reim K, Mansour M, Varoqueaux F, McMahon HT, Südhof TC, Brose N, Rosenmund C (2001) Complexins regulate a late step in Ca2+-dependent neurotransmitter release. Cell 104(1):71–81. https://doi.org/10.1016/s0092-8674(01)00192-1. PMID: 11163241
12. Rosenmund C, Feltz A, Westbrook GL (1995) Synaptic NMDA receptor channels have a low open probability. J Neurosci 15(4):2788–2795. https://doi.org/10.1523/JNEUROSCI.15-04-02788.1995. PMID: 7536820; PMCID: PMC6577776
13. Murthy VN, Stevens CF (1999) Reversal of synaptic vesicle docking at central synapses. Nat Neurosci 2(6):503–507. https://doi.org/10.1038/9149. PMID: 10448213
14. Basu J, Betz A, Brose N, Rosenmund C (2007) Munc13-1 C1 domain activation lowers the energy barrier for synaptic vesicle fusion. J Neurosci 27(5):1200–1210. https://doi.org/10.1523/JNEUROSCI.4908-06.2007. PMID: 17267576; PMCID: PMC6673179
15. Xu J, Pang ZP, Shin OH, Südhof TC (2009) Synaptotagmin-1 functions as a Ca2+ sensor for spontaneous release. Nat Neurosci 12(6):759–766. https://doi.org/10.1038/nn.2320. PMID: 19412166; PMCID: PMC2739891
16. Groffen AJ, Martens S, Díez Arazola R, Cornelisse LN, Lozovaya N, de Jong AP, Goriounova NA, Habets RL, Takai Y, Borst JG, Brose N, McMahon HT, Verhage M (2010) Doc2b is a high-affinity Ca2+ sensor for spontaneous neurotransmitter release. Science 327(5973):1614–1618. https://doi.org/10.1126/science.1183765. Epub 2010 Feb 11. Erratum in: Science. 2010 May 7;328(5979):690. PMID: 20150444; PMCID: PMC2846320
17. Schotten S, Meijer M, Walter AM, Huson V, Mamer L, Kalogreades L, ter Veer M, Ruiter M, Brose N, Rosenmund C, Sørensen JB, Verhage M, Cornelisse LN (2015) Additive effects on the energy barrier for synaptic vesicle fusion cause supralinear effects on the vesicle fusion rate. Elife 4:e05531. https://doi.org/10.7554/eLife.05531. Erratum in: Elife. 2015;4. https://doi.org/10.7554/eLife.09036. PMID: 25871846; PMCID: PMC4426983
18. Pyott SJ, Rosenmund C (2002) The effects of temperature on vesicular supply and release in autaptic cultures of rat and mouse

hippocampal neurons. J Physiol 539 (Pt 2):523–535. https://doi.org/10.1113/jphysiol.2001.013277. PMID: 11882684; PMCID: PMC2290147

19. Gerber SH, Rah JC, Min SW, Liu X, de Wit H, Dulubova I, Meyer AC, Rizo J, Arancillo M, Hammer RE, Verhage M, Rosenmund C, Südhof TC (2008) Conformational switch of syntaxin-1 controls synaptic vesicle fusion. Science 321(5895):1507–1510. https://doi.org/10.1126/science.1163174. Epub 2008 Aug 14. PMID: 18703708; PMCID: PMC3235364
20. Stevens CF, Wesseling JF (1999) Augmentation is a potentiation of the exocytotic process. Neuron 22(1):139–146. https://doi.org/10.1016/s0896-6273(00)80685-6. PMID: 10027296
21. Xue M, Craig TK, Xu J, Chao HT, Rizo J, Rosenmund C (2010) Binding of the complexin N terminus to the SNARE complex potentiates synaptic-vesicle fusogenicity. Nat Struct Mol Biol 17(5):568–575. https://doi.org/10.1038/nsmb.1791. Epub 2010 Apr 18. PMID: 20400951; PMCID: PMC3172005
22. Trimbuch T, Rosenmund C (2016) Should I stop or should I go? The role of complexin in neurotransmitter release. Nat Rev Neurosci 17(2):118–125. https://doi.org/10.1038/nrn.2015.16. PMID: 26806630
23. Xue M, Stradomska A, Chen H, Brose N, Zhang W, Rosenmund C, Reim K (2008) Complexins facilitate neurotransmitter release at excitatory and inhibitory synapses in mammalian central nervous system. Proc Natl Acad Sci U S A 105(22):7875–7880. https://doi.org/10.1073/pnas.0803012105. Epub 2008 May 27. PMID: 18505837; PMCID: PMC2409395
24. Trimbuch T, Xu J, Flaherty D, Tomchick DR, Rizo J, Rosenmund C (2014) Re-examining how complexin inhibits neurotransmitter release. Elife 3:e02391. https://doi.org/10.7554/eLife.02391. PMID: 24842998; PMCID: PMC4040926
25. Dobrunz LE (2002) Release probability is regulated by the size of the readily releasable vesicle pool at excitatory synapses in hippocampus. Int J Dev Neurosci 20(3–5):225–236. https://doi.org/10.1016/s0736-5748(02)00015-1. PMID: 12175858
26. Glasgow SD, McPhedrain R, Madranges JF, Kennedy TE, Ruthazer ES (2019) Approaches and limitations in the investigation of synaptic transmission and plasticity. Front Synaptic Neurosci 11:20. https://doi.org/10.3389/fnsyn.2019.00020. PMID: 31396073; PMCID: PMC6667546
27. Zucker RS, Regehr WG (2002) Short-term synaptic plasticity. Annu Rev Physiol 64:355–405. https://doi.org/10.1146/annurev.physiol.64.092501.114547. PMID: 11826273
28. Hennig MH (2013) Theoretical models of synaptic short term plasticity. Front Comput Neurosci 7:154. https://doi.org/10.3389/fncom.2013.00154. PMID: 24198783; PMCID: PMC3812535
29. Goda Y, Stevens CF (1998) Readily releasable pool size changes associated with long term depression. Proc Natl Acad Sci U S A 95(3):1283–1288. https://doi.org/10.1073/pnas.95.3.1283. PMID: 9448323; PMCID: PMC18746
30. Waites CL, Garner CC (2011) Presynaptic function in health and disease. Trends Neurosci 34(6):326–337. https://doi.org/10.1016/j.tins.2011.03.004. Epub 2011 May 17. PMID: 21596448
31. Amir RE, Van den Veyver IB, Wan M, Tran CQ, Francke U, Zoghbi HY (1999) Rett syndrome is caused by mutations in X-linked MECP2, encoding methyl-CpG-binding protein 2. Nat Genet 23(2):185–188. https://doi.org/10.1038/13810. PMID: 10508514
32. Bienvenu T, Carrié A, de Roux N, Vinet MC, Jonveaux P, Couvert P, Villard L, Arzimanoglou A, Beldjord C, Fontes M, Tardieu M, Chelly J (2000) MECP2 mutations account for most cases of typical forms of Rett syndrome. Hum Mol Genet 9(9):1377–1384. https://doi.org/10.1093/hmg/9.9.1377. PMID: 10814719
33. Na ES, Monteggia LM (2011) The role of MeCP2 in CNS development and function. Horm Behav 59(3):364–368. https://doi.org/10.1016/j.yhbeh.2010.05.014. Epub 2010 May 31. PMID: 20515694; PMCID: PMC3077534
34. Chao HT, Zoghbi HY, Rosenmund C (2007) MeCP2 controls excitatory synaptic strength by regulating glutamatergic synapse number. Neuron 56(1):58–65. https://doi.org/10.1016/j.neuron.2007.08.018. PMID: 17920015; PMCID: PMC2198899
35. Neher E, Marty A (1982) Discrete changes of cell membrane capacitance observed under conditions of enhanced secretion in bovine adrenal chromaffin cells. Proc Natl Acad Sci U S A 79(21):6712–6716. https://doi.org/10.1073/pnas.79.21.6712. PMID: 6959149; PMCID: PMC347199

36. Lindau M, Neher E (1988) Patch-clamp techniques for time-resolved capacitance measurements in single cells. Pflugers Arch 411(2):137–146. https://doi.org/10.1007/BF00582306. PMID: 3357753

37. Lindau M (1991) Time-resolved capacitance measurements: monitoring exocytosis in single cells. Q Rev Biophys 24(1):75–101. https://doi.org/10.1017/s0033583500003279. PMID: 2047522

38. Ritzau-Jost A, Delvendahl I, Rings A, Byczkowicz N, Harada H, Shigemoto R, Hirrlinger J, Eilers J, Hallermann S (2014) Ultrafast action potentials mediate kilohertz signaling at a central synapse. Neuron 84(1):152–163. https://doi.org/10.1016/j.neuron.2014.08.036. Epub 2014 Sep 11. PMID: 25220814

39. Gillespie JI (1979) The effect of repetitive stimulation on the passive electrical properties of the presynaptic terminal of the squid giant synapse. Proc R Soc Lond B Biol Sci 206(1164):293–306. https://doi.org/10.1098/rspb.1979.0106. PMID: 43975

40. von Gersdorff H, Matthews G (1994) Dynamics of synaptic vesicle fusion and membrane retrieval in synaptic terminals. Nature 367(6465):735–739. https://doi.org/10.1038/367735a0. PMID: 7906397

41. Sun JY, Wu LG (2001) Fast kinetics of exocytosis revealed by simultaneous measurements of presynaptic capacitance and postsynaptic currents at a central synapse. Neuron 30(1):171–182. https://doi.org/10.1016/s0896-6273(01)00271-9. PMID: 11343653

42. Sun JY, Wu XS, Wu LG (2002) Single and multiple vesicle fusion induce different rates of endocytosis at a central synapse. Nature 417(6888):555–559. https://doi.org/10.1038/417555a. PMID: 12037569

43. Geiger JR, Jonas P (2000) Dynamic control of presynaptic Ca(2+) inflow by fast-inactivating K(+) channels in hippocampal mossy fiber boutons. Neuron 28(3):927–939. https://doi.org/10.1016/s0896-6273(00)00164-1. PMID: 11163277

44. Hallermann S, Pawlu C, Jonas P, Heckmann M (2003) A large pool of releasable vesicles in a cortical glutamatergic synapse. Proc Natl Acad Sci U S A 100(15):8975–8980. https://doi.org/10.1073/pnas.1432836100. Epub 2003 Jun 18. PMID: 12815098; PMCID: PMC166423

45. Gillis KD (1995) Techniques for membrane capacitance measurements. In: Sakmann B, Neher E (eds) Single-channel recording. Springer, Boston. https://doi.org/10.1007/978-1-4419-1229-9_7

46. Oltedal L, Hartveit E (2010) Transient release kinetics of rod bipolar cells revealed by capacitance measurement of exocytosis from axon terminals in rat retinal slices. J Physiol 588 (Pt 9):1469–1487. https://doi.org/10.1113/jphysiol.2010.186916. Epub 2010 Mar 8. PMID: 20211976; PMCID: PMC2876803

47. Kaeser PS, Regehr WG (2017) The readily releasable pool of synaptic vesicles. Curr Opin Neurobiol 43:63–70. https://doi.org/10.1016/j.conb.2016.12.012. Epub 2017 Jan 16. PMID: 28103533; PMCID: PMC5447466

48. Zhou ZY, Wan QF, Thakur P, Heidelberger R (2006) Capacitance measurements in the mouse rod bipolar cell identify a pool of releasable synaptic vesicles. J Neurophysiol 96(5):2539–2548. https://doi.org/10.1152/jn.00688.2006. Epub 2006 Aug 16. PMID: 16914610

49. Balakrishnan V, Puthussery T, Kim MH, Taylor WR, von Gersdorff H (2015) Synaptic vesicle exocytosis at the dendritic lobules of an inhibitory interneuron in the mammalian retina. Neuron 87(3):563–575. https://doi.org/10.1016/j.neuron.2015.07.016. PMID: 26247863; PMCID: PMC4529965

50. Hartveit E, Veruki ML, Zandt BJ (2019) Capacitance measurement of dendritic exocytosis in an electrically coupled inhibitory retinal interneuron: an experimental and computational study. Physiol Rep 7(15):e14186. https://doi.org/10.14814/phy2.14186. PMID: 31379117; PMCID: PMC6680060

51. Sakaba T, Neher E (2001) Quantitative relationship between transmitter release and calcium current at the calyx of held synapse. J Neurosci 21(2):462–476. https://doi.org/10.1523/JNEUROSCI.21-02-00462.2001. PMID: 11160426; PMCID: PMC6763832

52. Wölfel M, Lou X, Schneggenburger R (2007) A mechanism intrinsic to the vesicle fusion machinery determines fast and slow transmitter release at a large CNS synapse. J Neurosci 27(12):3198–3210. https://doi.org/10.1523/JNEUROSCI.4471-06.2007. PMID: 17376981; PMCID: PMC6672471

53. Lou X, Korogod N, Brose N, Schneggenburger R (2008) Phorbol esters modulate spontaneous and Ca2+-evoked transmitter release via acting on both Munc13 and protein kinase C. J Neurosci 28(33):8257–8267. https://doi.org/10.1523/JNEUROSCI.0550-08.

2008. PMID: 18701688; PMCID: PMC6670566

54. Ariel P, Ryan TA (2010) Optical mapping of release properties in synapses. Front Neural Circuits 4:18. https://doi.org/10.3389/fncir.2010.00018. PMID: 20802854; PMCID: PMC2928663
55. Mathie A, Wooltorton JR, Watkins CS (1998) Voltage-activated potassium channels in mammalian neurons and their block by novel pharmacological agents. Gen Pharmacol 30(1):13–24. https://doi.org/10.1016/s0306-3623(97)00034-7. PMID: 9457476
56. Kaplan JH, Ellis-Davies GC (1988) Photolabile chelators for the rapid photorelease of divalent cations. Proc Natl Acad Sci U S A 85(17):6571–6575. https://doi.org/10.1073/pnas.85.17.6571. PMID: 3137570; PMCID: PMC282015
57. Delaney KR, Zucker RS (1990) Calcium released by photolysis of DM-nitrophen stimulates transmitter release at squid giant synapse. J Physiol 426:473–498. https://doi.org/10.1113/jphysiol.1990.sp018150. PMID: 1977904; PMCID: PMC1189900
58. Bollmann JH, Sakmann B, Borst JG (2000) Calcium sensitivity of glutamate release in a calyx-type terminal. Science 289(5481):953–957. https://doi.org/10.1126/science.289.5481.953. PMID: 10937999
59. Schneggenburger R, Neher E (2000) Intracellular calcium dependence of transmitter release rates at a fast central synapse. Nature 406(6798):889–893. https://doi.org/10.1038/35022702. PMID: 10972290
60. Sakaba T (2008) Two Ca(2+)-dependent steps controlling synaptic vesicle fusion and replenishment at the cerebellar basket cell terminal. Neuron 57(3):406–419. https://doi.org/10.1016/j.neuron.2007.11.029. PMID: 18255033
61. Trigo FF, Corrie JE, Ogden D (2009) Laser photolysis of caged compounds at 405 nm: photochemical advantages, localisation, phototoxicity and methods for calibration. J Neurosci Methods 180(1):9–21. https://doi.org/10.1016/j.jneumeth.2009.01.032. Epub 2009 Feb 7. PMID: 19427524
62. Burgalossi A, Jung S, Meyer G, Jockusch WJ, Jahn O, Taschenberger H, O'Connor VM, Nishiki T, Takahashi M, Brose N, Rhee JS (2010) SNARE protein recycling by αSNAP and βSNAP supports synaptic vesicle priming. Neuron 68(3):473–487. https://doi.org/10.1016/j.neuron.2010.09.019. Erratum in: Neuron. 2012 Feb 9;73(3):620. PMID: 21040848
63. Trigo FF, Sakaba T, Ogden D, Marty A (2012) Readily releasable pool of synaptic vesicles measured at single synaptic contacts. Proc Natl Acad Sci U S A 109(44):18138–18143. https://doi.org/10.1073/pnas.1209798109. Epub 2012 Oct 16. PMID: 23074252; PMCID: PMC3497746
64. Schneggenburger R, Meyer AC, Neher E (1999) Released fraction and total size of a pool of immediately available transmitter quanta at a calyx synapse. Neuron 23(2):399–409. https://doi.org/10.1016/s0896-6273(00)80789-8. PMID: 10399944
65. Thanawala MS, Regehr WG (2013) Presynaptic calcium influx controls neurotransmitter release in part by regulating the effective size of the readily releasable pool. J Neurosci 33(11):4625–4633. https://doi.org/10.1523/JNEUROSCI.4031-12.2013. PMID: 23486937; PMCID: PMC3711668
66. Elmqvist D, Quastel DM (1965) A quantitative study of end-plate potentials in isolated human muscle. J Physiol 178(3):505–529. https://doi.org/10.1113/jphysiol.1965.sp007639. PMID: 5827910; PMCID: PMC1357310
67. Thanawala MS, Regehr WG (2016) Determining synaptic parameters using high-frequency activation. J Neurosci Methods 264:136–152. https://doi.org/10.1016/j.jneumeth.2016.02.021. Epub 2016 Mar 10. PMID: 26972952; PMCID: PMC4833598
68. Verhage M, Sørensen JB (2008) Vesicle docking in regulated exocytosis. Traffic 9(9):1414–1424. https://doi.org/10.1111/j.1600-0854.2008.00759.x. Epub 2008 Apr 28. PMID: 18445120
69. Imig C, Min SW, Krinner S, Arancillo M, Rosenmund C, Südhof TC, Rhee J, Brose N, Cooper BH (2014) The morphological and molecular nature of synaptic vesicle priming at presynaptic active zones. Neuron 84(2):416–431. https://doi.org/10.1016/j.neuron.2014.10.009. Epub 2014 Oct 22. Erratum in: Neuron. 2014 Nov 19;84(4):882. PMID: 25374362
70. Harris KM, Sultan P (1995) Variation in the number, location and size of synaptic vesicles provides an anatomical basis for the nonuniform probability of release at hippocampal CA1 synapses. Neuropharmacology 34(11):1387–1395. https://doi.org/10.1016/0028-3908(95)00142-s. PMID: 8606788

71. Schikorski T, Stevens CF (1999) Quantitative fine-structural analysis of olfactory cortical synapses. Proc Natl Acad Sci U S A 96(7):4107–4112. https://doi.org/10.1073/pnas.96.7.4107. PMID: 10097171; PMCID: PMC22428
72. Xu-Friedman MA, Harris KM, Regehr WG (2001) Three-dimensional comparison of ultrastructural characteristics at depressing and facilitating synapses onto cerebellar Purkinje cells. J Neurosci 21(17):6666–6672. https://doi.org/10.1523/JNEUROSCI.21-17-06666.2001. PMID: 11517256; PMCID: PMC6763067
73. Richmond JE, Davis WS, Jorgensen EM (1999) UNC-13 is required for synaptic vesicle fusion in C. elegans. Nat Neurosci 2(11):959–964. https://doi.org/10.1038/14755. PMID: 10526333; PMCID: PMC2585767
74. Broadie K, Prokop A, Bellen HJ, O'Kane CJ, Schulze KL, Sweeney ST (1995) Syntaxin and synaptobrevin function downstream of vesicle docking in Drosophila. Neuron 15(3):663–673. https://doi.org/10.1016/0896-6273(95)90154-x. PMID: 7546745
75. Hess SD, Doroshenko PA, Augustine GJ (1993) A functional role for GTP-binding proteins in synaptic vesicle cycling. Science 259(5098):1169–1172. https://doi.org/10.1126/science.8438167. PMID: 8438167
76. Weimer RM (2006) Preservation of C. elegans tissue via high-pressure freezing and freeze-substitution for ultrastructural analysis and immunocytochemistry. In: Strange K (ed) C. elegans. Methods in molecular biology, vol 351. Humana Press. https://doi.org/10.1385/1-59745-151-7:203
77. Siksou L, Rostaing P, Lechaire JP, Boudier T, Ohtsuka T, Fejtová A, Kao HT, Greengard P, Gundelfinger ED, Triller A, Marty S (2007) Three-dimensional architecture of presynaptic terminal cytomatrix. J Neurosci 27(26):6868–6877. https://doi.org/10.1523/JNEUROSCI.1773-07.2007. PMID: 17596435; PMCID: PMC6672225
78. Wang SSH, Held RG, Wong MY, Liu C, Karakhanyan A, Kaeser PS (2016) Fusion competent synaptic vesicles persist upon active zone disruption and loss of vesicle docking. Neuron 91(4):777–791. https://doi.org/10.1016/j.neuron.2016.07.005. PMID: 27537483; PMCID: PMC4991631
79. Hammarlund M, Palfreyman MT, Watanabe S, Olsen S, Jorgensen EM (2007) Open syntaxin docks synaptic vesicles. PLoS Biol 5(8):e198. https://doi.org/10.1371/journal.pbio.0050198. PMID: 17645391; PMCID: PMC1914072
80. Siksou L, Varoqueaux F, Pascual O, Triller A, Brose N, Marty S (2009) A common molecular basis for membrane docking and functional priming of synaptic vesicles. Eur J Neurosci 30(1):49–56. https://doi.org/10.1111/j.1460-9568.2009.06811.x. Epub 2009 Jun 25. PMID: 19558619
81. Aravamudan B, Fergestad T, Davis WS, Rodesch CK, Broadie K (1999) Drosophila UNC-13 is essential for synaptic transmission. Nat Neurosci 2(11):965–971. https://doi.org/10.1038/14764. PMID: 10526334
82. Varoqueaux F, Sigler A, Rhee JS, Brose N, Enk C, Reim K, Rosenmund C (2002) Total arrest of spontaneous and evoked synaptic transmission but normal synaptogenesis in the absence of Munc13-mediated vesicle priming. Proc Natl Acad Sci U S A 99(13):9037–9042. https://doi.org/10.1073/pnas.122623799. Epub 2002 Jun 17. PMID: 12070347; PMCID: PMC124419
83. Watanabe S, Liu Q, Davis MW, Hollopeter G, Thomas N, Jorgensen NB, Jorgensen EM (2013) Ultrafast endocytosis at Caenorhabditis elegans neuromuscular junctions. Elife 2: e00723. https://doi.org/10.7554/eLife.00723. PMID: 24015355; PMCID: PMC3762212
84. Watanabe S, Rost BR, Camacho-Pérez M, Davis MW, Söhl-Kielczynski B, Rosenmund C, Jorgensen EM (2013) Ultrafast endocytosis at mouse hippocampal synapses. Nature 504(7479):242–247. https://doi.org/10.1038/nature12809. Epub 2013 Dec 4. PMID: 24305055; PMCID: PMC3957339
85. Vardar G, Salazar-Lázaro A, Brockmann M, Weber-Boyvat M, Zobel S, Kumbol VW, Trimbuch T, Rosenmund C (2021) Reexamination of N-terminal domains of syntaxin-1 in vesicle fusion from central murine synapses. Elife 10:e69498. https://doi.org/10.7554/eLife.69498. PMID: 34427183; PMCID: PMC8416022
86. Ariel P, Hoppa MB, Ryan TA (2013) Intrinsic variability in Pv, RRP size, Ca(2+) channel repertoire, and presynaptic potentiation in individual synaptic boutons. Front Synaptic Neurosci 4:9. https://doi.org/10.3389/fnsyn.2012.00009. PMID: 23335896; PMCID: PMC3542534
87. Zarebidaki F, Camacho M, Brockmann MM, Trimbuch T, Herman MA, Rosenmund C (2020) Disentangling the roles of RIM and Munc13 in synaptic vesicle localization and

neurotransmission. J Neurosci 40(49):9372–9385. https://doi.org/10.1523/JNEUROSCI.1922-20.2020. Epub 2020 Nov 2. PMID: 33139401; PMCID: PMC7724145

88. Cochilla AJ, Angleson JK, Betz WJ (1999) Monitoring secretory membrane with FM1-43 fluorescence. Annu Rev Neurosci 22:1–10. https://doi.org/10.1146/annurev.neuro.22.1.1. PMID: 10202529
89. Harata N, Pyle JL, Aravanis AM, Mozhayeva M, Kavalali ET, Tsien RW (2001) Limited numbers of recycling vesicles in small CNS nerve terminals: implications for neural signaling and vesicular cycling. Trends Neurosci 24(11):637–643. https://doi.org/10.1016/s0166-2236(00)02030-0. PMID: 11672807
90. Harata N, Ryan TA, Smith SJ, Buchanan J, Tsien RW (2001) Visualizing recycling synaptic vesicles in hippocampal neurons by FM 1-43 photoconversion. Proc Natl Acad Sci U S A 98(22):12748–12753. https://doi.org/10.1073/pnas.171442798. PMID: 11675506; PMCID: PMC60125
91. Pyle JL, Kavalali ET, Piedras-Rentería ES, Tsien RW (2000) Rapid reuse of readily releasable pool vesicles at hippocampal synapses. Neuron 28(1):221–231. https://doi.org/10.1016/s0896-6273(00)00098-2. PMID: 11086996
92. Stevens CF, Williams JH (2000) "Kiss and run" exocytosis at hippocampal synapses. Proc Natl Acad Sci U S A 97(23):12828–12833. https://doi.org/10.1073/pnas.230438697. PMID: 11050187; PMCID: PMC18849
93. Henkel AW, Lübke J, Betz WJ (1996) FM1-43 dye ultrastructural localization in and release from frog motor nerve terminals. Proc Natl Acad Sci U S A 93(5):1918–1923. https://doi.org/10.1073/pnas.93.5.1918. PMID: 8700859; PMCID: PMC39883
94. Rose T, Schoenenberger P, Jezek K, Oertner TG (2013) Developmental refinement of vesicle cycling at Schaffer collateral synapses. Neuron 77(6):1109–1121. https://doi.org/10.1016/j.neuron.2013.01.021. PMID: 23522046
95. Miesenböck G, De Angelis DA, Rothman JE (1998) Visualizing secretion and synaptic transmission with pH-sensitive green fluorescent proteins. Nature 394(6689):192–195. https://doi.org/10.1038/28190. PMID: 9671304
96. Bunt G, Wouters FS (2004) Visualization of molecular activities inside living cells with fluorescent labels. Int Rev Cytol 237:205–277. https://doi.org/10.1016/S0074-7696(04)37005-1. PMID: 15380669
97. Poskanzer KE, Davis GW (2004) Mobilization and fusion of a non-recycling pool of synaptic vesicles under conditions of endocytic blockade. Neuropharmacology 47(5):714–723. https://doi.org/10.1016/j.neuropharm.2004.07.026. PMID: 15458843
98. Fernandez-Alfonso T, Ryan TA (2008) A heterogeneous "resting" pool of synaptic vesicles that is dynamically interchanged across boutons in mammalian CNS synapses. Brain Cell Biol 36(1–4):87–100. https://doi.org/10.1007/s11068-008-9030-y. Epub 2008 Oct 22. PMID: 18941900; PMCID: PMC2924874
99. Kim SH, Ryan TA (2010) CDK5 serves as a major control point in neurotransmitter release. Neuron 67(5):797–809. https://doi.org/10.1016/j.neuron.2010.08.003. PMID: 20826311; PMCID: PMC2939042
100. Alivisatos AP, Gu W, Larabell C (2005) Quantum dots as cellular probes. Annu Rev Biomed Eng 7:55–76. https://doi.org/10.1146/annurev.bioeng.7.060804.100432. PMID: 16004566
101. Zhang Q, Cao YQ, Tsien RW (2007) Quantum dots provide an optical signal specific to full collapse fusion of synaptic vesicles. Proc Natl Acad Sci U S A 104(45):17843–17848. https://doi.org/10.1073/pnas.0706906104. Epub 2007 Oct 29. PMID: 17968015; PMCID: PMC2077028
102. Valtorta F, Meldolesi J, Fesce R (2001) Synaptic vesicles: is kissing a matter of competence? Trends Cell Biol 11(8):324–328. https://doi.org/10.1016/s0962-8924(01)02058-x. PMID: 11489637
103. Algar WR, Susumu K, Delehanty JB, Medintz IL (2011) Semiconductor quantum dots in bioanalysis: crossing the valley of death. Anal Chem 83(23):8826–8837. https://doi.org/10.1021/ac201331r. Epub 2011 Sep 19. PMID: 21928771
104. Rizzoli SO, Betz WJ (2004) The structural organization of the readily releasable pool of synaptic vesicles. Science 303(5666):2037–2039. https://doi.org/10.1126/science.1094682. PMID: 15044806

Part IV

Modelling

Chapter 15

Phenomenological Modeling of Diverse and Heterogeneous Synaptic Dynamics at Natural Density

Agnes Korcsak-Gorzo, Charl Linssen, Jasper Albers, Stefan Dasbach, Renato Duarte, Susanne Kunkel, Abigail Morrison, Johanna Senk, Jonas Stapmanns, Tom Tetzlaff, Markus Diesmann, and Sacha J. van Albada

Abstract

This chapter sheds light on the synaptic organization of the brain from the perspective of computational neuroscience. It provides an introductory overview on how to account for empirical data in mathematical models, implement such models in software, and perform simulations reflecting experiments. This path is demonstrated with respect to four key aspects of synaptic signaling: the connectivity of brain networks, synaptic transmission, synaptic plasticity, and the heterogeneity across synapses. Each step and aspect of the modeling and simulation workflow comes with its own challenges and pitfalls, which are highlighted and addressed.

Key words Simulation, Modeling, Synaptic organization, Spiking neural networks, Plasticity, Heterogeneity, Phenomenological model, Connectivity

1 Introduction

Creating mathematical models from experimental neurophysiological data has grown into an established and essential method for investigating the brain. Based on these mathematical models and exploiting the upswing of affordable and powerful computing architectures over the last few decades, a new sub-field concerned with the computational modeling of neurobiological systems has emerged. The discipline using mathematical modeling and analysis methods to understand principles of brain organization, dynamics,

Agnes Korcsak-Gorzo and Charl Linssen equally contributed in this chapter.

Supplementary Information The online version contains supplementary material available at [https://doi.org/10.1007/978-1-0716-4019-7_15].

Joachim H. R. Lübke and Astrid Rollenhagen (eds.), *New Aspects in Analyzing the Synaptic Organization of the Brain*, Neuromethods, vol. 212, https://doi.org/10.1007/978-1-0716-4019-7_15,

and function is called *computational neuroscience*. This discipline is also sometimes referred to as theoretical or mathematical neuroscience, each term having its own slightly different emphasis. One of the most challenging subjects of this comparatively young domain is the synaptic organization of the brain. This chapter reviews the status quo of synaptic modeling approaches. It is targeted primarily at experimentalists and aims to provide insight into ways data can be used to build mathematical or computational models.

The methods described in the previous chapters of this volume reveal diverse dynamical processes and heterogeneous components involved in synaptic signaling at various spatial and temporal scales. This variability is amplified by the size and complexity of neurobiological systems: both the density and the total number of synapses in mammalian brains are impressive, the former being on the order of 10^9 per cubic millimeter in the cerebral cortex [1] and the latter being estimated as roughly 5×10^{14} in the human brain [2]. Each cubic millimeter of the human brain contains on the order of 10^4–10^5 neurons adding up to about 10^{11} neurons in the brain as a whole [3].

How can we model such a heterogeneous, complex, and dense large-scale system? The process of modeling and simulation can be understood as a cycle, as depicted in Fig. 1. First, the experimental

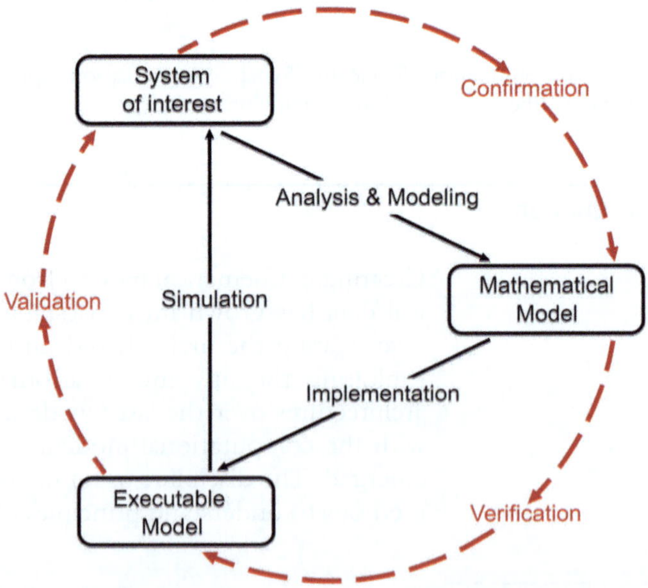

Fig. 1 Cycle of modeling and simulation. The empirical data from the brain structure under study (the "system of interest") is first mathematically modeled and then implemented in software. Closing the loop, simulation results can be compared to experimental recordings. (Reproduced with permission from Fig. 1 in Trensch et al. [4])

results recorded from the system of interest (here, the synaptic organization of the brain) are analyzed, and a mathematical model is formulated. Then, the mathematical model is translated into computer language, i.e., into an executable model that implements the mathematical operations needed to simulate the model. Finally, the model of the system is executed, whereby this simulation is the numerical analog to an experiment. This process yields results that can be compared with the experimental results. In turn, this comparison may deliver outcomes that can be used to improve the mathematical and computational models. Comparisons between the system of interest, the mathematical model, and the executable model ensure quality control. In general, three types of checks for correctness can be distinguished [4]: *Confirmation* ensures that the mathematical formulation applies to the system of interest, *verification* that the executable model sufficiently represents the mathematical model, and *validation* that the simulation outcome is consistent with and predictive of the system of interest. This chapter focuses on the inner triangle of arrows: the practical methods to formulate mathematical models in an informed way, to translate them into manageable and correct executable models, to run simulations, and to inform further modeling choices using the obtained data.

The challenge of computational neuroscience is to analyze the rich dynamics of neuronal systems and abstract their complexity into mathematical models that still capture essential characteristics of the experimental findings. At the same time, these models should be simple enough to be tractable and generalizable and thus reveal possible laws that govern the dynamical system. A fundamental question in this endeavor is what processes and variables are of interest and best describe the data. In synaptic organization, candidates include the connection strengths, synaptic time constants, delays, vesicle release characteristics, synaptic plasticity, and neuromodulation.

Developing this thought further, a modeler needs to decide on the scale the model represents and how to parameterize it. It is advisable to constrain the number of model parameters to a minimal set that answers a specific research question. Limiting the parameter space increases the tractability, mechanistic interpretability, and robustness of the model and reduces the risk of overfitting. However, capturing biological detail and enhancing the direct link between parameters and their biological counterparts can usually only be done with a large set of parameters. Heterogeneity can be represented by introducing parameter value distributions, leading to additional parameters characterizing the dispersion and possibly higher-order properties of the corresponding distributions. Overall, a suitable parameterization involves a tradeoff between the model's controllability and biological plausibility.

These decisions on which aspects of the system to express as variables and the choice of the corresponding model equations are abstraction steps: they formalize a hypothesis on which features are germane to the question at hand and which mathematical descriptions are appropriate for capturing the phenomena of interest (*see* Subheading 7.8). In general, this abstraction can be approached from two different directions. The ***bottom-up*** approach starts from the low-level properties of the neurons and synapses making up the system and models the complexity step by step in the hope of achieving realistic dynamical and functional properties. However, one major point of modeling is to improve our understanding of a system. Given that the starting point is a poor understanding, this approach suffers from the fundamental problem that essential features might be abstracted away or obfuscated by an abundance of less relevant details. Another point can be to provide accurate predictions, even if we do not understand the model. The opposite approach, ***top-down*** modeling, starts from the high-level dynamical, functional, or behavioral properties one would like to capture and then proposes concrete implementations. The drawback of this approach is that a model created in this way might not fully conform to biology, so it is difficult to draw conclusions about the brain. One solution is to use different degrees of abstraction at different scales to arrive at an understanding of the system, which is the motivation behind ***multi-scale modeling***. Ideally, biological realism is incrementally enhanced through cycles of data comparison and refinement (*see* Fig. 1 and Subheading 7.1).

Some neurophysiological observations can be modeled with analytically solvable equations, i.e., in an exact way and usually with pen and paper. However, various simplifying assumptions generally flow into such abstractions, and deriving an analytical solution to a model's equations becomes less feasible as its complexity increases. For such cases, numerical solutions can provide a useful alternative. This computational approach tends to be slower, but it can validate the analytical approach by requiring fewer simplifying assumptions and it may even provide novel theoretical insight.

Simple small network models frequently consist of equations that can be solved analytically or calculated numerically with few computational resources. However, both numerical and analytical approaches reach certain limits when attempting to replicate realistic neuron numbers in the volume of the brain region under consideration. As the number of neurons increases, the number of connections grows quadratically in networks without spatial dependence and linearly for distant neurons in models incorporating spatial dependence since most connections are local. To approximate ***natural density***, analytical techniques like mean-field theory sacrifice biological specificity. With sufficient computing power, numerical methods may solve model equations at natural density. A limitation is that, to date, this is only possible for small brains or small portions of larger brains.

We restrict the scope of this chapter to phenomenological models, which represent the empirical relationship between phenomena without explaining the reason for the interaction. We neglect the molecular level or ultrastructure, i.e., structures visible at magnifications higher than that provided by standard optical light microscopy. Furthermore, we address spiking neuron models, mainly so-called point or few-compartment neurons, which neglect the precise morphology of the neuron, as the effective dynamics of a morphologically complex neuron can often already be meaningfully captured by such models.

The models are usually translated to be executable by a computer, i.e., implemented in one of the various computer languages. One or an interconnected set of dynamical model components representing neurobiological entities such as neurons or synapses is *simulated*, i.e., evolved in time, for a specific duration with a set of parameters, initial conditions, and stimuli.

The generic framework that can numerically solve the dynamical equations of various models is called a *simulator*. It solves complex interactions with many coupled differential equations typically incrementally in discrete time steps. Spiking neuronal network simulators also communicate spike events from senders to receivers. The event-based nature of synaptic interactions in the form of action potentials is advantageous for the efficiency of the simulation. Furthermore, a simulator can record dynamical state variables in the network or other observables like connectivity and apply stimuli during simulation.

Since the simulator needs to organize and maintain the appropriate data structures in computer memory, a substantial amount of RAM may also be required, depending on the simulated neurobiological system. Biologically realistic network models are often large-scale to reflect or approach the natural density of neurons and synapses, have additional computational overhead due to heterogeneity, and are typically simulated for a long time, e.g., to gather statistics or study behavioral timescales. Thus, a simulator should be efficient and scale to high-performance architectures in terms of processing and memory usage.

In addition to these performance aspects, criteria for a good simulator include functional completeness, numerical accuracy, and reproducibility of results. Furthermore, a simulator increases its value for the community if the available models are relevant to many members and the documentation is comprehensive and easy to understand. Developing a simulator that fulfills all these criteria and supports diverse models is complex and time-intensive. Consequently, simulators with peer-reviewed collections of implemented models are continuously developed as a community effort and shared as software packages to benefit the field of computational neuroscience.

From the range of existing simulators for biologically inspired neuronal networks, this chapter focuses on NEST [5], an open-source software tool designed to simulate anywhere from small to large-scale networks of diverse spiking neuron models, and its associated domain-specific modeling language NESTML [6] that facilitates the creation of new neuron and synapse models. Other simulators with various scientific foci and special areas of application include NEURON [7], Brian [8], Nengo [9], Arbor [10], and ANNarchy [11]. Some common (simulator-agnostic) interfaces are provided by PyNN [12] and the modeling language NeuroML [13].

This chapter is structured as follows. Each section from Subheadings 2 to 6 presents a two-step recipe to go from experimental data on a specific feature or mechanism of the synaptic organization to simulations: first, how to mathematically model experimental data, and second, how to simulate this model. Subheading 2 starts with the most simplified view of brain circuitry, namely binary connections between neurons. This view is advanced in Subheading 3 to the notion of weighted connections. Then, dynamics is introduced to the existence of connections in Subheading 4 and the weights of those connections in Subheading 5. Finally, Subheading 6 discusses how to model the additional heterogeneity in all these features. Throughout the text, links to Subheading 7 highlight aspects that are particularly challenging or contain pitfalls. Furthermore, links to Subheading 9 provide URLs to NEST code snippets that help the reader develop an intuition on how the usage of the discussed models would look in code. The chapter ends with some concluding remarks in Subheading 8.

2 Connectivity

2.1 From Empirical Data to Mathematical Models

Anatomical and physiological experiments are yielding ever richer data sets on brain connectivity. Integration of these data into dynamical models can help gain insight into their implications for brain activity and function, besides identifying gaps in the data which can inform future experiments. The data cover diverse scales, ranging from electron microscopy at the sub-micron scale of synapses to light microscopy for neuronal morphology, paired recordings identifying fractions of connected neuron pairs, glutamate uncaging at the scale of tens to hundreds of microns, axonal tracing for long-range connectivity, and diffusion imaging for the whole-brain scale [14].

Despite the richness of the available data, none of these experimental approaches can specify full connectomes at the single-neuron resolution, especially in organisms with complex brain structures such as mammals. Therefore, we need to make predictions in order to complete the detailed connectivity data. One strategy is to find statistical regularities in the existing data and use these to extrapolate to missing data points. Of course, models do not need to be fully data-driven; various abstractions may be used to explore the influence of specific aspects of the connectivity. We illustrate the data-driven approach using the example of the cerebral cortex. In view of the incompleteness of the known cortical connectivity for any individual, we describe the connectivity in a probabilistic manner. A different strategy for generating the connectivity may be to grow connections according to developmental or other plasticity rules (for structural synaptic plasticity, *see* Subheading 4).

The cerebral cortex contains different types of excitatory and inhibitory neurons, distinguished by their morphology, electrophysiology, connectivity, and molecular make-up [15, 16]. We refer to the set of neurons of the same type in a given cortical area and layer as a *population* (*see* Fig. 2). Connection probabilities are specific to both source and target populations. Both within and

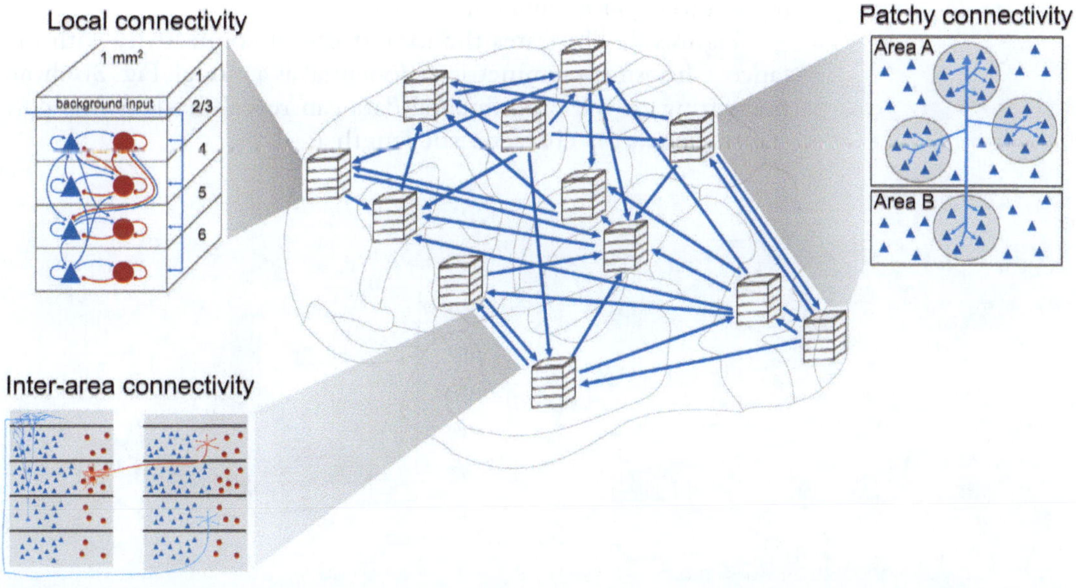

Fig. 2 Substructures of the cortex on different scales. The neurons in the cortex are organized into areas on the macroscale (middle). In each area, neurons are organized into layers with distinct connectivity (here, shown as a microcircuit under 1 mm^2 of cortical surface) and into populations of neurons with similar properties within each layer (upper left). Blue triangles and red circles represent excitatory and inhibitory neurons, respectively. Long-range projections connect the areas (lower left). Moreover, on an intermediate scale of millimeters, both intra-area and inter-area excitatory connections cluster into "patches" (right, only outgoing connections for one neuron shown). (Adapted with permission from Fig. 1 in Schmidt et al. [22] under license CC BY 4.0 originally from Fig. 1 in Potjans and Diesmann [23] and Fig. 1 in Kunkel et al. [24])

between areas, connectivity is also layer-specific. Furthermore, connection probability decays with the distance between neurons, both locally within areas and at longer ranges between areas [17–19]. A further organizing principle is that excitatory connectivity tends to form patches [20, 21], meaning that neurons establish additional synapses onto other nearby neurons resulting in spatial clusters (*see* Fig. 2).

When formalizing these properties into models, a number of subtleties are involved [25]. First, the term *connection probability* needs to be defined carefully. This could, for instance, refer to either the total number of synapses divided by the product of the source and target population sizes or the probability for any neuron pair to be connected via at least one synapse. The two definitions diverge in the case of *multapses*, multiple synapses between a given source and target neuron pair, often observed in reconstruction data [26]. Further, models can either allow self-connections, also called *autapses*, or prohibit them. Moreover, beyond a certain model size, the spatial decay of the connection probability becomes important. To capture this, simulated neurons are assigned spatial coordinates, and additional specifications are necessary, including boundary conditions and the choice of connectivity profile. Common choices for the local profile are Gaussian and exponential functions, where the latter generally appears to be a better approximation to experimental data [17, 18].

Figures 3a illustrates the local decay of connectivity with distance. Choosing a symmetric exponential as a model, Fig. 3b shows that fitting to the experimental data can reveal fundamental constants such as the characteristic length λ.

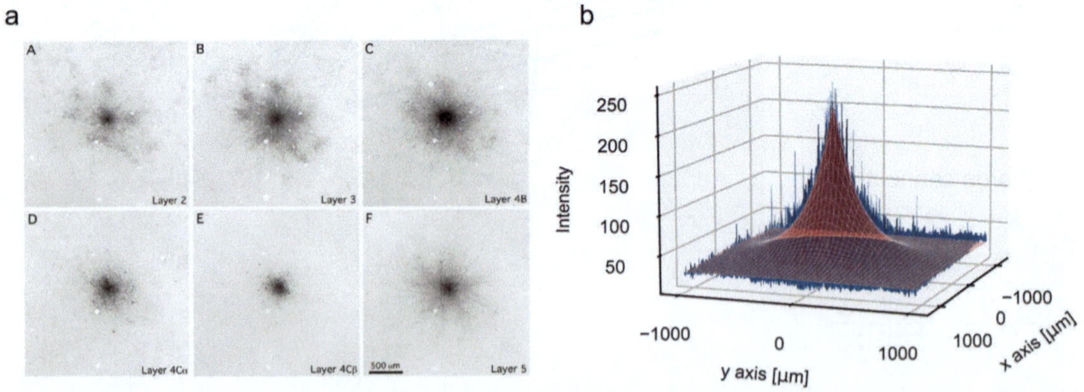

Fig. 3 Fitting a model of connectivity to observed data. (**a**) Layer-resolved axonal tracing data from V1 of New World monkeys. (Adapted from Sincich et al. [27], Fig. 3, Copyright 2001 Society for Neuroscience). Images show staining of cortical layers after an injection of biocytin into layer 3, anterogradely staining axons. (**b**) Fit of a 2D symmetric exponential function $f(r) = ae^{-\lambda/r} + b$ (red) to the axonal density distribution (blue) of layer 5. The model does not fully capture the connectivity profile for layers 2, 3, and 4B (**b**) A–C, which display patchy connectivity

When including *patchy connectivity*, the spatial position of the patches can be specified via a radial distance from a cell body and an angle [21]. Further possible parameters are the number of patches, the size of each patch, and the degree of overlap between patches. Layer-specific axonal tracing data, such as fractions of supragranular labeled neurons from retrograde tracing experiments [28], can inform the laminar inter-area patterns of cortical models. Here one should pay attention to the fact that, on the target side, axonal tracing tells us about axonal or synaptic locations but not about the locations of the target cell bodies. To a reasonable approximation, one can statistically estimate which synapses are established on which target neurons using morphological reconstructions, a method that assigns the number of synapses proportionally to the total length of dendritic elements in the vicinity of the synapses [22, 29].

This is only a tiny selection of data and features that can be included in neuronal network models. One can go into greater complexity and, for example, consider the higher-level organization of networks, such as hierarchical modularity or small-world properties. For a further discussion on model detail in general, *see* Subheading 8.

2.2 From Mathematical Models to Simulation

To simulate how the dynamics of a neuronal network model evolve, the mathematical model description needs to be translated into an executable one. This is preferably done using a dedicated simulator to avoid mistakes in the implementation and to enhance comparability and reproducibility of results (for the precise definition of the different forms of *reproducibility*, *see* Subheading 7.2). Executing a neuronal network simulation typically involves two successive phases: during the *build phase*, the network is set up on the machine by instantiating objects and data structures for neurons and synapses. The subsequent *simulation phase* propagates the network state for a specified biological model time. How fast a simulation runs, i.e., how the biological model time relates to the wall-clock time, crucially depends not only on the machine specifications but also on the representation of the network model on the machine. Parallel computing combines the computational power of many separate compute cores or nodes to enable large-scale simulations; to this end, NEST uses a hybrid approach with the Message Passing Interface (MPI) and Open Multi-Processing (OpenMP). The former enables parallel computing on multiple processors with distributed memory, while the latter enables parallel computing even on single processors with shared memory, referred to as *threading*. The total number of so-called *virtual processes* is determined as the product of the number of MPI processes and the number of OpenMP threads per process. A direct mapping between network structure and hardware is in general difficult to realize. Therefore, NEST uniformly distributes the neurons of each population across the available processors to balance the compute load (Subheading 7.3). The neurons are connected via synapses, which are assigned specific *weights* and *delays* reflecting conduction times. Synapse models are stored and updated on the same compute nodes that hold their postsynaptic partner neurons. Maintaining the complete network connectivity in computer memory enables the use of plasticity mechanisms that can modify synaptic strengths at runtime (for functional synaptic plasticity, *see* Subheading 5). The alternative *procedural connectivity* approach generates the required routing information on the fly and thereby requires fewer memory resources [30, 31].

Establishing synapses in a computational network model requires defining which neurons are connected. For specific data-driven models, the network structure can be loaded from a file, but simulators also provide built-in routines for generating connectivity. These routines (*see* Program 1) range from a primitive that just connects individual source and target neurons, to high-level connection rules acting on the neuron population level [25]. For example, the deterministic rule *all-to-all* connects each neuron of a source population to each neuron of a target population. Probabilistic rules account for the often statistically described sparse connectivity in biological neuronal networks. *Random, fixed in-degree* connectivity, for instance, specifies only the number of incoming connections per neuron but not which individual ones are selected as sources. If the connectivity is described as *pairwise Bernoulli*, each pair of neurons is connected with a given probability. The fixed in-degree rule needs to be combined with the specification of whether multapses are allowed, whereas the pairwise Bernoulli rule excludes them by definition as each pair of neurons is considered only once. High-level connection rules enable efficient low-level implementations such as parallelization of the network construction.

Pseudo-random number generators (pRNGs) are used for drawing connections according to a probabilistic rule and optionally also for setting neuron and synapse parameters (*see* Subheading 6). The resulting network realization will be identical if the same sequence of random numbers is sampled; this is achieved by fixing the pRNG seed. Random distributions sometimes have to be constrained in order to restrict the sign of a weight, e.g, according to Dale's law [32], or to enforce connection delays to be larger than the simulation time step, for which a typical value is 0.1 ms. A longer minimum delay, for instance, 1 ms, can furthermore be used to limit the necessary frequency of communication between virtual processes.

Large-scale neuronal network models require high-performance computing. Employing several compute nodes in parallel not only distributes the workload but also gives access to sufficient memory for storing the network connectivity. Storing a single synaptic weight costs 8 bytes in NEST [33] as it sums up the effects of a set of vesicles that may differ in size, as well as of potentially different amounts of receptors. Moreover, a typical neuron in the mammalian brain has on the order of 10^4 synapses [34]. This leads to a substantial amount of resources required for large models. Networks with reduced neuron and synapse numbers can preserve some characteristics (e.g., firing rates) of full-scale networks if the downscaling is compensated for with informed parameter adjustments [35]. The pairwise correlation structure of the neuronal activity, however, cannot be preserved simultaneously, rendering neuroscientific simulations at natural density a necessity where correlation structure is relevant. This may be the case, for instance, to ensure the correct network state: correlation changes may even shift a network between linearly stable and unstable regimes. Data structures that keep the memory usage per MPI process constant regardless of the total number of MPI processes used in the simulation [36] provide a potential solution, paving the way toward brain-size networks with realistic connectivity.

3 Synaptic Transmission

3.1 From Empirical Data to Mathematical Models

Electrochemical signaling between neurons is mediated by various receptor types, expressed post- and presynaptically. Different receptor types trigger different physiological responses. Ligand- or voltage-gated ion channels influence ionic flows through the membrane, with distinctive kinetics for each receptor type (examples include AMPA and NMDA). Metabotropic receptors trigger intracellular biochemical signaling cascades with downstream actions that are typically not instantaneously noticeable but mediate physiological adaptation processes. *Electrical synapses* (*gap junctions*) are transmembrane channels that form a direct electrical and biochemical coupling between the cytosol of two adjacent cells. Compared to *chemical synapses*, they provide increased speed as the signal does not need to be converted from electrical to chemical and back across a synaptic cleft. The composition of receptor types on a neuron's synapses, their spatial distribution on the dendritic tree and cell body, and their individual, instantaneous efficacy and response kinetics determine how and at which timescales the neuron filters and integrates its many presynaptic inputs. We will focus here on chemical synapses.

The amplitude of the postsynaptic response is proportional to the synapse's strength or *weight*, which depends on the amount and types of both neurotransmitters and receptors as well as the state of the postsynaptic neuron. Synaptic weights can be estimated from paired recordings, usually *in vitro*, to avoid background activity that confounds the measurements. However, here it should be taken into account that the synaptic weight obtained from paired-cell recordings is determined by a combination of biophysical properties, e.g., postsynaptic receptor density, amount of released neurotransmitters, reuptake kinetics, or existence of more than one connection between a pair of cells (multapses). Hence, the terms *strength* and *weight* refer to an effective, phenomenological quantity. Mapping the corresponding parameters to the *in vivo* condition is nontrivial because experimental conditions like temperature and extracellular fluid may differ, as well as a high-conductance network state affecting the measured quantities such as time constants [37, 38]. Change of synaptic strengths over time is discussed in Subheading 5.

When a presynaptic neuron has emitted an action potential, the signal that arrives at the postsynaptic neuron can be observed as a postsynaptic potential (PSP): the deflection in the somatic membrane voltage caused by the incoming spike. Alternatively, synaptic currents (PSCs) may be measured at different holding potentials using voltage-clamp recordings. From a PSC or PSP, the weight of the synapse can be derived. Synapses are often modeled as injecting a current into the postsynaptic cell or acting as a conductance, the current of which is proportional to the difference between the membrane potential and a synapse- or receptor-specific reversal potential. In the simplest approximation of the postsynaptic response kinetics, the time course of this current or conductance may be modeled as a Dirac delta function causing a step increase in the membrane potential or current, respectively. With increasing levels of complexity, the time course of the PSC (or postsynaptic conductance, PSG) can be approximated by an instantaneous rise followed by an exponential decay or by a double exponential with separate time constants for the rising and the decaying phase [39].

Beside spatially precise communication via synapses, the spatially more diffuse process of *neuromodulation* can alter the excitability of neurons and affect synaptic plasticity (*see* Subheading 5). Neuromodulation is achieved by the release of a neurotransmitter with less detail in the connectivity patterns than in typical synaptic (for instance, glutamatergic) neurotransmission. The neuromodulator, such as dopamine or serotonin, is typically released from a neuron whose cell body lies in a small, circumscribed nucleus in the brain but which projects broadly and affects many downstream targets simultaneously. The precise spatiotemporal profile of neurotransmitter concentration is often approximated in models by assuming the neuromodulator diffuses through extracellular space, referred to as *volume transmission*. Simulating the diffusion process entails solving the Laplace equation—an equation that involves the spatial gradient and divergence operators, requiring a different type of solver than those that solve the neuronal network system dynamics. Instead of a detailed representation of the geometry of extracellular space (for instance, based on the finite-element method), the medium may be assumed to be spatially homogeneous, and diffusion can even be assumed to occur instantaneously, considerably simplifying the model and its computational requirements [40].

Neurons have a spatial extent, and their dendrites often exhibit intricate branching patterns. Consequently, the spatial collocation of synapses on the dendrites has important consequences for the neuron's response to input. Dendritic responses are often nonlinear, as dendrites are studded with a high density of voltage-gated channels, which, combined with intracellular responses like calcium signaling, can cause a nonlinear interaction between nearby synaptic inputs. In addition, the dendrite itself can exhibit action potentials distinct from a somatic action potential, for instance, involving a local, intracellular calcium transient [see, e.g., 41]. The (local) change in membrane potential and conductance, in turn, affects the integration at adjacent synapses in the branch. The triggering of dendritic action potentials by co-activated and co-located synapses and their effects on the somatic dynamics can be accounted for in simple point neuron models by including nonlinearities in synaptic input currents [42, 43]. For a more fine-grained analysis, multicompartment models are commonly used. In these models, each neuron consists of dozens or hundreds of compartments, each equipped with a distinct type of dynamics and parameterization and coupled to neighboring compartments according to Ohm's law [39]. Multicompartment models permit integrating experimental data at a highly detailed level of description but are computationally and conceptually much more complex and demanding. On the other hand, some biophysical details like synaptic adaptation can be adequately modeled without the need to address the microscopic biophysics of synaptic vesicles but can be treated phenomenologically by adding one or a few extra continuous state variables to the model [e.g., 44, 45].

3.2 From Mathematical Models to Simulation

NEST integrates equations for neurons with linear subthreshold dynamics exactly [46] and uses standard numerical solvers for nonlinear neuron models. For synapses, an efficient approach is to specify the characteristic time evolution of some postsynaptic quantity, such as current or conductance, as a linear system of equations. If responses sum linearly across a neuron's synapses, they can be lumped together into a single or a few state variables and do not have to be stored and updated for each synapse separately. For this reason, the postsynaptic response is typically specified as part of the (postsynaptic) neuron model. As described in Subheading 3.1, the postsynaptic kernel could be, for example, a Dirac delta function (causing an instantaneous jump in the postsynaptic membrane potential), an instantaneous rise followed by an exponential decay, or a double-exponential function with a finite rise time. Furthermore, because they are linear, solving these equations does not require a numerical solver but only multiplication with a constant at each time step [47]. The reduction to a simple multiplication generally makes the solution much more precise: the computed values are closer to the mathematically "true" solution and more efficient to compute. Thus, simulations of networks with many synapses become feasible. Multiple types of synapses can be easily incorporated into this scheme by grouping them according to their kinetics, for instance, into a separate AMPA and NMDA group (*see* Program 2).

In simulations of large networks, the layout of data structures in memory and communication can become bottlenecks. Conceptual modeling decisions can interact with data layouts; for example, the synaptic delay can be chosen as a property of the synapses or the pre- or postsynaptic neurons. In the point-neuron framework, the delay is assigned to either a neuron's axonal or dendritic side, implying different biophysical interpretations and simulation outcomes. The biophysical object of a synapse is not necessarily represented in code by a specific software object but distributed into a presynaptic and a postsynaptic component. In the instantiation of a particular model, these components may not even live on the same compute node. As noted in Subheading 2.2, NEST stores synapses on the process containing the postsynaptic neuron.

In simulations using parallel computing, spike events and potentially other quantities such as synaptic weights have to be communicated between threads, processes, or across a computer network (Subheading 7.4). Parallel computing presents a set of unique design requirements because the evolution of the dynamical model needs to occur synchronously, lest the model's state becomes internally inconsistent when some parts of it have become desynchronized in time. This requirement can be addressed by instituting a minimum, nonzero transmission delay for each synaptic connection in the model. A delay between the presynaptic spike and the resulting postsynaptic response effectively decouples neurons for this time window so that events can be transmitted across the computer network in a regular cadence at the end of each window (Subheading 7.5). This decoupling allows simulations to scale to many compute nodes [48].

From a mathematical modeling point of view, gap junctions are much simpler than chemical synapses; their delay is negligible, and they do not filter the input. However, modeling gap junctions numerically can be challenging because they entail an instantaneous coupling between compartments. The *waveform relaxation* technique helps retain simulation efficiency when combining gap junctions with a numerical simulation method that takes advantage of a minimum, nonzero synaptic delay. Each neuron is considered a separate subsystem in this technique, and the gap junction coupling terms (current flowing from one neuron into the other) are solved iteratively. This solution requires exchanging data (in particular, membrane potentials) between the gap-junction coupled neurons only at the end of each minimum delay step, thus limiting the necessary communication frequency. In exchange, it requires only a modest increase in computation and the size of the communicated packets since solving the forward dynamics of each separately considered neuron needs to be repeated only once per iteration of the waveform relaxation algorithm [49, 50].

4 Structural Plasticity

The models introduced earlier in this chapter had static connectivity (*see* Subheading 2). However, macroscopic observations of the brain have revealed that the connections in cortical networks continually change as new synapses form and others dwindle and disappear [for a review, see 51]. This rewiring is a lifelong process to encode experiences but happens extensively during development and recovery from lesions in the brain tissue [for a review, see 52]. The underlying mechanisms introducing dynamics into the connectivity are summarized as *structural synaptic plasticity.*

4.1 From Empirical Data to Mathematical Models

Including structural plasticity mechanisms into a synapse model can increase its biological plausibility, e.g., regarding learning, development, reformation after lesions [53] or topographic map formation [54]. Structural plasticity might also be the basis for associative connections [55] and metaplasticity [56]. However, plasticity mechanisms capable of generating network connectivity in a principled fashion can also be helpful in other ways. First, they can help reduce dependence on cumbersome and expensive connectivity recordings in animals (*see* Subheading 2). Second, they can serve a range of functional purposes. For example, they can enhance learning performance [57] or increase the storage efficiency of long-term memories and, by that, prevent catastrophic forgetting [58]. Plasticity mechanisms also frequently serve a homeostatic function. In general, the term "homeostasis" refers to a range of vital physiological processes that assist organisms in maintaining internal states (such as body temperature, blood sugar levels, and heart rate) at optimal levels. Likewise, homeostatic plasticity maintains quantities such as spiking activity or numbers of connections at an energetically or computationally favorable set-point [59]. Efficient pruning of the connectivity and preserving sparse connectivity [60] can help save energy and optimize the usage of limited synaptic resources, which is particularly important in neuromorphic computing [57, 61, 62].

To understand the process of developing a comprehensive and accurate mathematical model of structural plasticity, the following paragraphs sketch the steps involved in creating the model suggested by Butz and Van Ooyen [53] as an example. This model is based on the observation that the creation and deletion of synapses can bring the postsynaptic neuron's firing rate into a certain physiological range. The authors consider synaptic elements, namely axonal boutons on the presynaptic side and dendritic spines on the postsynaptic side. When two such elements are combined, a synapse is created. The dynamics of the number of synaptic elements for each neuron depends on a readiness variable c (associated with the calcium concentration), which indicates the propensity of a neuron to grow synapses. The readiness variable is a low-pass filtered version of the spiking activity and thus approximates the neuron's instantaneous rate up to a scalar multiplier.

The algorithm comprises four steps, which are repeated until the connectivity converges. First, it continuously updates the spiking activity of the neurons since each neuron's mean firing rate influences the creation of synaptic elements. Second, it updates the readiness c for each neuron:

$$\frac{\mathrm{d}c}{\mathrm{d}t} = -\frac{c(t)}{\tau} + \beta\delta\left(t - t_j^{\mathrm{f}}\right), \tag{1}$$

i.e., c decays exponentially with the time constant τ and increases by a fixed amount β whenever the neuron j spikes at t_j^{f}, where $\delta(\cdot)$ denotes the Dirac delta function and f stands for "firing." Third, a homeostatic rule drives the neuron to reach and maintain a target activity by deleting postsynaptic elements if the instantaneous activity is higher than the target activity or creating synaptic elements if the current activity is lower than the target activity. A growth curve defines the speed of these modifications toward a target calcium concentration c_{target} by means of a growth rate ν and can be expressed as a linear function

$$\frac{\mathrm{d}z}{\mathrm{d}t} = \nu\left(1 - \frac{c(t)}{c_{\mathrm{target}}}\right). \tag{2}$$

Alternatively, a downward shifted Gaussian or other more complex function may be used, as long as it has a zero-crossing with a negative gradient that allows convergence. If the value of z increases or decreases by 1, the neuron grows or deletes a synaptic element, respectively. The algorithm creates new connections between randomly chosen synaptic elements from the available set in the fourth and last step. This set comprises synaptic elements generated in previous iterations that are not yet connected and the connection partners of deleted synaptic elements.

Beyond this specific example, there exists a range of different structural plasticity models: some have rules for deleting, some for forming synapses, and some for both [for a book, see 63]. Often the algorithm prunes synapses that do not have the chance to become active again [64], are the weakest according to a specific metric [65, 66], or have too little causal correlation between pre- and postsynaptic spikes [67]. Sometimes the mechanism's objective is to maintain a preset number of connections [57], preserve short-range connections [68], or prune until the connectivity has converged to the most efficient constellation [64]. In many algorithms, the determining factors for rewiring are the pre- and postsynaptic activity and the vicinity of other synapses. In general, one can divide structural plasticity mechanisms into two categories: Hebbian structural plasticity, which leads to an increase in the number of synapses during phases of high neuronal activity and, conversely, a decrease in phases of low neuronal activity; and homeostatic structural plasticity, which balances these changes by removing and adding synapses [69].

4.2 From Mathematical Models to Simulation

The NEST implementation of the discussed particular structural plasticity mechanism [70] updates the network connectivity at time intervals that are long compared to the computational time step used to update the neurons, based on experimental observations (*see* Program 3). This slow timescale makes the algorithm more efficient, as the available synaptic elements do not need to be calculated and communicated at every time step. However, when using structural plasticity to generate connections in a network, note that convergence is not guaranteed but determined by the growth rate, network connectivity, and network activity; thus, visual guidance is advised (*see* 71 and Subheading 7.6).

5 Functional Plasticity

The strength of a synapse is usually parameterized by a single static value, the synaptic efficacy or synaptic weight (*see* Subheading 3). However, existing synapses can grow stronger or weaker as an effect of a variety of biophysical mechanisms on both the pre- and postsynaptic side, phenomena collectively known as *functional synaptic plasticity*. These adjustments of synaptic efficacies are likely to form the basis of learning and memory processes in the brain. Thus, this section addresses the temporal evolution of synaptic efficacies and the underlying mechanisms.

5.1 From Empirical Data to Mathematical Models

Over 70 years ago, Hebb famously postulated that neurons that fire together wire together [72]. Since then, many phenomenological models of functional plasticity have been derived and developed. Introducing categories brings some order into the vast landscape of models, even if they do not have clear-cut boundaries and often overlap. Four categorizations are common. First, with respect to the timescale: while short-term plasticity models cover timescales from milliseconds to seconds, long-term plasticity models cover minutes to hours, and homeostatic plasticity models (e.g., synaptic scaling) even up to days [73, 74]. While structural plasticity can occur over the course of hours [75], timescales of functional plasticity are typically shorter. One needs structural plasticity to create new synapses (happening on a long timescale) which then grow stronger via functional plasticity (happening on a faster timescale). Vice versa, synapses that have grown weak are more likely to be pruned. To date, models usually include either functional or structural plasticity but not both together, and the effects of these mechanisms on synaptic learning are thus studied independently. A second categorization distinguishes functional plasticity according to the mechanisms involved: Unsupervised learning rules are based on unlabeled data, supervised learning rules involve a target signal, and reinforcement learning rules function via rewards [73, 74]. A third categorization considers the number of factors that constitute the update formula of the synaptic efficacy: Standard correlation-based rules usually involve two factors, the pre- and postsynaptic spiking, as opposed to three-factor models that involve an additional modulatory signal, e.g., neuromodulation. Fourth, based on activity dependence, plasticity mechanisms occur in two counteracting forms: in Hebbian-type mechanisms, higher activity levels of the pre- and postsynaptic neurons lead to strengthening of the (positive or negative) weights, whereas in homeostatic mechanisms, to their weakening [69]. Without further constraints, Hebbian-type plasticity may lead to a positive feedback loop and, consequently, substantial changes in synaptic weights and network activity. In contrast, homeostatic synaptic plasticity pushes the synaptic efficacy up if activities are low and down if neuronal activities are high, inducing a negative feedback loop and stabilizing the dynamics.

Experimental studies show that the efficacy of a synapse can change for a short time window of hundreds to thousands of milliseconds depending on the history of the presynaptic spikes [76–78]. This phenomenon is termed *short-term plasticity (STP)*, or more precisely, *short-term facilitation (STF)* if the efficacy is elevated, and *short-term depression (STD)* if the efficacy is decreased. The biophysical mechanism underlying STP is the dynamics of vesicle pools and spike-triggered exocytosis. On the one hand, after the generation of a spike, calcium accumulates in the presynaptic axon terminal, increasing the probability of neurotransmitter release, which enhances the synaptic efficacy and thus causes STF. On the other hand, repetitive firing leads to the depletion of vesicles and saturation of postsynaptic receptors, which decreases the efficacy and thus causes STD. The mechanisms for STF and STD are counteracting, and a combination of both can be present in the same synapse. Depending on the synapse or neuron type, one of them may be more pronounced. These phenomena form the basis for many STP models [for a review, see 79]. The following paragraph outlines one possible modeling approach by using the example of the model proposed by Tsodyks and Markram [76].

The starting point is the view introduced in Subheadings 2 and 3: a neuron k receives spikes from neuron j over a synapse with the weight w_{jk}. Now, to make the static synaptic weight a dynamical variable, w_{jk} is multiplied by a time-dependent scaling factor $f_{\mathrm{a}}(t)$, modeled by a set of three coupled differential equations:

$$\frac{\mathrm{d}f_{\mathrm{a}}}{\mathrm{d}t} = -\frac{f_{\mathrm{a}}}{\tau_{\mathrm{i}}} + u^{+}f_{\mathrm{r}}^{-}\,\delta\left(t - t_j^{\mathrm{f}}\right), \tag{3}$$

$$\frac{\mathrm{d}f_{\mathrm{r}}}{\mathrm{d}t} = \frac{f_{\mathrm{i}}}{\tau_{\mathrm{r}}} - u^{+}f_{\mathrm{r}}^{-}\,\delta\left(t - t_j^{\mathrm{f}}\right), \text{ and} \tag{4}$$

$$\frac{\mathrm{d}f_{\mathrm{i}}}{\mathrm{d}t} = \frac{f_{\mathrm{a}}}{\tau_{\mathrm{i}}} - \frac{f_{\mathrm{i}}}{\tau_{\mathrm{r}}}, \tag{5}$$

describing the utilization of synaptic resources by each presynaptic spike arriving at time t_j^{f}. Here, $f_{\mathrm{a}}(t)$, $f_{\mathrm{r}}(t)$, and $f_{\mathrm{i}}(t) = 1 - f_{\mathrm{r}}(t) - f_{\mathrm{a}}(t)$ denote the fractions of active, recovered, and inactive synaptic resources, respectively, and $\delta(\cdot)$ the Dirac delta function. The superscripts "$-$" and "$+$" refer to the values of the associated variables before and after their update. Each presynaptic spike increases the active synaptic resources, i.e., the synaptic weight, and simultaneously reduces the available (recovered) resources by an amount proportional to u^{+}. The utilization variable u^{+} represents the probability of vesicle release, controlled by the calcium concentration in the axon terminal. In the absence of STF, this utilization is constant. However, in facilitating synapses, it is dynamic and evolves according to

$$\frac{\mathrm{d}u}{\mathrm{d}t} = -\frac{u}{\tau_{\mathrm{fac}}} + U(1 - u^{-})\delta\left(t - t_j^{\mathrm{f}}\right), \tag{6}$$

where the parameter U corresponds to the amplitude of the postsynaptic current (the synaptic weight) in response to a single isolated presynaptic spike. The time constants τ_{r}, τ_i, and τ_{fac} describe the recovery time from synaptic depression, the decay of the postsynaptic currents, and the decay of the utilization, respectively. Despite its simplicity, this phenomenological model approximates experimental findings well [76].

The modifications of the synaptic efficacy by STP occur only during presynaptic firing and last for a few hundred milliseconds. After presynaptic firing has stopped, the synaptic resource variables, and hence the synaptic weight, return to their resting states $f_{\mathrm{a}} = 0$, $f_{\mathrm{r}} = 1$, $f_{\mathrm{i}} = 0$, $u = 0$, and w_{jk}. In contrast, *spike-timing-dependent plasticity (STDP)* has a prolonged effect on synaptic efficacy and thus constitutes a form of long-term plasticity. This form of plasticity was discovered in several spike pairing experiments where a pre- and a postsynaptic neuron were repetitively stimulated to emit spikes at a predefined interval [80, 81]. Reviews of the experimental findings can be found in Caporale and Dan [82], Markram et al. [83] and Brzosko et al. [84]. Although the results of these studies vary across cell types and pairing protocols, they all find that the induced change of the synaptic efficacy depends on the precise time difference $\Delta t = t_{\mathrm{post}} - t_{\mathrm{pre}}$ between a pair of pre- and postsynaptic spikes. The mapping between the sign and value of the weight changes and the time lags between pre- and postsynaptic spikes is described by the *STDP kernel*, which can take either a Hebbian or an anti-Hebbian form and is sometimes modeled as either symmetric or anti-symmetric [all forms illustrated by Fig. 2 in 85]. In nature, however, the size and shape of the potentiation and depression windows might differ, leading to an overall asymmetric window (*see* Fig. 4a). Generally, a postsynaptic spike occurring slightly after the presynaptic spike ($\Delta t > 0$) induces *long-term potentiation (LTP)*, whereas a postsynaptic spike occurring slightly before the presynaptic spike ($\Delta t < 0$) induces *long-term depression (LTD)*. Thus, STDP can encode a causal relationship between the firing of the pre- and postsynaptic neuron and the synaptic weight change. Since this finding follows Hebb's principle, this type of STDP belongs to the class of Hebbian plasticity rules.

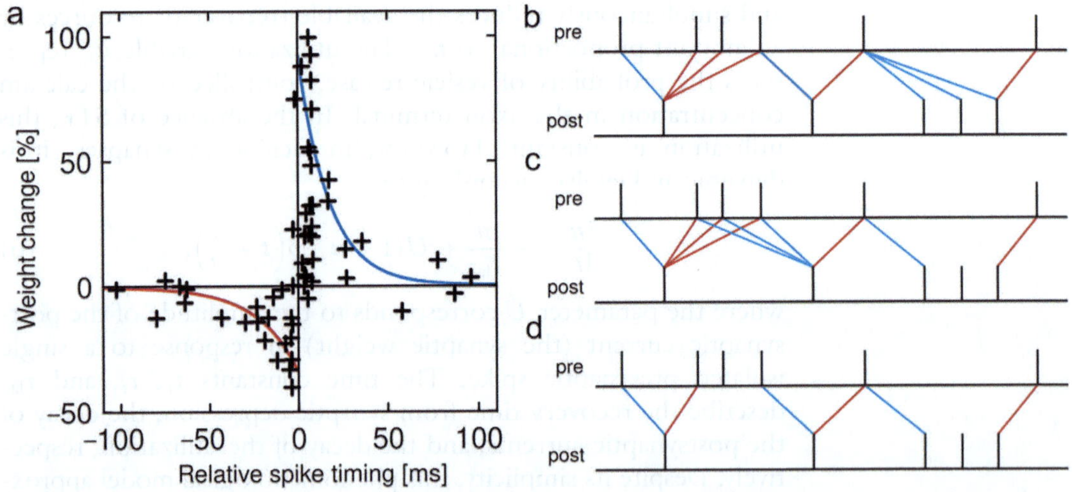

Fig. 4 Spike-timing-dependent plasticity (STDP). (**a**) Weight change expressed as a function of relative pre- and postsynaptic spike timing for an example of STDP with an anti-symmetric (and slightly asymmetric), Hebbian learning window. Markers correspond to empirical data from Bi and Poo [81]. Solid lines show exponential approximations used in the model (red indicates depression and blue indicates potentiation in all panels). An anti-Hebbian window would look similar but mirrored about the vertical axis. (**b**–**d**) Three nearest-neighbor spike pairing rule variants for STDP. (**b**) *Symmetric:* each presynaptic spike is paired with the last postsynaptic spike, and each postsynaptic spike is paired with the last presynaptic spike. (**c**) *Presynaptic centered:* each presynaptic spike is paired with the last postsynaptic spike and the next postsynaptic spike. (**d**) *Reduced symmetric:* as in panel (**c**), but only for closest pairs. (Adapted from Fig. 7 in Morrison et al. [74])

Formalizing this robust finding based on the above experimental data allows for mathematical treatment. Morrison et al. [74] developed a phenomenological model with only a few free parameters, which reproduces the experimental observations without referencing the underlying molecular mechanisms. They restricted the observables that can enter the plasticity rule to locally available ones because biologically plausible phenomenological models should only contain terms that are identifiable with mechanisms that exist in biology. In the case of STDP, the synaptic weight change depends on the pre- and postsynaptic spike times and potentially also on the current synaptic weight. Experiments demonstrating that action potentials back-propagating through the dendritic tree convey information about a postsynaptic spike support the fact that postsynaptic spikes can be available at the synapse [80].

The dependence of LTP and LTD on Δt is usually captured by exponential functions with decay times $\tau_{\pm}$. Morrison et al. [74] give a simple model of the weight change Δw for pair-based STDP:

$$\begin{aligned} \Delta w_{+} &= F_{+}(w)\, e^{-\frac{|\Delta t|}{\tau_{+}}} \quad \text{if } \Delta t > 0, \\ \Delta w_{-} &= -F_{-}(w)\, e^{-\frac{|\Delta t|}{\tau_{-}}} \quad \text{if } \Delta t \leq 0, \end{aligned} \tag{7}$$

where the functions $F_{\pm}$ capture the dependence on the current weight w and have to be specified further by fitting them to experimental data [86, 87] (Fig. 4a). A spike pair can be defined in different ways: for example, each presynaptic spike can be paired with the most recent preceding postsynaptic spike and vice versa, which is one variant of the class of *nearest-neighbor schemes* (see Fig. 4b–d), whereas in the *all-to-all scheme*, each presynaptic spike is paired with all preceding postsynaptic spikes [74, 88].

The STDP model described above can serve as a starting point for designing extended models to describe more nuanced experimental results, for example by including the postsynaptic membrane potential as an additional modulatory factor beyond the pre- and postsynaptic spikes. Along these lines, Clopath and Gerstner [89] and Clopath et al. [90] account for the effects of voltage-dependent receptors and channels. Their approach is based, among other findings, on experiments showing that the same spike pairing protocol can, depending on the postsynaptic membrane potential V_{m}, induce no change in synaptic weights at all, LTD, or LTP [91]. While a V_{m} smaller than an experimentally determined threshold potential Θ_{-} induces neither LTD nor LTP, an intermediate membrane potential $\Theta_{-} < V_{\mathrm{m}} < \Theta_{+}$ triggers LTD, and a high $V_{\mathrm{m}} > \Theta_{+}$ enables LTP. To capture this behavior, the mathematical description of the plasticity rule contains terms for facilitation and depression that are active based on these conditions of the membrane voltage, formally expressed as Heaviside step functions. With this mechanism, Clopath and Gerstner [89] were able to reproduce the complex frequency dependence of the synaptic weight changes in spike pairing experiments [92].

In Urbanczik and Senn [93], the postsynaptic membrane potential is included as a modulating factor. This plasticity rule, in particular, applies to synapses that connect to the dendrite of a postsynaptic neuron. Experiments show that presynaptic spikes that do not cause postsynaptic spikes lead to a depression of synaptic weights whose strength increases with increasing dendritic voltage [94]. From this observation, Urbanczik and Senn [93] conclude that the synaptic weights are adjusted such that the dendritic voltage assumes high values if and only if the soma of the postsynaptic neuron emits spikes. Therefore, in this rule, the difference between the dendritic voltage and the somatic activity drives the synaptic weight change.

Instead of the postsynaptic membrane potential, a third factor could also be a neuromodulator concentration, which is motivated by experimental studies [for a review, see 95] and by the fact that they provide a biologically plausible implementation of reward signals [96].

5.2 From Mathematical Models to Simulation

The STP implementation in NEST (*see* Programs 4 and 5) exploits several practical properties of the corresponding differential equations for their numerical integration. Since the synaptic resources are conserved (i.e., the fractions f_r, f_a, and f_i add up to 1), f_a can be eliminated from the system. Furthermore, thanks to its linear form, the system of coupled differential equations Eqs. (3) to (5) can be integrated exactly between two consecutive presynaptic spikes [46]. Concretely, the joint state of u, f_r, and f_a can be iteratively evolved by multiplying the state at the previous presynaptic spike with a *propagator matrix*.

To simulate STDP, Eq. (7) needs to be calculated efficiently [74]. Having restricted the model parameters to those locally available at the synapse facilitates the implementation in software (*see* Program 6). These constraints also improve the model's performance, since network simulators running on distributed systems take advantage of a limited need for global access to variables to reduce memory consumption and high-latency communication between compute nodes [97, 98]. The all-to-all pairing scheme can be efficiently implemented using a specific update scheme of the synaptic traces. These traces represent a fading memory of past spikes at the synapse without explicit knowledge of all past spike times [74, 86]. If a pre- or postsynaptic spike occurs, the corresponding trace and synaptic weight are updated, while no actions need to be performed in the periods in between. Defining the exact order of updates in a plasticity model, particularly with regard to pre- and postsynaptic spike timing, indicated by the "+" and " − " in Eqs. (3) to (5) is crucial and facilitated by the high-level language NESTML (*see* Program 7).

More complex learning scenarios, like reinforcement learning, are made possible by advanced plasticity rules, which, for example, depend on the postsynaptic membrane potential or neuromodulators [99]. However, these rules typically make it more difficult to discover an effective implementation. For example, neuromodulators (e.g., dopaminergic signals) affect several nearby synapses through *volume transmission* requiring a notion of physical 3D space (*see* Subheading 3.1). Moreover, the presence of time-continuous signals in some of these advanced plasticity rules necessitates the storage of the signal history if one wishes to keep the efficient *event-driven* scheme of updating the synaptic weights only at presynaptic spike times [97]. Depending on how the data structures are laid out in a simulator, accessing the continuous third-factor variables can be difficult or computationally costly because they need to be queried for every spike at every synapse. Generally, rules that require only spike times are more efficient in memory and compute time than rules that depend on the entire history of variables, like the membrane potential trace.

Given access to the synaptic weights, another form of functional plasticity, *weight normalization*, can be realized. It entails keeping the total sum, or a norm of all incoming synaptic strengths of a neuron constant by re-normalizing all its synaptic weights (*see* Program 8). Since, in the brain, these weight changes happen on timescales of several hundreds of milliseconds, the iterative re-normalization takes place on a coarse time grid, increasing the operation's efficiency.

Other advanced plasticity rules include, for example, a third-factor postsynaptic dendritic current [93] or inhibitory plasticity [100]. The discovery of new plasticity rules can be, to a certain degree, even automated [101]. Furthermore, state models with *synaptic tagging and capture* (STC), described, e.g., in Barrett et al. [102], incorporate even plasticity effects beyond synapse-specific ones. Ultimately, state-of-the-art computational plasticity models transcend the simple STP and STDP models [see, e.g., 103]. Algorithmically, however, these complicated models often use a combination of plasticity mechanisms and thus can be synthesized from such a base stack of simpler models.

6 Heterogeneity

Complexity and heterogeneity are ubiquitous and well-established design principles in neurobiological systems [104], covering a multitude of components and mechanisms at various spatial and temporal scales. From an information processing perspective, such variability is a fundamental component of the system, as it determines the types of computations a given circuit can perform and constrains the representational expressivity of its dynamics [105].

6.1 From Empirical Data to Mathematical Models

Biological synaptic connectivity is highly diverse in most of its constituent properties, including the type of neurotransmitter used, the composition of presynaptic vesicles and docking proteins (affecting release probability), the postsynaptic receptor composition (affecting efficacy and kinetics of the elicited response), transmitter reuptake and re-use, and the involvement of gliotransmission [see, e.g., 106], but also properties characterizing signal propagation such as axon diameter and conductance velocity [107–109]. These various types of diversity translate to a high degree of heterogeneity in phenomenological parameters characterizing mathematical models of the synaptic connectivity and dynamics, such as the synaptic weight [110–114], synaptic time constants [115, 116], response latencies (synaptic delays; 116, 117), and parameters specifying the plasticity dynamics [118]. In addition, biological neuronal networks exhibit a high degree of heterogeneity in the anatomical connectivity structure, such as the total number of inputs and outputs per neuron (in/out-degrees; 80, 119–123), and the composition of presynaptic source and postsynaptic target neuron populations.

Previous theoretical work on recurrent neuronal networks shows that heterogeneity in single-neuron properties or connectivity broadens the distribution of firing rates [116, 124] and affects the stability of asynchronous or oscillatory states as well as the level of synchrony [117, 123, 125–130]. A large number of theoretical and experimental studies point at the benefit of heterogeneity for the information processing capabilities of neuronal networks [105, 129, 131–140]. Therefore, modeling studies aiming at understanding the dynamical and functional principles of biological neuronal networks need to account for the synaptic (and other types of) heterogeneity.

Depending on the type of synaptic heterogeneity, its implementation in mathematical models may follow different strategies. Synaptic heterogeneity is expressed on local scales, such as in the connections between neurons in a given layer of a cortical column, and on large scales, such as in cortical inter-area connections. One form of this heterogeneity results from cell-type, layer, or area specificity. It reflects the anatomical and electrophysiological diversity of neurons in different brain regions and, in addition, emerges from specific interactions with other components of the nervous system or with the environment during brain development and learning. In mathematical models, this specificity is usually accounted for by subdividing the network into several populations representing different cell types or brain regions and applying distinct connectivity, synapse, and plasticity parameters for each pair of populations (Fig. 5a).

Another form of synaptic heterogeneity appears in an unspecific, quasi-random manner. It refers to variations in the synaptic characteristics across an ensemble of neuron pairs of seemingly

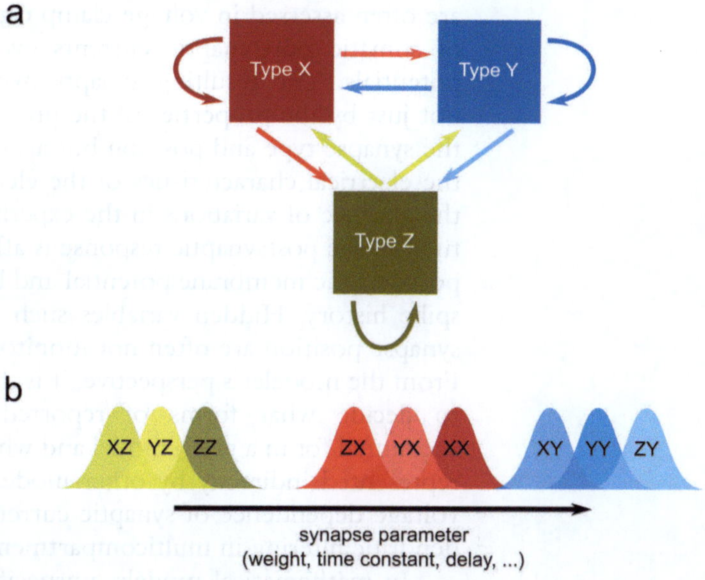

Fig. 5 Specific and unspecific heterogeneity in synaptic connectivity. (**a**) Sketch of a neuronal network comprising three populations of neurons of type X, Y, and Z (boxes). The properties of the different projections (arrows), such as the number of synapses, the synaptic weights, synaptic time constants, or synaptic delays, depend on the types of pre- and postsynaptic neurons. We refer to the resulting synapse-type specific diversity as *specific heterogeneity*. (**b**) For each type of projection $\{PQ\}$ from population Q to population P ($P, Q \in \{X, Y, Z\}$), the synaptic parameters are distributed (illustrated here with bell-shaped curves). We refer to this form of variability as *unspecific heterogeneity*. The parameters characterizing each distribution, such as the mean (horizontal position of each curve) or the variance, are usually synapse-type specific

identical type, for example, connections between a group of neurons with similar morphological and electrophysiological characteristics located in the same layer of a given cortical column (Fig. 5b). Similarly to the cell-type-, layer-, or area-specific diversity described above, the unspecific forms of heterogeneity are partly caused by synaptic plasticity, i.e., by adapting synaptic parameters during learning and development. In this respect, unspecific heterogeneity is not truly unspecific; it is, on the contrary, the result of fine-tuning, optimization, or specialization. Without knowing the details of these processes, the resulting diversity appears random or unspecific. Moreover, the distinction between type-specific and unspecific forms of heterogeneity relies on the assumption that different neuron types are distinguishable [141]. Without knowing the characteristics separating two neuronal phenotypes, these cell classes are treated as one type, and the observed diversity in neuron and synapse parameters appears unspecific.

To some extent, synaptic heterogeneity may also result from variations in experimental protocols and in unobserved variables affecting the synapse characteristics. Synaptic weights, for example, are often assessed in voltage-clamp experiments as the amplitudes of somatic postsynaptic currents evoked by presynaptic action potentials. The resulting synaptic weights are then determined not just by the properties of the pre- and postsynaptic cells or by the synapse type and position but also by the holding potential or the electrical characteristics of the electrode-cell contact. Even in the absence of variations in the experimental protocol, the amplitude of the postsynaptic response is affected by fluctuations in the postsynaptic membrane potential and by the pre- and postsynaptic spike history. Hidden variables such as the spike history or the synapse position are often not monitored in experimental studies. From the modeler's perspective, it is therefore not straightforward to decide what forms of reported heterogeneity should be accounted for in a given model and what forms are perhaps already represented indirectly by other model features (for example, the voltage dependence of synaptic currents, short-term plasticity, or dendritic filtering in multicompartment models).

In mathematical models, unspecific heterogeneity is typically accounted for in a probabilistic manner. Here, the parameters characterizing synaptic connectivity, such as synaptic weights, time constants, delays, in-degrees, etc., are randomly drawn from certain distributions. In particular, in the brain, many properties follow long-tailed distributions, often approximating the lognormal distribution [142–144]. These distributions, or the parameters characterizing them, such as the mean or the standard deviation, are extracted from experimental data. The rationale underlying this probabilistic modeling approach is twofold. First, it acknowledges that the synaptic parameters are typically not known for every single synapse in a given network. The majority of experimental studies provide data for small subsets of synapses, often pooled across different recording sessions or animals. Second, the probabilistic approach greatly simplifies the models, as the total number of parameters is substantially reduced. In probabilistic modeling approaches, the "model" is not defined by a single instantiation of a network and all its parameters but by the ensemble of many independent realizations generated from a given set of parameter distributions. Observations or findings obtained from a single network realization are meaningless unless they appear generically, i.e., frequently, for many different model realizations.

As described above, synaptic heterogeneity is often the result of an adaptation, development, or fine-tuning process. As demonstrated in a number of studies [e.g., *see* 145–147], such processes typically lead to dependencies between parameters. Certain plasticity processes, for example, lead to a competition (anti-correlation) between synapses, such that the strengthening of one synapse results in the weakening of another synapse [148, 149]. In a comprehensive probabilistic model, the set of parameters $\{\xi_1, \xi_2, \ldots, \xi_n\}$ for a specific network realization is generated according to a joint probability density function (pdf) $p_{1,\ldots,n}(x_1, x_2, \ldots, x_n)$, describing the probability of observing $\xi_1 = x_1$, $\xi_2 = x_2$, ..., and $\xi_n = x_n$. This joint pdf captures all parameter dependencies. Many modeling studies neglect parameter dependencies and assume that the joint pdf $p_{1,\ldots,n}(x_1, x_2, \ldots, x_n) = p_1(x_1)p_2(x_2)\ldots p_n(x_n)$ factorizes. In these studies, each parameter ξ_i is drawn from its respective marginal distribution $p_i(\cdot)$, independently of all other parameters. As before, this simplifying assumption typically reflects a lack of knowledge, as the available experimental data generally do not capture parameter dependencies. Theoretical studies show that this choice can have detrimental consequences for the dynamical and functional properties of the resulting system. Bahuguna et al. [147], for example, demonstrate that when the parameter dependencies are unknown, replacing all parameters by their respective mean (and thereby ignoring diversity) can be a better choice than drawing them from their marginal distributions.

A more direct approach toward modeling synaptic heterogeneity is the explicit account of known plasticity, learning, or developmental processes that dynamically lead to the observed diversity in synaptic parameters, including the dependencies described above [86, 149]. Similarly, multivariate parameter distributions may arise from optimization procedures or supervised learning methods fitting the model to some desired target dynamics or behavior [147, 150, 151]. While these top-down approaches are promising and commonly used in state-of-the-art computational neuroscience, they bear the risk that the underlying data or targets do not sufficiently constrain the model of the actual biological system and hence lead to a multitude of solutions that may not be realized in nature. A combination of bottom-up and top-down constraints appears to be the most reliable method to reduce this form of uncertainty.

6.2 From Mathematical Models to Simulation

Investigating the role of heterogeneity in synaptic connectivity by means of analytical mathematical methods is challenging [116, 117]. Therefore, theoretical studies often neglect heterogeneity to simplify the mathematical treatment and provide intuitive insight. A common strategy underlying many mathematical approaches is to reduce the dimensionality of the neuronal network dynamics by assuming that the network can be decomposed into homogeneous subpopulations, each of which comprises neurons with identical neuronal and synaptic parameters. While this approach can account for the specific heterogeneity described above to some extent, it can hardly describe the effects of unspecific heterogeneity. Simulation enables us to test whether the insights obtained under these homogeneity assumptions remain valid if heterogeneity in synaptic parameters is considered.

Even in simulation studies, however, accounting for synaptic heterogeneity is challenging. Acknowledging that every synapse is unique requires representing each synapse with the full individual set of parameters. In a homogeneous network where all synapses have identical properties, the connectivity is fully described by the adjacency matrix (which neuron is connected to which, and how often) and a small set of parameters describing the synapse characteristics, such as the synaptic weight w, the delay d, or the synaptic time constant τ. In a heterogeneous network where each synapse $\{j \rightarrow i\}$ is unique, in contrast, the individual weights w_{ij}, delays d_{ij}, and time constants τ_{ij} need to be stored for each connection. Therefore, representing the heterogeneous connectivity in simulations of neuronal networks at natural density imposes high memory demands for the underlying computing architecture (*see* Subheading 7.7).

In models of neuronal networks with heterogeneous synaptic connectivity, the heterogeneity is either implemented by drawing synapse parameters from predefined distributions or by a self-organization process driven by some plasticity or learning dynamics (*see* Subheading 6.1). Simulations based on the first, the probabilistic approach, require efficient methods of drawing random numbers from specified distributions during the network generation phase. The NEST simulator, for example, permits the high-level specification of probabilistic connection rules by the user (*see* Subheading 2), including distributions of synaptic weights, synaptic delays, or plasticity parameters. The task of generating a specific connectivity realization by drawing random numbers from these distributions is then delegated to fast low-level (C++) routines. The second approach relies on simulating plastic networks or on numerical optimization methods. Strategies for simulating different forms of synaptic plasticity are described in Subheadings 4 and 5. Simulating plastic networks with natural connection density is still a major challenge in computational neuroscience. Slow biological processes such as learning and development on timescales of hours, days, and years are presently inaccessible to simulation (or restricted to small and highly simplified models) because of the required wall-clock time. In this respect, dedicated neuromorphic computing architectures are particularly interesting as simulation platforms for neuroscience, as they offer the potential for faster-than-real-time simulations and hence, for an understanding of plasticity mechanisms on long timescales [152, 153].

7 Notes

In this section, we highlight some of the challenges and pitfalls that may be encountered during modeling and simulation.

7.1 Keeping Model Refinements Biologically Plausible

Developing a mathematical model that exhibits a dynamic behavior close to empirical biological data involves iterative optimization procedures, e.g., fitting the experimental data or calibrating the model parameters. While this optimization improves the model in some aspects, it may, at the same time, alter it in ways no longer motivated by biological observations [146]. For example, fitting is inherently biased in that, by definition, it improves the validation of certain model features at the cost of those not included in the fitting procedure. Thus, each optimization step should be conducted cautiously and checked for biological plausibility.

7.2 Reproducible Simulations

For small-scale simulations, sometimes custom simulation kernels are written. However, besides possibly duplicating published and established routines, these self-made frameworks are likely to contain bugs and lack documentation, for instance, on edge-case behavior. Therefore even for small networks, it helps to use standardized simulators. In particular, these simulators offer the benefit of being well characterized under different operating conditions using a diverse array of automated tests, being updated on a regular release cycle, and benefiting from the open-source model of iterative refinement [154]. All these factors increase the likelihood of long-term reproducible results.

An individual simulator should exhibit *replicable* behavior: repeated simulations of the same model should yield bitwise identical results, regardless of the number of threads or processing nodes used, due to the use of deterministic pseudo-random number generators. However, simulating the same model on a different platform or using a different numerical solver or time step size for ordinary differential equations (ODEs) may alter the results, especially in network models exhibiting chaotic and unstable dynamics. Nonetheless, results and conclusions should be *reproducible*, obtaining the same overall quantitative and qualitative conclusions [for a commentary on this terminology, see 155]. Reproducibility of results requires the original software to be available (including libraries and other dependencies) and, where applicable, the original (raw or pre-processed) dataset(s) and relevant metadata.

Similar to the model descriptions, it increases the reproducibility of methods and results [156] to use and contribute to existing simulation frameworks by reporting bugs, improving implemented methods, and developing and publishing custom modules of the respective framework, e.g., in NEST, in the form of *extension modules* (*see* Program 9).

7.3 Distribution of Compute Workload

It is beneficial to distribute the workload evenly across compute nodes, even for networks with complex connectivity and heterogeneous population properties. One way to achieve this is a *round-robin* distribution of neurons across compute nodes, i.e., in the case of M compute nodes, assigning neuron n to node ($n \bmod M$).

In general, the computing system's size affects the workload distribution. On small machines, the number of synapses per neuron is larger than the number of compute nodes the simulation runs on. Hence, each neuron typically has many targets on every compute node. However, with growing network size and the emergence of new supercomputer architectures over the last decades, the ratio between the numbers of synapses per neuron and compute nodes is in some cases reversed. On these new-generation supercomputers, the distribution of neurons over many nodes decreases the chance that a neuron shares a node with a connected partner, especially considering the sparsity of biological neuronal networks. Mitigating this issue, even more modern compute nodes follow the opposite trend: they possess more memory and cores per processor and thus more processing power, which reduces the number of nodes required for the simulation and brings the neurons and their targets closer together.

One of the main computational challenges remains the connectivity of neuronal networks. For example, representing each of the estimated 10^{14} synaptic connections in the brain individually by two double-precision numbers requires about 1.6 PB of main memory. Furthermore, neurons form connections with nerve cells not only in their vicinity but also in various remote areas. This feature distinguishes neuronal simulations from simulations of classical physical systems, that use, for example, finite-element methods exploiting the locality of physical interactions. However, memory and communication bandwidth, as well as cache efficiency, are more critical than floating-point performance of spike communication to local and distant targets [157].

Future work should address how computer network connectivity and the simulation distribution across compute nodes can follow (simulated) biological network organization, such as a modular organization on both small and large scales in the brain.

7.4 Scalability in Theory and Practice

Recording the simulation time under varying network or computing system sizes characterizes an implementation's *scalability*, which is essential to judging its efficiency [36]. The scenario of increasing the computing resources while keeping the network size fixed is called *strong scaling*, and the scenario of increasing the network size and computing resources proportionally is called *weak scaling*.

Ideally, when in strong scaling the hardware becomes twice as powerful, the simulation time is divided by two. However, in practice, the gains are usually less due to communication overhead and other bottlenecks in the system. Sometimes, a parallelization can even exhibit a super-linear speed-up [158], but this behavior might only become apparent at very large computing system scales.

Scaling up standard network models to test weak scaling can induce unrealistic activity patterns, for example, regarding the regularity and synchrony of spiking. Since synapses tend to vastly outnumber neurons, the number of synapses is an important determinant of the necessary computing resources. Thus, a reasonable approach is to keep the in-degrees constant when increasing the model size and thereby reduce the overall connection probability [see, e.g., 36]. This method tends to lessen activity correlations between neurons and hence diminish synchrony. In the case of ideal weak scaling, a network of twice the size should run for twice the time, but in practice, the performance is worse due to the same reasons as for strong scaling. Moreover, there is a complex dependence of the scaling behavior on network properties, such as the connectivity's modularity [157].

7.5 Precise Spike Times in Discrete-Time Simulation

A typical simulation of a continuous-time dynamical process runs in discrete steps of time Δt. However, exchanging events (spikes) on a grid can cause synchronization in the network as a pure simulation artifact. This effect disappears in the limit of $\Delta t \rightarrow 0$, but decreasing the time step increases the time necessary for the simulation to complete, so a tradeoff has to be made. A more computationally efficient solution is to store an extra offset value in spike events, which, in combination with a minimum synaptic delay and an algorithm that finds the precise time of spiking, decouples the simulation time step from the temporal precision with which spikes are exchanged [159].

7.6 Simulating until Convergence

Simulating neuroplasticity for too short of a period, especially longer-timescale processes like normalization and neuromodulation, is a common pitfall. Several of these dynamic processes have the potential to cause an abrupt bifurcation in the system late in the simulation. Additionally, some plasticity rules produce a long-tailed distribution of synaptic strengths [160], whereby the distribution reaches equilibrium again only after an extended simulation period. A limited measurement duration can also be a problem in empirical neuroscience, but simulations can, in principle, run as long as desired. The only limitations are the computing resources available (as these need to be shared with other users on high-performance computing systems) and the amount of time the simulation takes to complete, which depends on the simulator's efficiency. Simulating until the measure of learning performance has adequately converged can circumvent this pitfall to some degree.

7.7 Limited Synaptic Weight Resolution

Limiting the numerical resolution of synaptic parameters, such as the synaptic weight, appears to be an obvious strategy to reduce memory and compute load. To some extent, nature itself copes well with quantized synaptic weights: Transmission in chemical synapses is quantized due to the release of neurotransmitters in discrete packages from vesicles in the presynaptic axon terminals. The analysis of spontaneous (miniature) postsynaptic currents, i.e., postsynaptic responses to the release of neurotransmitters from single presynaptic vesicles, reveals that the resolution of synaptic weights is indeed finite for chemical synapses. As shown by Malkin et al. [161], the amplitudes of spontaneous excitatory postsynaptic currents recorded from different types of excitatory and inhibitory cortical neurons follow a unimodal distribution with a peak at about 20 pA and a lower bound at about 10 pA. Such a cut-off is present despite several factors that may wash out the discreteness of the synaptic transmission, such as variability in vesicle sizes, variability in the position of vesicle fusion zones, quasi-randomness in neurotransmitter diffusion across the synaptic cleft, and variability in postsynaptic receptor densities. However, the discreteness of synaptic strengths is obscured for evoked synaptic responses involving neurotransmitter release from many presynaptic vesicles and for superpositions of inputs from many synapses, and thus unlikely to play a particular role in the dynamics of the neuronal network as a whole.

Inspired by these observations, Dasbach et al. [162] systematically investigated the effects of a limited synaptic weight resolution on the dynamics of recurrent spiking neuronal networks resembling local cortical circuits. They show that a naive quantization of synaptic weights generally leads to a distortion of the firing statistics. However, in the example of one network type, they could demonstrate that the firing statistics remain unaffected under a weight discretization that preserves the mean and variance of the total synaptic input currents. In networks with sufficiently heterogeneous in-degrees, the firing statistics stay constant, even when replacing all synaptic weights with the mean of the weight distribution, i.e., entirely neglecting the unspecific form of heterogeneity in synaptic weights. Applying this finding in simulations reduces the memory demands substantially. The effect of discretized synaptic weights in networks undergoing different forms of synaptic plasticity has rarely been investigated [163] and remains a subject for future study.

7.8 Precise Specification of the Model

An incomplete and ambiguous description of a model without following conventions can impede understanding by the reader and thus the reproducibility of the respective study. The first step to avoiding this pitfall is to gain a clear picture of the requirements a model description must fulfill. For this, Nordlie et al. [164] propose that a neuronal network model description must contain a complete and detailed account of its architecture and the dynamics of its parts. In the formalization of model descriptions, computational neuroscience is less mature than other fields of science. Nonetheless, best practice guidelines are emerging that suggest the use of standardized tables of the model characteristics that cover the network architecture and connectivity, all neuron and synapse models used, the applied input stimuli, and the recorded data, described by a combination of text, equations, figures, subtables, and pseudocode [164]. Formalized sketches of the network and unified connectivity concepts [25] help computational neuroscientists to unequivocally convey their models and, consequently, readers to understand them.

It is advisable to use one of the numerous formal languages that facilitate such specifications and make them consistent, e.g., NeuroML [13], NESTML [6], or PyNN [12]. Most of such languages adhere to the class of either declarative or procedural languages. While a declarative language specifies the model's features, a procedural language specifies the series of commands or instructions needed for constructing the model. Best practices for either of these approaches include formally defined syntax and semantics or an API specification, both uniquely identified by version numbers.

It is good practice to keep the specification and implementation of a model separate. For example, implementation details such as the time resolution and the spike threshold detection method are essential for the reproducibility of the results but are not part of the model itself. Like the model description, the implementation specification should be complete and sufficient, as it can be challenging to reverse-engineer implementations and test the robustness of the results to implementation alterations [164]. However, the model should be robust to different choices in such implementation details.

Generating an executable representation of the model can be fully automated. For example, tools like NESTML specialize in processing the model descriptions of different complexity levels and in verbose formats like XML, creating executable implementations and visualizations, and facilitating debugging [165].

To summarize, it is advisable to follow standardized model descriptions, implementations, and generation procedures wherever possible for a project and ideally share the models with the community in dedicated databases for computational neuroscience models like ModelDB [166] or Open Source Brain [167].

8 Conclusions

Large-scale neuronal network simulations are a key tool for understanding brain processes. They make the complex nonlinear dynamics of neuronal activity, which are out of reach with analytic methods, accessible to inquiry. Moreover, continuous interaction between computational modeling and advances in empirical understanding have iteratively refined simulation approaches throughout the history of computational neuroscience. We hence outline a few promising directions for empirical and modeling approaches to work together.

New empirical data can bolster simulation studies in many ways. For example, long-term tracking of synapses *in vivo* will elucidate the relationship between synaptic plasticity and function

[168]. Besides, recordings of complete connectomes at single-neuron resolution are becoming feasible for ever-larger brains and may soon be available on the scale of a mouse brain [169]. Data-driven models based on such detailed and specific connectivity are complemented by models whose connectivity is generated with statistical approaches. The latter models also profit from more available data constraining their parameters; higher-level network organization can, for instance, be informed by data on hierarchical modularity and small-world properties. Sometimes, the conditions of data retrieval are inconsistent between experiments, or experiments only cover a small subset of the model system, and modeling could benefit from additional studies to fill the gaps.

How the available data are integrated into models depends on the particular research question and the level of abstraction appropriate for it. The modeler needs to decide (or find out) which features and phenomena of the natural system need to be represented in detail and which ones can be approximated. For instance, a biophysically detailed model with discrete vesicle release dynamics would be suitable when investigating how a compound influencing vesicle fusion to the membrane affects synaptic transmission. In contrast, investigating the compound's effects on large-scale network dynamics could necessitate approximation of the vesicle release by a simplified set of continuous quantities and differential equations. In essence, a good computational model should represent the relevant attributes of the studied biological structure, have explanatory power and simultaneously not necessitate extensive simulation time.

There are several ways in which the infrastructure of computer simulations itself can innovate, indicated by current trends and feature requests from the community. Simulation efficiency is still a bottleneck, especially in simulations involving plasticity, as they need to run for a comparatively long time. Furthermore, large-scale networks require powerful, high-performance compute clusters, which provide large amounts of RAM, and enable running the simulation in parallel and distributed across many CPUs or GPUs which are interconnected through a low-latency network. More advanced synaptic plasticity rules, for example involving tripartite synapses influenced by astrocytes or neuromodulators like dopamine, still lack software support for efficient simulations on a larger scale. Finally, simulations at cellular resolution could be extended toward multiphysics modeling by incorporating other physical phenomena. Such models could account for the volume diffusion of neuromodulators or the electric field in the neuropil to simulate ephaptic coupling for instance.

In general, what should one strive for in a model? The statistician George Box famously said, "All models are wrong, but some are useful." Although it helps to remind ourselves of the difficulties of modeling, to say that all models are wrong is not doing them justice, as put by Sir David R. Cox in a comment on Chatfield [170]: "The very word *model* implies simplification and idealization. The idea that complex physical, biological or sociological systems can be exactly described by a few formulae is patently absurd. The construction of idealized representations that capture important stable aspects of such systems is, however, a vital part of general scientific analysis and statistical models [...]". A model with unlimited parameters can fit the data perfectly but at the cost of generalizability and explanatory power. Ockham's razor, or the law of parsimony, provides a helpful heuristic in this context, guiding us to prefer the theory with fewer parameters between two competing theories. This objective is also formalized in Bayesian information criteria, which penalize models with larger numbers of parameters. In other words, the aim to reduce complexity should guide modeling choices to address a given research question.

However, condensing the biophysical details and terminology to arrive at a parsimonious, phenomenological formulation could impede testing such a minimal model experimentally. Moreover, the rigorous application of Ockham's razor leads to models optimized for single phenomena (e.g., connectivity, synaptic transmission, structural or functional plasticity) that are thus hard to combine. To solve this problem, after reducing the phenomena to their essential variables, we should express each model in a way that allows for their combination. Biophysical details could provide contact points between model concepts from different subdomains. Progress in integrating heterogeneous phenomena in large-scale models requires that models act as platforms that can be modified and extended over time. For example, after a minimal model has achieved a satisfactory performance, modelers could test whether the same results still hold for models of greater biophysical detail.

As a foreseeable trend, more and more computational neuroscientists will adopt procedures ensuring the reproducibility of their methods and results over the following years. This approach includes, where possible, data, model, and code sharing in open-access online repositories, adherence to open standards for model formats and software tools, active management of metadata, containerized distribution of dependencies, unit testing, and continuous integration instead of creating new in-house toolchains from the ground up. Last but not least, active contribution to an existing model database or open-source software, be it as small as a feature request, is a low-threshold action conducive to reproducible research everyone can take. Ultimately, the community's research interests in the form of these requests and contributions shape the landscape of available models in an open-access simulator.

9 Links to the Programs

1. https://nest-simulator.readthedocs.io/en/latest/tutorials/pynest_tutorial/part_3_connecting_networks_with_synapses.html
2. https://github.com/nest/nestml/blob/master/models/neurons/wb_cond_multisyn.nestml
3. https://www.nest-simulator.org/py_sample/structural_plasticity/
4. https://nest-simulator.readthedocs.io/en/latest/auto_examples/tsodyks_depressing.html
5. https://nest-simulator.readthedocs.io/en/latest/auto_examples/tsodyks_facilitating.html
6. https://nestml.readthedocs.io/en/latest/models_library/stdp.html
7. https://nestml.readthedocs.io/en/latest/nestml_language/synapses_in_nestml.html
8. https://nest-simulator.readthedocs.io/en/latest/guides/weight_normalization.html
9. https://nest-extension-module.readthedocs.io/en/latest/extension_modules.html

Acknowledgements

The authors thank Dennis Terhorst and Jessica Mitchell for technical help with the NEST documentation links, Angela Fischer for advice on the layout of the figures, and Rainer Waser for helpful feedback on the manuscript. Furthermore, they gratefully acknowledge funding from the European Union's Horizon 2020 Framework Programme for Research and Innovation under Specific Grant Agreement No. 945539 (Human Brain Project SGA3), from the Joint Lab "Supercomputing and Modeling for the Human Brain," the Helmholtz Association Initiative and Networking Fund under project number SO-092 (Advanced Computing Architectures), the Federal Ministry of Education and Research BMBF under grant ID 03ZU1106CB within NeuroSys as part of the initiative "Clusters4Future," the Helmholtz Metadata Collaboration (HMC) under funding ID ZT-I-PF-3-026, the Deutsche Forschungsgemeinschaft (DFG, German Research Foundation) under grant ID 368482240/GRK2416, HiRSE_PS, the Helmholtz Platform for Research Software Engineering—Preparatory Study, an innovation pool project of the Helmholtz Association, and the Jülich-Aachen Research Alliance Center for Simulation and Data Science (JARA-CSD) School for Simulation and Data Science

(SSD), as well as computing time granted by the JARA Vergabegremium and provided on the JARA Partition part of the supercomputer JURECA DC at Forschungszentrum Jülich (computation grant JINB33).

References

1. Alonso-Nanclares L, Gonzalez-Soriano J, Rodriguez JR, DeFelipe J (2008) Gender differences in human cortical synaptic density. Proc Natl Acad Sci 105(38):14615–14619. https://doi.org/10.1073/pnas.0803652105
2. Linden DJ (2018) Our human brain was not designed all at once by a genius inventor on a blank sheet of paper. In: Think tank: forty neuroscientists explore the biological roots of human experience. Yale University Press, pp 1–8. https://doi.org/10.12987/9780300235470-002
3. Azevedo FAC, Carvalho LRB, Grinberg LT, Farfel JM, Ferretti RE, Leite RE, Filho WJ, Lent R, Herculano-Houzel S (2009) Equal numbers of neuronal and nonneuronal cells make the human brain an isometrically scaled-up primate brain. J Comp Neurol 513(5): 532–541. https://doi.org/10.1002/cne.21974
4. Trensch G, Gutzen R, Blundell I, Denker M, Morrison A (2018) Rigorous neural network simulations: A model substantiation methodology for increasing the correctness of simulation results in the absence of experimental validation data. Front Neuroinform 12. https://doi.org/10.3389/fninf.2018.00081
5. Gewaltig MO, Diesmann M (2007) NEST (NEural Simulation Tool). Scholarpedia 2(4):1430. https://doi.org/10.4249/scholarpedia.1430
6. Plotnikov D, Rumpe B, Blundell I, Ippen T, Eppler JM, Morrison A (2016) NESTML: a modeling language for spiking neurons. arXiv 1606.02882. https://doi.org/10.48550/arXiv.1606.02882
7. Hines ML, Carnevale NT (2001) NEURON: a tool for neuroscientists. Neuroscientist 7(2): 123–135. https://doi.org/10.1177/107385840100700207
8. Stimberg M, Brette R, Goodman DFM (2019) Brian 2, an intuitive and efficient neural simulator. eLife 8:e47,314. https://doi.org/10.7554/eLife.47314
9. Bekolay T, Bergstra J, Hunsberger E, DeWolf T, Stewart TC, Rasmussen D, Choo X, Voelker A, Eliasmith C (2014) Nengo: a python tool for building large-scale functional brain models. Front Neuroinform 7:48. https://doi.org/10.3389/fninf.2013.00048
10. Abi Akar N, Cumming B, Karakasis V, Küsters A, Klijn W, Peyser A, Yates S (2019) Arbor — A morphologically-detailed neural network simulation library for contemporary high-performance computing architectures. In: 27th Euromicro International Conference on Parallel, Distributed and Network-Based Processing (PDP), pp 274–282. https://doi.org/10.1109/EMPDP.2019.8671560
11. Vitay J, Dinkelbach HÜ, Hamker FH (2015) ANNarchy: a code generation approach to neural simulations on parallel hardware. Front Neuroinform 9:19. https://doi.org/10.3389/fninf.2015.00019
12. Davison AP, Brüderle D, Eppler JM, Kremkow J, Muller E, Pecevski D, Perrinet L, Yger P (2009) Pynn: a common interface for neuronal network simulators. Front Neuroinf 2:11. https://doi.org/10.3389/neuro.11.011.2008
13. Gleeson P, Crook S, Cannon RC, Hines ML, Billings GO, Farinella M, Morse TM, Davison AP, Ray S, Bhalla US, Barnes SR, Dimitrova YD, Silver RA (2010) NeuroML: A language for describing data driven models of neurons and networks with a high degree of biological detail. PLOS Comput Biol 6. https://doi.org/10.1371/journal.pcbi.1000815
14. Van Albada SJ, Morales-Gregorio A, Dickscheid T, Goulas A, Bakker R, Bludau S, Palm G, Hilgetag CC, Diesmann M (2021) Bringing anatomical information into neuronal network models. In: Giugliano M, Negrello M, Linaro D (eds) Computational modelling of the brain, advances in experimental medicine and biology, vol 1359. Springer, pp 201–234. https://doi.org/10.1007/978-3-030-89439-9_9
15. Gouwens NW, Sorensen SA, Berg J, Lee C, Jarsky T, Ting J, Sunkin SM, Feng D, Anastassiou CA, Barkan E, et al. (2019) Classification of electrophysiological and morphological neuron types in the mouse visual cortex. Nature Neurosci 22(7):

1182–1195. https://doi.org/10.1038/s41593-019-0417-0
16. Hodge RD, Bakken TE, Miller JA, Smith KA, Barkan ER, Graybuck LT, Close JL, Long B, Johansen N, Penn O, et al. (2019) Conserved cell types with divergent features in human versus mouse cortex. Nature 573(7772): 61–68. https://doi.org/10.1038/s41586-019-1506-7
17. Packer AM, Yuste R (2011) Dense, unspecific connectivity of neocortical parvalbumin-positive interneurons: a canonical microcircuit for inhibition? J Neurosci 31(37): 13,260–13,271. https://doi.org/10.1523/JNEUROSCI.3131-11.2011
18. Perin R, Berger TK, Markram H (2011) A synaptic organizing principle for cortical neuronal groups. Proc Natl Acad Sci 108(13): 5419–5424. https://doi.org/10.1073/pnas.1016051108
19. Ercsey-Ravasz M, Markov NT, Lamy C, Van Essen DC, Knoblauch K, Toroczkai Z, Kennedy H (2013) A predictive network model of cerebral cortical connectivity based on a distance rule. Neuron 80(1):184–197. https://doi.org/10.1016/j.neuron.2013.07.036
20. Felleman DJ, Van Essen DC (1991) Distributed hierarchical processing in the primate cerebral cortex. Cereb Cortex 1(1): 1–47. https://doi.org/10.1093/cercor/1.1.1-a
21. Voges N, Schüz A, Aertsen A, Rotter S (2010) A modeler's view on the spatial structure of intrinsic horizontal connectivity in the neocortex. Prog Neurobiol 92(3):277–292. https://doi.org/10.1016/j.pneurobio.2010.05.001
22. Schmidt M, Bakker R, Hilgetag CC, Diesmann M, Van Albada SJ (2018) Multiscale account of the network structure of macaque visual cortex. Brain Struct Funct 223(3):1409–1435. https://doi.org/10.1007/s00429-017-1554-4
23. Potjans TC, Diesmann M (2014) The cell-type specific cortical microcircuit: relating structure and activity in a full-scale spiking network model. Cereb Cortex 24(3): 785–806
24. Kunkel S, Potjans TC, Morrison A, Diesmann M (2009) Simulating macroscale brain circuits with microscale resolution. Front Neuroinform. https://doi.org/10.3389/conf.neuro
25. Senk J, Kriener B, Djurfeldt M, Voges N, Jiang HJ, Schüttler L, Gramelsberger G, Diesmann M, Plesser HE, Van Albada SJ (2022) Connectivity concepts in neuronal network modeling. PLOS Comput Biol 18(9):e1010,086. https://doi.org/10.1371/journal.pcbi.1010086
26. Kasthuri N, Hayworth KJ, Berger DR, Schalek RL, Conchello JA, Knowles-Barley S, Lee D, Vázquez-Reina A, Kaynig V, Jones TR, et al. (2015) Saturated reconstruction of a volume of neocortex. Cell 162(3):648–661. https://doi.org/10.1016/j.cell.2015.06.054
27. Sincich LC, Blasdel GG (2001) Oriented axon projections in primary visual cortex of the monkey. J Neurosci 21(12):4416–4426. https://doi.org/10.1523/JNEUROSCI.21-12-04416.2001
28. Markov NT, Vezoli J, Chameau P, Falchier A, Quilodran R, Huissoud C, Lamy C, Misery P, Giroud P, Ullman S, et al. (2013) Anatomy of hierarchy: feedforward and feedback pathways in macaque visual cortex. J Comp Neurol 522(1):225–259. https://doi.org/10.1002/cne.23458
29. Rees CL, Moradi K, Ascoli GA (2017) Weighing the evidence in Peters' rule: does neuronal morphology predict connectivity? Trends Neurosci 40(2):63–71. https://doi.org/10.1016/j.tins.2016.11.007
30. Roth U, Eckardt F, Jahnke A, Klar H (1997) Efficient on-line computation of connectivity: Architecture of the connection unit of NESPINN. In: Proceedings of the MicroNeuro, Dresden, pp 31–39
31. Knight JC, Nowotny T (2021) Larger GPU-accelerated brain simulations with procedural connectivity. Nature Comput Sci 1(2):136–142. https://doi.org/10.1038/s43588-020-00022-7
32. Strata P, Harvey R (1999) Dale's principle. Brain Res Bull 50(5-6):349–350. https://doi.org/10.1016/S0361-9230(99)00100-8
33. Kunkel S, Schmidt M, Eppler JM, Plesser HE, Masumoto G, Igarashi J, Ishii S, Fukai T, Morrison A, Diesmann M, et al. (2014) Spiking network simulation code for petascale computers. Front Neuroinform 8:78. https://doi.org/10.3389/fninf.2014.00078
34. Kandel ER, Schwartz JH, Jessell TM, Siegelbaum SA, Hudspeth AJ (2012) Principles of neural science, 5th edn. McGraw-Hill, p 1472
35. van Albada SJ, Helias M, Diesmann M (2015) Scalability of asynchronous networks is limited by one-to-one mapping between effective connectivity and correlations. PLOS Comput Biol 11(9):e1004,490. https://doi.org/10.1371/journal.pcbi.1004490
36. Jordan J, Ippen T, Helias M, Kitayama I, Sato M, Igarashi J, Diesmann M, Kunkel S

(2018) Extremely scalable spiking neuronal network simulation code: From laptops to exascale computers. Front Neuroinform 12: 2. https://doi.org/10.3389/fninf.2018.00002

37. Destexhe A, Rudolph M, Paré D (2003) The high-conductance state of neocortical neurons in vivo. Nature Rev Neurosci 4(9): 739–751. https://doi.org/10.1038/nrn1198
38. Maksimov A, Diesmann M, Van Albada SJ (2018) Criteria on balance, stability, and excitability in cortical networks for constraining computational models. Front Comput Neurosci 12:44. https://doi.org/10.3389/fncom.2018.00044
39. Gerstner W, Kistler WM, Naud R, Paninski L (2014) Neuronal dynamics: from single neurons to networks and models of cognition. Cambridge University Press. https://doi.org/10.1017/CBO9781107447615
40. Potjans W, Morrison A, Diesmann M (2010) Enabling functional neural circuit simulations with distributed computing of neuromodulated plasticity. Front Comput Neurosci 4: 141. https://doi.org/10.3389/fncom.2010.00141
41. Larkum ME, Wu J, Duverdin SA, Gidon A (2022) The guide to dendritic spikes of the mammalian cortex in vitro and in vivo. Neuroscience 489:15–33. https://doi.org/10.1016/j.neuroscience.2022.02.009
42. Jahnke S, Timme M, Memmesheimer RM (2012) Guiding synchrony through random networks. Phys Rev X 2(4):041,016. https://doi.org/10.1103/physrevx.2.041016
43. Bouhadjar Y, Wouters DJ, Diesmann M, Tetzlaff T (2022) Sequence learning, prediction, and replay in networks of spiking neurons. PLOS Comput Biol 18(6):e1010,233. https://doi.org/10.1371/journal.pcbi.1010233
44. Tsodyks M, Pawelzik K, Markram H (1998) Neural networks with dynamic synapses. Neural Comput 10(4):821–835. https://doi.org/10.1162/089976698300017502
45. Brette R, Gerstner W (2005) Adaptive exponential integrate-and-fire model as an effective description of neuronal activity. J Neurophysiol 94(5):3637–3642. https://doi.org/10.1152/jn.00686.2005
46. Rotter S, Diesmann M (1999) Exact digital simulation of time-invariant linear systems with applications to neuronal modeling. Biol Cybern 81(5-6):381–402. https://doi.org/10.1007/s004220050570
47. Morrison A, Straube S, Plesser HE, Diesmann M (2007) Exact subthreshold integration with continuous spike times in discrete-time neural network simulations. Neural Comput 19(1):47–79. https://doi.org/10.1162/neco.2007.19.1.47
48. Morrison A, Diesmann M (2007) Maintaining causality in discrete time neuronal network simulations. In: Lectures in supercomputational neurosciences: dynamics in complex brain networks. Springer, pp 267–278. https://doi.org/10.1007/978-3-540-73159-7_10
49. Hahne J, Helias M, Kunkel S, Igarashi J, Bolten M, Frommer A, Diesmann M (2015) A unified framework for spiking and gap-junction interactions in distributed neuronal network simulations. Front Neuroinform 9:22. https://doi.org/10.3389/fninf.2015.00022
50. Jordan J, Helias M, Diesmann M, Kunkel S (2020) Efficient communication in distributed simulations of spiking neuronal networks with gap junctions. Front Neuroinform 14:12. https://doi.org/10.3389/fninf.2020.00012
51. Stettler DD, Yamahachi H, Li W, Denk W, Gilbert CD (2006) Axons and synaptic boutons are highly dynamic in adult visual cortex. Neuron 49(6):877–887. https://doi.org/10.1016/j.neuron.2006.02.018
52. Butz M, Wörgötter F, Van Ooyen A (2009) Activity-dependent structural plasticity. Brain Res Rev 60(2):287–305. https://doi.org/10.1016/j.brainresrev.2008.12.023
53. Butz M, Van Ooyen A (2013) A simple rule for dendritic spine and axonal bouton formation can account for cortical reorganization after focal retinal lesions. PLOS Comput Biol 9(10):e1003,259. https://doi.org/10.1371/journal.pcbi.1003259
54. Bamford SA, Murray AF, Willshaw DJ (2010) Synaptic rewiring for topographic mapping and receptive field development. Neural Netw 23(4):517–527. https://doi.org/10.1016/j.neunet.2010.01.005
55. Gallinaro JV, Rotter S (2018) Associative properties of structural plasticity based on firing rate homeostasis in recurrent neuronal networks. Sci Rep 8(1):1–13. https://doi.org/10.1038/s41598-018-22077-3
56. Kalantzis G, Shouval HZ (2009) Structural plasticity can produce metaplasticity. PLOS One 4(11):e8062. https://doi.org/10.1371/journal.pone.0008062
57. Bellec G, Kappel D, Maass W, Legenstein R (2017) Deep rewiring: Training very sparse

deep networks. arXiv 1711.05136. https://doi.org/10.48550/arXiv.1711.05136

58. Knoblauch A (2017) Impact of structural plasticity on memory formation and decline. In: The rewiring brain. Elsevier, pp 361–386. https://doi.org/10.1016/B978-0-12-803784-3.00017-2
59. Turrigiano G (2012) Homeostatic synaptic plasticity: Local and global mechanisms for stabilizing neuronal function. Cold Spring Harb Perspect Biol 4(1):a005,736–a005,736. https://doi.org/10.1101/cshperspect.a005736
60. Kappel D, Habenschuss S, Legenstein R, Maass W (2015) Network plasticity as Bayesian inference. PLOS Comput Biol 11(11): e1004,485. https://doi.org/10.1371/journal.pcbi.1004485
61. Billaudelle S, Cramer B, Petrovici MA, Schreiber K, Kappel D, Schemmel J, Meier K (2021) Structural plasticity on an accelerated analog neuromorphic hardware system. Neural Netw 133:11–20. https://doi.org/10.1016/j.neunet.2020.09.024
62. George R, Indiveri G, Vassanelli S (2017) Activity dependent structural plasticity in neuromorphic systems. In: IEEE Biomedical Circuits and Systems Conference (BioCAS). IEEE, pp 1–4. https://doi.org/10.1109/BIOCAS.2017.8325074
63. Van Ooyen A, Butz-Ostendorf M (2017) The rewiring brain: a computational approach to structural plasticity in the adult brain. Academic Press. https://doi.org/10.1016/B978-0-12-803784-3.00033-0
64. Iglesias J, Eriksson J, Grize F, Tomassini M, Villa AEP (2005) Dynamics of pruning in simulated large-scale spiking neural networks. Biosystems 79(1-3):11–20. https://doi.org/10.1016/j.biosystems.2004.09.016
65. Hawkins J, Ahmad S (2016) Why neurons have thousands of synapses, a theory of sequence memory in neocortex. Front Neural Circuits, 23. https://doi.org/10.3389/fncir.2016.00023
66. Roy S, Banerjee A, Basu A (2014) Liquid state machine with dendritically enhanced readout for low-power, neuromorphic vlsi implementations. IEEE Trans Biomed Circuits Syst 8(5):681–695. https://doi.org/10.1109/TBCAS.2014.2362969
67. Bourjaily MA, Miller P (2011) Synaptic plasticity and connectivity requirements to produce stimulus-pair specific responses in recurrent networks of spiking neurons. PLOS Comput Biol 7(2):e1001,091. https://doi.org/10.1371/journal.pcbi.1001091
68. Butz M, Steenbuck ID, Van Ooyen A (2014) Homeostatic structural plasticity increases the efficiency of small-world networks. Front Synaptic Neurosci 6:7. https://doi.org/10.3389/fnsyn.2014.00007
69. Fauth M, Tetzlaff C (2016) Opposing effects of neuronal activity on structural plasticity. Front Neuroanat 10. https://doi.org/10.3389/fnana.2016.00075
70. Diaz-Pier S, Naveau M, Butz-Ostendorf M, Morrison A (2016) Automatic generation of connectivity for large-scale neuronal network models through structural plasticity. Front Neuroanat 10:57. https://doi.org/10.3389/fnana.2016.00057
71. Nowke C, Diaz-Pier S, Weyers B, Hentschel B, Morrison A, Kuhlen TW, Peyser A (2018) Toward rigorous parameterization of underconstrained neural network models through interactive visualization and steering of connectivity generation. Front Neuroinform 12:32. https://doi.org/10.3389/fninf.2018.00032
72. Hebb DO (1949) The organisation of behaviour: A neuropsychological theory. Wiley. https://doi.org/10.4324/9781410612403
73. Magee JC, Grienberger C (2020) Synaptic plasticity forms and functions. Annu Rev Neurosci 43(1):95–117. https://doi.org/10.1146/annurev-neuro-090919-022842
74. Morrison A, Diesmann M, Gerstner W (2008) Phenomenological models of synaptic plasticity based on spike timing. Biol Cybern 98(6):459–478. https://doi.org/10.1007/s00422-008-0233-1
75. Okabe S, Kim HD, Miwa A, Kuriu T, Okado H (1999) Continual remodeling of postsynaptic density and its regulation by synaptic activity. Nature Neurosci 2(9):804–811
76. Tsodyks MV, Markram H (1997) The neural code between neocortical pyramidal neurons depends on neurotransmitter release probability. Proc Natl Acad Sci 94:719–723. https://doi.org/10.1073/pnas.94.2.719
77. Markram H, Wang Y, Tsodyks M (1998) Differential signaling via the same axon of neocortical pyramidal neurons. Proc Natl Acad Sci 95(9):5323–5328. https://doi.org/10.1073/pnas.95.9.5323
78. Gupta A, Wang Y, Markram H (2000) Organizing principles for a diversity of GABAergic interneurons and synapses in the neocortex. Science 287(5451):273–278. https://doi.org/10.1126/science.287.5451.273

79. Zucker RS, Regehr WG (2002) Short-term synaptic plasticity. Annu Rev Physiol 64(1): 355–405. https://doi.org/10.1146/annurev.physiol.64.092501.114547
80. Markram H, Lübke J, Frotscher M, Roth A, Sakmann B (1997) Physiology and anatomy of synaptic connections between thick tufted pyramidal neurones in the developing rat neocortex. J Physiol 500(2):409–440. https://doi.org/10.1113/jphysiol.1997.sp022031
81. Bi Gq, Poo Mm (1998) Synaptic modifications in cultured hippocampal neurons: dependence on spike timing, synaptic strength, and postsynaptic cell type. J Neurosci 18(24):10,464–10,472. https://doi.org/10.1523/JNEUROSCI.18-24-10464.1998
82. Caporale N, Dan Y (2008) Spike timing–dependent plasticity: a Hebbian learning rule. Annu Rev Neurosci 31:25–46. https://doi.org/10.1146/annurev.neuro.31.060407.125639
83. Markram H, Gerstner W, Sjöström PJ (2012) Spike-timing-dependent plasticity: a comprehensive overview. Front Synaptic Neurosci 4:2. https://doi.org/10.3389/fnsyn.2012.00002
84. Brzosko Z, Mierau SB, Paulsen O (2019) Neuromodulation of spike-timing-dependent plasticity: past, present, and future. Neuron 103(4):563–581. https://doi.org/10.1016/j.neuron.2019.05.041
85. Li Y, Zhong Y, Zhang J, Xu L, Wang Q, Sun H, Tong H, Cheng X, Miao X (2014) Activity-dependent synaptic plasticity of a chalcogenide electronic synapse for neuromorphic systems. Sci Rep 4(1):4906. https://doi.org/10.1038/srep04906
86. Morrison A, Aertsen A, Diesmann M (2007) Spike-timing-dependent plasticity in balanced random networks. Neural Comput 19(6): 1437–1467. https://doi.org/10.1162/neco.2007.19.6.1437
87. Van Rossum MC, Bi GQ, Turrigiano GG (2000) Stable Hebbian learning from spike timing-dependent plasticity. J Neurosci 20(23):8812–8821. https://doi.org/10.1523/JNEUROSCI.20-23-08812.2000
88. Burkitt AN, Meffin H, Grayden DB (2004) Spike-timing-dependent plasticity: the relationship to rate-based learning for models with weight dynamics determined by a stable fixed point. Neural Comput 16(5):885–940. https://doi.org/10.1162/089976604773135041
89. Clopath C, Gerstner W (2010) Voltage and spike timing interact in STDP–a unified model. Front Synaptic Neurosci 2:25. https://doi.org/10.3389/fnsyn.2010.00025
90. Clopath C, Büsing L, Vasilaki E, Gerstner W (2010) Connectivity reflects coding: a model of voltage-based STDP with homeostasis. Nature Neurosci 13:344–352. https://doi.org/10.1038/nn.2479
91. Ngezahayo A, Schachner M, Artola A (2000) Synaptic activity modulates the induction of bidirectional synaptic changes in adult mouse hippocampus. J Neurosci 20(7):2451–2458. https://doi.org/10.1523/JNEUROSCI.20-07-02451.2000
92. Sjöström PJ, Turrigiano GG, Nelson SB (2001) Rate, timing, and cooperativity jointly determine cortical synaptic plasticity. Neuron 32(6):1149–1164. https://doi.org/10.1016/S0896-6273(01)00542-6
93. Urbanczik R, Senn W (2014) Learning by the dendritic prediction of somatic spiking. Neuron 81(3):521–528. https://doi.org/10.1016/j.neuron.2013.11.030
94. Artola A, Bröcher S, Singer W (1990) Different voltage dependent thresholds for inducing long-term depression and long-term potentiation in slices of rat visual cortex. Nature 347:69–72. https://doi.org/10.1038/347069a0
95. Pawlak V, Wickens JR, Kirkwood A, Kerr JND (2010) Timing is not everything: neuromodulation opens the STDP gate. Front Synaptic Neurosci 2:146. https://doi.org/10.3389/fnsyn.2010.00146
96. Wörgötter F, Porr B (2005) Temporal sequence learning, prediction, and control: a review of different models and their relation to biological mechanisms. Neural Comput 17(2):245–319. https://doi.org/10.1162/0899766053011555
97. Stapmanns J, Hahne J, Helias M, Bolten M, Diesmann M, Dahmen D (2021) Event-based update of synapses in voltage-based learning rules. Front Neuroinform 15. https://doi.org/10.3389/fninf.2021.609147
98. Morrison A, Mehring C, Geisel T, Aertsen A, Diesmann M (2005) Advancing the boundaries of high-connectivity network simulation with distributed computing. Neural Comput 17(8):1776–1801. https://doi.org/10.1162/0899766054026648
99. Weidel P, Duarte R, Morrison A (2021) Unsupervised learning and clustered connectivity enhance reinforcement learning in spiking neural networks. Front Comput Neurosci 15:18. https://doi.org/10.3389/fncom.2021.543872

100. Vogels TP, Sprekeler H, Zenke F, Clopath C, Gerstner W (2011) Inhibitory plasticity balances excitation and inhibition in sensory pathways and memory networks. Science 334(6062):1569–1573. https://doi.org/10.1126/science.1211095
101. Jordan J, Schmidt M, Senn W, Petrovici MA (2021) Evolving interpretable plasticity for spiking networks. eLife 10:e66,273. https://doi.org/10.7554/eLife.66273
102. Barrett AB, Billings GO, Morris RGM, Van Rossum MCW (2009) State based model of long-term potentiation and synaptic tagging and capture. PLOS Comput Biol 5(1): e1000,259. https://doi.org/10.1371/journal.pcbi.1000259
103. Mongillo G, Barak O, Tsodyks M (2008) Synaptic theory of working memory. Science 319(5869):1543–1546. https://doi.org/10.1126/science.1150769
104. Koch C, Laurent G (1999) Complexity and the nervous system. Science 284(5411): 96–98. https://doi.org/10.1126/science.284.5411.96
105. Duarte R, Morrison A (2019) Leveraging heterogeneity for neural computation with fading memory in layer 2/3 cortical microcircuits. PLOS Comput Biol 15(4): e1006,781. https://doi.org/10.1371/journal.pcbi.1006781
106. Parpura V, Zorec R (2010) Gliotransmission: exocytotic release from astrocytes. Brain Res Rev 63(1-2):83–92. https://doi.org/10.1016/j.brainresrev.2009.11.008
107. Girard P, Hupé JM, Bullier J (2001) Feedforward and feedback connections between areas V1 and V2 of the monkey have similar rapid conduction velocities. J Neurophysiol 85(3): 1328–1331. https://doi.org/10.1152/jn.2001.85.3.1328
108. Liewald D, Miller R, Logothetis N, Wagner HJ, Schüz A (2014) Distribution of axon diameters in cortical white matter: an electron-microscopic study on three human brains and a macaque. Biol Cybern 108(5): 541–557. https://doi.org/10.1007/s00422-014-0626-2
109. Muller L, Chavane F, Reynolds J, Sejnowski TJ (2018) Cortical travelling waves: mechanisms and computational principles. Nature Rev Neurosci 19(5):255–268. https://doi.org/10.1038/nrn.2018.20
110. Song S, Sjöström P, Reigl M, Nelson S, Chklovskii D (2005) Highly nonrandom features of synaptic connectivity in local cortical circuits. PLOS Comput Biol 3(3):e68. https://doi.org/10.1371/journal.pbio.0030068
111. Lefort S, Tomm C, Sarria JCF, Petersen CCH (2009) The excitatory neuronal network of the C2 barrel column in mouse primary somatosensory cortex. Neuron 61:301–316. https://doi.org/10.1016/j.neuron.2008.12.020
112. Koulakov AA, Hromadka T, Zador AM (2009) Correlated connectivity and the distribution of firing rates in the neocortex. J Neurosci 29(12):3685–3694. https://doi.org/10.1523/JNEUROSCI.4500-08.2009
113. Avermann M, Tomm C, Mateo C, Gerstner W, Petersen CCH (2012) Microcircuits of excitatory and inhibitory neurons in layer 2/3 of mouse barrel cortex. J Neurophysiol 107(11):3116–3134. https://doi.org/10.1152/jn.00917.2011
114. Ikegaya Y, Sasaki T, Ishikawa D, Honma N, Tao K, Takahashi N, Minamisawa G, Ujita S, Matsuki N (2013) Interpyramid spike transmission stabilizes the sparseness of recurrent network activity. Cereb Cortex 23(2): 293–304. https://doi.org/10.1093/cercor/bhs006
115. Kuhn A, Aertsen A, Rotter S (2004) Neuronal integration of synaptic input in the fluctuation-driven regime. J Neurosci 24(10):2345–2356. https://doi.org/10.1523/JNEUROSCI.3349-03.2004
116. Roxin A, Brunel N, Hansel D, Mongillo G, Van Vreeswijk C (2011) On the distribution of firing rates in networks of cortical neurons. J Neurosci 31(45):16,217–16,226. https://doi.org/10.1523/jneurosci.1677-11.2011
117. Brunel N, Hakim V (1999) Fast global oscillations in networks of integrate-and-fire neurons with low firing rates. Neural Comput 11(7):1621–1671. https://doi.org/10.1162/089976699300016179
118. Kampa BM, Letzkus JJ, Stuart GJ (2007) Dendritic mechanisms controlling spike-timing-dependent synaptic plasticity. Trends Neurosci 30(9):456–463. https://doi.org/10.1016/j.tins.2007.06.010
119. Feldmeyer D, Egger V, Lübke J, Sakmann B (1999) Reliable synaptic connections between pairs of excitatory layer 4 neurones within a single 'barrel' of developing rat somatosensory cortex. J Physiol 521(1): 169–190. https://doi.org/10.1111/j.1469-7793.1999.00169.x
120. Feldmeyer D, Lübke J, Silver RA, Sakmann B (2002) Synaptic connections between layer 4 spiny neurone-layer 2/3 pyramidal cell pairs in juvenile rat barrel cortex: physiology and anatomy of interlaminar signalling within a cortical column. J Physiol 538(3):803–822.

https://doi.org/10.1113/jphysiol.2001.012959

121. Feldmeyer D, Lübke J, Sakmann B (2006) Efficacy and connectivity of intracolumnar pairs of layer 2/3 pyramidal cells in the barrel cortex of juvenile rats. J Physiol 575:583–602. https://doi.org/10.1113/jphysiol.2006.105106
122. Stepanyants A, Hirsch J, Martinez LM, Kisvarday ZF, Ferecsko AS, Chklovskii DB (2008) Local potential connectivity in cat primary visual cortex. Cereb Cortex 18(1): 13–28. https://doi.org/10.1093/cercor/bhm027
123. Roxin A (2011) The role of degree distribution in shaping the dynamics in networks of sparsely connected spiking neurons. Front Comput Neurosci 5(8). https://doi.org/10.3389/fncom.2011.00008
124. Van Vreeswijk C, Sompolinsky H (1998) Chaotic balanced state in a model of cortical circuits. Neural Comput 10(6):1321–1371. https://doi.org/10.1162/089976698300017214
125. Tsodyks M, Mitkov I, Sompolinsky H (1993) Pattern of synchrony in inhomogeneous networks of oscillators with pulse interactions. Phys Rev Lett 71(8). https://doi.org/10.1103/PhysRevLett.71.1280
126. Golomb D, Rinzel J (1993) Dynamics of globally coupled inhibitory neurons with heterogeneity. Phys Rev Lett 48(6):4810–4814. https://doi.org/10.1103/PhysRevE.48.4810
127. Neltner L, Hansel D, Mato G, Meunier C (2000) Synchrony in heterogeneous networks of spiking neurons. Neural Comput 12(7): 1607–1641. https://doi.org/10.1162/089976600300015286
128. Denker M, Timme M, Diesmann M, Wolf F, Geisel T (2004) Breaking synchrony by heterogeneity in complex networks. Phys Rev Lett 92(7):074,103–1–074,103–4. https://doi.org/10.1103/PhysRevLett.92.074103
129. Mejias JF, Longtin A (2012) Optimal heterogeneity for coding in spiking neural networks. Phys Rev Lett 108. https://doi.org/10.1103/PhysRevLett.108.228102
130. Pfeil T, Jordan J, Tetzlaff T, Grübl A, Schemmel J, Diesmann M, Meier K (2016) Effect of heterogeneity on decorrelation mechanisms in spiking neural networks: A neuromorphic-hardware study. Phys Rev X 6:021,023. https://doi.org/10.1103/PhysRevX.6.021023
131. Stocks NG (2000) Suprathreshold stochastic resonance in multilevel threshold systems. Phys Rev Lett 84(11):2310. https://doi.org/10.1103/PhysRevLett.84.2310
132. Shamir M, Sompolinsky H (2006) Implications of neuronal diversity on population coding. Neural Comput 18(8):1951–1986. https://doi.org/10.1162/neco.2006.18.8.1951
133. Chelaru MI, Dragoi V (2008) Efficient coding in heterogeneous neuronal populations. Proc Natl Acad Sci 105(42). https://doi.org/10.1073/pnas.0807744105
134. Osborne LC, Palmer SE, Lisberger SG, Bialek W (2008) The neural basis for combinatorial coding in a cortical population response. J Neurosci 28(50):13,522–13,531. https://doi.org/10.1523/JNEUROSCI.4390-08.2008
135. Padmanabhan K, Urban NN (2010) Intrinsic biophysical diversity decorrelates neuronal firing while increasing information content. Nature Neurosci 13(10):1276–1282. https://doi.org/10.1038/nn.2630
136. Marsat G, Maler L (2010) Neural heterogeneity and efficient population codes for communication signals. J Neurophysiol 104(5): 2543–2555. https://doi.org/10.1152/jn.00256.2010
137. Holmstrom LA, Eeuwes LBM, Roberts PD, Portfors CV (2010) Efficient encoding of vocalizations in the auditory midbrain. J Neurosci 30(3):802–819. https://doi.org/10.1523/JNEUROSCI.1964-09.2010
138. Yim MY, Aertsen A, Rotter S (2013) Impact of intrinsic biophysical diversity on the activity of spiking neurons. Phys Rev E 87. https://doi.org/10.1103/PhysRevE.87.032710
139. Lengler J, Jug F, Steger A (2013) Reliable neuronal systems: The importance of heterogeneity. PLOS One https://doi.org/10.1371/journal.pone.0080694
140. Mejias JF, Longtin A (2014) Differential effects of excitatory and inhibitory heterogeneity on the gain and asynchronous state of sparse cortical networks. Front Comput Neurosci 8. https://doi.org/10.3389/fncom.2014.00107
141. Battaglia D, Karagiannis A, Gallopin T, Gutch HW, Cauli B (2013) Beyond the frontiers of neuronal types. Front Neural Circuits 7:13. https://doi.org/10.3389/fncir.2013.00013
142. Buzsáki G, Mizuseki K (2014) The log-dynamic brain: how skewed distributions affect network operations. Nature Rev Neurosci 15(4):264–278. https://doi.org/10.1038/nrn3687
143. Robinson PA, Gao X, Han Y (2021) Relationships between lognormal distributions of

neural properties, activity, criticality, and connectivity. Biol Cybern 115(2):121–130. https://doi.org/10.1007/s00422-021-00871-z

144. Morales-Gregorio A, Van Meegen A, Van Albada SJ (2022) Ubiquitous lognormal distribution of neuron densities across mammalian cerebral cortex. bioRxiv. https://doi.org/10.1101/2022.03.17.480842

145. Prinz AA, Bucher D, Marder E (2004) Similar network activity from disparate circuit parameters. Nature Neurosci 7:1345–1352. https://doi.org/10.1038/nn1352

146. Achard P, De Schutter E (2006) Complex parameter landscape for a complex neuron model. PLOS Comput Biol 2(7):e94. https://doi.org/10.1371/journal.pcbi.0020094

147. Bahuguna J, Tetzlaff T, Kumar A, Kotaleski JH, Morrison A (2017) Homologous basal ganglia network models in physiological and Parkinsonian conditions. Front Comput Neurosci 11:79. https://doi.org/10.3389/fncom.2017.00079

148. Abbott LF, Nelson SB (2000) Synaptic plasticity: taming the beast. Nature Neurosci 3:1178–1183. https://doi.org/10.1038/81453

149. Tetzlaff C, Kolodziejski C, Timme M, Wörgötter F (2011) Synaptic scaling in combination with many generic plasticity mechanisms stabilizes circuit connectivity. Front Comput Neurosci 5:47. https://doi.org/10.3389/fncom.2011.00047

150. Eliasmith C, Anderson CH (2002) Neural engineering: computation, representation, and dynamics in neurobiological systems. MIT Press. https://mitpress.mit.edu/9780262050715/

151. Bellec G, Scherr F, Subramoney A, Hajek E, Salaj D, Legenstein R, Maass W (2020) A solution to the learning dilemma for recurrent networks of spiking neurons. Nature Commun, 3625. https://doi.org/10.1038/s41467-020-17236-y

152. Furber S (2016) Large-scale neuromorphic computing systems. J Neural Eng 13(5):051,001. https://doi.org/10.1088/1741-2560/13/5/051001

153. Wunderlich T, Kungl AF, Müller E, Hartel A, Stradmann Y, Aamir SA, Grübl A, Heimbrecht A, Schreiber K, Stöckel D, Pehle C, Billaudelle S, Kiene G, Mauch C, Schemmel J, Meier K, Petrovici MA (2019) Demonstrating advantages of neuromorphic computation: A pilot study. Front Neurosci 13. https://doi.org/10.3389/fnins.2019.00260

154. Zaytsev YV, Morrison A (2013) Increasing quality and managing complexity in neuroinformatics software development with continuous integration. Front Neuroinform 6:31. https://doi.org/10.3389/fninf.2012.00031

155. Plesser HE (2018) Reproducibility vs. replicability: a brief history of a confused terminology. Front Neuroinform 11:76. https://doi.org/10.3389/fninf.2017.00076

156. Goodman SN, Fanelli D, Ioannidis JPA (2016) What does research reproducibility mean? Sci Transl Med 8(341). https://doi.org/10.1126/scitranslmed.aaf5027

157. Van Albada SJ, Kunkel S, Morrison A, Diesmann M (2014) Integrating brain structure and dynamics on supercomputers. In: Brain-inspired computing, pp 22–32. https://doi.org/10.1007/978-3-319-12084-3_3

158. Kurth AC, Senk J, Terhorst D, Finnerty J, Diesmann M (2022) Sub-realtime simulation of a neuronal network of natural density. Neuromorph Comput Eng 2(2):021,001. https://doi.org/10.1088/2634-4386/ac55fc

159. Hanuschkin A, Kunkel S, Helias M, Morrison A, Diesmann M (2010) A general and efficient method for incorporating precise spike times in globally time-driven simulations. Front Neuroinform 4. https://doi.org/10.3389/fninf.2010.00113

160. Teramae JN, Tsubo Y, Fukai T (2012) Optimal spike-based communication in excitable networks with strong-sparse and weak-dense links. Sci Rep 2:485. https://doi.org/10.1038/srep00485

161. Malkin SL, Kim KK, Tikhonov DB, Zaitsev AV (2014) Properties of spontaneous and miniature excitatory postsynaptic currents in neurons of the rat prefrontal cortex. J Evol Biochem Physiol 50(6):506–514. https://doi.org/10.1134/S0022093014060052

162. Dasbach S, Tetzlaff T, Diesmann M, Senk J (2021) Dynamical characteristics of recurrent neuronal networks are robust against low synaptic weight resolution. Font Neurosci 15. https://doi.org/10.3389/fnins.2021.757790

163. Pfeil T, Potjans TC, Schrader S, Potjans W, Schemmel J, Diesmann M, Meier K (2012) Is a 4-bit synaptic weight resolution enough?–constraints on enabling spike-timing dependent plasticity in neuromorphic hardware.

Front Neurosci 6:90. https://doi.org/10.3389/fnins.2012.00090

164. Nordlie E, Gewaltig MO, Plesser HE (2009) Towards reproducible descriptions of neuronal network models. PLOS Comput Biol 5(8):1–18. https://doi.org/10.1371/journal.pcbi.1000456

165. Blundell I, Brette R, Cleland TA, Close TG, Coca D, Davison AP, Diaz-Pier S, Fernandez Musoles C, Gleeson P, Goodman DFM, et al (2018) Code generation in computational neuroscience: a review of tools and techniques. Front Neuroinform 12:68. https://doi.org/10.3389/fninf.2018.00068

166. McDougal RA, Morse TM, Carnevale T, Marenco L, Wang R, Migliore M, Miller PL, Shepherd GM, Hines ML (2016) Twenty years of ModelDB and beyond: building essential modeling tools for the future of neuroscience. J Comput Neurosci 42(1):1–10. https://doi.org/10.1007/s10827-016-0623-7

167. Gleeson P, Cantarelli M, Marin B, Quintana A, Earnshaw M, Sadeh S, Piasini E, Birgiolas J, Cannon RC, Cayco-Gajic NA, et al (2019) Open Source Brain: a collaborative resource for visualizing, analyzing, simulating, and developing standardized models of neurons and circuits. Neuron 103(3):395–411. https://doi.org/10.1016/j.neuron.2019.05.019

168. Holtmaat A, Bonhoeffer T, Chow DK, Chuckowree J, De Paola V, Hofer SB, Hübener M, Keck T, Knott G, Lee WCA, et al. (2009) Long-term, high-resolution imaging in the mouse neocortex through a chronic cranial window. Nat Protoc 4(8): 1128–1144. https://doi.org/10.1038/nprot.2009.89

169. Foxley S, Sampathkumar V, De Andrade V, Trinkle S, Sorokina A, Norwood K, La Riviere P, Kasthuri N (2021) Multi-modal imaging of a single mouse brain over five orders of magnitude of resolution. NeuroImage 238:118,250. https://doi.org/10.1016/j.neuroimage.2021.118250

170. Chatfield C (1995) Model uncertainty, data mining and statistical inference. J Roy Stat Soc A (Stat Soc) 158(3):419. https://doi.org/10.2307/2983440

Chapter 16

Computational Methods and Algorithms to Segment and Model Recurrent Bursting Events in Long-Time Series Springer Nature Book about Neuromethods

L. Zonca, E. Dossi, N. Rouach, and D. Holcman

Abstract

Long-time series of neuronal recordings are resulting from the activity of connected neuronal networks. Yet how neuronal properties can be extracted remains empirical. We review here the data analysis based on network models to recover physiological parameters from electrophysiological and calcium recordings in neurons and astrocytes. After, we present the recording techniques and activation events, such as burst and interburst and Up and Down states. We then describe time-series segmentation methods developed to detect and to segment these events. To interpret the statistics extracted from time series, we present computational models of neuronal populations based on synaptic short-term plasticity and after hyperpolarization. We discuss how these models are calibrated so that they can reproduce the statistics observed in the experimental time series. They serve to extract specific parameters by comparing numerical and experimental statistical moment or entire distributions. Finally, we discus cases where calibrated models are used to predict the selective impact of some parameters on the circuit behavior, properties that would otherwise be difficult to dissect experimentally.

Key words Signal segmentation, Algorithms, Computational models, Deconvolution, Nano-pipette, Bursting, Correlation, AfterHyperPolarization, Synaptic short-term plasticity, Neuronal network, Parameter estimation, Optimization

1 Introduction

Neuronal recordings reveal complex network activation patterns. Although these patterns are clearly appearing in the data, we still need to segment them. This is the preliminary step for analyzing, decoding, and thus better clarifying how neuronal networks impact the brain function [1]. Electrophysiological recordings with patch-clamp [2] provide a direct access to the voltage dynamics of the soma or dendrites of a single neuron (Fig. 1). On the one hand, this technique revealed transient behaviors, such as spiking [3, 4], persistent fluctuations of the membrane potential called Up and Down

Joachim H. R. Lübke and Astrid Rollenhagen (eds.), *New Aspects in Analyzing the Synaptic Organization of the Brain*, Neuromethods, vol. 212, https://doi.org/10.1007/978-1-0716-4019-7_16,

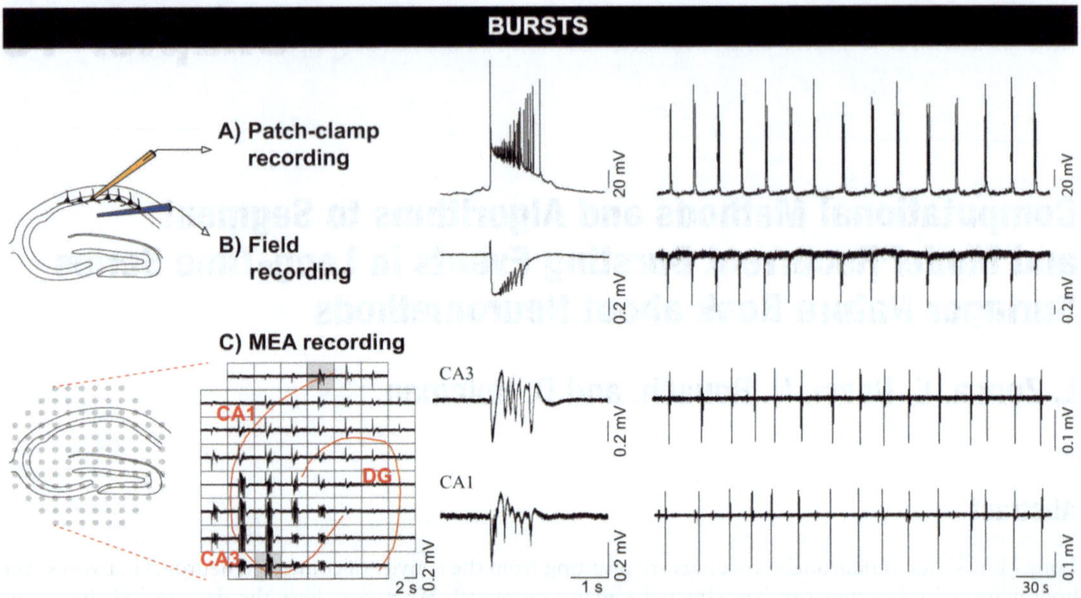

Fig. 1 Bursting activity recorded in mouse hippocampal slices with different electrophysiology techniques. (**a**) Patch-clamp recording for a pipette penetrating the cell. (**b**) Local field potential (LFP) when the tip of the pipette is immersed in the bath. (**c**) Multi-electrode array (MEA) recordings consisting of multiple electrodes recording the sum for the activity of small and large neuronal networks. This recording allows to follow the center of an epileptic focus. Left, schematic representations of patch-clamp and field pipette positions for recording and MEA array. Middle, enlarged view of single bursts recorded with the three techniques. Right, representative continuous recordings of recurrent bursts (5-minutes time windows). (Modified from [20], and unpublished data)

states [5] or bursting [6]. On the other hand, indirect recordings of neuronal activity such as local field potentials or multiple electrode array (MEA), are used to record population ensembles containing tens to millions of neurons and thus providing access to global activity (Fig. 1).

Another method of probing network activity consists in recording transient calcium events based on recording fluorescent dyes. Although with a much coarser time-resolution, calcium recordings reveal spatial synchrony between neurons [7, 8]. The neuronal code embedded into the signal from single recordings contains several components, mixing intrinsic background intrinsic fluctuations and instrumental noise. In that context, spontaneous events, resulting from the background fluctuations, can be difficult to differentiate from evoked events [9]. To overcome theses difficulties, methods are combining signal processing to extract statistics from the data with the help of mathematical modeling to interpret data. We propose to review here some of these approaches.

The present review is organized as follows: first, we describe briefly the electrophysiological and calcium fluorescence recording techniques, as well as the signals they provide. Second, we present time segmentation procedures to distinguish large amplitude events, characterized by sharp transitions, from the background noise. We recall how to construct and interpret the statistics obtained from the segmentation, with a focus on the bursting events but also on the inter-events intervals. The present approach is generic as it is used to analyze spontaneous or evoked activity. Ongoing cortical neuronal activity in the absence of sensory inputs or with sensory stimulation often reveals similar distributions [9]. Finally, in the third part, we review parsimonious models based on short-term synaptic plasticity, offering a reduced number of variables. These models are used to interpret the distributions of burst and interburst intervals and Up and Down states. Once calibrated, these models allow to evaluate role of specific variables, which could be otherwise difficult to disentangle experimentally.

2 Electrophysiological and Calcium Time Series Recordings

2.1 Patch-Clamp, Local Field Potential, and MEA

The activity of the brain spans over multiple temporal and spatial scales, that require an adequate set of technologies to be addressed. In the past decade, robust recording approaches have revealed transient events in individual neurons and their organized network. Electrophysiology recordings capture a wide spectrum of neuronal phenomena, from single neuron spiking to slow and fast network oscillations. Different electrophysiological techniques can be used to study brain neuronal activity, among which patch-clamp, local field potential (LFP) or multi-electrode array (MEA) recordings.

The patch-clamp technique is based on the use of a glass micropipette to form high resistance seals on tiny patches of cell membrane [10], allowing high-resolution recording of currents from single ionic channels, or from the entire population of ionic channels, on a cell membrane [11]. This powerful technique thus provides intracellular recordings of neuronal activity at the single cell level, but its use is limited to a few neurons per experiments. During patch-clamp, an electrode is inserted in the soma of a chosen neuron, and is now routinely used to obtain long-time stable recordings for hours. Examples of double recordings in the hippocampus show a first patch electrode inside a neuron (Fig. 1a) and a field recording, called local field potential (LFP), where the pipette is simply immersed in the tissue (Fig. 1b). We recall that LFP recording is an electrophysiology technique that focuses on the activity of large populations of cells. Indeed, LFP data mirror the spatially weighted electrical activity in the proximity of the recording electrode [12, 13], including excitatory and inhibitory post-synaptic potentials, non-synaptic calcium spikes as well as afterpotential of somatodendritic spikes and voltage-gated membrane oscillations [14–16]. Thus, LFP measurements detect a variety of subthreshold activity close to the electrode and provide a view of local extracellular network activity at a single site [12]. From long-time recordings, it is possible to sample short bursting events of few seconds, with different electric signatures depending on the recording techniques, but with similar time scale.

The key advantage of multi-electrode array (MEA) recording compared to techniques mentioned above is the capability to record (and stimulate) neuronal activity (extracellular local field potentials and extracellular action potentials) at multiple sites simultaneously [15], thus allowing the study of neuronal network activity pattern and propagation. Microelectrodes embedded in large arrays detect changes in the extracellular field caused by ionic flows through the closest neurons and nearby cells [16, 17], and the amplitude and shape of detected signals depend on the orientation and distance between the neuronal sources and the recording electrodes [15, 18] (Fig. 1c). Based on the spatial precision gained by the MEA, events with decaying amplitudes can be tracked, enabling to find regions where a burst can spread and thus and to study for example epileptic focus. MEAs have a similar time-resolution as patch-clamp and LFP but the burst dynamics results from the summation of many groups of neurons. Other improved recording technics include tetrodes to record extracellular field potentials invivo, multiple electrodes, such as the neuropixels probe features using 960 recording electrodes, allowing to record more that 700 neurons simultaneously from brain regions in an awake mouse. Similarly, neuroSeeker probes can also record from 1000 of electrodes in vivo [19]. To conclude transient patterns such as bursting can vary depending on the recording methods, but not the associated time scale, allowing to define a precise time interval and thus to recover duration statistics.

2.2 Bursting Events Revealed by Calcium Dyes

Contrary to the fast time-resolved electrophysiological recordings presented in Fig. 1, bursting events are also revealed by slower calcium transient events. This approach is well discussed in [21] for calcium imaging and also most recently for voltage dyes [22–24]. The time courses $f_{dyes}(t)$ of fast events results from the convolution between the dyes response time scale $g_{intrin}(t)$ and the signal itself $C(t)$, which could last much longer than the event. This slower response leads to possible temporal imprecision in the beginning and the end of a burst or a spike, as shown in Fig. 2. The general problem is often to recover the concentration or voltage from the slow dyes signal. This question is well known in signal processing and is equivalent to finding the kernel C of the following equation

$$f_{dyes}(t) = \int_0^t C(s) g_{intrin}(t-s)\,ds + N(t), \tag{1}$$

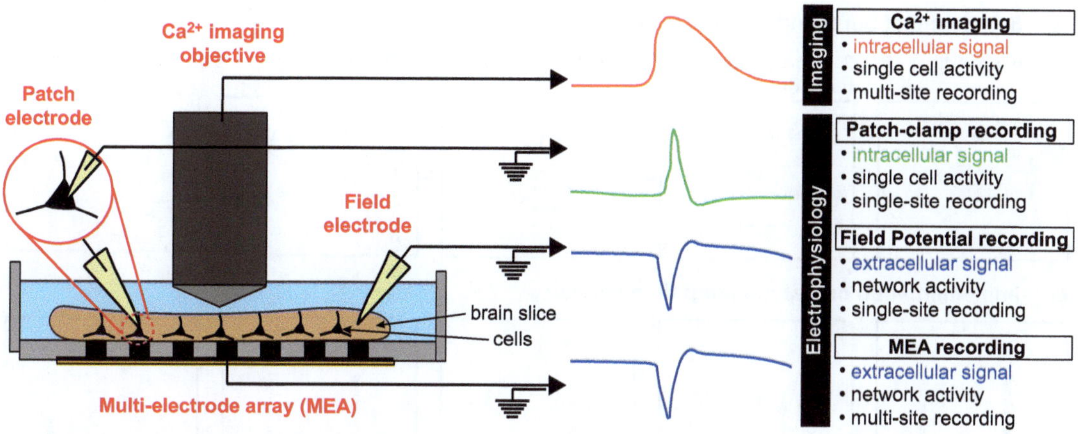

Fig. 2 Comparison between voltage and dyes recording techniques. Patch-clamp, field potential and MEA recordings versus the slow Ca^{2+} imaging event

where $N(t)$ is a noise source. A solution based on fitting the empirical curve by an analytical function was proposed in [25]. This allows to recover a fast voltage from slower dye responses.

3 Segmentating, Correlation, and Causal Analysis of Long Time-Series

Segmenting transient bursting events automatically in electrophysiological recordings is the first step for collecting large amount of samples and for building the associated statistics. Segmentation methods [26] consist in isolating transient events of interest such as spikes, bursts, or interburst intervals. This segmentation is possible due to the structure of a burst made of sharp transients with a significant in amplitude change compared to the background signal. Segmenting events with an amplitude similar to the background signal would be quite difficult, unless a difference can be identified in the frequency domain, which is the case in ripples [16]. Several examples of fast transients with large amplitudes that differentiate bursting from the background signal and also hyperpolarization from resting phases are shown in Fig. 1. We shall now describe the successive steps to segment bursting events in patch-clamp and MEA time series.

3.1 Burst Detection and Segmentation in Patch-Clamp Recordings

The segmentation procedure is to detect burst, AHP, and interburst intervals in neuronal patch-clamp recordings.

3.1.1 Initial Low-Pass Filtering

The first step of burst segmentation consists in removing from the membrane potential $s(t)$ the high frequency components often generated by fast opening and closing of voltage activated channels

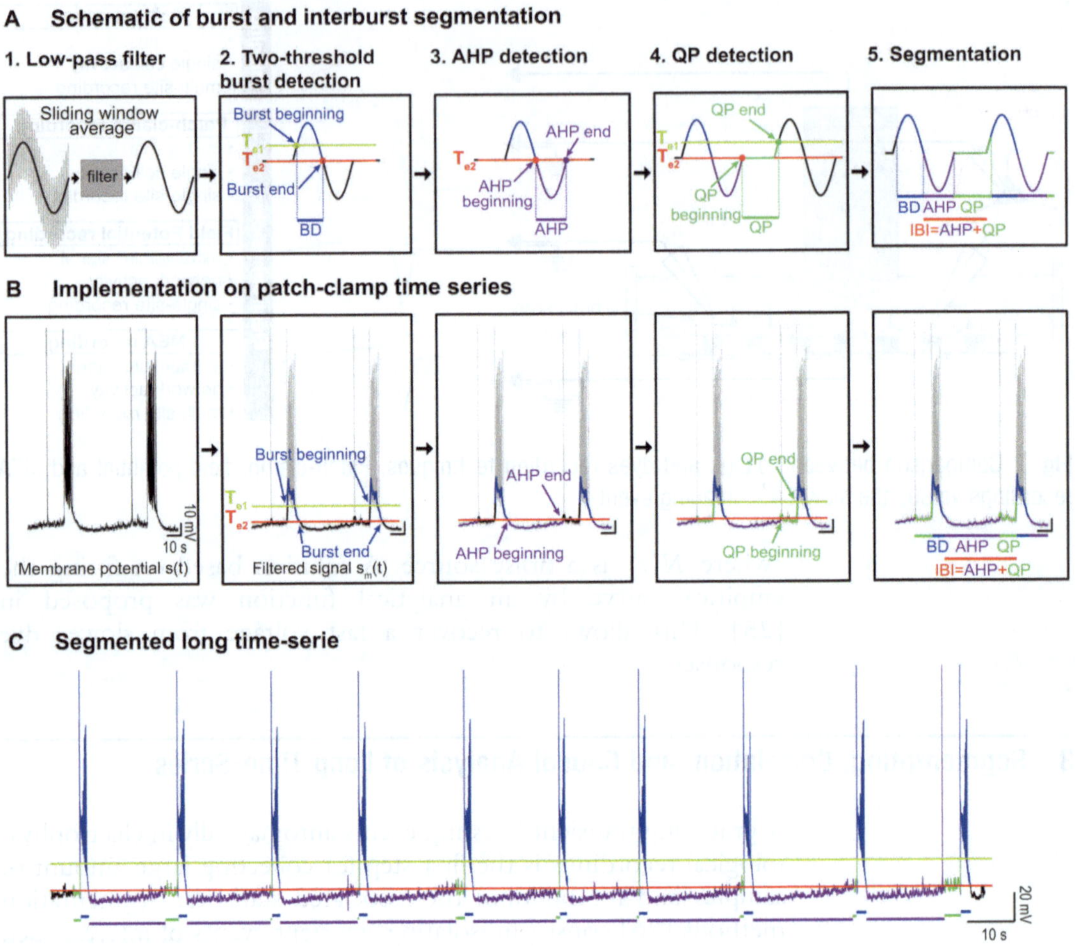

Fig. 3 Burst detection principle. (**a**) Burst and interburst segmentation algorithm. 1. The membrane potential signal is low-pass filtered using a sliding window of length $T_W = 1s$. 2. A burst is detected using the two thresholds $T_{e,1}$ (beginning) and $T_{e,2}$ (end) on the filtered signal s_m. 3. AHP begins at the end of the burst and lasts until s_m is back above resting membrane potential ($= T_{e,2}$) 4. QP begins at the end of AHP until the next burst. 5. Extraction of burst durations, (BD,blue), AHP duration (magenta) and quiescent phases (QP, green). IBI (red) is composed of AHP and QP. (**b**) Implementation on patch-clamp time-series. Step by step application of the algorithm described in A on electrophysiological data. (**c**) 200s electrophysiological recording segmented into burst (blue) AHP (magenta) and QP (green) periods using the algorithm described in A

[2, 27]. This step is achieved by applying a low-pass filter to $s(t)$, using a sliding window average: for each time point t, the filtered signal $s_m(t)$ (Fig. 3a1) is obtained by taking the average of the unfiltered signal s over a window of size $T_w = 1s$ centered at time t:

$$s_m(t) = \frac{1}{T_w} \sum_{i=t-T_w/2}^{t+T_w/2} s(i). \tag{2}$$

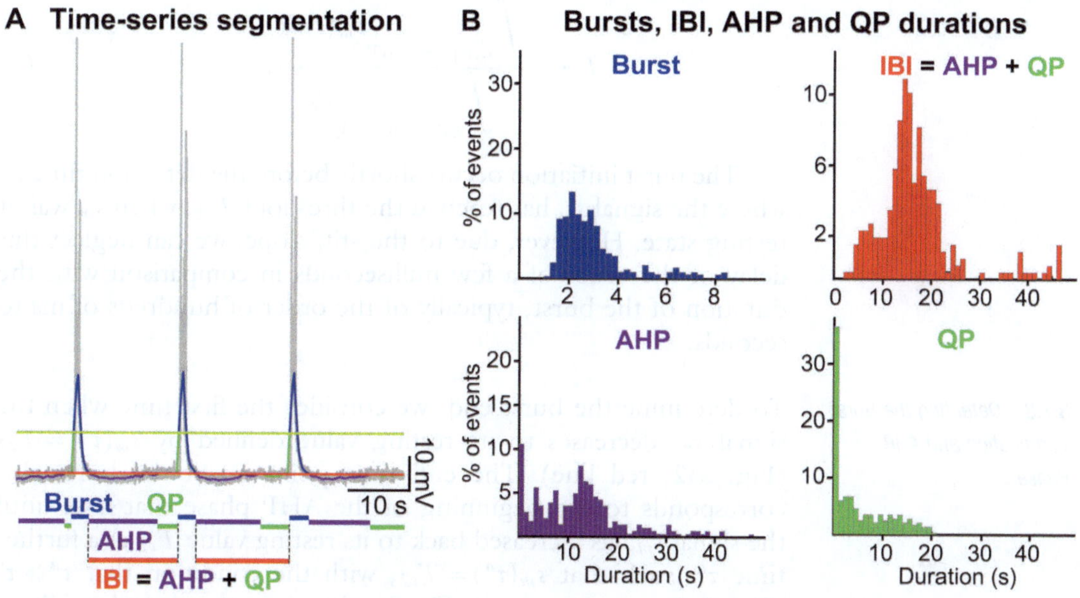

Fig. 4 (**a**) Segmentation of the traces into burst (blue), AHP (pink) and QP (green), scalebar: 10 s, 10 mV. (**b**) Distributions of bursts (blue), IBI = AHP+QP (red), AHP (pink) and QP (green) durations

This procedure is easy to use, fast at low computational cost. Other averaging methods include SavitzkyGolay or kernel filters. In addition, increasing the filter frequency is simply obtained by increasing the size of the window. This is an advantage compared to classical low-pass filters such as Butterworth versus Chebichev, Hermite [28, 29], where the order of the filter has to be specified, as well as a cutting frequency. These filters also introduce some distortions in the signal.

3.1.2 Detecting the Burst Initiation

The burst initiation is detected by finding the first time τ^i when the filtered signal s_m reaches a threshold T_{e1} (Fig. 4a2, light green), i.e.,

$$s_m(\tau^i) = T_{e1}. \tag{3}$$

The threshold T_{e1} is chosen relatively to the amplitude of the signal variations as

$$T_{e1} = \frac{T_{e2} + \max_{t\in[0,T]} s_m(t)}{2}, \tag{4}$$

which uses the average between the maximum value of the filtered signal and the resting membrane potential T_{e2} (red line). In general, there is no automatic procedure to detect the membrane resting potential, which can vary over different traces.

When the voltage contains mostly bursting and AHP events, with a baseline in the range $R_{mp} = [-55 - 65]$mV, one possibility is to impose that the threshold T_{e2} is obtained by averaging the signal inside the range R_{mp} that is

$$T_{e2} = \frac{\int_{s_m \in [-55\ -65]} s_m(s) ds}{\int_{s_m \in [-55\ -65]} ds}. \tag{5}$$

The burst initiation occurs shortly before the detection time τ^i, where the signal s_m has reached the threshold T_{e1}, when s_m was at resting state. However, due to the stiff slope, we can neglect this delay of the order of a few milliseconds in comparison with the duration of the burst, typically of the order of hundreds of ms to seconds.

3.1.3 Detecting the Burst Termination and AHP Phase

To determine the burst end, we consider the first time when the signal s_m decreases to its resting value defined by $s_m(\tau^e) = T_{e2}$ (Fig. 3a2, red line). The end time point τ^e of the burst also corresponds to the beginning of the AHP phase that lasts until the signal s_m has increased back to its resting value T_{e2} at a further time τ^a, such that $s_m(\tau^a) = T_{e2}$, with the condition that $\tau^a > \tau^e$ (Fig. 3a3, magenta trace). Finally, this time segmentation allows to define a quiescence phase QP, between the end of the AHP at time τ^a and the initiation of the next burst (Fig. 3a4, green trace). In summary, the nth burst duration is the difference between the termination and initiation times

$$BD_n = \tau_n^e - \tau_n^i. \tag{6}$$

The collection of durations $(BD_1, ..BD_n)$ (blue), the AHP durations $AHPD_n = \tau_n^a - \tau_n^e$, (magenta), and the quiescent phases $QP_n = \tau_{n+1}^i - \tau_n^a$ (green) define the empirical distributions of segmented traces (Figs. 3c and 4a). These distributions are obtained from segmentating path-clamp recordings, as shown in Fig. 4b.

3.2 Segmenting Bursting in MEA Recordings

The method described above can also be applied with little modifications to MEA recordings. However, the AHP cannot be identified in these recordings, allowing to simplify the segmentation procedure. We explain briefly how the steps have to be adjusted: The starting point is to apply a low-pass filter to the input field recordings using a sliding time window with a length of $T_w = 0.4s$ leading to an output signal $s_{m,MEA}$. The bursts and interburst intervals are then detected using the absolute value of the filtered signal $|s_{m,MEA}|$. Since the MEA baseline is zero, the burst initiation time τ^i is obtained when the signal $|s_{m,MEA}|$ reaches one third of its maximum value

$$|s_{m,MEA}|(\tau^i) = \frac{\max_{[0,T]}(|s_{m,MEA}|)}{3}. \tag{7}$$

Similarly, the end time τ^e of the burst is found under the condition

$$|s_{m,MEA}|(\tau^e) = T_{end} = \frac{max(|s_{m,MEA}|)}{15}. \quad (8)$$

This empirical threshold T_{end} allows to identify the end of the burst. It corresponds to the time at which the signal enters into a fluctuation regime around zero, where the noise is dominant in the interburst interval. In general, it would be possible to define this threshold by considering the standard deviation of the signal in between bursts:

$$\sigma_{IBI} = \frac{\int_{s_{m,MEA}(s) \in IBI_i} (s_{m,MEA}(s) - \overline{s_{m,MEA}})^2 ds}{\int_{s_{m,MEA}(s) \in IBI_i} ds}, \quad (9)$$

where IBI_i are the interburst intervals. This standard deviation can be computed by a simple thresholding with one threshold to detect the burst events. The complementary of these events define the signal where the standard deviation should be computed. In practice, we can replace $\frac{max(|s_{m,MEA}|)}{15}$ by σ_{IBI}. The Matlab code for the above segmentation algorithms is available for download at bionewmetrics.segmentation_tools. Each algorithm is provided as a Matlab function that takes the data time-series as an input and where the threshold parameters and filtering can be adjusted.

3.3 Segmenting Intracellular Recordings into Up and Down States

We present here another application of the segmentation method associated with the changes of the membrane potential of connected neurons, that can oscillate between two dominant values called Up (depolarized) and a Down (hyperpolarized) states. These two-state oscillation is due to intrinsic channel properties combined with synaptic dynamics [30]. In the past 20 years, neuronal information processing in the cortex has been associated with these Up and Down state dynamics [8, 9]: Up-state is often associated with a direct processing of an external stimulation response such as auditory, visual, or somatosensory (whisker deviation), but could also appear spontaneously [7] in the form of an organized activity [31]. Up and Down states are recognized by the two peaks in the voltage distribution. Interestingly, segmenting intracellular recordings into periods of Up and Down states falls into the method described above. This segmentation allows to build the statistics for their durations, as shown for the auditory and barrel cortex (Fig. 5a–b). Interestingly, the segmentation of the intracellular recordings for various neuronal types revealed that the distributions of time in the Up state shows multiple peaks [32, 33], a sign of a non-Poissonian distribution (Fig. 5c–d left). This is however not the case for the distribution of times in the Down state (right). Extracting from the statistics of these distribution relevant

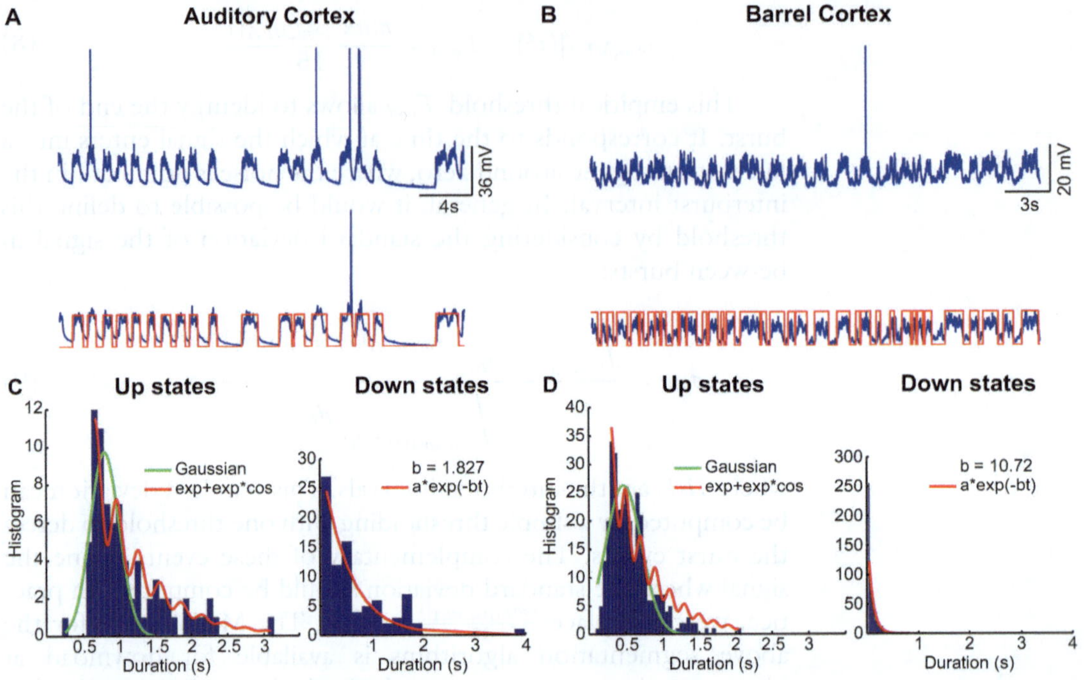

Fig. 5 (**a**–**b**) Segmentation between Up and Down states (red) of intracellular recordings (blue) in the auditory (**a**) and barrel (**b**) cortex, data from [9]. (**c**–**d**) Distributions of the Up (left) and Down (right) states durations fitted with an analytical curve for both datasets

Table 1
Experimental time-series segmentation parameters

	Patch-clamp	MEA
Low-pass filter window T_w	1s	0.4s
Burst detection threshold T_{e1}	$\frac{T_{e2}+\max_{t\in[0,T]} s_m(t)}{2}$	$\frac{\max_{[0,T]}(\lvert s_{m,MEA}\rvert)}{3}$
End of burst & AHP threshold T_{e2}	$\frac{\int_{s_m\in[-55-65]} s_m(s)ds}{\int_{s_m\in[-55-65]} ds}$	$\frac{max(\lvert s_{m,MEA}\rvert)}{15}$

parameter for neuronal activity will be discuss in Subheading 4. To interpret the fast oscillations in the Up state, a modeling approach is needed as we shall see in Subheading 4 (Table 1).

3.4 Time Correlation Between Induced and Spontaneous Burst and Interburst Events

In this section, we discuss how burst segmentation is used to study the correlation between successive events. We start with burst reverberation [34, 35], which consists in the activation of a single neuron that can trigger the firing of an entire well connected network for seconds due to synaptic facilitation and depression dynamics (Fig. 6a–b). This effect was first reported in cultured islands [36] and then found in slices. Interestingly, the bursting

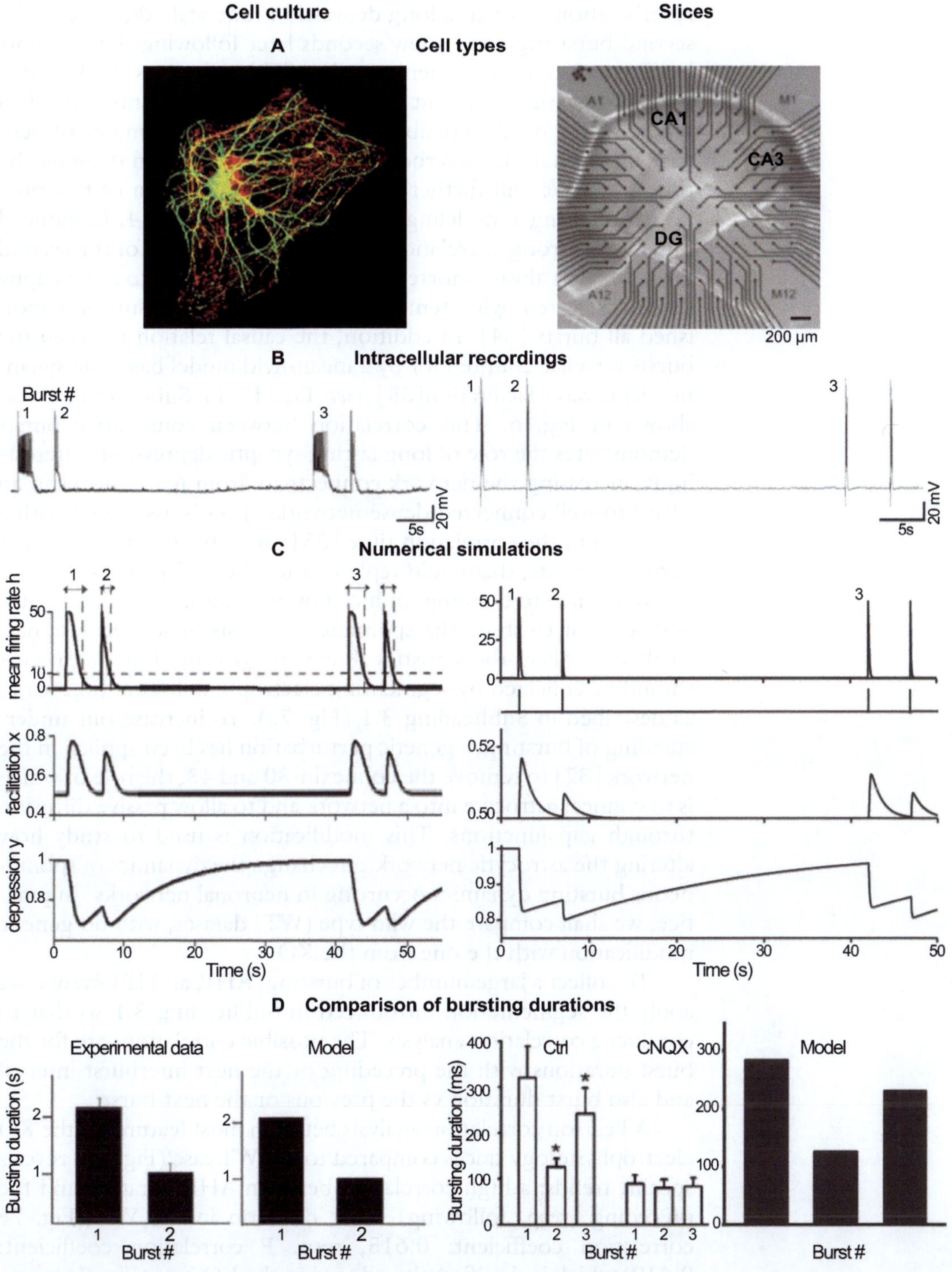

Fig. 6 Bursting reverberation times in neuronal cultures and slices. (**a**) Cultured neurons (left) and mice hippocampal slices (right). (**b**) Intracellular recordings with evoked bursts 5 s apart (bursts 1 and 2) and then 30 s and 35 s later (bursts 3 and 4) in both cell types (cultures and slice). (**c**) Simulations of model 17 reproducing the successive bursts calibrated on both datasets, mean firing rate h (upper), facilitation x (center) and depression y (lower). (**d**) Comparison of the successive burst durations in data and model experiments

reverberation revealed a long depression time scale that affects the second burst triggered a few seconds later following the first one, leading to a short-term memory effect. This effect was also found in slices, but with a different time scale (hundreds of ms instead of seconds, Fig. 6c–d), possibly suggesting that the number of neurons involved in the reverberation could play a role in defining the time scale. We will further discuss the interpretation of this phenomenon using a modeling approach in Subheading 4. In induced bursts, the strong correlation between the durations of the second burst which is always shorter than the first one is due to the synaptic dynamics, as removing temporarily the synaptic transmission abolished all bursts [34]. In addition, the causal relation between the bursts is well accounted for by a mean-field model based on synaptic depression-facilitation[35] (*see* Eq. 17 in Subheading 4) as shown in Fig. 6. This correlation between consecutive bursts demonstrates the role of long lasting synaptic depression. Interestingly, increasing the network connectivity from few neurons in an island to well connected dense networks slices is associated with a reduction of the correlation time [35], probably due to a stronger network activity, that could replete faster the available vesicles.

In parallel to bursting, induced by an external stimulation, it is also relevant to study the spontaneous bursts generated in a network and collect the statistics. For that goal, the bursting events should be collected by segmenting electrophysiological time series as described in Subheading 3.1 (Fig. 7a). To increase our understanding of bursting, a genetic perturbation has been applied in the network [37] to remove the connexin-30 and 43, the role of which is to connect astrocyte into a network and to allow passive diffusion through gap junctions. This modification is used to study how altering the astrocytic network can change the dynamics of spontaneous bursting dynamics occurring in neuronal networks. In practice, we shall compare the wild type (WT) dataset, with no genetic modification with the one from the KO.

To collect a large number of bursting, AHP, and IBI events, we apply the segmentation introduced in Subheading 3.1 so that to conduct a correlation analysis. The possible correlations are for the burst durations with the preceding or the next interburst interval and also burst duration vs the previous or the next burst.

A Pearson correlation analysis between most features in the KO electrophysiology traces compared to the WT case (Fig. 7b) reveals specific trends: a high correlation between AHP duration and the preceding (resp. following) burst duration in the WT (Fig. 7e correlation coefficient: 0.615, resp. F correlation coefficient: 0.619) which is significantly reduced in the KO case (Fig. 7e correlation coefficient: – 0.154, resp. F correlation coefficient: 0.099). Regarding the total interburst duration vs burst durations, the correlation is also reduced in the KO (preceding burst: fig. 7C correlation coefficient: 0.766 for the WT and 0.052 for the KO,

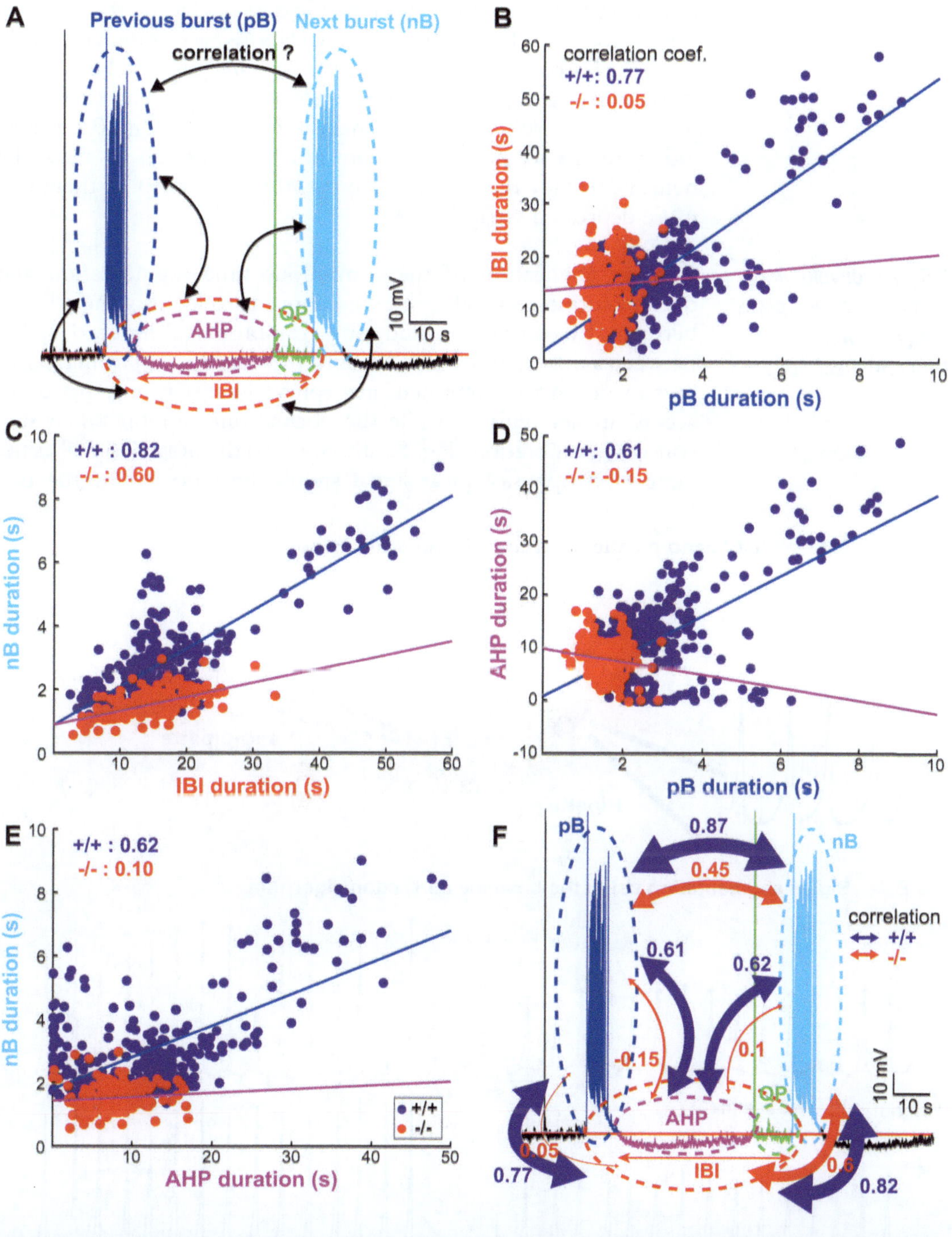

Fig. 7 Correlations between and across burst and interburst intervals. (**a**) Segmented time-series showing the time epochs used for the correlation study. (**b–e**) Correlation plots for Burst occurring in the WT (+/+, purple) and in Connexin KO (−/−, red) between Interburst Interval (IBI) vs previous Burst (pB), next Burst (nB) vs IBI, AHP vs pB and nB vs AHP. (**f**) Correlation coefficients (arrows) between the different epochs for (+/+, purple) and (−/−, red), the arrow width represents the correlation strength

following burst: D correlation coefficient: 0.815 for the WT and 0.596 for the KO). All the correlations are summarized in Fig. 7f. To conclude, the segmentation analysis can be used to study correlation between bursting events, suggesting here that possibly a control feedback mechanism could be lost in KO, resulting in refractory periods that are not long enough to allow the neurons to fully recover after a burst, therefore keeping them in a more depressed synaptic state.

3.5 Deconvolution Kernel Between Spikes Recorded from Different Electrodes

Another application of the continuous time segmentation and deconvolution methods presented above is to identify the relation between the amplitude and time responses of induced spikes between a normal pipette and a nano-pipette [38, 39] from recordings made on the same neuronal soma. Since a nano-pipette can access smaller regions, while the normal one cannot, a first step consists in calibration (Fig. 8a) the small to the normal signal as the same electrophysiological signal should be reported by the two

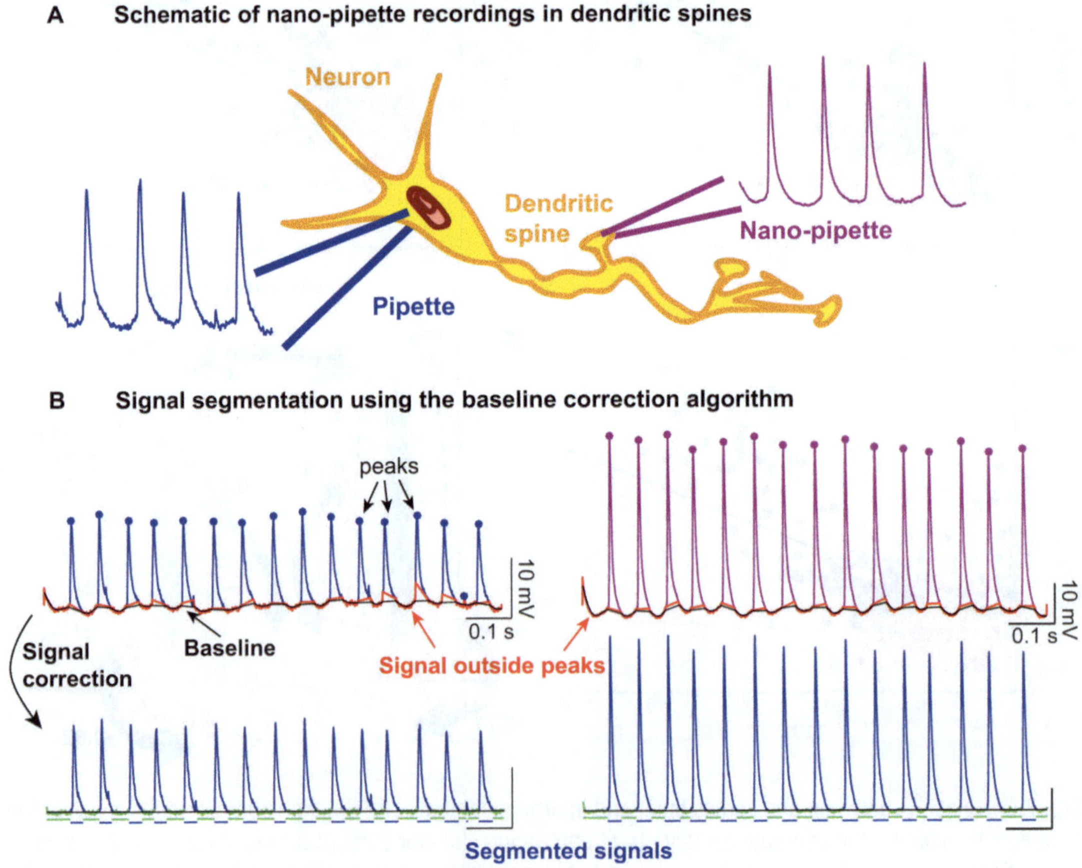

Fig. 8 Recording and segmenting spikes from normal and nano-pipettes. (**a**) Schematic representation of dual recordings with pipettes of different sizes. (**b**) Pre-processing step to remove the low component followed by a segmentation to isolate the spikes

pipettes. We use a causal convolution transformation to map the response of the nano-pipette to the one of the normal pipette. This mapping requires first a pre-processing of the signal to remove the low component and thus align the baselines, which can be done using the method described in Subheading ?? and applying the same segmentation procedure to isolate each spike (Fig. 8b).

The deconvolution procedure consists in finding the kernel function $K(t)$ that sends the response $p_{sm}(t)$ of the nano-pipette to the response of the larger pipette $p_{la}(t)$, so that we have the causal relation

$$p_{la}(t) = \int_0^t K(t-s)p_{sm}(s)ds. \tag{10}$$

The direct Fourier transform can be used here so that for each spike k, we get an approximated kernel K_k obtained by inverting the Fourier transform of relation 10 (Fig. 9a)

$$\hat{K}_k(f) = \frac{\hat{p}_{la}(f)}{\hat{p}_{sm}(f)}. \tag{11}$$

We use the mean of all estimates as an approximation of a stationary kernel:

$$\tilde{K}(t) \sim \frac{1}{N}\sum_{1}^{N} K_k(t). \tag{12}$$

In the case of a non-stationary signal, another possibility would be to average the last $N = 10$ events before, so that the kernel can evolve with time. Finally, the inverse Fourier transform allows to recover the time-dependent kernel.

The method is validated on spikes that did not contribute to find the kernel (Fig. 9b). Once the kernel is found, it can be used to deconvolve the signal recorded by the nano-pipette in nanoscale regions, where the large pipette cannot access. The baseline correction and spike detection were performed using the calcium data analysis toolbox described in the previous section. The code for the deconvolution step and kernel reconstruction is provided here bionewmetrics.deconvolution.

3.6 Conclusion About Segmentation and Toward the Need of Modeling for Interpretation

In the past two decades, automatized electrophysiological time series segmentation of long recordings has been used to reveal various type of voltage, timing, and many more distributions. In addition, the focus of studying fluctuations of the membrane potential revealed how complex dynamics such as Up and Down states are associated with the first step of neuronal encoding of an external stimulation. When the neuronal tissue is studied in the limit where population burst emerges spontaneously, usually by

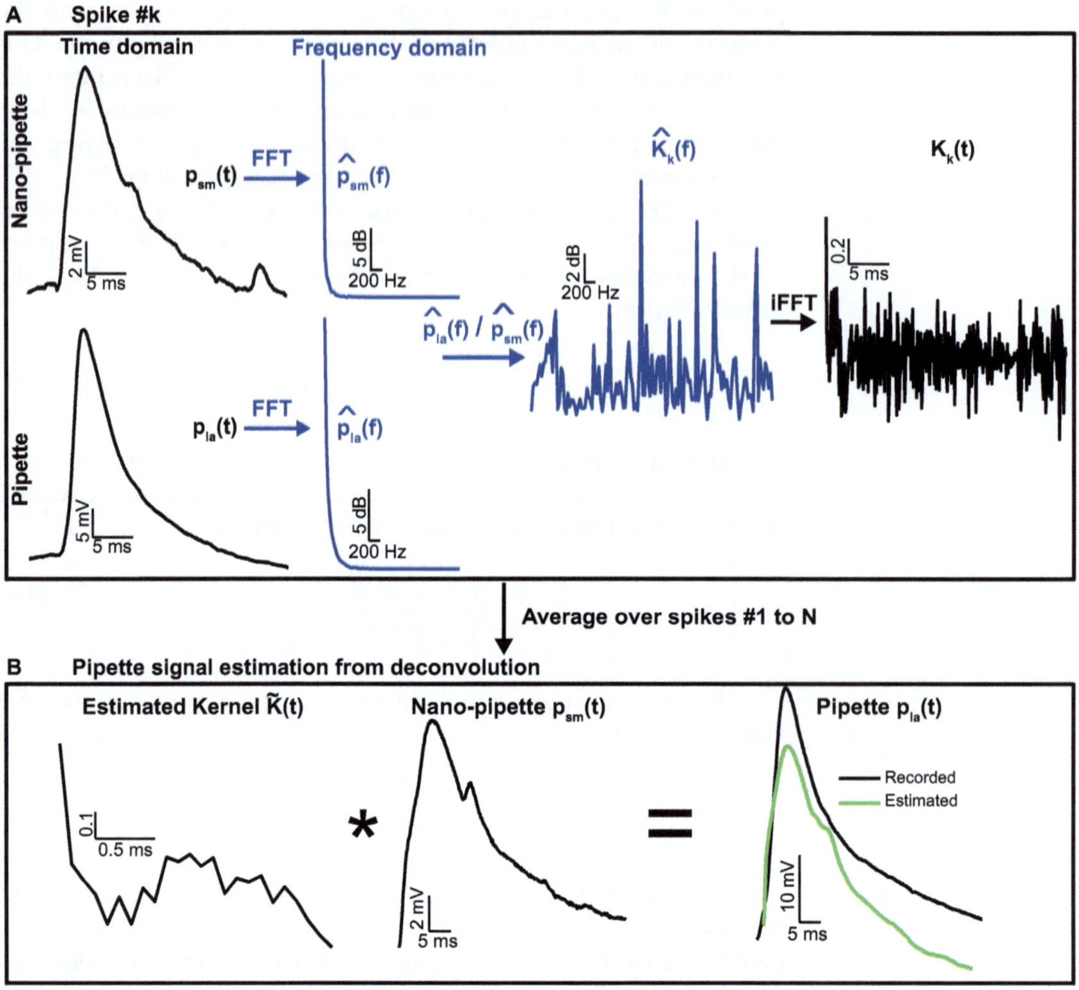

Fig. 9 Deconvolution of segmented spikes between normal and nano-pipettes. (**a**) Estimation of the local convolution kernel K_k from the pipette and nano-pipette signal during spike k. (**b**) Estimated kernel $\tilde{K}(t)$ obtained from averaging the spike 1 to N (left). convolution with the a nano-pipette spike $p_{sm}(t)$ (center) allows to recover the pipette signal (right, green)

switching off the inhibition, their dynamics can reveal some inherent properties of the underlying neuronal mechanism that could be prevalent in pathological conditions.

Collecting local transient duration distributions already revealed several physiological functions: (1) busting activity is more frequent but shorter in epileptic conditions after altering the astrocyte network [20] compared to normal condition. (2) The similarity between Up states responses in vivo induced spontaneously or evoked suggest profound [9] mechanisms of brain representation and replay [7].

Interpreting the statistical properties of bursting, duration of Up states, etc... is often difficult as they result from the combination of multiple mechanisms such as channel activation, neuronal connectivity and synaptic dynamics. The specific role of each of these components can be analyzed with a modeling approach, numerical simulations and also, when possible, by deriving the appropriate formulas that relate the multiple parameters together.

Furthermore, time responses or empirical distributions can provide serious constraints for the model ability to capture the appropriate physiological phenomenology. This is often revealed by the fitting step of analytical or simulated curves derived from the model. An optimal fit procedure should then be used to extract specific parameters of the model, as we shall explain in the next section.

4 Parsimonious Neuronal Network Models Based on Short-Term Synaptic Plasticity

A large class of models have been developed to account for neuronal networks activity, using channels dynamics [2] but also synaptic short-term properties [40, 41]. These models can account for neuronal networks that exhibit a large variety of time-frequency behaviors during spontaneous or induced responses by external stimulations. The time-frequency transient responses are often complex as explored by the time-frequency spectrogram [42] or correlation analysis. They can span a large frequency range, from 0 to hundreds of Hz, as well as time scales from a few milliseconds to several seconds.

Reduced models started with Hodgkin-Huxley [43] based on channel conductances, that summarize the neuronal activity as a single voltage dynamics representing an ensemble of connected neurons. Other models have followed with refined the dynamics according to the physiological properties they needed to account for. Morris-Lecar [44] and Li-Rinzel focus on calcium dynamics and many other variants are summarized in the book [45].

Interestingly, models that account for synaptic properties can already account for a large diversity of physiological behaviors such as synaptic responses, Up-Down states [46], bursting dynamics [35, 47], transient network response to an external stimulation [48], memory capacity retrieval [49] or theta-paced flickering between place-cell-maps in the hippocampus [50]. We present in this section coarse-grained models used to characterize the statistics of bursting, as discussed in the sections above. The models are based on mean-field approximation of neuronal network driven by synaptic short-term plasticity [51].

4.1 Modeling Short-Term Synaptic Depression-Facilitation Between Two Neurons

The initial neuronal synaptic model is based on the biophysical mechanism associated with short-term synaptic depression which results from a lack of releasable presynaptic vesicles. Lack of vesicles prevents the transmission between two connected neurons from occurring [52]. Specifically, each action potential (AP) of a presynaptic neuron triggers a vesicular release with a given probability, usually around $p \sim 0.2$. Thus, to describe the effect of an AP on the post-synaptic neuron, the change in the fraction of available resources (vesicles) is monitored by splitting the resources into three states: an effective state E (vesicle are available), an inactive I state (nothing is left) and a recovered R state (which accounts for the time to reconstitute a pool of readily releasable vesicles) [52]. The equations associated to the synaptic response are

$$\begin{aligned} \frac{dR}{dt} &= \frac{I}{\tau_{rec}} - U_{SE}R\delta(t - t_{AP}) \\ \frac{dE}{dt} &= \frac{E}{\tau_{inact}} + U_{SE}R\delta(t - t_{AP}) \\ I &= 1 - R - E, \end{aligned} \tag{13}$$

where τ_{rec} and τ_{inact} are the time scales for recovery and inactivation respectively and U_{SE} reflects the fraction of used resources at each action potential occurring at time t_{AP}.

4.2 Modeling Short-Term Synaptic Plasticity in a Single Neuron and at the Network Level

The model equations 13 developed for a synaptic interaction can be extended to account for the dynamics of a homogeneously and sufficiently connected neuronal ensemble [40]. In that case, the system becomes

$$\begin{aligned} \tau\frac{dM}{dt} &= -M + JxuM + I_{ext}(t) + \sigma\dot{\xi} \\ \frac{dx}{dt} &= \frac{1-x}{\tau_D} - uxM \\ \frac{du}{dt} &= \frac{U-u}{\tau_f} + U(1-u)M, \end{aligned} \tag{14}$$

where the variable M represents the mean firing rate of the neuronal population, the synaptic efficacy is modulated by the fraction of available synaptic vesicles x and the release probability u can be interpreted as presynaptic calcium accumulation [35]. The external input $I_{ext}(t)$ can vary in time and the additive noise $\sigma\dot{\xi}$ is approximated as a Gaussian.

The release probability u is increased (facilitated) every time a spike is generated and increases the presynaptic firing rate, while synaptic resources decrease. In the absence of any presynaptic firing, the variables x, u, and M recover to their resting values, 1, U and 0 with time constants τ_D, τ_F and τ. The synaptic dynamics can be monitor by the product Jux. Instead of a linear neuronal input, the second term in the first equation of 14 can be replaced by $g(JxuM + I_{ext})$, where $g(z) = a\log(1 + \exp z/b)$, where a and b are constants [49].

4.3 Modeling Short-Term Synaptic Depression to Account for Up and Down State Dynamics

Accounting for depression only, and thus keeping the facilitation constant in the mean-field firing rate eq. 14, with an additive Brownian noise leads to the simplified equations

$$\begin{aligned} \tau\dot{V} &= -V + \mu U w_T \alpha (V-T)^+ + \sqrt{\tau}\sigma\dot{\omega} \\ \dot{\mu} &= \frac{1-\mu}{t_r} - U\mu\alpha(V-T)^+, \end{aligned} \tag{15}$$

where V is the average synaptic input (in mV), $(V-T)^+ = max(V-T, 0)$ the mean firing rate, $\alpha = 1Hz/mV$ a conversion factor, w_T the average synaptic strength, U the vesicle utilization parameter, and t_r the recovery time constant of the synaptic depression. Finally, $\dot{\omega}$ is the normalized white noise and σ its amplitude. Under the condition that the synaptic weight is high enough $w_T > w_c$, where w_c is a critical value above which the dynamics exhibits a bistability between two attractors (a focus and a center) separated by a separatrix and an unstable limit cycle (Fig. 10). The stochastic dynamics allows to create random switches between these states. This dynamics recapitulates Up and Down states driven by short-term plasticity and synaptic noise in cortical neurons [46]. The parameter values used in [46] are summarized in Table 2.

Analysis of model 15 reveals that the time distribution in Up states f_{Up} is not Poissonian [53], but consists in a sum of oscillating exponentials, but a truncation to the first two terms is acceptable:

$$f_{Up}(t) = \sum_{k\geq 0} e^{-\lambda_k t}\cos(\omega_k t + \phi_k) \approx Ae^{-\lambda_0 t} + Be^{-\lambda_1 t}\cos(\omega t + \phi), \tag{16}$$

where A, B are constants and the parameters $\lambda_0, \lambda_1, \omega, \phi$ can be related to the dynamics in the phase-space of the focus attractor in Fig. 10 [54]. A similar oscillation phenomenon was also reported in [55–57]. For the fitting procedure and the range of used parameters, we refer to the procedure described below in Subheading 5 (*see* also Table 3). The analysis of this dynamics further reveals that during Up states, a dominant frequency emerges, which depends on the parameters. These oscillations can be located in the θ or α bands [58], confirming that the Up-state activity generates oscillations, a process that thus could be simply explained by synaptic dynamics.

Table 2
Parameters of model 15

τ	U	J	σ	T	t_r
0.05s	0.5	12.6mV/Hz	2.2mV	2mV	0.8s

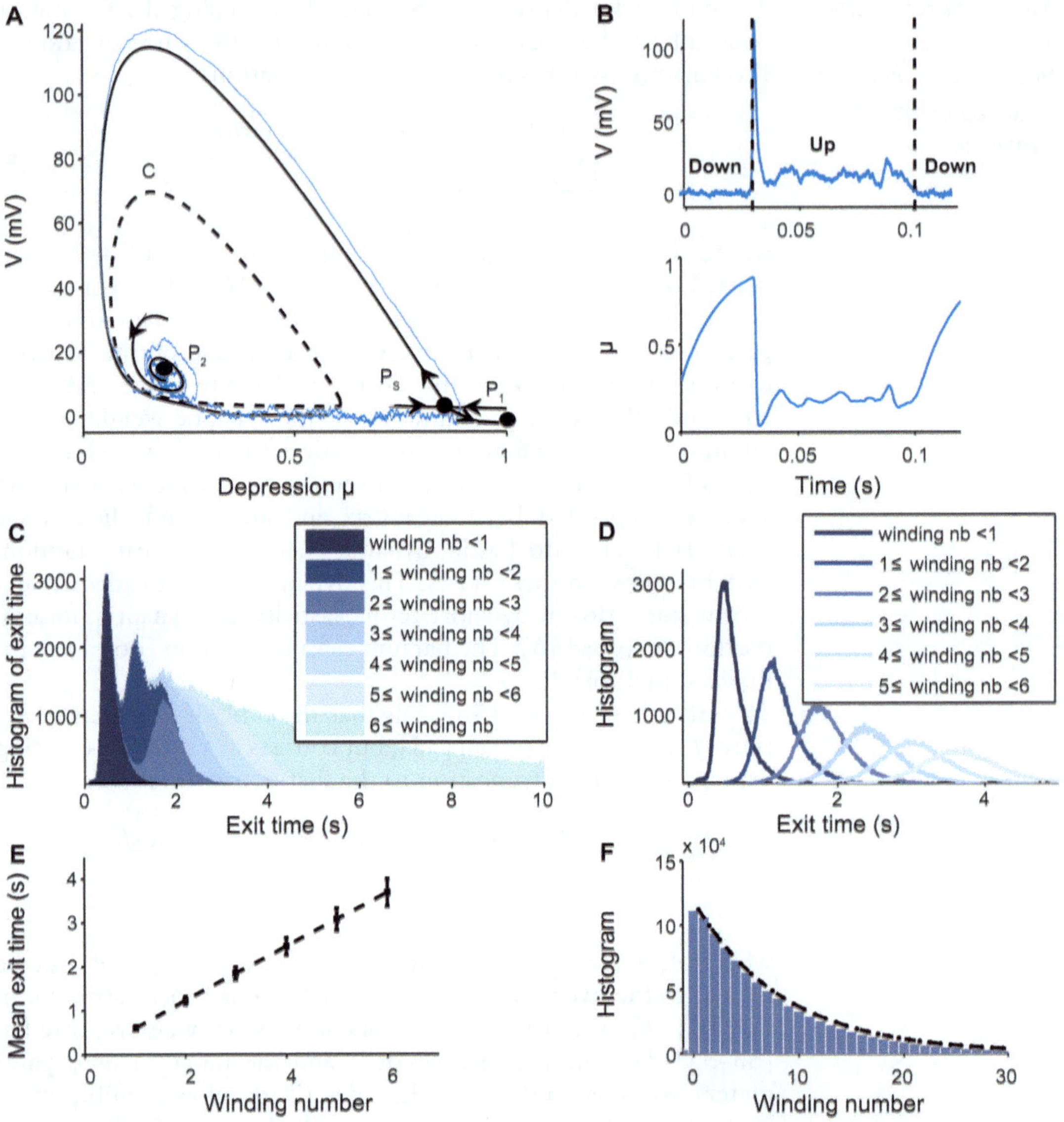

Fig. 10 Up-Down state model: (**a**) Two-dimensional phase-space of model 15 with two attractors (P_1 and P_2) and one saddle-point (P_S). Stochastic trajectories (blue) can escape from P_1 pushed by noise, they follow the unstable manifold of P_S (black curve) and circle around P_2 until they cross the limit cycle (dotted line)and fall back to P_1. (**b**) Stochastic time-series of the two variables V (upper) and μ (lower) exhibiting Up and Down states separated by vertical dotted lines. (**c**) Histogram of the exit times from the attractor point P_2 color coded by the number of revolutions around P_2 called winding number obtained from numerical simulations. (**d**) Histogram of exit times from P_2 conditioned on the winding number. (**e**) Mean exit time as a function of the winding number from numerical simulations. (**f**) Distribution of the total winding number before exit

Table 3
Equation 16 parameters

Recordings	A	B	λ_0	λ_1	ω (Frequency)	ϕ(Phase)
Barrel cortex	54.4438	13.100	1.92	1.99	31.41	-3
Auditory cortex	23.9135	10	1.48	1.87	21.22	-4.3

4.4 Synaptic Short-Term Depression with Facilitation to Model Neuronal Population Bursts

The previous system of Eqs. 15 used to model depressed synapses associated in a well connected neuronal network is insufficient to account for bursting events, which usually require an amplification phase to propagate a single spiking event initiated by one cell into a population response. The facilitation property plays exactly such a role. System 14 can be extended by considering the three variables: the mean firing rate h, the facilitation x, and the depression y, reflecting the fraction of available vesicles. This new system satisfies

$$\begin{aligned} \tau\dot{h} &= -h + Jxyh^{+} + I_{ext}(t) + \sqrt{\tau}\sigma\dot{\omega} \\ \dot{x} &= \frac{X - x}{t_f} + K(1 - x)h^{+} \\ \dot{y} &= \frac{1 - y}{t_r} - Lxyh^{+}, \end{aligned} \tag{17}$$

where $h^{+} = max(h, 0)$ is a linear threshold function of the synaptic current which represents the firing rate. The first equation describes the response of the network ensemble to an external stimulation $I_{ext}(t)$, the term $\dot{\omega}$ is an additive Gaussian noise and σ its amplitude. This term represents fluctuations in the synaptic current (synaptic noise). The second equation describes the facilitation response, and the third one is the depression, similar to system 15. The parameters K and L describe how the firing rate modulates the depression and facilitation variables. The time scales t_f and t_r define the recovery of a synapse from the network activity. Finally, the combined effect of synaptic short-term facilitation and depression on the mean voltage is described by the term Jxy where the parameter J represents the network connectivity (mean number of synapses per neuron) [59].

The dynamical system 17 has been used to study the bursting dynamics observed in an island of cultured neurons when inhibitory neurons are suppressed [35]. The model accounts for short-time memory effects: while a direct stimulation can generate a 2s burst, a second stimulation at 5s interval lead to a reduced 1s burst. However, when the second stimulation occurs 30s later, instead of a reduced second burst, the duration is back to 2s. What is the cause for such effect? can this model explains this property?

System 17 has been used to answer positively this question and to compute the bursting reverberation time T_R [35]. The model could precisely account for this short-term adaptation and showed that the shortest second burst was the result of long recovery of the depression parameter: the second burst was initiated while the depression was still in the recovery phase after 5s, but not after 30s. Fitting the experimental data allowed to estimate the parameters reported in Table 2.

4.5 Estimation of the Burst Duration Based on an Approximated Synaptic Short-Term Depression-Facilitation Model

Although model 17 is nonlinear, it can be used to estimate the bursting duration with respect to the model parameters, and in particular to the synaptic connectivity J. The first step consists in defining the bursting profile using the firing rate variable h: the duration of bursting is defined as the time during which the firing rare h is above a threshold h_{th}. During this reverberation period, the firing rate remains approximately constant in the initial phase of the response, which allows a partial decoupling of the synaptic equations. The approximated system is

$$\begin{aligned} \tau \dot{h} &= -h + Jxyh + H\delta(t - t_{stim}) \\ \dot{x} &= \frac{X - x}{t_f} + K(1 - x)H \\ \dot{y} &= \frac{1 - y}{t_r} - LxyH. \end{aligned} \tag{18}$$

During the early bursting time, we approximate Eqs. 2 and 3 of system 18, by considering the approximation $h(t) \approx H$ at its initial value, leading to this approximated system. This system can be used to estimate the burst duration T_R with respect to the threshold h_{th} by a direct linear integration.

Although this approximation could affect drastically the dynamics, it decouples the equations of system 18 and by resolving them with respect to the variable h, lead to the approximation procedure (supplementary information of [35]). Using the notation $T_R = T_0 = \tau \ln\left(\frac{H}{h_{th}}\right)$ for $J = 0$, the derived expression for the bursting time is

$$T_R(J) = \frac{1 - JX\tau - \sqrt{(J\tau X - 1)^2 - 2J\tau XH(K - LX)\tau \ln\left(\frac{H}{h_{th}}\right)}}{J\tau XH(K - LX)}. \tag{19}$$

Interestingly, comparing simulations with the plot of expression 19 shows that it is a good approximation for $J < 1.9$. However, when $J > 1.9$, the reverberation time is larger than 1s. Thus, the approximation made in Eq. 18 is not valid anymore. Other approximations are discussed in [35]. Interestingly, the formula reveals how each parameter contributes to the bursting time and the analysis further revealed a maximum for the bursting time: when the network is not enough connected, the bursting time increases with J due to the facilitation that triggers a response in a large ensemble of neurons. However, in a too densely connected network, depression dominates, thus decreasing the burst duration.

Finally, it is possible to relate the abstract degree of synaptic connectivity J used in Eqs. 17 to the actual number of synapses by calibrating the model with experimental data: this procedure has been applied in neuronal cultures, where the connectivity $J = 40$ (for $\tau = 0.05$) accounted for approximately 3000 synapses per neuron in hippocampal islands containing 5 to 30 neurons homogeneously connected [35]. Finally, the present analysis based on the depression-facilitation model differs significantly from fitting hidden-Markov model approach [60], where several exponential distributions including the gamma, inverse Gaussian and geometric distributions are used to fit the experimental histograms of bursting time.

To conclude, the present coarse-grained neuronal model provides the basis to fit data and to extract parameters, as opposed to a fitting procedure based on machine learning, where the very large amount of parameters usually prohibit a biophysical interpretation.

4.6 Modeling Burst and Interburst Dynamics Using AfterHyperPolarization and Short-Term Synaptic Plasticity

While the short-term synaptic plasticity model 17 could reproduce with a satisfactory precision the changes in the bursting patterns over tens of seconds, the model could not account for the long interburst durations. To account for long time scale and the transient return of the membrane potential to the resting potential, another mechanism had to be taken into account: indeed this long transient recovery of hundreds of milliseconds is known to result from the AfterHyperpolarization (AHP) mechanism. It is not mediated by synaptic properties, but rather by a combination of calcium and potassium channels with medium to slow timescales [61]. While the bursting time depends on the reverberation events involving depletion of vesicular pools and the refilling time was determinant for the next burst generation, yet the membrane potential has its own independent dynamics. Thus, a new mechanism needed to be incorporated to account for bursting and AHP periods. With such mechanism, the model would then be able to predict not only bursting, interburst interval but also possible correlations that could occur between successive bursts, burst and interburst intervals and between successive interbursts.

Bursting events have been traditional modeled by several abstract variables in a phase-space [45]. Another possibility is to use Na^+ and delayed rectifier K^+ currents to produce basic spiking and add Ca^{2+}-dependent currents to generate bursting. Here we will present an alternative approach that does not account for the direct channels biophysical properties but rather models the phenomenology of AHP, leading to hyperpolarization and a long transient return to equilibrium. Thus instead of adding the channel mechanisms, the effect of AHP directly on the short-term synaptic model 17 is taken into account [62]. This approach requires two parsimonious modifications to account for two features:

1. Decreasing the mean voltage value after bursting. This effect is achieved by modifying the transient behavior: in the model 17, a new equilibrium state is added, representing hyperpolarization (see below).
2. Modifying the voltage timescales to account for the slower AHP dynamics recovery. Two new time constants have to be considered: The slow τ_{sAHP} and the medium τ_{mAHP}. During the period characterized by the medium scale τ_{mAHP}, the voltage is forced to hyperpolarize. While the slow recovery to the resting membrane potential is dominated by the slow τ_{sAHP}.

The new coarse-grained AHP-short-term-plasticity model equations are summarized as follows:

$$\begin{aligned} \tau_0 \dot{h} &= -(h - T_0) + Jxy(h - T_0)^+ + \sqrt{\tau_0}\sigma\dot{\omega} \\ \dot{x} &= \frac{X - x}{\tau_f} + K(1 - x)(h - T_0)^+ \\ \dot{y} &= \frac{1 - y}{\tau_r} - Lxy(h - T_0)^+, \end{aligned} \tag{20}$$

where the variables τ_0 and T_0 are now piecewise constant, depending on the burst decomposition into four phases, illustrated in Fig. 11a. We shall describe the four phases of a bursting-Interburst cycle within such a model.

4.6.1 Step 1: Burst

To emulate a population burst (Fig. 11a, blue) we use model 17. A burst is associated with a decay of the depression variable y and an increase of the facilitation variable x. These processes are governed by the fast time constant τ_0 for the mean voltage variable h which is $\tau_0 = \tau$ while the resting value of h is $T_0 = T = 0$ (*see* Table 5 for the parameters values). In practice, this behavior is defined for a certain configuration of the parameter space: $\{y > \Upsilon_{AHP}$ and $h \geq H_{AHP}$ or $\dot{y} \leq 0\}$ (Fig. 11b, above the purple and orange surfaces).

4.6.2 Step 2: Hyperpolarization

The hyperpolarization phase (Fig. 11a, orange) starts at the end of the burst as the depression variable y reaches its minimum. In the model, this condition is represented in the equation by $y < \frac{1}{1+Lx(h-T_0)}$. To force the voltage to hyperpolarize, the resting value of h has to change transiently to the value $T_0 = T_{AHP} < T$. Finally, the time constant of h changes to a medium value $\tau_0 = \tau_{mAHP}$. This phase lasts as long as $y < \Upsilon_h$ (Fig. 11b, below

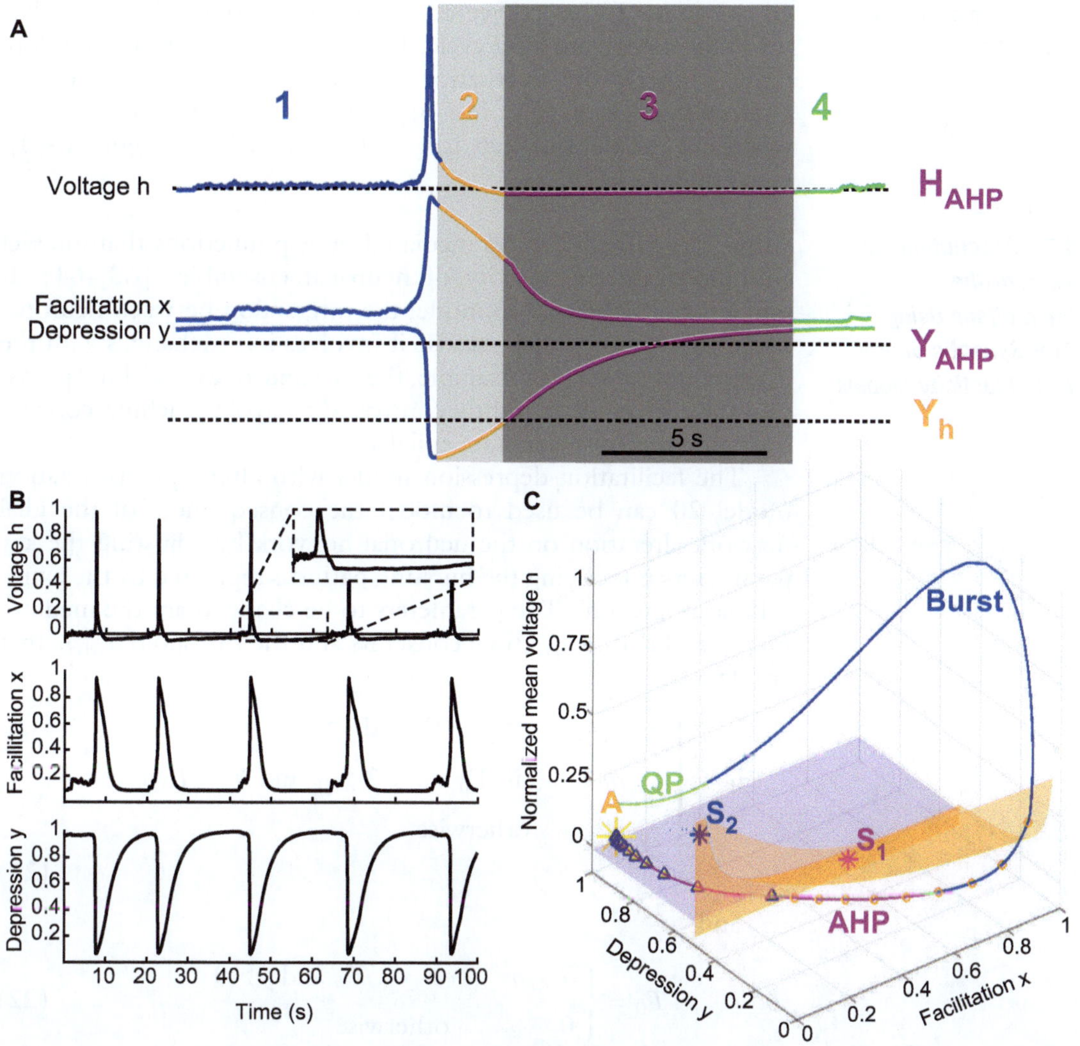

Fig. 11 (**a**) *h*, *x*, and *y* time-series decomposed over the four phases of a burst. (**b**) Stochastic time-series of *h*, *x*, and *y* exhibiting spontaneous bursts followed by AHP periods (inset). (**c**) Phase-space of model 20 with a deterministic burst trajectory decomposed according to the four phases

the orange surface), which is a threshold value chosen empirically to provide enough time for the mean voltage variable h to reach its maximum hyperpolarization value T_{AHP}.

4.6.3 Step 3: Slow Recovery Period

At the end of the hyperpolarization period, to account for the slow recovery phase of h (Fig. 11a–b, purple), the time constant has to change to its slowest value $\tau_0 = \tau_{sAHP}$. Finally, the mean voltage h recovers from hyperpolarization by setting back the resting value to the initial value $T_0 = 0$. These two modifications allow the slow recovery of h to its resting membrane potential. In practice, during this phase, the variable y is still increasing. To determine the end of this phase, each variable h and y must return close to their resting value, that is $\{\Upsilon_{AHP} < y \text{ or } h < H_{AHP}\}$.

4.6.4 Step 4: Quiescent Phase (QP)

Once system 20 has recovered to a its steady state, the network is ready for a next bursting cycle. However, this occurs at a random time, triggered by fluctuations around equilibrium due to the random fluctuation (Fig. 11a–b, green). In this state, the parameter values are the ones of step 1: $\{y > \Upsilon_{AHP}$ and $h \geq H_{AHP}$ or $\dot{y} \leq 0\}$, $\tau_0 = \tau$ and $T_0 = T$.

4.7 Accounting for Neuron-Glia Interactions Using AHP-Synaptic Short-Term Plasticity Models

Astrocytes form networks mediated by gap junctions that interact and modulate the activity of neuronal ensembles [63, 64]. It remains quite difficult to model the interaction between neuronal and astrocyte networks because it involves ion exchanges with the extracellular space. For example, the dynamic of extracellular potassium levels has to be tightly controlled to avoid reaching concentrations that could generate epilepsy.

The facilitation-depression model with afterhyperpolarization model 20 can be used to model the consequences of the glial network alteration on the neuronal network by adjusting the key parameters underlying the bursting patterns regulated by the extracellular potassium. The parameters to be changed are the medium τ_{mAHP} and slow τ_{sAHP} time constants and the threshold T_{AHP} that are defined by

$$\tau_0 = \begin{cases} \tau_{mAHP} & \text{if } y < \Upsilon_h \text{ and } \dot{y} > 0 \\ \tau_{sAHP} & \text{if } \Upsilon_0 < y < \Upsilon_{AHP} \text{ and } h < H_{AHP} \\ \tau_0 & \text{otherwise} \end{cases} \tag{21}$$

and

$$T_0 = \begin{cases} T_{AHP} & \text{if } y < \Upsilon_h \text{ and } \dot{y} > 0 \\ 0 & \text{otherwise} \end{cases}. \tag{22}$$

The adjustment procedure of these parameters is defined as follows: for each parameter set, the histograms of burst and interburst durations are generated by running numerical simulations and by segmenting the simulated time-series as described in Subheading 4.8. Then an optimization procedure is used to explore the parameter space and to determine the optimal set of parameters. This can be done on a wild type (WT) but also on a genetically perturbed mice.

Once the parameters are calibrated, this approach can be used to determine the relative contribution of the changes in either AHP, basal membrane depolarization, synaptic noise or depression to a connexin-deficient astroglial network (i.e., without gap junctions between the astrocytes). This approach allows to quantify the role of each of these parameters on the neuronal activity, by assessing its effect on the neuronal bursting pattern.

This paradigm is thus to run bursting simulations with parameters obtained in the fit of the WT distributions, except for some tested factor (AHP, noise amplitude or depolarization threshold), for which we can use the parameters obtained from the fit of a connexin-deficient condition. To conclude, simulations based on data reveals how a calibrated mean-field model can be used to identify the contribution of astrocytes on a neuronal network by controlling the AHP dynamics.

4.8 Variant of the Segmenting Algorithm Applied to Simulated Time-Series

A variant of the segmentation procedure of the burst and interburst interval adapted for the time-series generated by system 20 is in contrast with the segmentation we presented for experimental time-series in Subheading 3.1. Simulated time-series are generated from Eqs. 20 and the dynamics of the firing rate is driven by the small Brownian noise. Such a model can reproduce the spontaneous bursting dynamics, followed by long AHP refractory periods and a quiescent phase, characterized by local fluctuations of the firing rate around the resting state (Fig. 11c). The adjusted segmentation algorithm, leading to bursting, AHP, and quiescent phase periods is as follows:

- **Burst initiation:** A bursting event is first detected when the variable h reaches a threshold $h(\tau_1) = T_1$, chosen sufficiently high to avoid detecting the small fluctuations due to the noise term. However, contrary electrophysiological time-series (Subheading 3.1), where a burst initial rise is very fast, here we cannot neglect the time delay between the burst initiation and the time τ_1 of burst detection. Thus, the time τ^{ini} of burst initiation is set equal to the last time previous to τ_1, where the variable h equals its resting value, that is $h(\tau^i) = T$ with $\tau^i < \tau_1$.
- **Burst termination:** The termination is detected by finding the time τ_2 such that the variable h passes a second threshold $T_2 < T$, i.e., $h(\tau_2) = T_2$. We apply a similar correction to this time by taking the last time τ^e before τ_2 where $h(\tau^e) = T$ with $\tau^e < \tau_2$.
- **AHP detection:** Similar to the procedure used for patch-clamp recordings (Subheading 3.1), the AHP period starts immediately at the end of the burst at time τ^e, and lasts until time τ_e^a when the voltage h reaches its equilibrium value, i.e., $h(\tau_e^a) = T$ and $\tau_e^a > \tau^e$.
- **QP detection:** The QP period starts at the AHP end time τ_e^a and lasts until the next burst is initiated.

We obtain the relevant statistics for burst, AHP, and QP duration distribution by collecting the durations generated by the simulation based on segmentation procedure describe above. We shall

Table 4
Simulated time-series segmentation parameters 15

Burst detection threshold T_{e1}	$100 + T$
End of burst & AHP threshold T_{e2}	$T - 1$

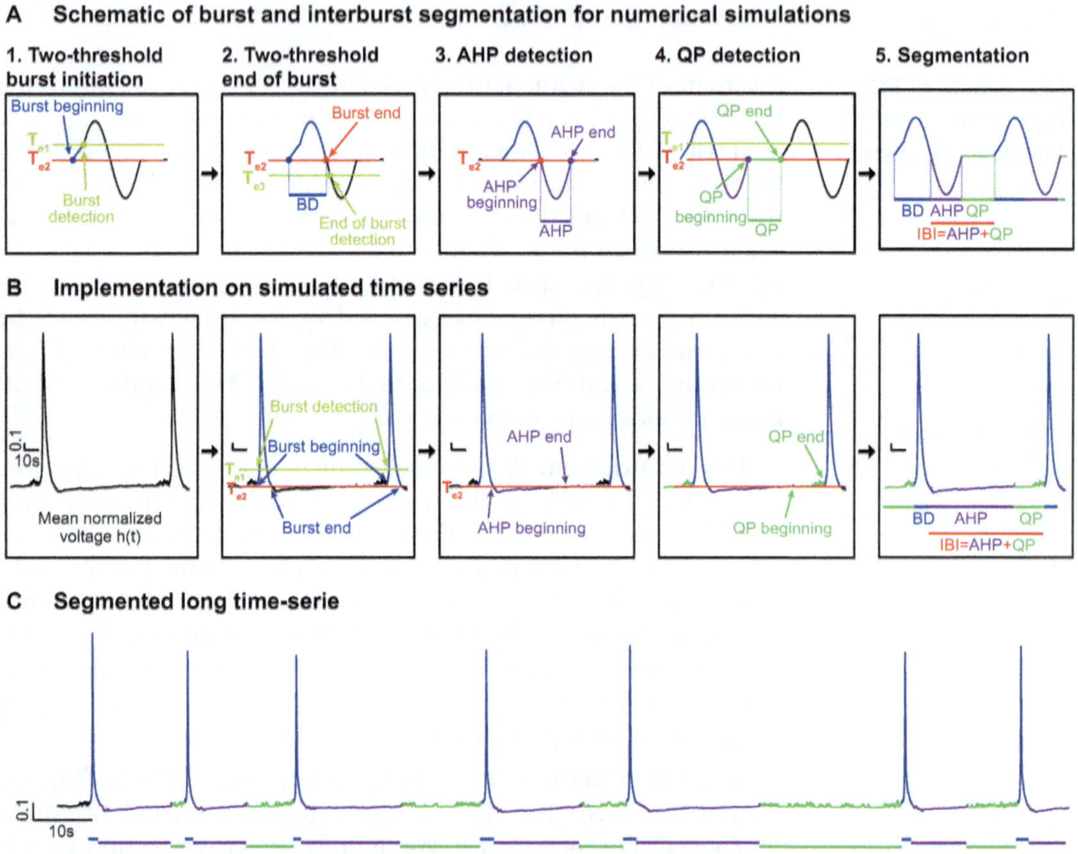

Fig. 12 Segmentation pipeline for the simulated time-series (**a**) Burst and interburst segmentation algorithm. 1. Burst initiation is detected using the two thresholds $T_{e,1}$ (detection) and $T_{e,2}$ (real burst beginning) on the h time-series. 2. End of burst is detected using the two thresholds $T_{e,3}$ (detection) and $T_{e,2}$ (real burst end) on h. 3. AHP begins at the end of the burst and lasts until h is back above resting membrane potential ($= T_{e,2}$) 4. QP begins at the end of AHP until the next burst beginning. 5. Extraction of burst durations, (BD,blue), AHP duration (magenta) and quiescent phases (QP, green). IBI (red) is composed of AHP and QP. (**b**) Implementation on simulated time-series. Step by step application of the algorithm described in A on simulated time-series. (**c**) 200s simulated time-serie of the mean normalized voltage h segmented into burst (blue) AHP (magenta) and QP (green) periods using the algorithm described in A

use them in Subheading 5.1 to calibrate the model parameters, a necessary procedure to evaluate the role of each parameter separately (Table 4). The code to simulated the model 20 and the segmentation of the simulated time-series is available at bionewmetrics.ahp_model (Fig. 12).

4.9 Modeling Slow Rhythmic Oscillations with Two Short-Depression Time Scales for the preBötzinger Complex

Another variant of the depression-facilitation model has served to study the bursting activity generated by the preBötzinger complex (preBtC), which gives rise to the respiratory rhythm [65–67]. The preBötC generates a spontaneous respiratory-like rhythm even in slices. The model developed in [68] predicted that a neuronal network model based on two successive synaptic depression mechanisms (Fig. 13a) allows a synchronized activity in the network (bursts) with a population refractory period (silence) that prevents an immediate bursting activity. In addition, combining electrophysiological recordings and the model suggested that the rhythm emerges through stochastic activation of intrinsic currents conveyed by recurrent synaptic connections (Fig. 13b), without the need of rhythmic pacemaker neurons.

Bursting in this model (Fig. 13c), is obtained by four principal state variables of a neuron (Fig. 13d–e): the voltage V, the facilitation variable x, and the two normalized depression-related variables Υ_{free} and Υ_{dock}. Synaptic fluctuations are the driving forces that can initiate population activities through recurrent excitation. The system of equation is given for each neuron i by

$$
\begin{aligned}
C\dot{V}_i &= -I_{Na} - I_K - I_L + \sum_{j \text{ connectedto } i} I_{syn,j} + \sigma \dot{W}_i \\
\dot{x}_i &= \frac{X - x_i}{\tau_f} + k(1 - x_i)H(V_i - T) \\
\dot{y}_{\text{free},i} &= \frac{1 - y_{\text{free},i} - y_{\text{dock},i}}{\tau_{\text{rec}}} - \frac{1}{\tau_{\text{dock}}}\left(y_{\text{dock}}^{\max} - y_{\text{dock}}\right) y_{\text{free},i} \\
&\quad \times \left[1 + \frac{x_i - X}{X} H(V - T)\right] \\
\dot{y}_{\text{dock}} &= \frac{1}{\tau_{\text{dock}}}\left(y_{\text{dock}}^{\max} - y_{\text{dock},i}\right) y_{\text{free},i}\left[1 + \frac{x_i - X}{X} H(V_i - T)\right] \\
&\quad - \frac{1}{\tau_{\text{rel}}} y_{\text{dock}} \frac{x_i - X}{X} H(V_i - T),
\end{aligned}
\tag{23}
$$

where the variables and parameters are described below. When a presynaptic AP arrives at a time t_0, which does not fall into the refractory period window, the synaptic current is

$$
i_{t_0}(t) = K_I(y_{\text{dock}}(t) - y_{\text{dock}}(t_0))H(V - T)H\left(y_{\text{dock}} - y_{\text{dock}}^{\min}\right), \tag{24}
$$

where K_I is a constant, which converts the variable accounting for the proportion of fused vesicles during an AP, into a post-synaptic current. For spikes arriving at times t_k, the synaptic current is for $t \in [t_k, t_{k+1}]$,

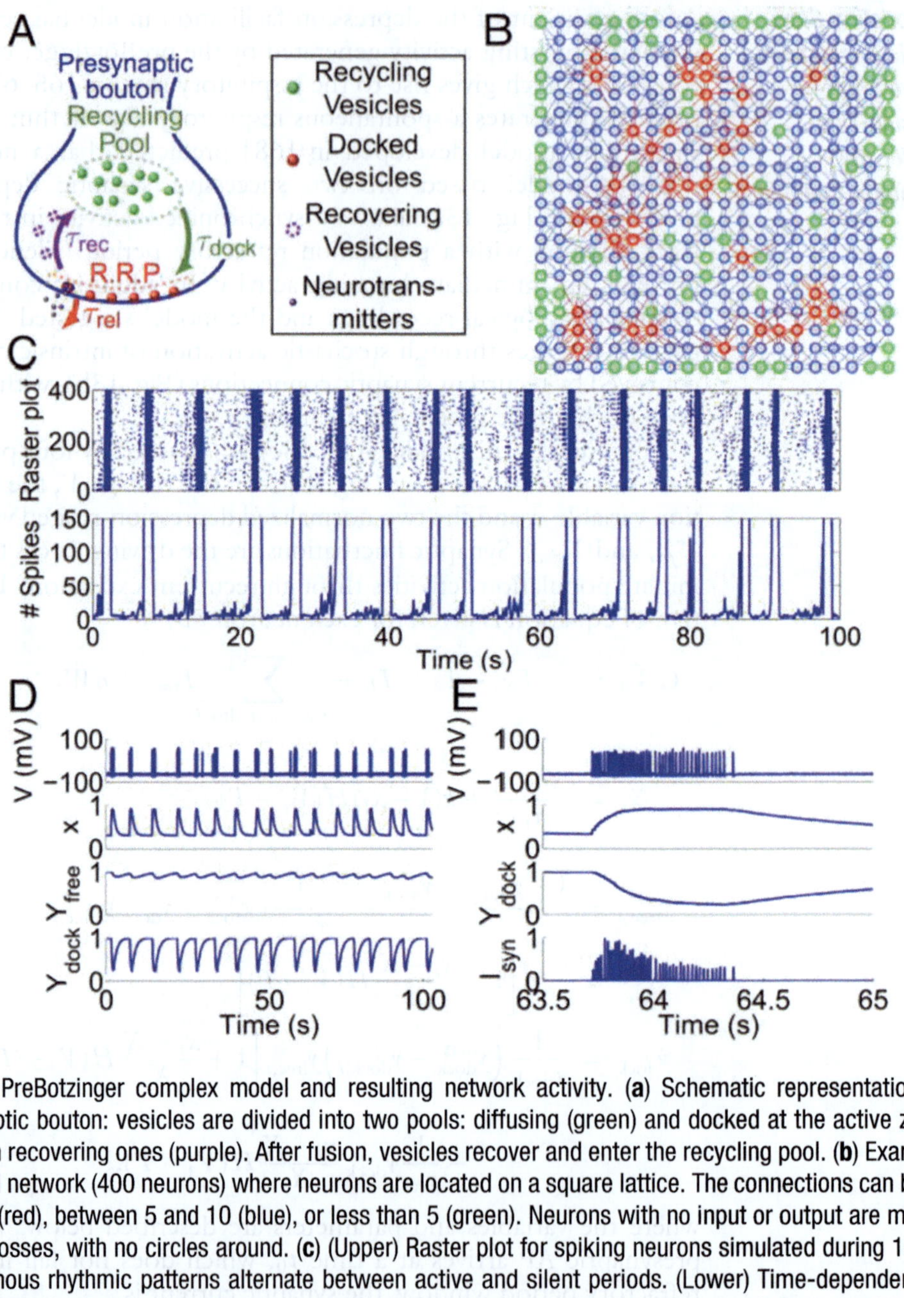

Fig. 13 PreBotzinger complex model and resulting network activity. (**a**) Schematic representation of the presynaptic bouton: vesicles are divided into two pools: diffusing (green) and docked at the active zone (AZ, red) with recovering ones (purple). After fusion, vesicles recover and enter the recycling pool. (**b**) Example of a neuronal network (400 neurons) where neurons are located on a square lattice. The connections can be higher than 10 (red), between 5 and 10 (blue), or less than 5 (green). Neurons with no input or output are marked by green crosses, with no circles around. (**c**) (Upper) Raster plot for spiking neurons simulated during 100 s, the synchronous rhythmic patterns alternate between active and silent periods. (Lower) Time-dependent plot of the number of spikes in the network, computed in time windows of 100 ms. The *y* axis is zoomed to observe the preburst increase in the number of spikes. (**d**) Time-dependent plot of the voltage V, the facilitation variable x, and the scaled variables $Y_{\text{free}} = y_{\text{free}}/y_{\text{free}}^{max}$, $Y_{\text{dock}} = y_{\text{dock}}/y_{\text{dock}}^{max}$ for a single neuron chosen randomly from the network (the mean bursting duration is 777 ± 98 ms, and the mean interburst interval is 5.2 ± 1.0 s). (**e**) Magnification of V, x, Y_{dock}, and I_{syn} for the neuron in D, during 1.5 s (same simulation as C and D) copyright from [68]

$$I_{syn}(t) = \begin{cases} 0 & \text{duringarefractoryperiod} \\ i_{t_k}(t - \tau_{del}), & \text{else} \end{cases}, \tag{25}$$

where the rest of the currents are given by the classical Hodgkin-Huxley model [2]

$$\begin{aligned} I_{Na} &= g_{Na} m_\infty^3 h(V - E_{Na}) \\ I_K &= g_k n^4 (V - E_K) \\ I_L &= g_L (V - E_L) \\ m_\infty &= \frac{\alpha_m}{\alpha_m + \beta_m} \\ \alpha_k &= \frac{1}{\tau_k} \frac{\theta_k - V}{e^{\frac{\theta_k - V}{\tau_k}} - 1} \quad \text{for } k = n, m \\ \beta_k &= \eta_k e^{-\frac{V+65}{\sigma_k}}. \end{aligned}$$

The variables m and n represent the opening of the Na^+ and K^+ channels respectively. For the closing of the Na^+ channel, $h = (0.89 - 1.1n)$. The synaptic current $\sum_j I_{syn,j}$ integrates the sum over all connecting neurons [68].

The synaptic dynamics depends on a two pool model that accounted for two different time scales of depression. Indeed, synaptic depression results from the depletion of the readily releasable pool (RRP) (Fig. 13a), where synaptic vesicles are gathered at the membrane before fusion. The other pool of recycling vesicles (recycling pool, RP) are diffusing. Finally after fusion, vesicles are not participating in any of the two previous pools. This state is described as recovering. In the present model, the total amount of vesicles in a synapse is constant. Thus the fraction y_{free} (resp. y_{dock}) of vesicles in the RP (resp. RRP), with the fraction of recovering vesicles y_{rec} satisfies the conservation equation

$$y_{\text{free}} + y_{\text{dock}} + y_{\text{rec}} = 1. \tag{26}$$

Finally, the synaptic facilitation variable x reflects all possible mechanisms that enhance vesicular release, and thus is associated with an increase in the release probability [52]. It was already described in Eq. 17 where τ_f is the facilitation rate, X its value at equilibrium, and H is the Heaviside function. The facilitation x increases due to the term $k(1 - x)$ during an AP, when the membrane potential V is above a threshold T ($H(V - T) = 1$), and relaxes back to equilibrium when V is below $T(H(V - T) = 0)$.

Simulations of the model revealed that for the chosen set of parameters, the fraction of vesicles in the RP (Υ_{free}) is quite stable, oscillating between 80% and 98% of its maximum value, whereas the fraction of docked vesicles in the RRP (Υ_{dock}) fluctuates between 20% and 100% of its filled state. The simulations suggested that the RRP alternates between an empty and full state, where the mean maximum number of vesicles is 4.5. When the percentage of vesicles in the RRP Υ_{dock} reaches its minimal value, the bursting period ends, leading to a decrease in the facilitation variable x that relaxes back to equilibrium (Fig. 13e). On the contrary, x increases exponentially when each population burst begins, which, associated with the decrease in Υ_{dock}, is reflected by a transient increase and subsequent decrease in the amplitude of synaptic currents during the bursts (I_{syn}, Fig. 13e).

Each neuron within the network that reaches its minimum y^{min}_{dock} enters a refractory state that shuts down its synaptic transmission. This minimum value leads to burst termination across the population because recurrent excitation ceases. As synapses recover, a subsequent burst can begin when several connected neurons spike in a short time interval to facilitate post-synaptic neuronal activation. This nonlinearly promotes spiking activity to spread through neighboring neurons and invade the entire network. These features were stable with respect to the network topology with a cycle periods (CPs), burst durations (BDs), and interburst intervals (IBIs) of 5.5 ± 1 s, 694 ± 138 ms, and 4.8 ± 1 s, respectively ($n = 10$ networks), which is in the range observed in vitro.

To conclude, this model could account for a large range of in vitro and in vivo data as well as small events called small burstlets, that could be observed before the emergence of a large burst [69]. The codes used in this project are available for download at bionewmetrics.preBotz.

4.10 Short-Term Plasticity Model Predicts the Capacity of Working Memory

Another application of synaptic short-term plasticity neuronal network concerns the capacity of working memory. How many random tasks or words can be transiently stored in the human memory? This number is related to the capacity of working memory (WM) [49], limited for most human to four.

To study the neuronal computation underlying such a number, a network model was proposed in [49], consisting in combining excitatory networks with synaptic currents. The mean-field short-term plasticity model consists in combining P excitatory networks, as presented in Eqs. 18 with a common inhibitory one, that could inhibit all of them simultaneously. Using Eqs. 17, for $\mu = 1..P$, the model can be written as

$$\begin{aligned}
\tau\frac{dh_\mu}{dt} &= -h_\mu + J_{EE}u_\mu x_\mu R_\mu - J_{EI}RI + I_b + I_e(t),\\
\frac{du_\mu}{dt} &= \frac{U - u_\mu}{\tau_f} + U(1 - u_\mu)R_\mu,\\
\frac{dx_\mu}{dt} &= \frac{1 - x_\mu}{\tau_d} - u\mu x_\mu R_\mu,\\
\tau\frac{dh_I}{dt} &= -h_I + J_{IE}\textstyle\sum_\nu R_\nu,
\end{aligned} \tag{27}$$

where the variables are the firing rate h_I of the inhibitory network, the time constant τ of excitatory and inhibitory neurons. h and R are the synaptic current and firing rate of excitatory and inhibitory neurons, respectively. The conversion to firing rate uses the function $R(h) = \alpha \ln(1 + \exp(h\alpha))$, which is a neuronal gain chosen in the form of a smoothed threshold-linear function.

The current I_b represents the background excitation input and I_e is the external input. As described in the previous subsections, u and x refer to the short-term facilitation and depression effects, respectively. The neuronal connectivities are encoded in the parameters J_{EE}, J_{EI} and J_{IE} that should be tuned.

When several items are generated by applying transient external excitation inputs to the several networks, a sub-number can be maintained successfully in the form of brief reactivation called population spikes. This number is around 4 and thus the interpretation is that the capacity of WM of four (see figure 1 of [49]).

The model allowed to estimate the capacity number N_c as the ratio of the duration T_{max} of the cycle between two consecutive recalls for a given network to the total duration t_s of separation between population spike generated by all neuronal networks.

The cycle duration can be approximated as the time it takes for the synaptic efficacy curve to reach a peak while neglecting the firing rate, thus from Eqs. 2 and 3 of the system 27, we have $T_{max} = \tau_d \frac{\tau_f/\tau_d}{1-U}$. The duration t_s depends on (1) the width of the population spike of the previous item, the delay and the width of the inhibitory pulse triggered by this population spike, and the time for a next cluster to recover from inhibition and initiate a new population spike. A refined analysis leads to the expression $t_s = \tau \ln(\frac{h_0}{I_b - I_c}) + C$, where C is a constant, the negative initial value h_0 is determined by the strength of inhibition triggered by a population spike of a previously activated network and I_c is a critical value parameter. This leads to the number of memory items

$$N_c = \frac{\tau_d \dfrac{\tau_f/\tau_d}{1-U}}{\tau \ln\left(\dfrac{h_0}{I_b - I_c}\right) + C}. \tag{28}$$

To conclude, an elementary network configuration exhibiting short-term depression and facilitation where all excitatory networks are connected to the same inhibitory network can generate a memory replay where the mean capacity is N_c. This number depends on the intrinsic time scale of the synaptic dynamics as well as the amplitude of some input currents.

4.11 Coupling Synaptic Short-Term Plasticity Neuronal Networks to Generate Brain Oscillations

Brain rhythms observed in EEG recordings often originate from multiple neuronal projections among different regions. For example the α-band [8–12] Hz, observed in the prefrontal cortex during general anesthesia [70, 71], involves thalamo-cortical projections [72–75]. We discuss briefly in this subsection how the α and the δ bands can be generated by connecting different neuronal networks characterized by depression-facilitation when all external sensory stimuli are shut down (Fig. 14a–b) [58].

General anesthesia on the propofol anesthetic is characterized by a dominant α-band but also low frequency delta-oscillations resulting from the switching between Up and Down states (δ-band). In the Up-state, the model has to reproduce the α-oscillation, as possible using Eq. 15, the dynamics of which produce a focus attractor [35, 53]. Indeed, the dynamics is driven by random fluctuations of the membrane potential. Switching between Up and Down state is associated with the slow δ-band oscillation. These oscillations can be reproduced by adding AHP to short-term synaptic plasticity with the use of Eqs. 20 (Fig. 14a). In this case, the overall network connectivity modulates the time spent in the Up-state.

When an inhibitory neuronal network is connected to an excitatory one with AHP, with a constant input current on the inhibitory population, then the network exhibits switching between Up and Down states. Furthermore, the injected current modulates the fraction of time spent in Up vs Down states. When this input current reaches a threshold, the excitatory population can be locked in a Down state, silencing the network. However, without AHP the inhibitory network cannot induce switching between Up and Down states, and the dominant α-band generated by the excitatory population remains locked in the Up state (Fig. 14b, right).

By further coupling two excitatory networks with an inhibitory one (Fig. 14c, left), α-oscillations are generated associated with a slow switching between Up and Down states (δ-band) thus reproducing the variety of behaviors that can be generated by the thalamo-cortical loop: the excitatory subsystem contains two populations: (1) an α-oscillation generator and (2) AHP-short-term plasticity network that generates Up-Down state switching. These two populations are connected by reciprocal connections and receive an inhibitory input from the inhibitory subsystem (Fig. 14c, left). This inhibitory network can also be activated by an external stimulation I_i that models the injection of propofol.

To conclude, by connecting three neuronal populations, Up and Down states can coexist with an α-band [58]: a stable α-band emerges once external stimuli are suppressed by the anesthetic agent (Fig. 14b, right). This behavior results from the interactions between the three networks. The amplitude of inhibitory inputs as well as the connectivity level are fundamental parameters that define the proportion of Up states and thus the persistence of the α-

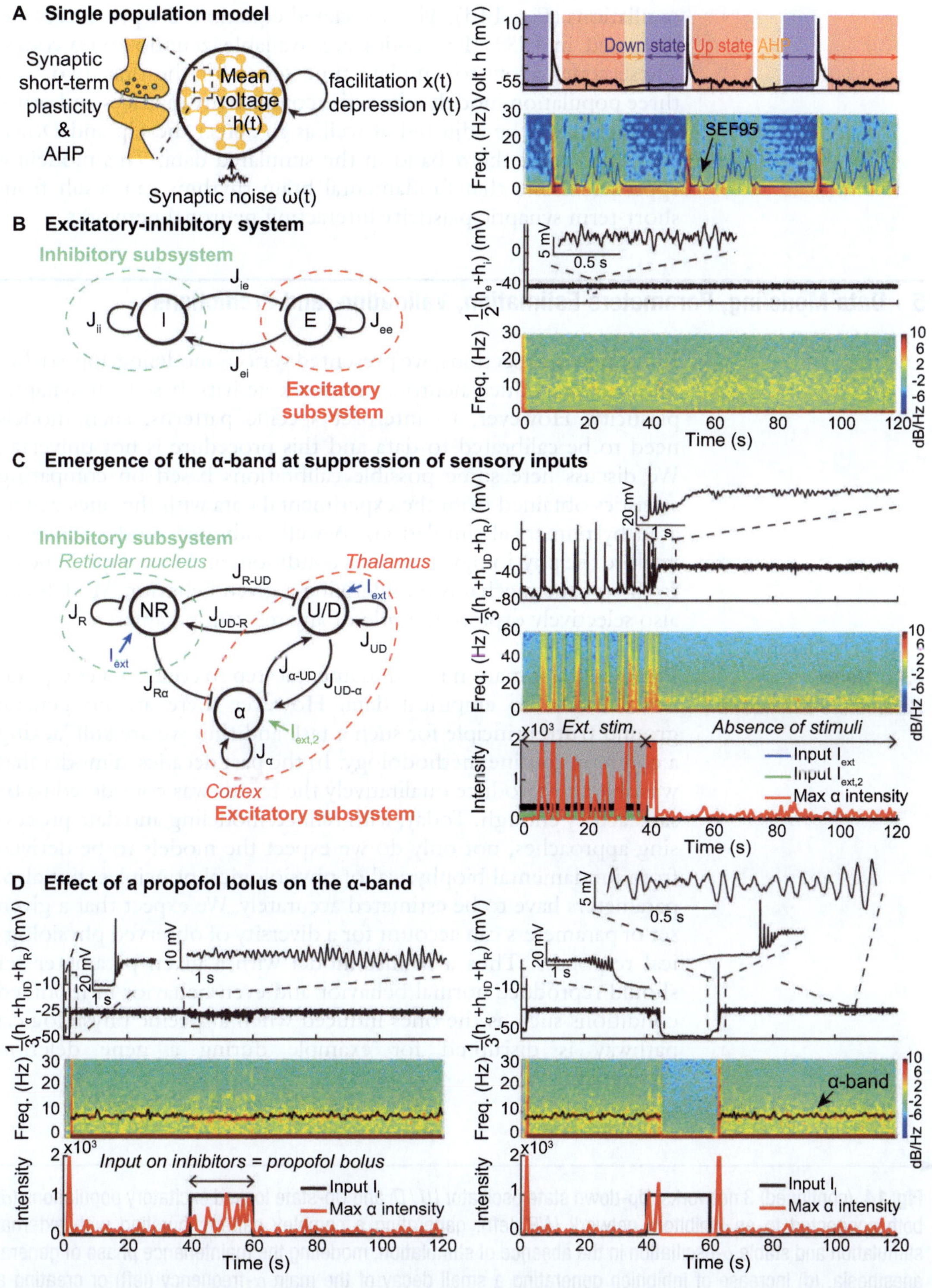

Fig. 14 Modeling oscillatory rhythms using 3 network models. **(a)** Single population with short-term facilitation-depression and AHP (left) producing slow oscillations between Up (red) and Dow (purple) states (right). **(b)** Excitatory population connected to an inhibitory one (left) generating a fragmented α-band (right) **(c)**

oscillations (Fig. 14d). The associated equations are presented and discussed in [58]. The codes are available zenodo.zonca.codes. These codes allow to simulate time series for the one, two, and three population models where the connectivity matrices and other parameters can be adjusted as well as to detect the Up and Down states and track the α-band in the simulated data. This modeling approach shows that fundamental brain rhythms can result from short-term synaptic plasticity interacting neuronal networks.

5 Data Modeling, Parameters Estimation, Validation, and Predictions

In the previous sections, we presented various modeling approaches that can reproduce neuronal network activity based on synaptic plasticity. However, to interpret specific patterns, such models need to be calibrated to data and this procedure is not universal. We discuss here some possible calibrations based on comparing statistics obtained from the experimental data with the ones generated by numerical simulations. A well calibrated model allows to explore the physiology in various conditions and to comprehend or to dissect the mechanism underlying a given behavior. Models can also selectively explore the role of any relevant parameter.

5.1 Model Parameters Estimation

Parameters estimation is a fundamental step to connect a computational model to empirical data. However there are no general ground truth principle for such a task and thus we are still lacking a common routine methodology. In the past decades, a model that was able to reproduce qualitatively the trends was considered to be satisfactory enough. Today, with refined modeling and data processing approaches, not only do we expect the models to be derived from fundamental biophysical of physiological principles, but also, parameters have to be estimated accurately. We expect that a given set of parameters can account for a diversity of observed physiological responses. Thus a robust model with a given parameter set should reproduce normal behavior and even behavior in modified conditions such as the ones induced when a specific physiological pathway is disrupted for example during a gene deletion experiment.

Fig. 14 (continued) 3 networks: Up-down state oscillator (*U*/*D*) and Up-state locked excitatory populations (α) both connected to an inhibitory network (*NR*, left), generating a complex pattern: bursting with external stimulation and stable α-oscillation in the absence of stimulation, modeling the maintenance phase of general anesthesia. (**d**) Increase of inhibition generating a small decay of the main α-frequency (left) or creating a complete suppression (right) adapted from [58]

5.2 Lack of a Principle in Parameter Estimation Procedures

5.2.1 Small Dimensional Space of Parameters

For a model with few parameters, known as low dimensional space, it is satisfactory to define a set of admissible parameters based on comparing mean and variance between simulations and experimental data. For modeling the bursting time, fixing the mean and variance of the response at 5 and 35 s was sufficient to determine the entire parameter space (Fig. 6) [35].

We shall now describe this case in more detail: there is a temporal correlation between bursts generated by small island of neurons [36]: a burst can be induced by a single neuronal stimulation that reverberate during few seconds into the ensemble of interconnected neurons (Fig. 6). Interestingly, the same stimulation generated few seconds after the first one lead to a decrease of almost half the duration of the first burst, while another stimulation generated 30 s later triggers a burst of duration comparable to the first one. The modeling approach revealed that such bursting temporal correlation is due to the short-term synaptic depression property, as confirmed experimentally by blocking the synaptic transmission [35].

To do so, the model has to be calibrated to reproduce the mean and the variance of the of burst dynamics as well as the decrease in the duration of the second burst. Specifically, the model contains 6 parameters, 4 were already known, leaving 2 to be calibrated. These constraints are quite strong and such success in reproducing the observed bursting dynamics suggested that the model captures the main mechanism underlying these bursting events. In particular, the long recovery timescale of the depression $\tau_r \approx 3s$, could explain that it takes tens of seconds for the synapse to be reactivated after a burst. Interestingly, the same model was used to reproduce the same phenomenology in hippocampal slices instead of islands by increasing further the depression timescale to $\tau_r \approx 20s$. Finally, having a calibrated model allows to explore the conditions of neuronal network physiology. In particular, the model revealed that the physiology conditions are exactly those that sets the longest burst duration. Further, as the mean number of synapses increases, which could reproduce development condition, the burst duration passes by a maximum before decreasing.

5.2.2 Large Dimensional Space of Parameters

When a model contains a larger number of parameters (more than 10), stringent criteria are needed to determine the parameter values. This can be achieved by comparing distributions instead of restricting to first and second statistical moments. In physiological models, parameters have a specific interpretation, which is not necessarily the case for machine-learning approaches, where the number of parameters have an other order of magnitude, up to millions [76]. The goal of a large parameter set is often to perform pattern recognition, the goal of which is different from studying the mechanism underlying a physiological function.

We discuss now a possible unified method to calibrate model parameters. In principle, this task is model dependent because each model has a specific purpose depending on the spatial and temporal scales it has been elaborated for. For mean-field network models, the goal is often to study neuronal properties from elementary processes such as synaptic-short term plasticity or channels behavior. For these models, the behavior or phase-space often depends on the range of the parameters. Thus finding the appropriate parameters is critical to obtain the relevant behavior and thus to clarify the complex dynamics.

For example, to obtain the emergence of Up and Down states with the synaptic depression Eqs. 15, the synaptic strength parameter J, associated with the average number of synaptic connections has to be large enough [46]. This qualitative description is certainly not sufficient nowadays and a comparison to data is requested. This demands can be answered by introducing a comparison criteria: for example, can the model or generated distribution predict the distribution of times in the Up states compared to other models [35]? The model equation 18 [53] predicts that the distribution does not follow a Poissonian or Gamma distribution, but rather could be explained by a sum of oscillating exponentials. This oscillating behavior is obtained by the probability distribution function derived from the associated Fokker-Planck equations [35, 53, 54]. Indeed, the distribution of times in the Up state is approximated as

$$f_{Up}(t) = A\exp(-\lambda_0 t) + B\exp(-\lambda_1 t)\cos(\omega_1 t + \phi_1). \tag{29}$$

This distribution contains 6 parameters: $A, B, \lambda_0, \lambda_1, \omega_1, \phi_1$, with $\lambda_1 > \lambda_0$. The constants A and B represent the relative contribution of the two main exponentials. The main decay is obtained by fitting the distribution of times in the Up state by the first exponential, leading to a value for the parameter λ_0. The distance between the two consecutive peaks is equal to $\frac{2\pi}{\omega_1}$, while the phase ϕ_1 accounts for a possible shift in the position of the maximum [53]. Finally, the second exponential decay term λ_1 can be found by multiplying the empirical distribution by $\exp(\lambda_0 t)$ and then by fitting the decay tail by an exponential. To conclude short-term synaptic models are used to derive curves to fit time distributions (Fig. 5).

5.3 Identifying Parameters Using Distributions

The models presented in the previous section based on synaptic short-term plasticity are insufficient to account for electrophysiological mechanisms such as after hyperpolarization (AHP), that requires modeling the voltage properties. Voltage modeling has remained difficult and empirical, due to the difficulty of simplifying and adapting the Maxwell equations. Voltage conduction based on channels has been approached by the phenomenological and empirical Hodgking-Huxley models and their successors where the voltage equation is mostly driven by capacitance and resistance in the form $CdV/dt = \sum I_{\text{ions}}$, where the sum account for various ionic exchanges across the membrane ionic channels [2].

To account for recent recordings of bursting obtained in slices, where the astrocyte network was disrupted, the phenomenological model 20 based on synaptic short-term plasticity and AHP was introduced with a total of 14 parameters, as already discussed above. The model accounts for the refractory period generated by the potassium channels, leading to a hyperpolarization phase. In the absence of a systematic approach to test how this model could reproduce the electrophysiological time series, it is possible to use the entire distributions of bursting and interbursting durations. This criteria turns out to be sufficient to identify all parameters. Fitting a distribution is quite constraining as it is a one dimensional curve. For example, many statistical distributions such as Gaussian, Gamma, or Theta functions are characterized by two parameters. Thus for a model containing $2N$ parameters, one possibility is to collect N distributions from experimental data.

In the present case, the model 20 had 14 parameters, 6 of which had already been calibrated [35, 46, 48, 52] leaving 8 parameters to identify. Accordingly, 4 distributions should be fitted. We chose the distributions of the bursting and AHP durations in a WT and in KO cases.

While each individual distribution could be fitted by two parameters, fitting the four histograms of bursting and AHP durations in WT and KO at the same time gave a much more robust constrain on 8 parameters. We shall now describe this fitting procedure to estimate the parameters summarized in the form of a pipeline (Fig. 15). The pipeline consists of a model equation and four histograms obtained by segmenting the electrophysiological time series of both burst and AHP durations in WT and KO. The steps are as follows:

1. 8 unknown parameters are represented in a vector $\boldsymbol{\theta}$ (bottom half of Table 5). For the initial condition, select a random choice inside the authorized ranges for each parameter value (Table 5 right column).
2. At step k, new values of the 8 parameters in θ_k are chosen randomly in their authorized range $[A, B]$ (Table 5 right column). Once the parameters are chosen, we run 5000 simulations from the model 20, leading to time series that are then segmented using the procedure described in Subheading 4.8. The statistics for three epochs burst, AHP, and QP durations are then generated.

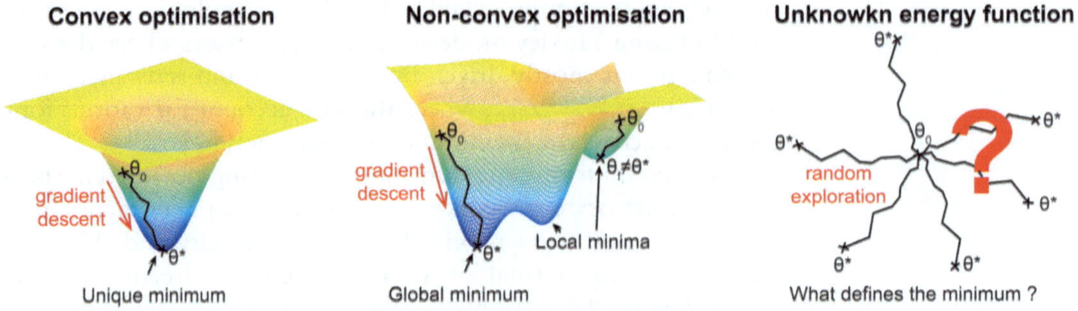

Fig. 15 Parameter fitting pipeline. (**1**) Model with 8 free parameters (vector θ) to be fitted. (**2**) At each step k, run and segment (ex. 5000s) simulations to generate time series with the parameters θ_k. (**3**) Build the cumulative histograms such as bursting and AHP durations from the numerical simulations. (**4–5**) Compare the simulated histograms to the experimental ones by computing the Kolmogorov-Smirnov distance $d_{ks}(k)$ between the experimental and simulated distributions. (**6**) Iterate at step (2) (ex: $K = 1000$ times) or (**7**) Select the optimal realization k^*. (**8**) The model with optimal parameters is used to explore the role any parameter of interest

Table 5
Model 20 parameters

	Fixed parameters	**Values**	
τ	Fast time constant for h	0.05s [46]	
K	Facilitation rate	0.037 Hz	(*modified: 0.04Hz in* [35])
L	Depression rate	0.028 Hz	(*modified: 0.037Hz in* [35])
τ_r	Depression time rate	2.9s	(*modified: 2–20s in* [35])
τ_f	Facilitation time rate	0.9s	(*modified: 1.3s in* [35])
T	Depolarization parameter	0	
H_{AHP}	End of recovery phase threshold on h	-7.5	
	Parameters to fit	Fitted values	Authorized range
τ_{mAHP}	Medium time constants for h	0.15s	$[0.05, 1]$ s
τ_{sAHP}	Slow time constants for h	5s	$[1, 20]$ s
J	Synaptic connectivity	4.21	$[3, 5]$ [48]
X	Facilitation resting value	0.08825	$[0, 0.2]$
σ	Noise amplitude	3	$[0.1, 10]$
T_{AHP}	Undershoot threshold	-30	$[-40, -5]$
Υ_{AHP}	Recovery phase threshold	0.85	$[0.75, 0.95]$
Υ_h	end of AHP phase threshold	0.5	$[0.45, 0.55]$

3. The segmentation of step 2 allows to build the histograms of burst and AHP durations from the numerical simulations.
4. To find the optimal parameters, generate histograms $f_{Sim}^{(k)}(t)$ from simulations that will be compared to the experimental ones $f_{Exp}(t)$. The method consist in computing the error associated with the Kolmogorov-Smirnov distance

$$\epsilon_k = d_{ks}(f_{Exp}(t), f_{Sim}^{(k)}(t)). \tag{30}$$

With four histograms to fit burst and AHP for the WT and KO datasets, it is possible to average over the four d_{ks} distances. The empirical two-sample Kolmogorov-Smirnov distance is given by the maximum of the difference between the cumulative distribution functions $F_{1,2}$ of the two empirical distributions

$$d_{ks}(F_1, F_2) = sup_x |F_1(x) - F_2(x)|. \tag{31}$$

5. When the error d_{ks} is higher than a threshold T_{fit}, then the iterate restarts at step 2, otherwise the fitting algorithm terminates. To ensure that the algorithm had converged, another solution would be to fix the number of iterations at step 2 (e.g $N_{iter} = 1000$) and to pick the best of all realizations:

$$k^* = Argmin_{k \in [1, N_{iter}]}(d_{ks,k}), \tag{32}$$

in that case,

$$\theta^* = \theta_{k^*}. \tag{33}$$

For step 2, this approach was used in the absence of a deterministic strategy to explore the parameter space. Other strategy could be a gradient descent if an energy function was known (Fig. 16). This procedure is quite generic and could be adapted to other models and any number parameters. The pipeline implementation in Matlab can be found in bionewmetrics.pipeline.parameter.calibration. The model used in this code is the model 20 but it can be replaced by any other model implemented as a Matlab function returning time-series with large amplitude activation events such as bursts.

5.4 Possible Improvements of the Fitting Pipeline

There are several possible improvements for the fitting pipeline: first, it will be soon possible to use fully automatized softwares as described in Subheading 5.3. Second, the fit quality between the data and the simulation depends on the chosen metric. Here we used the Kolmogorov-Smirnov distance but there are other distances that could be used to better capture similarities and differences between the histograms. One possibility would be to use the Wasserstein distance [77, 78], which is defined in optimal transport theory [79, 80]. This distance is also called the earth mover's

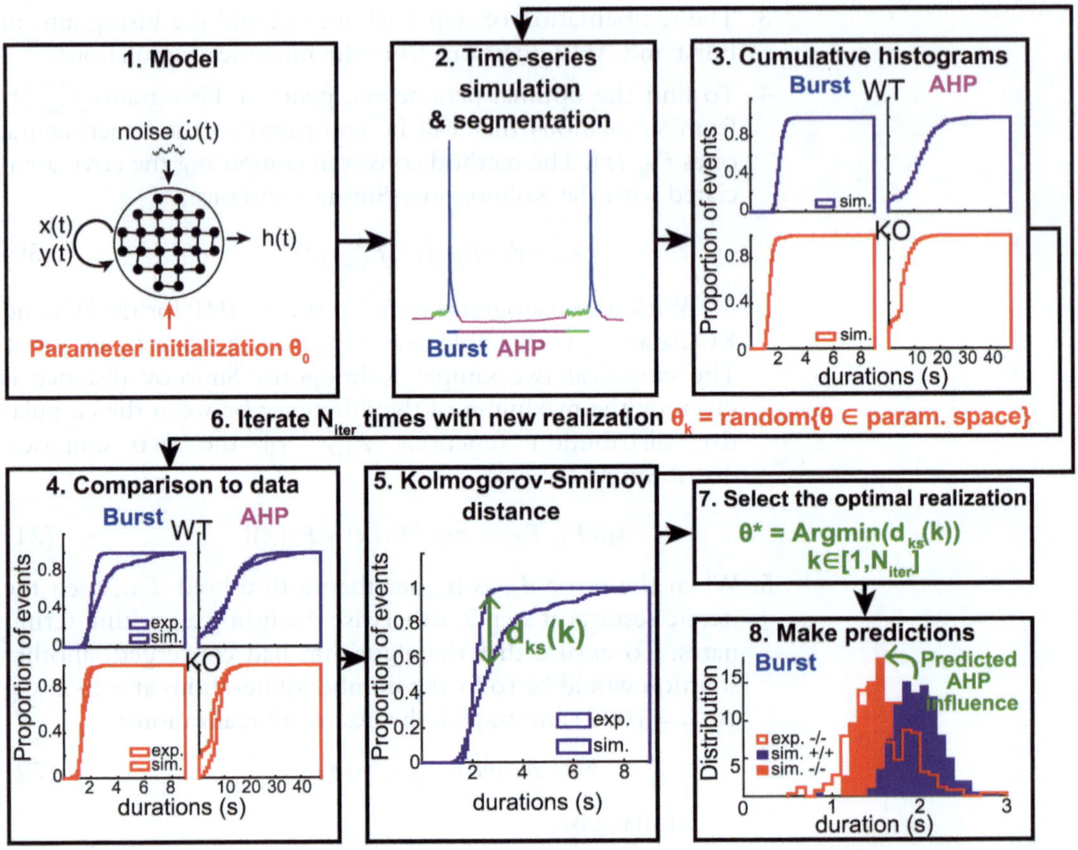

Fig. 16 Optimization procedures. Illustration of an heuristic to select the optimal parameters: *Left* Convex optimization allows to find iteratively the optimal parameter values by converging to the global minimum θ^* of an energy function (yellow surface) starting from any point θ_0 of the parameter space. The fastest path follows the steepest slope using a gradient descent algorithm (red arrow). *Middle* Non-convex optimization can lead to multiple local minima ($\theta_f \neq \theta^*$) and thus the selection of the optimal parameters depends on the initial point. *Right* When no energy functional is known, there are standard procedure to known to find the best parameter values. A possible approach is to use a random exploration with multiple scales

distance because can be seen as the minimal workload required to move one pile of soil to another one with specific shapes. The piles of soil being metaphors for abstract distributions, allowing to compar histograms. Since this method is already used in various theoretical fields, several implemented versions of this metric are openly available online (see for example [81] for a Matlab implementation, the scipy library in Python also provides an implementation of this distance, see scipy.stats.wasserstein_distance).

Finally, it is often difficult to find an optimal strategy for the exploration of the parameter space especially in high dimension. A random exploration is greedy and not efficient. When the space is the produce of intervals, the unit space parameter is a cube in high dimension. One possibility would be to use a uniform point distribution with a resolution r. Then refining the exploration around the minimum point is possible by a subdivision with a resolution

$r/2$. This could lead to an accelerated search of the optimal parameters. It remains an open question in general to know whether there is a unique solution for such complex dynamical systems. The optimal solution could also be located in a shallow minimum where neighboring points provide as well good results.

5.5 Network Models to Decipher the Role of Each Parameter

After the calibration step, network model such as the one presented by Eqs. 20 can be used to decipher the respective influence of AHP, synaptic noise or membrane depolarization on neuronal bursting activity, that are all affected by the disruption of the astrocyte network. Specifically, this step can be performed by running simulations with all the parameters obtained in the case of the WT fit and then changed to the KO values only the parameters corresponding to the condition to be explored.

The distribution of the simulated burst and IBI durations can be compared to ones of the KO data. Using this approach, it turns our that AHP was sufficient to explain the change in burst and IBI durations observed in the KO data, while the increase of synaptic noise had some effect on burst durations but not much on IBI and the membrane depolarization had no effect on either duration. To conclude, these results clarify the physiological mechanisms underlying the regulation of neuronal bursting by astrocytes in a way that would not have been possible experimentally.

6 Discussion and Perspectives

We illustrated in this book chapter how data analysis can benefit from neuronal network modeling based on synaptic short-term plasticity. We presented a generic signal processing approach and the associated algorithms to segment large amplitude bursting events from a background, characterized by a smaller amplitude. Indeed, this approach requires that burst-like events should be well separated and differ sufficiently from the background signal. Although this segmentation algorithm is quite elementary, when the baseline can vary or when multiple bursting event can coexist, several adjustment are necessary, as we saw for the case of calcium dynamics in astrocytes.

We did not review here amplitude events that need to be segmented and replaced in real time: this could be the case for artifacts located in the EEG channels [82, 83]. For example, during general anesthesia, large artifacts can be generated due to eye motion (EOG), muscle contraction (EMG) or simply because electrodes are transiently moved. Removing these artifacts require not only to segment the exact epoch, but also to replace them by a more physiological signal. The method presented in this chapter is insufficient for this task and other approaches based on wavelet transform and coefficient transportation have been developed for real-time brain monitoring [83].

A new generation of segmentation problems involve time-frequency dynamics, relevant to detect short high frequencies responses such as ripples [84, 85]. It is also the case for the segmentation of transient epileptoid signs during children anesthesia [86, 87]. This approach requires different techniques such as comparison with a pre-existing alphabet of motifs using more elaborated projection methods such as independent component analysis.

Finally, computational models have now matured enough so that they play a key role in our understanding, quantification and prediction of neuronal network dynamics, allowing to decipher for example the contribution of short-term synaptic plasticity from other biophysical parameters. They also served to study the emergence of oscillation regimes from non-synchronous oscillators, or to quantify the contribution of membrane fluctuation in rhythm genesis [68, 88].

Probably, the next generation of models will become avatars of physiological processes. To that end, they should have well calibrated parameters to reproduce the statistics of physiological realizations. Obtaining such calibrations remains difficult for medical applications such as real-time brain prediction during performing tasks, anesthesia, coma or simply to predict an epileptic crisis sufficiently in advance. Another difficulty to move from neuronal network representation to models involving brain regions is that this complexity remains difficult to handle due to large amount of parameters, but once such calibrations would be possible by comparing distribution, we can foresee real-time applications and progress in managing or controlling the brain under various conditions.

Acknowledgements

DH research is supported by a grant ANR NEUC-0001, a Memolife grant and the European Research Council (ERC) under the European Unions Horizon 2020 research and innovation programme (grant agreement No 882673). European Research Council (Consolidator grant 683154) (NR) European Unions Horizon 2020 research and innovation program (Marie Sklodowska-Curie Innovative Training Networks, grant 722053, EU-GliaPhD) (NR) Ligue Francaise contre lEpilepsie (ED) Aviesan (ED). French Research Ministry (ED386, Ecole Doctorale de Sciences Mathmatiques Paris centre). LZ fellowship was supported Fondation pour la Recherche Medicale (FRM FDT202012010690). D. H and N. R are suppported by ANR AstroXcite.

Competing Interests The authors declare no competing financial interests.

References

1. Kandel ER, Schwartz JH, Jessell TM, of Biochemistry D, Jessell MBT, Siegelbaum S, Hudspeth A (2000) Principles of neural science, vol 4. McGraw-Hill, New York
2. Hille B (1978) Ionic channels in excitable membranes. current problems and biophysical approaches. Biophys J 22(2):283–294
3. Stuart G, Schiller J, Sakmann B (1997) Action potential initiation and propagation in rat neocortical pyramidal neurons. J Physiol 505(3): 617–632
4. Markram H, Lübke J, Frotscher M, Sakmann B (1997) Regulation of synaptic efficacy by coincidence of postsynaptic aps and epsps. Science 275(5297):213–215
5. Garaschuk O, Milos RI, Grienberger C, Marandi N, Adelsberger H, Konnerth A (2006) Optical monitoring of brain function in vivo: from neurons to networks. Pflügers Archiv 453(3):385–396
6. Llinás RR, Steriade M (2006) Bursting of thalamic neurons and states of vigilance. J Neurophysiol 95(6):3297–3308
7. Kenet T, Bibitchkov D, Tsodyks M, Grinvald A, Arieli A (2003) Spontaneously emerging cortical representations of visual attributes. Nature 425(6961):954–956
8. Cossart R, Aronov D, Yuste R (2003) Attractor dynamics of network up states in the neocortex. Nature 423(6937):283–288
9. Chen X, Rochefort NL, Sakmann B, Konnerth A (2013) Reactivation of the same synapses during spontaneous up states and sensory stimuli. Cell Rep 4(1):31–39
10. Kornreich BG (2007) The patch clamp technique: principles and technical considerations. J Vet Cardiol 9(1):25–37
11. Sachs F, Auerbach A (1983) [9] single-channel electrophysiology: use of the patch clamp. In: Methods in enzymology, vol 103. Elsevier, Amsterdam, pp 147–176
12. Mazzoni A, Logothetis NK, Panzeri S (2013) Information content of local field potentials. In Principles of Neural Coding, pp 411–429
13. Pesaran B (2009) Uncovering the mysterious origins of local field potentials. Neuron 61(1): 1–2
14. Oren I, Paulsen O (2010) Currents in space: understanding inhibitory field potentials. J Physiol 588(Pt 12):2015
15. Obien MEJ, Deligkaris K, Bullmann T, Bakkum DJ, Frey U (2015) Revealing neuronal function through microelectrode array recordings. Front Neurosci 8:423
16. Buzsáki G, Watson BO (2012) Brain rhythms and neural syntax: implications for efficient coding of cognitive content and neuropsychiatric disease. Dialog Clin Neurosci 14(4): 345–367
17. Anastassiou CA, Buzsáki G, Koch C, Quiroga R, Panzeri S (2013) Biophysics of extracellular spikes. Principles Neural Coding 15:146
18. Nunez PL, Srinivasan R et al (2006) Electric fields of the brain: the neurophysics of EEG. Oxford University Press, Oxford
19. Hong G, Lieber CM (2019) Novel electrode technologies for neural recordings. Nat Rev Neurosci 20(6):330–345
20. Chever O, Dossi E, Pannasch U, Derangeon M, Rouach N (2016) Astroglial networks promote neuronal coordination. Sci Signaling 9(410):ra6
21. Savtchenko LP, Bard L, Jensen TP, Reynolds JP, Kraev I, Medvedev N, Stewart MG, Henneberger C, Rusakov DA (2018) Disentangling astroglial physiology with a realistic cell model in silico. Nat Commun 9(1):1–15
22. Holcman D, Yuste R (2015) The new nanophysiology: regulation of ionic flow in neuronal subcompartments. Nat Rev Neurosci 16(11): 685–692
23. Zhang H, Reichert E, Cohen AE (2016) Optical electrophysiology for probing function and pharmacology of voltage-gated ion channels. Elife 5:e15202
24. Liu Y, Lu Y, Chen G, Wang Q (2020) Recent progress of hybrid optical probes for neural membrane potential imaging. Biotechnol J 15(12):2000086
25. Cartailler J, Kwon T, Yuste R, Holcman D (2018) Deconvolution of voltage sensor time series and electro-diffusion modeling reveal the role of spine geometry in controlling synaptic strength. Neuron 97(5):1126–1136
26. Cohen MX (2014) Analyzing neural time series data: theory and practice. MIT Press, Cambridge
27. Gribkoff VK, Kaczmarek LK (2008) Structure, function, and modulation of neuronal voltage-gated ion channels. John Wiley & Sons, New York
28. Jaffard S, Meyer Y, Ryan RD, Boggess A, Narcowich FJ. Wavelets: tools for science and technology, p 3
29. Schuss Z (2011) Nonlinear filtering and optimal phase tracking, vol 180. Springer, Berlin

30. Destexhe A, Hughes SW, Rudolph M, Crunelli V (2007) Are corticothalamic upstates fragments of wakefulness? Trends Neurosci 30(7): 334–342
31. Yang W, Carrillo-Reid L, Bando Y, Peterka DS, Yuste R (2018) Simultaneous two-photon imaging and two-photon optogenetics of cortical circuits in three dimensions. elife 7: e32671
32. Holcman D, Schuss Z (2014) Oscillatory survival probability and eigenvalues of the non-self-adjoint Fokker–Planck operator. Multiscale Model Simul 12(3):1294–1308
33. Dao Duc K, Parutto P, Chen X, Epsztein J, Konnerth A, Holcman D (2015) Synaptic dynamics and neuronal network connectivity are reflected in the distribution of times in up states. Front Comput Neurosci 9:96
34. Cohen D, Segal M (2011) Network bursts in hippocampal microcultures are terminated by exhaustion of vesicle pools. J Neurophysiol 106(5):2314–2321
35. Dao Duc K, Lee CY, Parutto P, Cohen D, Segal M, Rouach N, Holcman D (2015) Bursting reverberation as a multiscale neuronal network process driven by synaptic depression-facilitation. PLoS One 10(5):e0124694
36. Cohen D, Segal M (2009) Homeostatic presynaptic suppression of neuronal network bursts. J Neurophysiol 101:2077–2088
37. Chever O, Dossi E, Pannasch U, Derangeon M, Rouach N (2016) Astroglial networks promote neuronal coordination. Sci Signaling 9(410):ra6
38. Jayant K, Hirtz JJ, Plante IJL, Tsai DM, De Boer WD, Semonche A, Peterka DS, Owen JS, Sahin O, Shepard KL et al (2017) Targeted intracellular voltage recordings from dendritic spines using quantum-dot-coated nanopipettes. Nat Nanotechnol 12(4):335–342
39. Jayant K, Wenzel M, Bando Y, Hamm JP, Mandriota N, Rabinowitz JH, Jen-La Plante I, Owen JS, Sahin O, Shepard KL et al (2019) Flexible nanopipettes for minimally invasive intracellular electrophysiology in vivo. Cell Rep 26(1):266–278
40. Tsodyks M, Pawelzik K, Markram H (1998) Neural networks with dynamic synapses. Neural Comput 10(4):821–835
41. Mongillo G, Barak O, Tsodyks M (2008) Synaptic theory of working memory. Science 319(5869):1543–1546
42. Buzsaki G (2006) Rhythms of the Brain. Oxford University Press, Oxford
43. Hodgkin AL, Huxley AF (1952) A quantitative description of membrane current and its application to conduction and excitation in nerve. J Physiol 117:500–544
44. Morris C, Lecar H (1981) Voltage oscillations in the barnacle giant muscle fiber. Biophys J 35(1):193–213
45. Izhikevich EM (2007) Dynamical Systems in Neuroscience: the geometry of excitability and bursting. MIT Press, Cambridege
46. Holcman D, Tsodyks M (2006) The emergence of up and down states in cortical networks. PLoS Comput Biol 2(3):174–181
47. Dao Duc K, Parutto P, Chen X, Epsztein J, Konnerth A, Holcman D (2015) Synaptic dynamics and neuronal network connectivity are reflected in the distribution of times in up states. Front Comput Neurosci 9:96
48. Barak O, Tsodyks M (2007) Persistent activity in neural networks with dynamic synapses. PLoS Comput Biol 3(2):e35
49. Mi Y, Katkov M, Tsodyks M (2017) Synaptic correlates of working memory capacity. Neuron 93(2):323–330
50. Mark S, Romani S, Jezek K, Tsodyks M (2017) Theta-paced flickering between place-cell maps in the hippocampus: a model based on short-term synaptic plasticity. Hippocampus 27(9): 959–970
51. Varela JA, Sen K, Gibson J, Fost J, Abbott L, Nelson SB (1997) A quantitative description of short-term plasticity at excitatory synapses in layer 2/3 of rat primary visual cortex. J Neurosci 17(20):7926–7940
52. Tsodyks MV, Markram H (1997) The neural code between neocortical pyramidal neurons depends on neurotransmitter release probability. Proc Natl Acad Sci USA 94:719–723
53. Dao Duc K, Schuss Z, Holcman D (2016) Oscillatory survival probability: analytical and numerical study of a non-poissonian exit time. Multiscale Model Simul 14(2):772–798
54. Duc KD, Schuss Z, Holcman D (2014) Oscillatory decay of the survival probability of activated diffusion across a limit cycle. Phys Rev E 89(3):030101
55. Verechtchaguina T, Sokolov I, Schimansky-Geier L (2007) Interspike interval densities of resonate and fire neurons. Biosystems 89(1–3): 63–68
56. Verechtchaguina T, Sokolov IM, Schimansky-Geier L (2006) First passage time densities in resonate-and-fire models. Phys Rev E 73(3): 031108
57. Verechtchaguina T, Sokolov I, Schimansky-Geier L (2006) First passage time densities in non-Markovian models with subthreshold oscillations. Europhys Lett 73(5):691

58. Zonca L, Holcman D (2021) Emergence and fragmentation of the alpha-band driven by neuronal network dynamics. PLoS Comput Biol 17(12):e1009639
59. Bart E, Bao S, Holcman D (2005) Modeling the spontaneous activity of the auditory cortex. J Comput Neurosci 19(3):357–378
60. McFarland JM, Hahn TT, Mehta MR (2011) Explicit-duration hidden Markov model inference of up-down states from continuous signals. PLoS One 6(6):e21606
61. McKiernan EC, Marrone DF (2017) Ca1 pyramidal cells have diverse biophysical properties, affected by development, experience, and aging. PeerJ 5:e3638
62. Zonca L, Holcman D (2021) Modeling bursting in neuronal networks using facilitation-depression and afterhyperpolarization. Commun Nonlinear Sci Numer Simul 94:105555
63. Giaume C, Koulakoff A, Roux L, Holcman D, Rouach N (2010) Astroglial networks: a step further in neuroglial and gliovascular interactions. Nat Rev Neurosci 11(2):87–99
64. Dallérac G, Chever O, Rouach N (2013) How do astrocytes shape synaptic transmission? Insights from electrophysiology. Front Cellular Neurosci 7:159
65. Smith J, Ellenberger H, Ballanyi K, Richter D, Feldman J (1991) Pre-bötzinger complex: a brainstem region that may generate respiratory rhythm in mammals. Science 254(5032): 726–729
66. Cui Y, Kam K, Sherman D, Janczewski WA, Zheng Y, Feldman JL (2016) Defining pre--bötzinger complex rhythm- and pattern-generating neural microcircuits in vivo. Neuron 91(3):602–614
67. Ashhad S, Kam K, Negro CAD, Feldman JL (2022) Breathing rhythm and pattern and their influence on emotion. Ann Rev Neurosci 45: 223–247
68. Guerrier C, Hayes JA, Fortin G, Holcman D (2015) Robust network oscillations during mammalian respiratory rhythm generation driven by synaptic dynamics. Proc Natl Acad Sci 112(31):9728–9733
69. Cui Y, Kam K, Sherman D, Janczewski WA, Zheng Y, Feldman JL (2016) Defining preböttzinger complex rhythm-and pattern-generating neural microcircuits in vivo. Neuron 91(3):602–614
70. Başar E, Güntekin B (2012) A short review of alpha activity in cognitive processes and in cognitive impairment. Int J Psychophysiol 86(1): 25–38
71. Başar E (2012) A review of alpha activity in integrative brain function: fundamental physiology, sensory coding, cognition and pathology. Int J Psychophysiol 86(1):1–24
72. Dossi RC, Nunez A, Steriade M (1992) Electrophysiology of a slow (0.5–4 Hz) intrinsic oscillation of cat thalamocortical neurones in vivo. J Physiol 447(1):215–234
73. Steriade M, Amzica F, Contreras D (1994) Cortical and thalamic cellular correlates of electroencephalographic burst-suppression. Electroencephalogr Clin Neurophysiol 90(1): 1–16
74. Contreras D, Steriade M (1995) Cellular basis of eeg slow rhythms: a study of dynamic corticothalamic relationships. J Neurosci 15(1): 604–622
75. Steriade M, Amzica F, Contreras D (1996) Synchronization of fast (30-40 Hz) spontaneous cortical rhythms during brain activation. J Neurosci 16(1):392–417
76. Gómez-de Mariscal E, García-López-de Haro C, Ouyang W, Donati L, Lundberg E, Unser M, Muñoz-Barrutia A, Sage D (2021) Deepimagej: a user-friendly environment to run deep learning models in imagej. Nat Methods 18(10):1192–1195
77. Rüschendorf L (1985) The wasserstein distance and approximation theorems. Probab Theory Related Fields 70(1):117–129
78. Vallender S (1974) Calculation of the Wasserstein distance between probability distributions on the line. Theory Probab Appl 18(4): 784–786
79. Ollivier Y, Pajot H, Villani C (2014) Optimal transport: theory and applications, vol 413. Cambridge University Press, Cambridge
80. Villani C (2008) Optimal transport: old and new, vol 338. Springer, Berlin
81. Kolbe N (2021) Wasserstein distance
82. Chavez M, Grosselin F, Bussalb A, Fallani FDV, Navarro-Sune X (2018) Surrogate-based artifact removal from single-channel eeg. IEEE Trans Neural Syst Rehab Eng 26(3):540–550
83. Dora M, Holcman D (2022) Adaptive single-channel EEG artifact removal for real-time clinical monitoring. IEEE Trans Neural Syst Rehab Eng
84. Buzsáki G (2006) Rhythms of the brain. Oxford University Press, Oxford
85. Weiss SA, Song I, Leng M, Pastore T, Slezak D, Waldman Z, Orosz I, Gorniak R, Donmez M, Sharan A et al (2020) Ripples have distinct spectral properties and phase-amplitude coupling with slow waves, but indistinct unit firing, in human epileptogenic hippocampus. Front Neurol 11:174

86. Rigouzzo A, Girault L, Louvet N, Servin F, De-Smet T, Piat V, Seeman R, Murat I, Constant I (2008) The relationship between bispectral index and propofol during target-controlled infusion anesthesia: a comparative study between children and young adults. Anesthesia Analgesia 106(4):1109–1116
87. Rigouzzo A, Khoy-Ear L, Laude D, Louvet N, Moutard ML, Sabourdin N, Constant I (2019) EEG profiles during general anesthesia in children: A comparative study between sevoflurane and propofol. Pediatric Anesthesia 29(3): 250–257
88. Del Negro CA, Funk GD, Feldman JL (2018) Breathing matters. Nature Reviews Neuroscience 19(6):351–367

Part V

Dendritic Spines

Chapter 17

Probing Dendritic Function with Holographic Structured Illumination

Vincent R. Daria and Hans A. Bachor

Abstract

Holographic structured illumination has revolutionized techniques to investigate how neurons process information. Neurons receive synaptic inputs via its dendrites, and understanding the processing of those inputs and the influence of the dendrite's nonlinear properties are key components to decode the neuron's computing power. Recent breakthroughs to understand dendritic function make use of structured illumination to stimulate and/or record dendritic activity either via photochemical or optogenetic tools. As a stimulator, patterned light resembles multiple synaptic inputs onto dendrites, which can be used to investigate the nonlinear properties of dendrites and the input–output transfer function of neurons. For optical recording, light patterns can excite fluorogenic indicators (calcium or voltage) at multiple sites, and the fluorescence can be collected simultaneously using an image sensor array (e.g., camera). Simultaneous recording of the changes in the fluorescence from multiple sites enables the probing of critical spatiotemporal phenomena occurring in dendrites. The future points to combining the advantages of holographic photostimulation, optical recording, and three-photon excitation with adaptive wavefront correction to enable the visualization of dendritic activity at high spatiotemporal resolution in deep regions of the intact brain. Probing the integrative and nonlinear properties of dendrites will get us closer to understanding of how these neurons function as tiny computers in the brain.

Key words Holographic projection, Two-photon excitation, Functional calcium imaging, Two-photon photolysis, Wavefront correction, Light scattering through tissues

1 Introduction

In the past decade, light has gradually replaced electrodes for probing the activity of neurons. Laser-based techniques to probe neuronal activity have been instrumental in major breakthroughs in neuroscience [1–3]. In vivo experiments benefit from recent developments, as deeper sections of the intact brain can now be probed using new laser technologies [4–6]. When probing neurons embedded in optically thick brain tissues, light must be able to penetrate through the tissue. Moreover, light's interaction with cells should be confined within a localized volume so as not to

Joachim H. R. Lübke and Astrid Rollenhagen (eds.), *New Aspects in Analyzing the Synaptic Organization of the Brain*, Neuromethods, vol. 212, https://doi.org/10.1007/978-1-0716-4019-7_17,

excite other cells unnecessarily. Such localized interaction can be achieved using a nonlinear multi-photon excitation, which is now an established technique for stimulating and imaging neurons [5, 7–9].

In a conventional multi-photon microscope, imaging is performed by raster scanning a focused near-infrared ultrafast laser and imaging the fluorescence via a single-channel detector (e.g., photomultiplier tube). Apart from their use in imaging, laser scanning microscopes have been used for quasi-simultaneous multi-site photostimulation along the dendrites of neurons, by rapidly switching the position of the focus from one site to another [10–12]. Temporal patterns or sequences of photo-stimulated synaptic inputs can be implemented, which mimic correlated firing patterns from presynaptic neurons [12]. However, when using the scanning approach, the scanning speed defines the temporal resolution for imaging or for multi-site photostimulation. Such limitation depends on the speed of the beam scanning device. For example, the settling time of moving a galvanometer- controlled scanning mirror is limited by the size of the mirror and inertia. Scanning speeds have significantly improved with the use of acousto-optic deflectors (AODs) [13–15], and random-access multi-photon microscopes using AODs are now commonly used for optical recording of Ca^{2+} activity [16].

An alternative to the scanning beam approach is to simultaneously project three-dimensional (3D) light patterns, which was originally introduced for optical manipulation of microscopic particles [17–22]. Holograms encoded on a spatial light modulator (SLM) can split a single laser into multiple laser beams, which has now found applications for probing neuronal activity [23–28]. The laser beams can be focused, and each focus can function as multiple light probes for stimulation or recording of neuronal activity. These light probes can induce targeted photochemical reactions or excite genetically encoded opsins that stimulate cells at multiple locations simultaneously. Moreover, these light probes can also be used to excite fluorogenic calcium or voltage indicators enabling simultaneous optical recording from multiple sites.

Figure 1a summarizes and compares the capabilities of the three different optical approaches for studying neuronal function using patterned illumination. The *y*-axis represents the temporal response and corresponding neuronal responses (synaptic events, dendritic spikes, and Ca^{2+} responses), while the *x*-axis shows the number of stimulation or recording sites (N). A beam scanner using galvanometer mirrors (GM) is able to change the position of a focus within ~300 μs, while an AOD can deflect a laser beam within ~10 μs. On the other hand, holographic projection is able to provide simultaneous probing of up to a certain number of sites depending on the available laser power. GM scanners offer the highest optical throughput and can be useful for probing slow

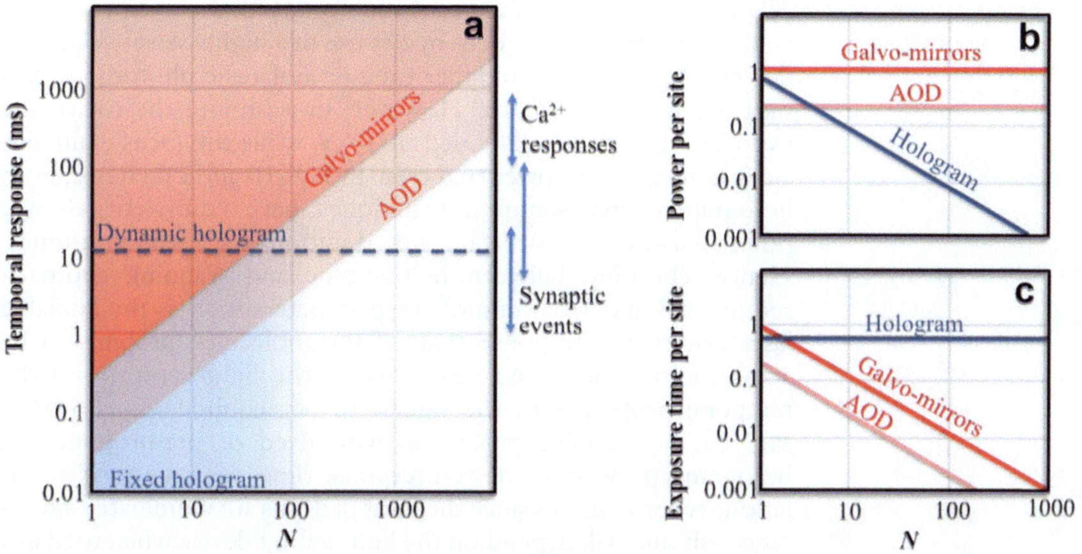

Fig. 1 Comparison between optical techniques. (**a**) Three techniques to produce patterned illumination: beam scanners using either (1) galvanometer-controlled mirrors (dark red) or (2) acousto-optic deflectors (AODs) (light red) and (3) holographic projection (blue). Temporal responses (left *y*-axis) can be associated with neuronal events (right *y*-axis) when used for studying neuronal circuits. (**b**) For *N* sites, the normalized exposure time per site for either scanning (red) technique is scaled by 1/*N*, while the holographic projection (bluo) can park thc illumination (constant) for the entire duration of the sampling time. (**c**) The normalized power per site is also scaled by 1/*N* in holographic projection (blue trace) but maintained in either scanning techniques (red). Hence, the average power per site over the sampling time is nearly constant for all systems

events (~1 s), while AOD scanners offer the fastest response but with low optical throughput (~50% per AOD). A two-dimensional (2D) scanning system uses two cascaded AODs with total throughput of ~25%, while a three-dimensional (3D) system uses four cascaded AODs with a total optical throughput of ~6%. Hence, the low throughput of AODs makes it a less preferred approach for multi-site photostimulation. Nonetheless, AODs have been used to perform random-access photostimulation of neurons with photolysis of caged neurotransmitters [11, 29] and genetically encoded opsins [30].

Holograms offer simultaneous projection of multiple foci. Holograms can be pre-programmed to project a fixed arrangement of multiple foci, which can be temporally gated using high-speed mechanical shutters [26], an AOD, or a digital micromirror device (DMD) [27, 31]. With high-speed computing algorithms implemented on a graphics processor unit (GPU), holograms can also be dynamically changed with temporal response limited by the refresh rates of the SLMs (up to ~100 Hz) [32]. While the power per site is scaled by $1/N$ in holographic projection, each focus can be parked at a longer exposure time, therefore improving the signal-to-noise ratio (SNR) when sensing the fluorescence simultaneously using a

2D image sensor (see blue traces in Fig. 1b and c). On the other hand, scanning a single focus makes use of a highly sensitive photodetector (e.g., photomultiplier tube or avalanche photodiodes) to measure the fluorescence. However, by scanning the focus, the exposure time per site is scaled by $1/N$, while the focus maintains a constant average power (red trace in Fig. 1b and c). Hence, both holographic and scanning techniques have equivalent average power per site over a certain time duration (e.g., sampling time). Hence, choosing between holographic and scanning approach requires an understanding of the parameters such as the available laser power, the response time of the molecules/proteins to be actuated, the speed and sensitivity of the light sensor, and the temporal properties of the signals to be acquired. As shown in Fig. 1a, holographic projection with fixed or pre-programmed holograms provides the fastest response times for a range of stimulation/recording sites since the light patterns are illuminated simultaneously and will depend on the light gating device when used as a stimulator or the camera frame rate when used for recording.

The key requirement in the projection of light patterns is to efficiently provide enough power suitable for molecular excitation. Optimal optical throughput can be achieved by spatially modulating the phase of the incident light. The phase-encoded light is optically transformed to project the desired spatial patterns at the output. Holographic projection systems use phase-only SLMs to shape the incident beam into different light patterns (e.g., multiple foci or user-defined light patterns). However, apart from the use of diffractive optics, there are factors that affect the optical throughput such as light scattering through biological tissues. Tissues reduce the number of photons arriving at the target. Tissues also cause distortions to the wavefront of the incident light that broaden the focus and degrade the resolution. These effects are detrimental in techniques requiring focal stimulation. Wavefront correction has been demonstrated to compensate for these distortions and can be implemented in existing holographic projection systems. Rectifying distortions of the wavefront of the light probes could provide optimal use of holographic systems for studying the brain. Other factors that affect the optical throughput of phase-only holograms are the truncation of its amplitude modulation component as well as poorly calculated holograms.

In this chapter, we discuss the applications of holographic projection systems for photostimulation and optical recording of dendritic activity in cortical neurons. We start by discussing basic holographic projection and how to incorporate wavefront correction for optimal light projection through brain tissues. We discuss its applications to photostimulation via multi-site two-photon photolysis of caged neurotransmitters. Multi-site photostimulation of synaptic inputs can be used to analyze the input/output transfer function of cortical pyramidal neurons. We also discuss how to use multi-foci patterns for optical recording of the membrane potential and Ca^{2+} activity along dendrites. As an optical recording technique, the fluorescence emitted from multiple excitation sites can be recorded simultaneously using a multi-channel detector such as a 2D image sensor (e.g., CCD camera). Parking the probe beam allows for longer dwell time (as compared to a scanned beam), thereby yielding a better signal-to-noise ratio (SNR). Most importantly, simultaneous illumination using multiple foci onto the sample enables high-speed detection of fluorescence with temporal resolution limited by the frame rate of the camera. With adaptive wavefront correction readily implementable in holographic setups, the future points to incorporating means to optimize light transmission to brain tissues to perform holographic stimulation and/or recording at deeper regions and targeting highly resolved dendrites and spines of the intact living brain. The ability to probe functioning neurons within their physiological environment sets the context of holographic projection to decode the role of dendrites and how single neurons function as tiny computers in our brain.

2 Structured Illumination via Holographic Projection

Structured illumination via holographic projection is achieved by calculating a computer-generated hologram (CGH) and encoding the hologram on an SLM (*see* Fig. 2a). When the SLM is illuminated, the optical transformation at the focus of a lens projects light patterns that corresponds to the encoded hologram. The expanded output in Fig. 2a shows a 3D light pattern of eight foci distributed in 3D. To calculate the hologram, several techniques have been proposed [33–36]. A CGH can be calculated by the superposition of prism and lens phase functions [36] or by an iterative optimization approach based on the Gerchberg–Saxton algorithm [33].

An analytic approach to calculate holograms is achieved by superposing prism and lens functions to calculate a hologram that projects arbitrary foci patterns in 3D. For N holographically projected multiple foci distributed around a volume of the focus of a lens, the input field representing the hologram, $h(u, v)$, is described by

$$h(u,v) = \sum_j^{N-1} \exp\left(i\frac{2\pi}{\lambda}\left(\frac{A}{R}\left(ux_j + vy_j\right) + \left(\frac{2Ar}{R}\right)Z_j + \Phi(u,v)\right)\right) \tag{1}$$

where (u, v) are transverse hologram coordinates, A is the numerical aperture of the lens, is the wavelength of the laser, and R is the radius of the operating region at the hologram, while $r = \sqrt{u^2 + v^2}$ is the radial coordinate from the optical axis. The output, on the

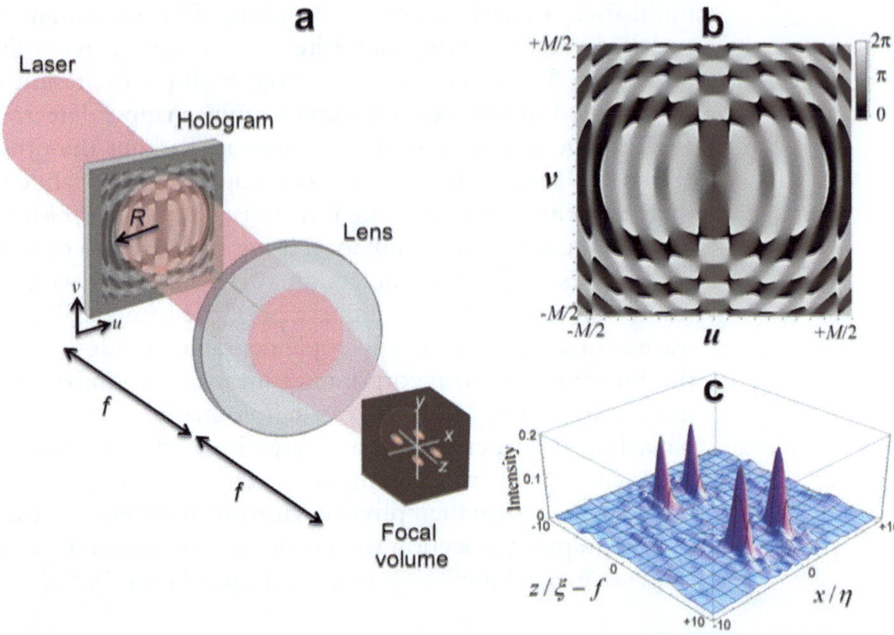

Fig. 2 Holographic projection of light patterns. (**a**) Schematic for phase-only holographic projection. (**b**) Representative hologram for producing four foci at the focal volume of a lens. (**c**) Numerically evaluated intensity distribution of four foci along *xz* plane using a laser encoded with holograms shown in **b**. Pseudo-grayscale of the phase holograms. Here, black and white are used for 0 and 2π phase shifts, respectively

other hand, is characterized by the 3D spatial coordinates (x, y, z), and the jth position of a focus has coordinates (x_j, y_j, z_j).

Equation 1 yields a complex field $h(u, v) = a(u, v)\exp(i\varphi(u, v))$, defined by an amplitude $a(u, v)$ and phase $\varphi(u, v)$. However, in practice, only the phase is encoded onto the SLM while setting $a(u, v) = 1$, which leads to the projection of spurious diffraction orders at the output [37]. The desired multiple foci with positions defined by (x_j, y_j, z_j) are associated with the first-order diffraction of light by the hologram. Moreover, a disturbing focus at the optical axis, also referred to as zero order, is projected from non-phase modulating areas of the SLM such as the region around its pixels. The focus at the zero order can be removed by introducing another out-of-phase focus at the center in addition to the desired multiple foci pattern [38]. On the other hand, a full complex field hologram can produce optimal 3D multiple foci array without spurious diffraction orders, but entails the use of two SLMs [39]. New methods to calculate holograms are able to produce uniform intensity distributions between the multiple foci [40]. Graphics processing units (GPU) can also be used to provide high-speed calculation and optimization of holograms [32, 34, 40].

Using the phase-only holograms generated, the 3D intensity distribution of the holographically projected multiple foci can be calculated using the Fresnel diffraction integral [32]. Figure 2b and c shows examples of phase-only CGHs and their corresponding intensity distributions along the *xz* plane (*see* Fig. 2d and e). The physical transverse and axial coordinates are scaled by $\eta = A \cdot \lambda^{-1}$ and $\xi = A^2 \cdot \lambda^{-1}$, respectively. Each diffraction-limited focus is characterized by the convolution of *N* delta functions with the 3D point-spread function of the optical system.

Apart from the phase information that yields structured illumination, Eq. 1 incorporates a third phase term ($\Phi(u, v)$), which refers to a wavefront correction term. $\Phi(u, v)$ acts as a global wavefront correction term that can be used to counter light scattering through biological tissues. $\Phi(u, v)$ may be derived via iterative optimization algorithms [41] or can be substituted with Zernike polynomials ($Z_n^m(u, v)$) [42]. Using Zernike modes to correct for scattering through biological tissues has been proposed previously by Rueckel et al. [43] and Zeng et al. [44], where they established tissue-specific aberration optimization techniques. Rahman and Booth [45] also used wavefront information retrieved from confocal back scattering to map optical distortion in tissue.

3 Photostimulation of Dendritic Activity

The cortex of the mammalian brain processes sensory inputs and generates motor commands among other functions. It has a laminar structure (*see* Fig. 3a) with six layers populated with pyramidal cells, which play a major role in information processing. Pyramidal neurons in layer 2/3 and layer 5 are characterized as having two major dendritic domains: apical and basal dendrites. The apical dendrites branch out from a single apical trunk, which extends from the apex of the soma opposite the axon. The basal dendrites, on the other hand, extend directly from the base of the soma.

Synaptic inputs onto the dendrites are processed by the neuron, which eventually fires an action potential via its axon. The axon's terminal boutons form synapses with dendritic spines of another neuron. Figure 3a illustrates a layer 2/3 pyramidal neuron whose terminal bouton is connected to an oblique dendrite of a layer 5 pyramidal neuron. Synaptic inputs onto spines of the receiving neuron are distributed along its entire dendritic tree. With the neuron's dendritic tree extending spatially in 3D and its inputs coming in temporal sequences, each neuron therefore receives multiple synaptic inputs that are spatiotemporally distributed in four dimensions (3D space and time). As the neuron processes these inputs, it fires an action potential, thereby propagating its processed information to other neurons in the network.

To understand how cortical circuits process information, we need to study the input–output transfer function of neurons, especially those of pyramidal neurons. Pyramidal neurons respond differently depending on the location of the synapse, whether it is

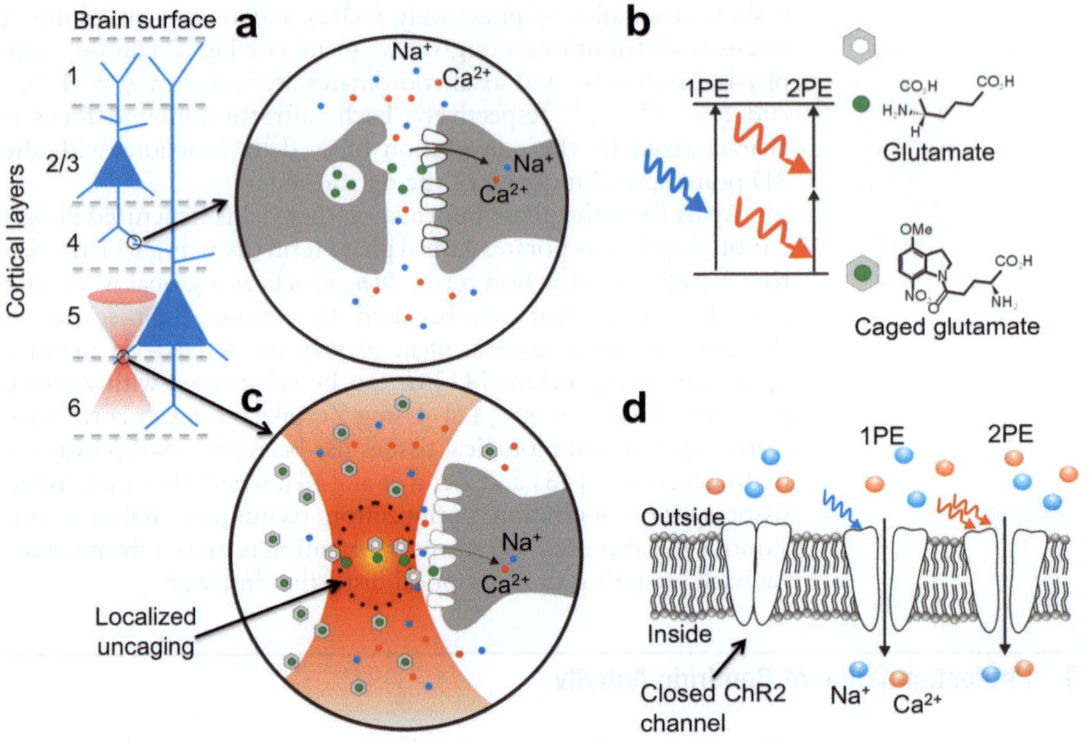

Fig. 3 Emulating a synaptic input using light. (**a**) Two pyramidal neurons in the cortex connected via a synapse. The inset shows a layer 2/3 neuron whose terminal bouton releases neurotransmitters (glutamate) to a spine of a layer 5 pyramidal neuron (recipient). (**b**) Photochemical reaction for photolysis of caged glutamate. (**c**) Single-photon (1P) and two-photon (2P) photolysis of caged neurotransmitters (MNI-caged-L-glutamate) for emulating a synaptic input. (**d**) Illustration of the gating of ions (Ca^{2+} and Na^{+}) into the neuron by genetically expressed light-activated channels (e.g., ChR2)

proximal or distal with respect to the soma. The variety of connections indicates that the various dendritic domains may play different roles in the processing of synaptic inputs. To understand these roles, we need to systematically analyze synaptic inputs from different dendritic locations of pyramidal neurons.

During a synapse, neurotransmitters are released from a sending neuron's terminal boutons and bind to receptors of the receiving neuron (*see* Fig. 3a). A neurotransmitter can bind to an α-amino-3-hydroxy-5-methyl-4-isoxazolepropionic acid (AMPA) receptor, which is an ion channel that opens up and allows entry of ions that cause the neuron to depolarize. As the neuron receives more synaptic inputs, it further depolarizes to reach a threshold and eventually fires an action potential consequently sending off neurotransmitters to other neighboring neurons in the network.

Photochemical tools have facilitated a systematic analysis of how neurons respond to synaptic inputs. For example, the synaptic input to a neuron can be emulated via photolysis of caged neurotransmitters. Photolysis can emulate neurotransmitter release from presynaptic cells. Figure 3b illustrates the energy diagram whereby a chemically caged neurotransmitter (e.g., caged glutamate) is freed from its cage. Depending on the chemical cage, the energy required to photolyze a caged neurotransmitter typically requires ultraviolet light to achieve single-photon (1P) photolysis. On the other hand, two-photon (2P) photolysis provides for a localized uncaging within the focus of the near-infrared ultrafast laser (*see* Fig. 3c). When released, the uncaged neurotransmitters bind to AMPA receptors causing the neuron to depolarize. The resulting voltage response can be measured via a patch-clamp recording from the neuron.

Aside from photolysis of caged neurotransmitters, the membrane of neurons can be encoded with “opsins” or light-activated ion channels (e.g., channelrhodopsin-2, ChR2), which when illuminated by light, by either 1P or 2P excitation, opens up and facilitates ion exchange similar to synaptic inputs (*see* Fig. 3d) [46]. These channels can be expressed onto the membrane using either transfection or transgenic techniques [47, 48].

Holographic projection can be used to trigger synaptic inputs and analyze the response of the cell [23, 24, 26–28, 49, 50]. However, in order to effectively study the input–output properties of neurons, the excitation light needs to be shaped and accurately positioned along spines of the neuron’s dendritic tree. Hence, it is important to render a 3D image of neurons prior to photostimulation. While other imaging modalities can be used, the laser scanning 2P microscope [7] facilitates rendering of the 3D morphology of neurons providing accurate maps of the dendritic arborizations. Once the image of the neuron is acquired, an appropriate hologram can be calculated to target photostimulation sites along the neuron’s dendritic tree.

From the image of the cell, a map of the dendritic tree can be used for locating stimulation sites and shaping the light. Once the stimulation sites are identified, the hologram can be calculated to position light at specific sites along the dendrite. To perform photolysis, caged neurotransmitters are applied into the bath of the brain slice. Uncaging of MNI-caged-L-glutamate is achieved by illuminating the sample for about 2 ms with laser wavelength set to λ ~ 360 nm for single-photon (1P) or λ ~ 720 nm for 2P photolysis. The neuron’s electrical response to photolysis can be recorded using the same electrode used to deliver the imaging dye into the neuron.

Figure 4a illustrates how structured illumination can be used to stimulate synapses at particular sites along the dendritic tree of cortical neurons. The neuron depolarizes in response to uncaging, and the excitatory postsynaptic potential (EPSP) or current (EPSC) depends on the total uncaging laser energy and number of uncaging sites. Figure 4b is adapted (with permission) from Lutz et al. [23], where a holographically shaped light is used for 1P photolysis

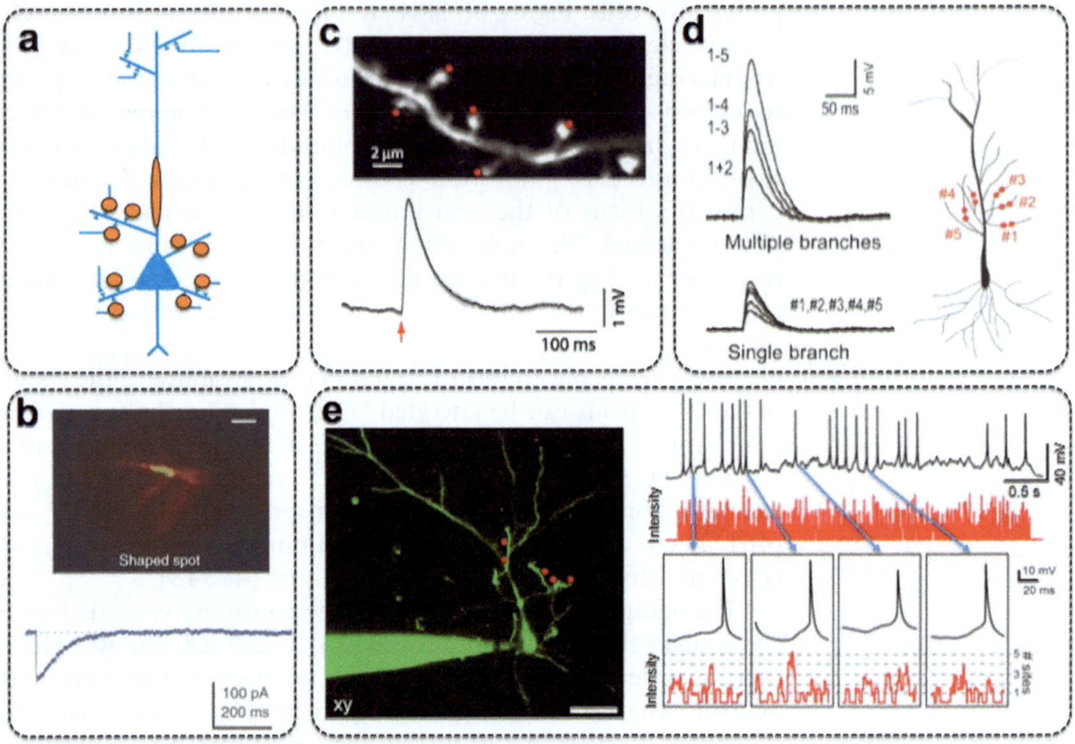

Fig. 4 Holographic photolysis of caged neurotransmitters. (**a**) Schematic of a pyramidal neuron indicated with uncaging sites for studying apical oblique and basal dendrites. (**b**) Adapted with permission from Lutz et al. [23]. (top) Light is shaped for 1P uncaging of a dendrite of a stellate cell. (bottom) EPSC response evoked by the cell during uncaging. (**c**) Adapted with permission from Nikolenko et al. [24]. (top) 2P image of a basal dendrite showing spines stimulated by 2P uncaging. The red spots indicate sites of simultaneous holographic 2P uncaging. (bottom) EPSP averaged from several trials. (**d**) Adapted with permission from Yang et al. [49]. Neuron trace of a hippocampal pyramidal neuron where it depolarizes during multi-site 1P uncaging along a single dendrite and from multiple dendrites. (**e**) Adapted with permission from Go et al. [27]. (left) 2P image of a layer 2/3 pyramidal neuron. (top left) Firing of action potentials by randomized synaptic inputs, where the intensity plot is related to the number of simultaneous inputs. (bottom left) Magnified traces show that when the neuron receives at least three simultaneous inputs, the neuron fires an action potential

($\lambda = 405$ nm) of MNI-caged-L-glutamate at a dendrite of a stellate cell in a cerebellar slice. An image of the cell loaded with Alexa Fluor 594 dye is acquired using widefield epifluorescence. Bottom trace shows the EPSC response evoked by the cell using the shaped light stimulation. Figure 4c is adapted (with permission) from Nikolenko et al. [24], showing spines of a basal dendrite from a layer 5 pyramidal neuron are stimulated by 2P photolysis (725 nm) of MNI-caged-L-glutamate. The cell is loaded with Alexa Fluor 488, and a high-contrast image of the spines is obtained via a 2P microscope. The red spots indicate sites of simultaneous uncaging via holographic projection of multiple foci. Neuronal response was measured via whole-cell patch-clamp recording, and the EPSP averaged from several trials is shown at the bottom trace of Fig. 4c.

Introducing multiple foci around the neuron's dendritic tree allows for studying the neuron's input–output transfer function [26, 27, 49]. Figure 4d is adapted (with permission) from Yang et al. [49], showing how a hippocampal pyramidal neuron depolarizes following multi-site single-photon photolysis along a single dendrite and from multiple dendrites. Simultaneous stimulation at multiple sites produces larger depolarization. With enough depolarization from multiple uncaging sites, neurons may fire action potentials. Figure 4e (left) is adapted from Go et al. [27], which shows a flattened image (from a 3D render) of a cortical layer 2/3 pyramidal neuron where red spots indicate five uncaging sites. The average power per site is set to 20 mW. The five uncaging sites are distributed among the basal and apical oblique dendrites in one plane and are randomly gated for uncaging pulses of 4 ms. Go et al. [27] used a digital micromirror device (DMD) at the image conjugate plane of the sample to gate each focus into temporal sequences. Figure 4e (right) shows EPSPs evoked from randomly turning on and off the uncaging at the sites. The spatiotemporal uncaging of five sites is sufficient to elicit action potentials in cortical neurons. The level of light gated by the DMD (see intensity plot in Fig. 4e) gives information on the number of stimuli active at any uncaging event. The DMD allows for changing of the photostimulation pattern within 750 μs. Using the DMD also has the added advantage of cleaning up the multi-site photostimulation pattern produced by the SLM. Micromirrors that are switched off redirect the zero and high diffraction orders off the region of interest and therefore eliminate the risk of unintended uncaging [27].

Figure 5 is adapted from Go et al. [28], which shows a 2P image of a layer 2/3 pyramidal neuron labeled with calcium indicator (OGB-1) loaded via the patch pipette (*see* Fig. 5a). Inset shows a high-magnification flattened multi-stack image of a basal dendrite with visible spines. Figure 5b (bottom left) shows the volume of interest displayed as an xz image with three representative image planes (i, $z = -5$ μm; ii, $z = 0$; and iii, $z = +5$ μm. Scale bar: 5 μm). Figure 5c shows somatic EPSPs of individual uncaging events from ten uncaging sites, while Fig. 5d shows the response when the number of simultaneous inputs is increased from 1 to 10. Figure 5e summarizes the dendrite's nonlinear response to increasing synaptic inputs as it plots the measured EPSP as compared to the sum of EPSP amplitudes for a given number of inputs.

4 Functional Imaging of Dendritic Activity

Aside from using light for stimulation, it can also be used to record neuronal activity via either functional calcium or voltage imaging. Neurons can be loaded or expressed with fluorogenic calcium or

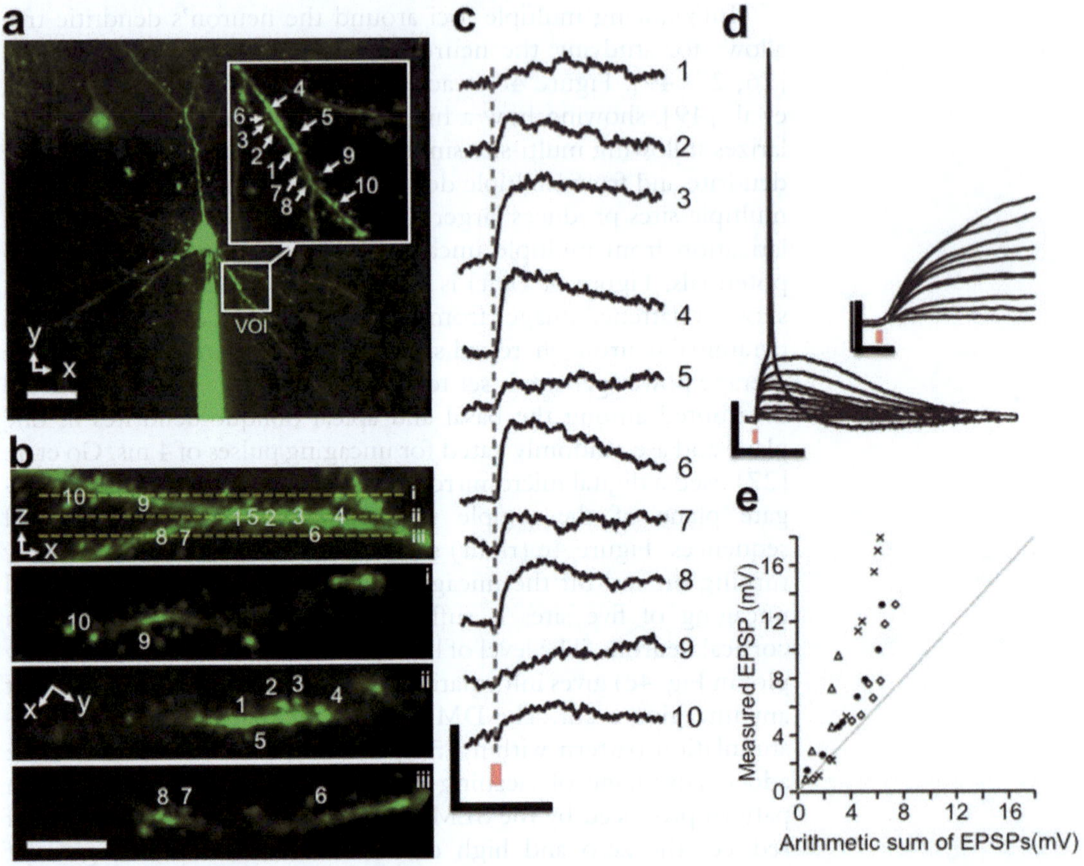

Fig. 5 Adapted with permission from Go et al. [28]. (**a**) Two-photon image of a layer 2/3 pyramidal neuron with high magnification of the dendrite shown as inset. (**b**) Volume of interest displayed as an xz image with three representative image planes (i, $z = -5$ μm; ii, $z = 0$; and iii, $z = +5$ μm). (**c**) Somatic EPSPs of individual uncaging events. Scale bars: 1 mV, 40 ms. (**d**) Uncaging responses with increasing number of simultaneous uncaging sites. Scale bars: 20 mV, 50 ms. Inset: magnified EPSP rise times for increasing number of uncaging sites. (**e**) Integration plot of measured multi-site uncaging response against the arithmetic sum of individual uncaging events. The dashed line indicates linear summation

voltage indicators, and the changes in the fluorescence are correlated with neuronal activity. Hence, optical recording via functional imaging can provide spatiotemporal analysis on the function of individual neurons as well as how populations of neurons communicate with each other.

In functional calcium imaging, the changes in intracellular calcium concentration are correlated with neuronal activity especially when the neuron fires action potentials. Within a single neuron, functional calcium imaging has been used to study Ca^{2+} influx into dendrites [51–53]. Membrane-permeable forms of Ca^{2+} indicators (acetoxymethyl (AM) ester) and genetically encoded Ca^{2+} indicators (GECI) can be used to analyze network activity of populations of neurons in an intact living brain (in vivo). The most

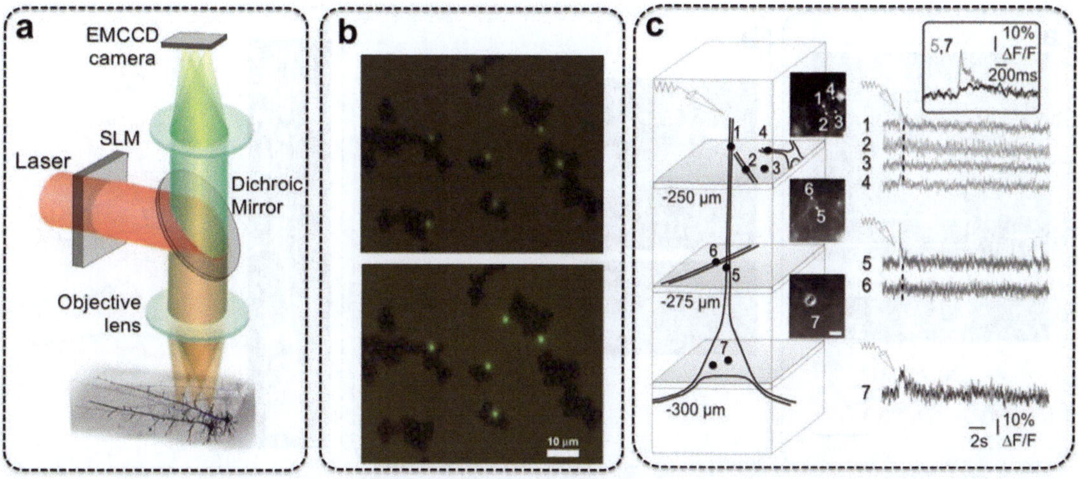

Fig. 6 Functional calcium imaging. (**a**) During an action potential (top trace), the intracellular Ca^{2+} concentration (bottom trace) increases with Ca^{2+} ions that enter via Ca^{2+} channels (green). Extrusion of Ca^{2+} by Ca^{2+} pumps (blue) allows intracellular Ca^{2+} to return back to the resting state. (**b**) Holographic projection of multiple foci along a plane of dispersed fluorescent microbeads. (**c**) Schematic for applying multi-site two-photon excitation of Ca^{2+} indicators along different regions of the neuron's dendritic tree for multi-site functional calcium imaging

common and recently used GECI are the GCaMPs and their derivatives [54].

While calcium imaging has been instrumental in studying the activity of neuronal populations, it lacks the required temporal resolution to analyze fast nonlinear responses along dendrites of neurons. Using light to sense the changes in membrane voltage offers a promising solution but still remains a challenge. Voltage imaging is achieved by measuring the electric field across the thin plasma membrane using voltage sensitive dyes (VSDs) or genetically encoded voltage indicators (GEVIs). However, the signal-to-noise ratio (SNR) is very low, and fast neuronal responses require exposure times that are approaching the acquisition limits of multi-pixel image sensors (e.g., camera).

Holographic projection can be used for simultaneous detection of 2P excited fluorescence from either calcium or voltage indicators. Figure 6a shows the optical layout of the holographic projection system for multi-site functional imaging of dendritic activity. Structured illumination targets different sites of interest, and the fluorescence from the various sites can be simultaneously recorded using a highly sensitive camera. Figure 6b is adapted from Daria et al. [32], which shows simultaneous detection of fluorescence excited by multiple foci along the transverse plane. Each focus targets a fluorescently labeled bead (2 µm diameter), and the fluorescence is detected onto a 2D sensor array (e.g., camera). Apart from multiple foci, shaped light can also be used to improve fluorescence signal detection, especially with 1P excitation.

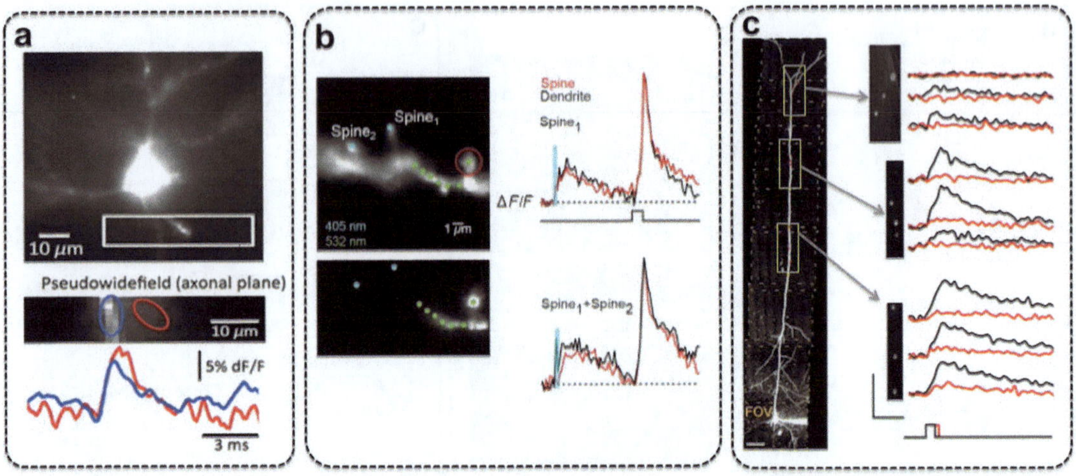

Fig. 7 (**a**) Schematic for holographic projection of multiple foci for simultaneous optical recording of neuronal activity. (**b**) Reconstructed image of a L5 pyramidal neuron loaded with calcium indicator. Scale bar is 50 μm. (**c**) Cross-sectional images of the apical dendrite with holographically projected multiple foci (red circles) and the fluorescence from the multiple foci as viewed under the EMCCD camera with limited region of interest (ROI). (**d**) The fluorescence signal from each focus plotted as a function of time in response to current pulse trains with 25 Hz (gray), 70 Hz (black), and 95 Hz (red)

While multi-site fluorescence detection can be achieved using a camera, the sensitivity of single-channel detectors (e.g., PMTs and avalanche photodiodes) is more superior than 2D sensor arrays. Ducros et al [31] used a PMT to perform multi-site calcium imaging via structured illumination. To distinguish fluorescence from multiple sites, each focus is gated temporally using a DMD, and the sequence of gated fluorescence represents spatially distributed recording sites. Figure 6c is adapted from Ducros et al. [31], where signals from different dendritic regions are detected using a single PMT. The DMD encodes each focus with a specific binary amplitude modulation sequence, and the fluorescence signals from all independently targeted sites are collected simultaneously onto the PMT and site-specifically decoded. Figure 6c (right) shows calcium signals from several locations (dendrites and soma) of a layer 2/3 pyramidal cell, which was loaded with calcium indicator (OGB1). The recorded signals are analyzed as fractional changes of the fluorescence (ΔF/F) and plotted as a function of time.

Using cameras with highly sensitive sensors (e.g., electron-multiplying charged coupled device (EMCCD) or scientific complementary metal-oxide (sCMOS) 2D sensors) allows for high-speed sensing of fast neuronal activity (~1 ms). Figure 7a is adapted from Foust et al. [55], where voltage imaging using holographically sculpted light enables discrimination of spike kinetic properties from neighboring axons and dendrites imaged using a high-speed camera (NeuroCCD). Fluorescence from the neuron loaded with VSD (JPW3028) shows voltage transients from axons (blue) and basal dendrites (red) regions with a holographically shaped illumination. Figure 7a (top) shows an image of the neuron indicated with a region of the axon and basal dendrite where voltage imaging was performed. The middle image shows section where the dendrite is in focus, while the axon (deeper) is out focus. Light is holographically targeted to excite fluorescence and record voltage transients only in dendritic tip (red trace), without exciting fluorescence in the adjacent axon in an out-of-focus deeper axial plane (blue trace).

Figure 7b is adapted (with permission) from Tanese et al. [56] where they combined voltage imaging with glutamate uncaging using holographic structured illumination. They performed simultaneous voltage imaging and scanless glutamate uncaging from individual spines using two independently controlled and synchronized CGH patterns of illumination. Figure 7b (left-top) shows the fluorescence image of a section of basal dendrite (taken via widefield illumination) used to generate illumination pattern for voltage imaging (green) and glutamate uncaging (blue). Figure 7b (left-bottom) shows the same dendritic section with patterned illumination indicated by the spots. Figure 7b (right) shows the optical signals corresponding to a subthreshold signal evoked by glutamate uncaging onto one spine (top) and two spines (bottom) succeeded by a supra threshold back propagating action potential evoked by depolarizing current pulse. Signals from spine head (red) and from parent dendrite (black) are shown.

Despite the slow response, fluorogenic calcium indicators facilitate optical recording of the changes in intracellular Ca^{2+} concentration and correlated with dendritic activity. Castanares et al. [57] used the holographic approach to analyze layer 5 pyramidal neurons in the cortex. These neurons have complex dendritic branches and are known to propagate dendritic spikes due to nonlinear conductance of voltage-gated ion channels. The neuron was loaded with Ca^{2+} indicator (Cal-520) via the patch pipette, and after about 30 min of loading the indicator, a 2P image was rendered (*see* Fig. 7c, left). From the 3D image, an accurate map of the dendritic tree can be obtained for locating regions of interest for observing changes in calcium concentration. To induce activity, current is injected onto the soma via a whole-cell patched electrode operating in current-clamp mode. The holographic structured illumination was used to perform multi-site Ca^{2+} imaging of dendritic spikes using the critical frequency method [53]. Fluorescence emanating from the various foci is simultaneously acquired using the EMCCD camera. By limiting the region of interest (ROI) and pixel binning of the camera to acquire only the region where the various foci are located, the speed of acquisition can be increased. Castanares et al. [57] were able to achieve up to 400 fps using an EMCCD camera (iXon Ultra 897, Andor). Figure 7c (right) shows the fluorescence from the multiple foci positioned along apical dendrites as captured by the camera. The traces are single-trial Ca^{2+} responses for each focus as a function of time [57]. A 25 Hz pulse train (gray traces) evoked a train of four action potentials, which resulted in small Ca^{2+} responses. However, 70 Hz (black traces) and 95 Hz (red traces) pulse trains resulted in significantly higher Ca^{2+} events. The increased Ca^{2+} responses represent dendritic Ca^{2+} spikes evoked at the nexus of the apical tuft dendrites [53].

5 Optimal Light Projection Through Brain Tissues

When using light to stimulate or record neuronal activity in brain tissue, scattering degrades the focus and hampers delivery of light unto neurons embedded deep within the tissue. Figure 8a demonstrates the effect of scattering as the focus is pushed deeper into the brain tissue to target certain dendritic structures. The point-spread function (PSF) of a focused laser broadens as the focus is positioned

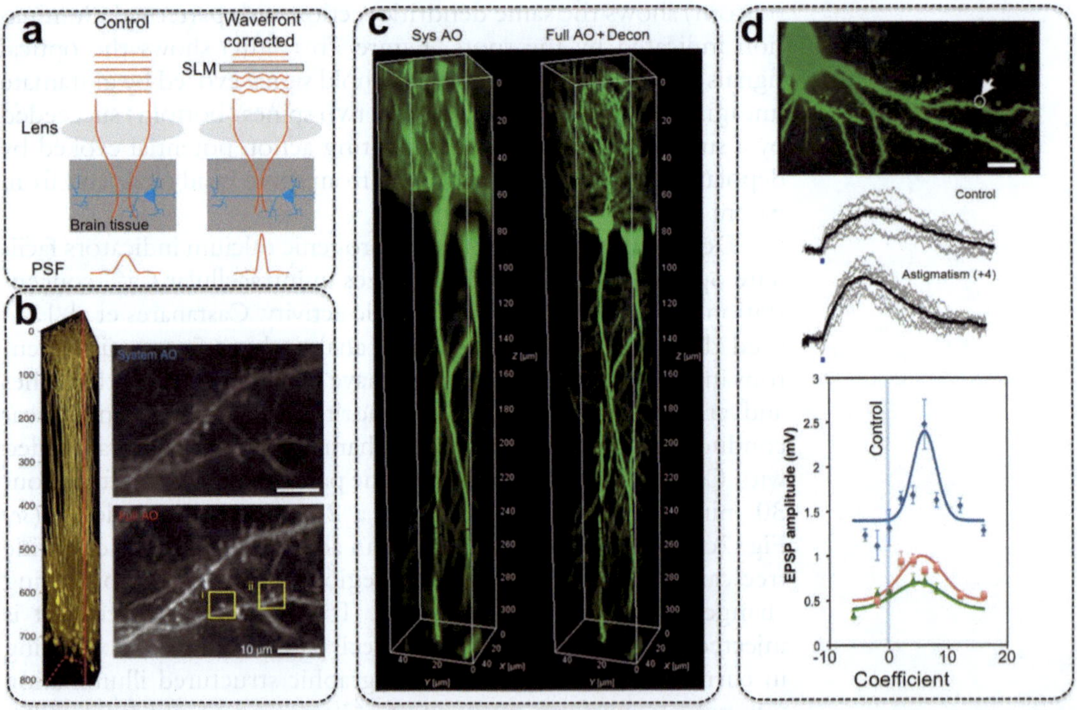

Fig. 8 Optimal 2P photolysis of caged glutamate with wavefront correction. (**a**) Cartoon of how light is distorted through biological media that broadens the focus at deeper regions. (**b**) Representative 2P image of a layer 5 pyramidal neuron showing location of the dendritic spine targeted for 2P photolysis. (**c**) Light-evoked EPSP responses (black traces) each superimposed on ten single episodes (gray traces) without (top) and with correction for astigmatism (bottom). (**d**) Plot of the EPSP amplitude as a function of strength of astigmatism correction (coefficient) from three different neurons (red, green, and blue). (**e**) Similarly, light-evoked EPSP responses without (top) and with correction for spherical aberration (bottom). (**f**) Plot of the EPSP amplitude as a function of strength of spherical aberration correction. Scale bar 20 μm

deeper into the tissue. The path of the incident light is scattered by the tissue, and for some types of media, the scattered paths can be treated as roughly deterministic and may be possible to build a correcting wavefront to counter the effect of scattering [58, 59].

Using the same holographic setup described in the previous section, the hologram can be incorporated with a correcting term (third phase term in Eq. 1) to adapt the wavefront of the incident beam with the light distortions caused by the brain tissue. The process is referred to as adaptive optics (AO), which introduces a corrective term to effectively reverse the distortions in the wavefront to produce an optimal focus. Several techniques have been proposed to measure and reverse the optical distortion in tissues [60, 61] based on photo-conjugation [58], sensorless imaging [62, 63], artificial guide star [64], and wavefront sensors [65].

Figure 8b is adapted from Liu et al. [66] where adaptive wavefront control was implemented with two-photon microscopy to perform in vivo imaging of the column in the visual cortex. Figure 8b (left) shows mRuby2-labeled layer 5 pyramidal neurons in a $150 \times 150 \times 810\,\mu m^3$ volume of vS1 cortex. Full AO corrections were applied at every 50 μm of depth. Figure 8b (right) shows a section showing dendritic spines at 710 μm below the pia in the same mouse as in Fig. 8e with system (top) and full (bottom) AO correction. Imaging deeper into brain regions allows for probing of dendritic activity necessary for understanding neuronal computation. 2P endomicroscopes allow for less invasive imaging through the skull as it uses a graded index lens that can be implanted in the skull. Figure 8c is adapted from Qin et al. [67], where they incorporated AO to correct for tissue aberrations and facilitate diffraction-limited deep brain imaging of dendritic structures in the hippocampus in mice. Qin et al. [67] used their 2P endomicroscope with AO to reconstruct a column of hippocampal CA1 pyramidal neurons imaged with system AO (left) and with their novel full correction plus deconvolution (right) to investigate the relationship between somatic and dendritic activity.

Deriving an adaptive wavefront to correct for arbitrary distortions caused by the tissue may be time-consuming. Instead of deriving an adaptive wavefront, light distortions may be corrected by pre-deriving a particular wavefront correction pattern identified for specific regions in the brain tissue. Such technique corrects for light distortions without using wavefront sensing [59]. Figure 8c (top) shows a representative image of a cortical pyramidal neuron showing a position of an uncaging site along its basal dendrite [42]. The basal dendrites of the neuron are located around 50–100 μm from the top surface of the parasagittal brain slice. Figure 8c (middle) shows the corresponding EPSP responses following 2P photolysis at the target spine indicated on the top figure. The EPSP responses are optimized by encoding a Zernike mode correction (astigmatism) onto the SLM via Eq. 1. Control plots are traces with no wavefront correction or when the Zernike coefficient for astigmatism is zero. Figures 8c (bottom) summarizes the changes in EPSP amplitude as a function of Zernike coefficient. Note that the coefficient refers to the strength of correction and may vary depending on the location of stimulation. The use of Zernike modes limits the optimization to 15 modes as compared to adjusting the phase of an array of $M \times M$ phase-shifting pixels of the SLM [41]. Depending on the spatial resolution of the SLM, typical number of pixels of an SLM ranges from $M = 600$ to 2048. Wavefront corrections can hence be made using this finite set of Zernike modes. Furthermore, selecting specific regions within the brain slice (e.g., cortex, hippocampus, etc.), the complexity can be reduced to about three to four Zernike modes, effectively reducing a $M \times M$-dimensional problem to three or four dimensions.

6 Conclusion

In this book chapter, we discussed how holographically projected multi-foci patterns can be used to stimulate and record the activity of single neurons. The hologram can be programmed to arbitrarily project multiple foci anywhere around the neuron in three dimensions. Each focus can be used to trigger synaptic inputs by two-photon photolysis of caged neurotransmitters. Hence, the input–output transfer function of different neurons can be studied either by simultaneous or temporally gated projection of multiple foci.

In two-photon excitation, the interaction is localized within the focus, which can be positioned to target dendritic spines without triggering other spines unnecessarily. When targeting spines in the brain tissue, light is scattered, and this hampers the delivery of light for stimulation. To counter the effect of tissue scattering, the hologram may include a correcting wavefront to improve focal stimulation. The use of Zernike modes limits the optimization to three to four modes as compared to adjusting the phase of an array of $M \times M$ phase-shifting pixels of the SLM. Lengthy optimization algorithms to find a more suitable wavefront correction are not applicable to time-critical experiments, such as those involving acute brain slices. Certain Zernike modes have been identified to correct for specific brain regions such as the cortex and the hippocampus of a rat brain in a parasagittal slice preparation.

The multiple foci can also be used for two-photon excitation of fluorogenic indicators, and the fluorescence emanating from the different foci can be collected simultaneously using a camera. Using a camera to monitor the temporal changes in the fluorescence enables observation of time-critical phenomena such as dendritic activity within a single neuron.

These techniques discussed in this book chapter allow probing of the integrative and nonlinear properties of the dendrites, which could provide us with a better understanding of how these neurons function. While recent experiments involving all-optical interrogation have already partially combined holographic with other optical techniques (e.g., scanning) [24, 28, 50, 68–73], the future points to the combination of these holographic techniques to exploit their benefits when used for photostimulation and optical recording. Holographic projection provides simultaneous illumination over all sites setting an exposure time that is equivalent to the sampling time. In contrast, high-speed scanning techniques need to subdivide the sampling time with the number of sites in order to achieve patterned stimulation, thereby setting a very short exposure time for both stimulation and recording.

Finding the right tools to probe neuronal activity is key to understanding the complex mechanisms involved during information processing within neurons. The pros and cons of the different techniques need to be assessed and matched with specific experimental requirements for answering certain neuroscience questions. The spatiotemporal resolution afforded by holographic projection has been useful for probing dendritic activity.

Ongoing improvements of the SLM technology (e.g., optical throughput, diffraction efficiency, and high refresh rate), of com puting efficacy for high-speed calculation of optimal holograms, and of photochemical/optogenetic tools (e.g., less cross talk, sensitivity, and speed) will enable holographic techniques to be applied to in vivo experiments and for real-time investigation of information processing between neurons. This enhanced visualization can be the basis for a better understanding of the integrative properties of dendrites and their functional role in the overall neuronal function.

Acknowledgments

Experimental results presented in this chapter were reproduced and slightly modified from published works of former colleagues: Mary Ann Go, Julian Choy, Sharmila Sáne, and Michael Castañares. This work was supported by the Australian Research Council Discovery Project (DP140101555) and the National Health and Medical Research Council (PG1105944).

References

1. Daria VR, Bachor HA (2015) Using light to probe neuronal function. Europhys Lett 111(3):38001
2. Go MA, Daria VR (2017) Light-neuron interactions: key to understanding the brain. J Opt 19(2):023002

3. Daria VR, Castanares ML, Bachor H (2021) Spatio-temporal parameters for optical probing of neuronal activity. Biophys Rev 13:13–33
4. Mittmann W et al (2011) Two-photon calcium imaging of evoked activity from L5 somatosensory neurons in vivo. Nat Neurosci 14(8): 1089–1093
5. Wang TY, Xu C (2020) Three-photon neuronal imaging in deep mouse brain. Optica 7(8): 947–960
6. Wang T et al (2020) Quantitative analysis of 1300-nm three-photon calcium imaging in the mouse brain. elife 9:e53205
7. Denk W, Strickler JH, Webb WW (1990) Two-photon laser scanning fluorescence microscopy. Science 248(4951):73–76
8. Svoboda K et al (1997) In vivo dendritic calcium dynamics in neocortical pyramidal neurons. Nature 385(6612):161–165
9. Helmchen F, Denk W (2005) Deep tissue two-photon microscopy. Nat Methods 2(12): 932–940
10. Denk W (1994) Two-photon scanning photochemical microscopy: mapping ligand-gated ion channel distributions. Proc Natl Acad Sci USA 91(14):6629–6633
11. Shoham S et al (2005) Rapid neurotransmitter uncaging in spatially defined patterns. Nat Methods 2(11):837–843
12. Branco T, Clark BA, Hausser M (2010) Dendritic discrimination of temporal input sequences in cortical neurons. Science 329(5999):1671–1675
13. Reddy GD, Saggau P (2005) Fast three-dimensional laser scanning scheme using acousto-optic deflectors. J Biomed Opt 10(6): 064038
14. Reddy GD et al (2008) Three-dimensional random access multiphoton microscopy for functional imaging of neuronal activity. Nat Neurosci 11(6):713–720
15. Katona G et al (2012) Fast two-photon in vivo imaging with three-dimensional random-access scanning in large tissue volumes. Nat Methods 9(2):201–208
16. Nadella KM et al (2016) Random-access scanning microscopy for 3D imaging in awake behaving animals. Nat Methods 13(12): 1001–1004
17. Eriksen RL, Daria VR, Gluckstad J (2002) Fully dynamic multiple-beam optical tweezers. Opt Express 10:597–603
18. Daria VR, Eriksen R, Glückstad J (2003) Dynamic manipulation of colloidal structures using a spatial light modulator. J Mod Opt 50: 1601–1614
19. Rodrigo PJ, Daria VR, Glückstad J (2005) Four-dimensional optical manipulation of colloidal particles. Appl Phys Lett 86:074103
20. Curtis JE, Koss A, Grier DG (2002) Dynamic holographic optical tweezers. Opt Commun 207:169–175
21. Dufresne ER, Grier DG (1998) Optical tweezer arrays and optical substrates created with diffractive optics. Rev Sci Instrum 69(5): 1974–1977
22. Daria VR, Go MA, Bachor HA (2011) Simultaneous transfer of linear and orbital angular momentum to multiple low-index particles. J Opt 13:044004
23. Lutz C et al (2008) Holographic photolysis of caged neurotransmitters. Nat Methods 5(9): 821–827
24. Nikolenko V et al (2008) SLM microscopy: scanless two-photon imaging and photostimulation with spatial light modulators. Front Neural Circuits 2:393
25. Papagiakoumou E et al (2010) Scanless two-photon excitation of channelrhodopsin-2. Nat Methods 7(10):848–U117
26. Go MA et al (2012) Simultaneous multi-site two-photon photostimulation in three dimensions. J Biophotonics 5(10):745–753
27. Go MA et al (2013) Four-dimensional multi-site photolysis of caged neurotransmitters. Front Cell Neurosci 7:231
28. Go MA et al (2019) A compact holographic projector module for high-resolution 3D multi-site two-photon photostimulation. PLoS One 14(1):e0210564
29. Losavio BE et al (2010) Acousto-optic laser scanning for multi-site photo-stimulation of single neurons in vitro. J Neural Eng 7(4): 045002
30. Liu Y et al (2012) Instrumentation of a compact random-access photostimulator based on acousto-optic deflectors. Rev Sci Instrum 83(2):025116
31. Ducros M et al (2013) Encoded multisite two-photon microscopy. Proc Natl Acad Sci USA 110(32):13138–13143
32. Daria VR et al (2009) Arbitrary multisite two-photon excitation in four dimensions. Appl Phys Lett 95(9)
33. Gerchberg RW, Saxton WO (1972) Practical algorithm for determination of phase from image and diffraction plane pictures. Optik 35(2):237
34. Reicherter M et al (2006) Fast digital hologram generation and adaptive force measurement in liquid-crystal-display-based holographic tweezers. Appl Opt 45(5): 888–896

35. Zhao Y et al (2015) Accurate calculation of computer-generated holograms using angular-spectrum layer-oriented method. Opt Express 23(20):25440–25449
36. Liesener J et al (2000) Multi-functional optical tweezers using computer generated holograms. Opt Commun 185:77–82
37. Palima D, Daria VR (2006) Effect of spurious diffraction orders in arbitrary multi-foci patterns produced via phase-only holograms. Appl Opt 45:6689–6693
38. Palima D, Daria VR (2007) Holographic projection of arbitrary light patterns with a suppressed zero-order beam. Appl Opt 46(20): 4197–4201
39. Go MA et al (2011) Optimal complex field holographic projection. Opt Lett 36(16): 3073–3075
40. Pozzi P, Mapelli J (2021) Real time generation of three dimensional patterns for multiphoton stimulation. Front Cell Neurosci 15:609505
41. Vellekoop IM, Mosk AP (2007) Focusing coherent light through opaque strongly scattering media. Opt Lett 32(16):2309–2311
42. Choy JMC et al (2017) Improving focal photostimulation of cortical neurons with pre-derived wavefront correction. Front Cell Neurosci 11:105
43. Rueckel M, Mack-Bucher JA, Denk W (2006) Adaptive wavefront correction in two-photon microscopy using coherence-gated wavefront sensing. Proc Natl Acad Sci USA 103(46): 17137–17142
44. Zeng J et al (2012) 3D resolved mapping of optical aberrations in thick tissues. Biomed Optics Express 3(8):1898–1913
45. Rahman SA, Booth MJ (2013) Direct wavefront sensing in adaptive optical microscopy using backscattered light. Appl Opt 52(22): 5523–5532
46. Nagel G et al (2003) Channelrhodopsin-2, a directly light-gated cation-selective membrane channel. Proc Natl Acad Sci USA 100(24): 13940–13945
47. Zhang F et al (2010) Optogenetic interrogation of neural circuits: technology for probing mammalian brain structures. Nat Protoc 5(3): 439–456
48. Zeng HK, Madisen L (2012) Mouse transgenic approaches in optogenetics. Prog Brain Res 196:193–213
49. Yang S et al (2011) Three-dimensional holographic photostimulation of the dendritic arbor. J Neural Eng 8(4):046002
50. Emiliani V et al (2015) All-optical interrogation of neural circuits. J Neurosci 35(41): 13917–13926
51. Tank DW et al (1988) Spatially resolved calcium dynamics of mammalian Purkinje-cells in cerebellar slice. Science 242(4879):773–777
52. Yuste R et al (1994) Ca2+ Accumulations in Dendrites of Neocortical Pyramidal Neurons: an Apical Band and Evidence for 2 Functional Compartments. Neuron 13(1):23–43
53. Larkum ME, Kaiser KM, Sakmann B (1999) Calcium electrogenesis in distal apical dendrites of layer 5 pyramidal cells at a critical frequency of back-propagating action potentials. Proc Natl Acad Sci USA 96(25):14600–14604
54. Nakai J, Ohkura M, Imoto K (2001) A high signal-to-noise Ca2+ probe composed of a single green fluorescent protein. Nat Biotechnol 19(2):137–141
55. Foust AJ et al (2015) Computer-generated holography enhances voltage dye fluorescence discrimination in adjacent neuronal structures. Neurophotonics 2(2)
56. Tanese D et al (2017) Imaging membrane potential changes from dendritic spines using computer-generated holography. Neurophotonics 4(3):031211
57. Castanares ML, Stuart G, Daria VR (2018) Holographic functional calcium imaging of neuronal circuit activity. In: Kao F-J, KEiser G, Gogoi A (eds) Advanced optical methods for brain imaging. Springer Nature, Switzerland, pp 143–165
58. Yaqoob Z et al (2008) Optical phase conjugation for turbidity suppression in biological samples. Nat Photonics 2(2):110–115
59. Schwertner M, Booth MJ, Wilson T (2004) Characterizing specimen induced aberrations for high NA adaptive optical microscopy. Opt Express 12(26):6540–6552
60. Booth MJ (1861) Adaptive optics in microscopy. Philos Trans R Soc A Math Phys Eng Sci 2007(365):2829–2843
61. Ji N (2017) Adaptive optical fluorescence microscopy. Nat Methods 14(4):374–380
62. Wang K et al (2014) Rapid adaptive optical recovery of optimal resolution over large volumes. Nat Methods 11(6):625
63. Ji N, Milkie DE, Betzig E (2010) Adaptive optics via pupil segmentation for high-resolution imaging in biological tissues. Nat Methods 7(2):141–U84
64. Tao XD et al (2011) Adaptive optics microscopy with direct wavefront sensing using fluorescent protein guide stars. Opt Lett 36(17): 3389–3391

65. Tao XD et al (2011) Adaptive optics confocal microscopy using direct wavefront sensing. Opt Lett 36(7):1062–1064
66. Liu R et al (2019) Direct wavefront sensing enables functional imaging of infragranular axons and spines. Nat Methods 16(7):615–618
67. Qin ZY et al (2020) Adaptive optics two-photon endomicroscopy enables deep-brain imaging at synaptic resolution over large volumes. Sci Adv 6(40)
68. Zhang Z et al (2018) Closed-loop all-optical interrogation of neural circuits in vivo. Nat Methods 15(12):1037–1040
69. Quan X et al (2018) Three-dimensional stimulation and imaging-based functional optical microscopy of biological cells. Opt Lett 43(21):5447–5450
70. Bovetti S et al (2017) Simultaneous high-speed imaging and optogenetic inhibition in the intact mouse brain (vol 7, 2017). Sci Rep 7: 40041
71. Bovetti S, Fellin T (2015) Optical dissection of brain circuits with patterned illumination through the phase modulation of light. J Neurosci Methods 241:66–77
72. Anselmi F et al (2011) Three-dimensional imaging and photostimulation by remote-focusing and holographic light patterning. Proc Natl Acad Sci USA 108(49): 19504–19509
73. Dal Maschio M et al (2010) Simultaneous two-photon imaging and photo-stimulation with structured light illumination. Opt Express 18(18):18720–18731

Chapter 18

Intravital Imaging and Analysis of the Structural Plasticity of Dendritic Spines in Multiple Brain Regions

Yuan Shi, Mochen Cui, Tanja Blume, and Jochen Herms

Abstract

The structural plasticity of dendritic spines is associated with changes in various cognitive performances across physiological and pathological conditions. With the rapid development of imaging and labelling methods, in vivo study of dendritic spine dynamics has been made possible from superficial to deep layers in many brain regions. However, depending on the brain region, there are multiple variants of surgical, imaging, and analyzing procedures, making them, as a whole, technically difficult to comprehend. Here, we introduce an open-skull coverslip-based surgical procedure of cranial window implantation, which is suitable for monitoring dendritic spine dynamics in the superficial cortical layer and can be easily adapted for monitoring brain regions of various depths and locations with minor modifications. Following these surgical procedures, we further describe intravital two-photon microscopy-based imaging approaches and methods for processing and analyzing acquired imaging data.

Key words In vivo two-photon imaging, Cranial window, Microprism, Hippocampal window, Structural plasticity of dendritic spines

1 Introduction

Dendritic spines, first described by Santiago Ramón y Cajal [1–3], are small protrusions of neuronal dendrite membrane. With neighboring axonal boutons, the vast majority of dendritic spines form synapses—the basic computational unit of brain information processing, receiving informational input from presynaptic terminals [4, 5]. With the advances in imaging and labelling techniques during the past decades, the structural properties of dendritic spines have been unveiled highly dynamic [6–9]. These dynamic alterations, reflecting the establishment or destruction of synapses, occur continuously throughout development, juvenile period, and adulthood, although their extent may vary between different brain regions [10, 11]. Emerging evidence suggests that such structural alterations in synaptic connectivity are likely the underpinning of cognitive changes across a variety of physiological responses to

Joachim H. R. Lübke and Astrid Rollenhagen (eds.), *New Aspects in Analyzing the Synaptic Organization of the Brain*, Neuromethods, vol. 212, https://doi.org/10.1007/978-1-0716-4019-7_18,

various stimuli [1, 4], such as sensory experience changes [12, 13], environmental enrichment [14, 15], drug administration [16, 17], hypoxia [18], and many, if not all, nervous system disorders, for instance, neurological [19, 20], neurodevelopmental [21, 22], neuropsychiatric [23–25], and neurodegenerative diseases [26–30] and many others. Given the structural plasticity of dendritic spines' involvement in and their diverse responses to such a variety of physiological and pathological processes, it has become one of the most intensively studied subjects in neuroscience. Using recently developed sophisticated imaging techniques—in vivo two-photon microscopy, with genetically labelled cellular targets in living animals—scientists are able to visualize individual dendritic spines in vivo at high temporal and spatial resolution.

In vivo two-photon microscopy is one of the most widely used fluorescence imaging methods to monitor the structural plasticity of dendritic spines in live animals through an implanted imaging window [31]. Unlike the conventional one-photon confocal microscope, the laser source of a two-photon microscope generates two near-infrared (NIR) excitation photons within a single femtosecond [32]. Once the two photons reach the target fluorophore, their combined energy promotes the target fluorophore to its excited state [32]; the target fluorophore then emits fluorescence that is to be received by the photomultipliers of the microscopic detectors [33] (*see* Fig. 1). One of the advantages of using NIR excitation is that NIR penetrates more deeply in scattering tissue, allowing the two-photon microscope imaging of the dendritic spines in living tissue to a depth of approximately 1 mm, which is 2–3 times deeper compared to the imaging range of one-photon confocal microscopes [34]. This advantage makes it possible to study the dendritic spines located in deep layers of the brain in their original living environment. On top of that, NIR excitation light has low phototoxicity [35], thus minimizing its effect on dendritic spines under repeated imaging. Therefore, the two-photon microscopy is suitable for longitudinal imaging of the structural plasticity of dendritic spines without affecting them.

In order to visualize dendritic spines using in vivo two-photon microscopy, the spines have to be labelled with fluorescence. The labelling of fluorescent reporter proteins in live animals can be achieved by several approaches, including in utero electroporation, viral vectors, and transgenesis. In utero electroporation allows transfection of fluorescent protein-encoding gene into stem/progenitor cells, which are in immediate proximity to the ventricle, and their descendant cells, allowing prominent expression of fluorescent signal in a relatively large population of cells [36, 37]. Fluorescently labelled neurons and dendritic spines could be detected even months after in utero electroporation [38, 39]. Apart from in utero electroporation, viral vectors, such as the adeno-associated virus (AAV), lentivirus, etc., are commonly employed to achieve precise

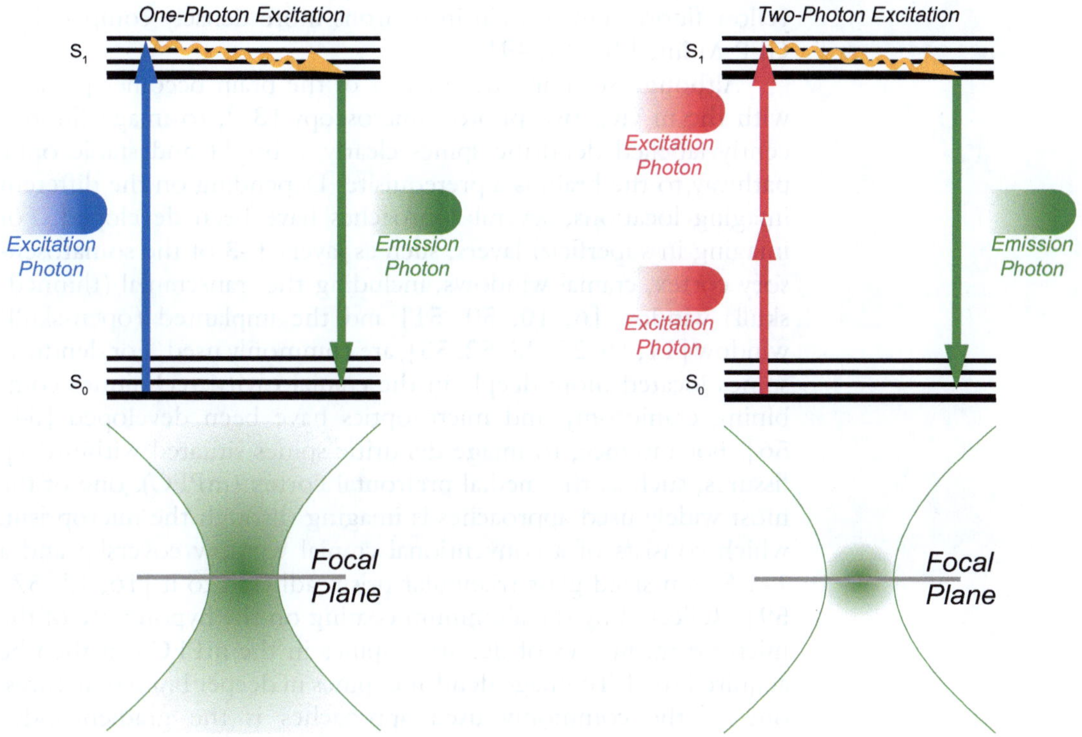

Fig. 1 Schematic diagram of one- and two-photon excitations. In one-photon excitation, a single photon with sufficient energy is absorbed to excite the fluorophore from the ground state (S_0) to the excited higher-energy state (S_1) and emit a photon with slightly lower energy when transiting back to the initial ground state S_0. In two-photon excitation, two lower-energy photons are simultaneously absorbed into the fluorophore, emitting a photon in the visible wavelength. Since sufficient energy for emission is provided only on one spot where the two photons hit simultaneously, the emitting light of the two-photon excitation is far more restrained around the focal plane compared to the one-photon excitation

region-specific and/or cell type-specific fluorescent labelling [40, 41]. Here, the transfected gene is assembled into the viral genome and expressed under the control of certain promoters or Cre-lox recombination systems, driving designated gene expression in specific cell types [42–45]. Another well-established approach is to genetically encode and express fluorescent proteins in dendritic spines in living animals [46]. Unlike in utero electroporation and viral vectors, transgenic animals express the hereditary stable and consistent pattern of fluorescent protein labeling in all stages of life. One of the most prevalently employed examples is the GFP-M (*Thy1-eGFP*) transgenic mouse line. In GFP-M mice, driven by the mouse thymus cell antigen 1 (Thy1) promoter [47], the enhanced green fluorescent protein (eGFP) is sparsely produced in a subset of cortical and hippocampal neurons, facilitating acquiring images of dendritic spines in vivo [46, 48]. There are other transgenic mouse lines with different fluorescent proteins and expression patterns. For instance, the YFP-H mouse line expresses yellow fluorescent protein in neurons more densely compared to GFP-M line [46, 48, 49].

Although scanning deep layers of the brain becomes possible with the in vivo two-photon microscopy [31], to image fluorescently labelled dendritic spines clearly, a bright and stable optic pathway to the brain is a prerequisite. Depending on the different imaging locations, several approaches have been developed. For imaging in superficial layers, such as layers 1–3 of the somatosensory cortex, cranial windows, including the transcranial (thinned-skull) window [6, 10, 50, 51] and the implanted (open-skull) window [11, 16, 27, 28, 52, 53], are commonly used. For dendritic spines located more deeply in the cranial cavity, techniques combining craniotomy and micro-optics have been developed [54–56]. For instance, to image dendritic spines situated within deep fissures, such as the medial prefrontal cortex (mPFC), one of the most widely used approaches is imaging through the microprism, which consists of a conventional cranial window coverslip and a 1–1.5-mm-sized glass triangular prism adhered to it [16, 17, 57–59]. Reflected by the aluminum coating on the hypotenuse of the microprism, images of dendritic spines in the mPFC can then be acquired [60]. To image dendritic spines in deeper brain structures, one of the commonly used approaches is the gradient-index (GRIN) lens, a cylindrical optic made with a microendoscope [54, 61], penetrating from the cortex to the hippocampus [56], allowing imaging of dendritic spines of the hippocampal neurons [62]. A simplified approach for hippocampal dendritic spine imaging can be achieved using a metal cannula sealed with a coverslip on the bottom, though to acquire images of dendritic spines through the cannula, a long working distance objective is required. Using the latter approach, dendrites of CA1 hippocampal pyramidal neurons in adult mice can be successfully traced over months [8, 63, 64].

In the following, we describe details about an open-skull coverslip-based surgical procedure of cranial window implantation and its modifications, which are suitable for monitoring dendritic spines in different regions of the mouse brain. We also describe corresponding imaging procedures and analyzing methods. Through all this information, we hope readers could acquire a comprehensive impression of in vivo imaging of dendritic spines and be inspired to explore further on spine dynamics and a broader spectrum of brain functions.

2 Materials

In this section, we introduce materials required for surgically establishing optical access to the brain and subsequent longitudinal imaging of dendritic spines using in vivo two-photon microscopy.

Notably, the following descriptions are but principles of the surgery and the imaging process. Depending on the region of the target, as well as the microscopic system to be employed, specific adjustments to the procedure and individual customization of certain devices should be considered.

2.1 Animals

GFP-M transgenic mice (*Thy1-eGFP*), originally obtained from the Jackson Laboratory and bred to heterozygosity on C57BL/6 genetic background, express the enhanced green fluorescent protein (eGFP) in sparse (*see* **Note 1**) and random subsets of neurons within specific populations [46]. All animal studies must conform to national and institutional regulations (*see* **Note 2**).

2.2 Reagents, Consumables, and Equipment for Surgical Procedures

1. Midazolam.
2. Fentanyl.
3. Medetomidine.
4. 2% lidocaine.
5. Carprofen.
6. Metamizole.
7. Sterile saline solution 0.9%.
8. Ethanol 70% (w/v).
9. Biopsy punch (Kai)—for carving and marking the skull.
10. Surgical instruments (fine scissors, fine curved and straight forceps, and scalpel).
11. Hot bead sterilizer (Fine Science Tools)—for sterilizing surgical instruments.
12. Optical glue (Norland Products)—for fixing the microprism on a circular coverslip (Norland Products).
13. Histoacryl blue tissue adhesive (B. Braun)—for sealing the margin around the coverslip.
14. Cyano fast glue (Hager)—for fixing the "z"-shaped holder on the skull.
15. UV-curable dental cement (Tetric EvoFlow)—for fixing the implant and head holder on the skull.
16. Demi Plus LED curing light system—for curing the UV-curable dental cement and the optical glue.
17. Eye ointment (Bepanthen).
18. Circular coverslip (diameter: 3 mm and 4 mm, no. 1).
19. Sterile cotton swabs with applicator.
20. Sugi pointed suction swab (Kettenbach)—for absorbing extra fluid during the operation.

21. Gelita-Spon hemostatic absorbable gelatin—for stopping bleeding during durotomy.
22. Bone wax—for stopping bleeding on the nasal bone while removing the periosteum and roughening the bone surface.
23. Microprism: right-angle shape (thickness, 1.5 mm; right-angle edge, 1.5 mm) with the hypotenuse surface coated with reflective aluminum (IMM Photonics).
24. UV light source (Opticure LED 200, Norland)—for curing the optical glue.
25. Hippocampal window—a customized metal cannula (diameter, 3 mm; height, 1.5 mm) sealed at one end with a circular glass coverslip (diameter, 3 mm, no. 1).
26. Dental drill with drill burrs (tip diameter, 0.5 mm; Meisinger).
27. Head holder (*see* **Note 3**)—for fixing the head of the mouse on the framework attached to the microscope stage for imaging.
28. 27 G syringe needle attached to a vacuum suction—for aspirating the cortex for hippocampal window implantation.
29. Pattex superglue—for stabilizing the cannula of the hippocampal window to the skull.
30. Stereotaxic frame (Neurostar) equipped with a heating pad.
31. Stereomicroscope with LED illumination (Olympus)—for window implantation.

2.3 Reagents, Equipment, and Software for In Vivo Two-Photon Imaging

1. Isoflurane.
2. 95% oxygen/5% carbon dioxide.
3. Double-distilled water (ddH_2O)—immersion medium for the water immersion objective.
4. Eye ointment (Bepanthen).
5. Stereoscope with LED illumination (Zeiss)—for checking and cleaning of the window coverslip.
6. LSM 880 microscope (Zeiss) or equivalent.
7. Chameleon Vision-S femtosecond laser (Coherent).
8. W Plan-Apochromat 20×/1.0 NA water-immersion objective (Zeiss).
9. Long working distance water-dipping objective (16×/0.8 NA water-immersion objective with a working distance of 3 mm) (Nikon).
10. Imaging framework equipped with a heating pad.
11. Zen software (Zeiss)—software for controlling microscope.

2.4 Software for Image Processing and Analysis

1. Zen software (Zeiss)—software for annotating individual dendritic spines.
2. Imaris 9.8 (Bitplane)—software for 3D reconstruction and quantification of dendritic spines.
3. Matlab (The MathWorks)—for Imaris-based dendritic spine categorization.

3 Methods

3.1 Surgical Procedures

In this section, we introduce surgical procedures for establishing optical access to superficial cortical layers, brain regions situated in deep fissures, and subcortical regions. We choose the somatosensory cortex, mPFC, and hippocampus as representatives, respectively. Properties of optical windows for these procedures are listed in Table 1.

Prior to surgery, sterilize all surgical instruments using a hot bead sterilizer (Fine Science Tools) (*see* **Note 4**) and sterilize the workspace with alcohol 70% (w/v) thoroughly. In the surgical procedures described here, surgical anesthesia is established using a general formula: an intraperitoneal injection with a mixture of 5.0 mg/kg midazolam, 0.05 mg/kg fentanyl, and 0.5 mg/kg medetomidine. To monitor the depth of anesthesia and analgesia, the toe-pinch withdrawal reflex is tested by pressing the mouse toes with forceps. After being anaesthetized, the mouse's body temperature is maintained using a heating pad, and mouse eyes are

Table 1
Properties of different types of optical windows used in this chapter

Type of optical window	Imaging depth	Window diameter	Recovery duration post surgeries	Duration of access
Glass coverslip-based cranial window	~900 μm under the brain surface [65, 66]	3–4 mm	4 weeks [16, 27, 28]	Up to months [16, 28]
Prism-based cranial window	~900 μm under the brain surface[a]	3–4 mm	1–3 weeks [16, 59][c]	Several weeks [16, 59]
Hippocampal cranial window	~900 μm under the brain surface[b]	3 mm	8–12 days [67] to 4 weeks [64]	Several weeks [8, 67]

Note: [a]Due to the long distance between the front end of the objective and the brain surface (through the microprism, 1–1.5 mm) and the dental acrylic around the window, the actual imaging depth highly depends on the working distance of the objective. For an objective with a 2 mm working distance, the most distant focal plane may not be able to reach 1000 μm under the brain surface. We use an objective with a 3 mm working distance; with properly conducted implantation, 1000 μm imaging depth can be achieved

[b]Due to the same reason, the actual imaging depth highly depends on the working distance of the objective

[c]Previous studies reported the inflammation around the implantation site reduces significantly 1 week after the surgery and does not hinder the in vivo imaging of dendritic spines. We usually wait for more days so the inflammation diminishes completely and the imaging goes more smoothly

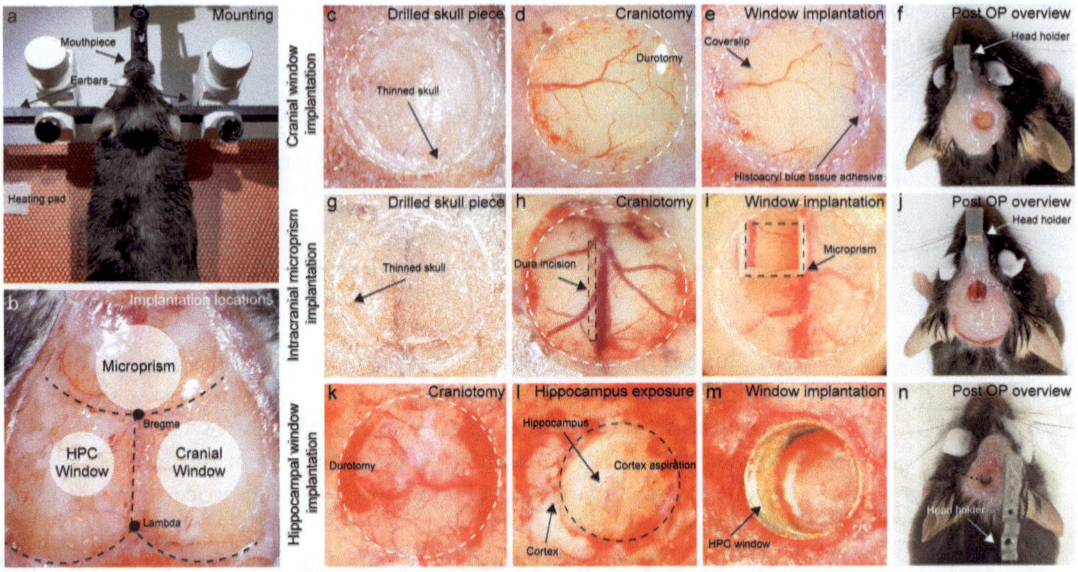

Fig. 2 Surgery procedures of implantations for cranial window, microprism, and hippocampal window. (**a**) Mounting of an anesthetized mouse on a stereotaxic frame. (**b**) Implantation locations of the cranial window, microprism, and HPC window, relative to bregma and lambda. (**c–f**) For cranial window implantation, the marked piece of the skull is thinned using a dental drill (**c**), and after the removal of the skull piece, the dura is carefully removed (**d**), and the craniotomy is sealed using a 4-mm-diameter coverslip (**e**), and with all implants secured by cured dental acrylic, the mouse is released for post-op recovery (**f**). (**g–j**) For intracranial microprism implantation, the marked skull is thinned (**g**), and the dura incision is performed in parallel with the superior sagittal sinus on the side of intended microprism implantation (**h**), the craniotomy is sealed with 4-mm-diameter microprism-attached coverslip (**i**), and with all implants secured by cured dental acrylic, the mouse is released for post-op recovery (**j**). (**k–n**) For hippocampal window implantation, the durotomy is conducted after the skull piece removal (**k**), the cortex is aspirated to expose the underneath dorsal hippocampus (**l**), and the coverslip-attached cannula is inserted into the cortex-free cylinder cavity to seal the craniotomy (**m**), and with all implants secured by cured dental acrylic, the mouse is released for post-op recovery (**n**). HPC hippocampal, OP operation

protected from dehydration and strong light by applying eye ointment (Bepanthen). Detailed procedures of different implantations are described in the following sections.

3.1.1 Cranial Window Implantation

This section describes the steps for cranial window implantation above the somatosensory cortex. The detailed surgical procedures are as follows:

1. Mount the anaesthetized mouse onto a stereotaxic frame: insert ear bars symmetrically (scales of the ear bars on both sides should be the same once fixed) into ear canals so that the head is flat, straight, and not moveable; the midline of the mouse's head should coincide with the midline of the mouthpiece; put the mouthpiece into the animal's mouth until its maxillary incisors are hooked to the round hole on the mouthpiece (*see* **Note 5**) (*see* Fig. 2a).

2. Clean the scalp with ethanol 70% (w/v) to sterilize the area, and rub it in with a cotton swab.
3. Apply local analgesic 2% lidocaine subcutaneously at the intended incision site on the scalp and wait for 2–3 min for the anesthesia to take effect.
4. Work under a stereomicroscope (Olympus) with adequate lighting, and lift the skin with forceps and make a straight single incision with scissors along the midline from the front (between eyes) to the back (~between ears) to expose the skull. Extend the incision if necessary.
5. Clean the hair from the margin of the cut skins with cotton swabs, remove all the residual hairs with fine-tipped forceps, and glue the border of the remaining skin with Histoacryl blue tissue adhesive to prevent inflammation (*see* **Note 6**).
6. Scrape off the skull periosteum with a scalpel.
7. Roughen the skull surface using the scalpel tip, to make the UV-curable dental cement bond to the skull more firmly (*see* **Note 7**), and use the bone wax to stop the hemorrhage in case of bleeding.
8. Detach muscles on the occipital bone as far as needed and make sure the bone surface is dry (*see* **Note 8**).
9. Mark bregma and set coordinates AP −2–2.5 mm, ML +2 as the center of window implantation with the 4-mm-diameter coverslip, and ensure the center marked on the parietal bone is suitable to accommodate the window.
10. Use a 4-mm-diameter biopsy punch (Kai) to carve and mark the skull piece to be replaced by the coverslip (*see* Fig. 2b).
11. Use a dental drill with a 005 drill bit (Fine Science Tools) to deepen the punch mark by gently pressing and tracing the drill bur along the mark until a mere leakage of cerebrospinal fluid is visible (*see* **Note 9**) (*see* Fig. 2c).
12. Use forceps to lift the loose skull piece and detach it from the rest of the skull. Do not damage the dura and brain tissue underneath. If small amounts of bleeding occur during removing the skull piece, absorb the blood using Sugi pointed suction swab, and flush the craniotomy with normal saline. Usually, the bleeding stops relatively fast if no major vessels are damaged (*see* Fig. 2d).
13. If the experimental design requires, the dura can be removed using forceps. During the durotomy, make sure the blood vessels on the brain surface are not damaged (*see* **Note 10**); details of dura removal are listed in **steps 14** and **15** (*see* Fig. 2d).

14. To remove the dura, nip and lift a small piece of it distant from the superior sagittal sinus using forceps and teal it off carefully. To prevent the brain surface to become dry and the dura to start sticking to the brain, drop some normal saline on the brain surface, so the stub end of the dura floats and becomes easy to be removed.
15. For places close to the superior sagittal sinus, where blood vessels are frequently attached to the dura, nip and lift the dura from the vessel-free side, and remove it piece by piece, to avoid rupturing the vessels and bleeding.
16. Close the craniotomy immediately with the 4 mm coverslip (*see* **Note 11**) (*see* Fig. 2e).
17. Dry the edges of the implant with Sugi pointed suction swab, and use the Histoacryl blue tissue adhesive to seal the margin between the circular coverslip and the skull.
18. Place the "z"-shaped head holder in the middle of nasal bones and use Cyano fast glue to fix the mouse on the framework attached to the microscope stage (*see* **Note 12**).
19. Apply a thin layer of UV-curable dental cement (*see* **Note 13**) around the coverslip and on bilateral parietal, frontal, and nasal bones (also cover the base part of the "z"-shaped holder), and cure the dental cement with UV curing light (Demi) for 30 s (*see* Fig. 2f).
20. Once the dental cement is cured, release the mouse from the framework.
21. The mouse is administered with carprofen (4 mg/kg, s.c.) and metamizole (200 mg/kg, i.g.) to limit inflammation and reduce pain and, additionally, receives one subcutaneous injection of prophylactic antibiotic enrofloxacin (5 mg/kg, s.c.). The mouse is then awakened by subcutaneous injection with a mixture of 1.2 mg/kg naloxone, 2.5 mg/kg atipamezole, and 0.5 mg/kg flumazenil.
22. Let the mouse recover in a clean cage under a heat lamp, or put the mouse in a 37 °C incubator with ventilation for 1–2 h. Once the mouse is fully conscious and mobile, it can be removed from the heating source and back to the regular cage for further recovery.

3.1.2 Intracranial Microprism Implantation

This section describes the steps for intracranial microprism implantation besides the mPFC.

Microprism assemble (at least 24 h in advance):

1. Work under a stereoscope (Zeiss) with sufficient light.
2. Clean the 3-mm-diameter coverslip (no. 1) with canned air and then ethanol, and wait for the coverslip to dry completely.

3. Clean the microprism with canned air and then ethanol, wait for the coverslip to dry completely, and make sure not to scratch the aluminum coating during cleaning.
4. Put a small drop of optical glue (Norland NOA 71) onto one of the right angle surfaces of the microprism, and make sure the glue does not overflow.
5. Adhere the microprism onto the coverslip by putting the glued side of the microprism on the coverslip (*see* **Note 14**).
6. Once the microprism is positioned as desired, apply UV light for 1 min on the assembled microprism.
7. Store the assembled microprism in a dark, dry, sterile place for subsequent use.

Properly prepared microprism is the prerequisite of the implantation and subsequent imaging. The detailed surgical procedures for intracranial microprism implantation are as follows:

1. *Repeat Subheading 3.1.1* **steps 1–8**.
2. To target the mPFC, the skull overlying the area of interest is located by the digital stereotaxic device with center coordinates at AP 0–2, ML 0 (*see* **Note 15**).
3. Use a 3-mm-diameter biopsy punch (Kai) to carve and mark the skull piece to be replaced by the coverslip based on the four coordinates acquired from **step 2** (*see* Fig. 2b).
4. Remove the marked skull piece using a dental drill with a 005 drill bit (Fine Science Tools). Conduct this step with extreme care to avoid damaging the superior sagittal sinus (*see* **Note 16**) (*see* Fig. 2g).
5. Make an incision on the dura (>1.5 mm) along the superior sagittal sinus on the contralateral brain surface of the hemisphere for imaging, and make sure the blood vessels on the brain surface are not damaged (*see* **Note 17**) (*see* Fig. 2h).
6. Close the craniotomy immediately with the 3 mm coverslip (*see* **Note 11**) by slipping the microprism into the incision (*see* **Note 18**). To make sure the reflection comes from the target region (medial prefrontal area), the contact surface of the microprism with the contralateral hemisphere was kept parallel with the midline of the mouse skull, while the front edge of the surface was placed at ~1.4–1.5 mm anterior to bregma (*see* Fig. 2i).
7. *Repeat Subheading 3.1.1* **steps 17** and **18**.
8. Apply an extremely thin layer (*see* **Note 19**) of UV-curable dental cement around the coverslip and on bilateral parietal, frontal, and nasal bones (also cover the base part of the "z"-shaped holder with a slightly thicker layer of UV-curable

dental cement to prevent potential detachment during imaging), and cure the dental cement with UV curing light (Demi) for 30 s (*see* Fig. 2j).

9. *Repeat Subheading 3.1.1* **steps 20–22**.

3.1.3 Hippocampal Window Implantation

This section describes steps for hippocampal window assembly and subsequent hippocampal window implantation.

Hippocampal window assembly (at least 24 h in advance):

1. Use the customized cannula, which consists of a stainless steel tube that measures 1.7 mm in length, 3 mm outer diameter, and 2.5 mm inner diameter, attached on one end with a 0.5-mm-wide ring of the same. The ring helps hold the cannula firmly on the skull without sinking or moving.
2. Apply UV-curable adhesives (Norland Product) at the other ringless end of the cannula, and carefully cover the end with a 3-mm-diameter coverslip (no. 1).
3. Apply UV light for 1–2 min on the assembled cannula with the coverslip for the adhesive to fully cure.

Hippocampal window surgery steps are detailed as follows:

1. *Repeat Subheading 3.1.1* **steps 1–8**.
2. To target the hippocampal CA1 region, the skull overlying the area of interest is located by the digital stereotaxic device at the coordinates: center at AP 2.2 mm, ML 1.8 mm.
3. Use a 3-mm-diameter biopsy punch (Kai) to carve and mark the skull piece to be replaced by the coverslip based on the central coordinate as above (*see* Fig. 2b).
4. *Repeat Subheading 3.1.1* **steps 11–15**.
5. After the dura is removed (*see* Fig. 2k), aspirate the cortex within the craniotomy region using a 27 G syringe needle attached to vacuum suction (*see* **Note 20**). Stop the aspiration when the alveus CA1 is reached. Then carefully peel away the surrounding cortical tissue and the top layers of the alveus (*see* Fig. 2l).
6. (*Optional*) A silicone (Kwik-Sil) layer could be applied onto the alveus after its top layers are removed to reduce the artifacts due to the movement of the hippocampus during the imaging process (*see* **Note 21**).
7. Insert a disinfected custom-made metal cannula (outer diameter, 3 mm; height, 1.5 mm) with one end sealed with a no. 1 3-mm-diameter coverslip into the cylinder cavity where the cortical tissue is aspirated. Make sure all bleedings are stopped, and no air bubbles are under the coverslip. If there are bleedings in the insertion area, stop them firstly using Sugi pointed suction swab and PBS irrigation (*see* Fig. 2m).

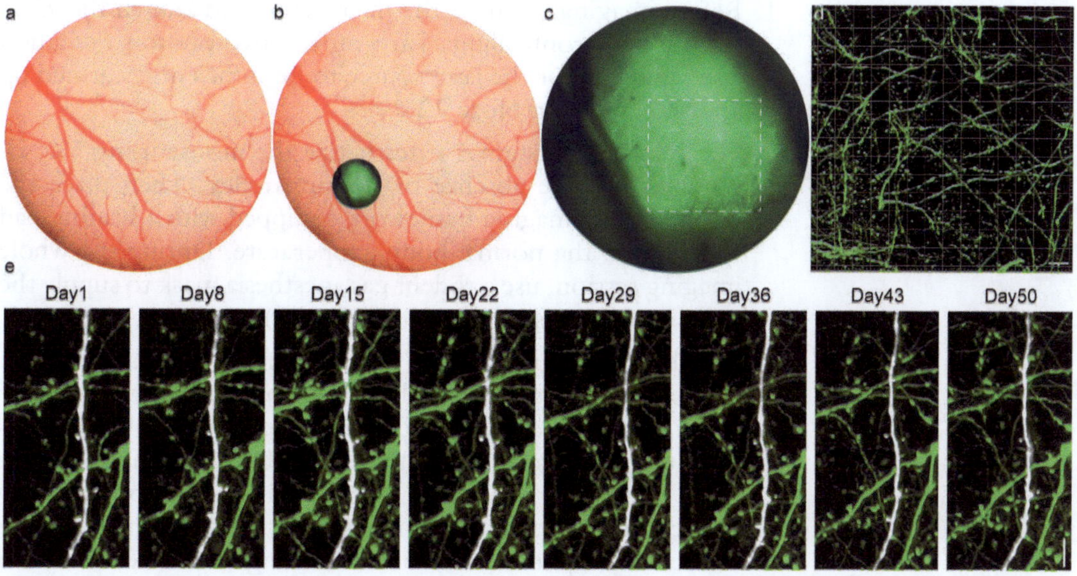

Fig. 3 Representative micrographs showing in vivo imaging area locating and GFP-labelled dendrites in the somatosensory cortex. (**a**) Overview of the window-covered region under the bright field. (**b** and **c**) A smaller region is selected from the window-covered region through the ocular (**b**), and under the GFP-exciting fluorescent light source, dendrites in the somatosensory cortical region are visible (**c**). (**d**) Dendrites in the square with white dash line in **c** are further imaged using the in vivo two-photon microscope. (**e**) The same dendrite is imaged longitudinally for several weeks to months. Scale bars: **d** = 20 μm; **e** = 5 μm

8. Dry the edges of the implant with Sugi pointed suction swab, use the Histoacryl to seal the margin between the cannula and the skull, and then fix the cannula also with superglue (Pattex) to the skull to minimize movement artifacts during imaging sessions.
9. Place the customized head holder in the middle of nasal bones and use Cyano fast glue to fix the mouse on the framework attached to the microscope stage (*see* **Note 12**) (*see* Fig. 2n).
10. *Repeat Subheading 3.1.1* **steps 19–22**.

3.2 Longitudinal In Vivo Two-Photon Imaging

3.2.1 Imaging Dendritic Spines Through the Cranial Window

This section describes steps for longitudinal in vivo two-photon imaging of dendrites and dendritic spines through the cranial window. The cranial window can be used for monitoring brain regions that are directly accessible by removing frontal, parietal, and occipital bone of mouse skull, e.g., motor cortex, somatosensory cortex, etc. Here we use the somatosensory cortex as a representative. The consecutive in vivo two-photon micrographs of eGFP-labelled layer-V-pyramidal-neurons-origin apical dendritic tufts in the somatosensory cortex are acquired on a regular basis through a 20× water-immersion objective on an LSM 880 microscope (Zeiss). The detailed imaging procedures are as follows:

1. Before imaging, turn on the microscope and launch the ZEN software for controlling. Turn on the two-photon excitation laser and wait for the laser source to be sufficiently warmed up and stabilized (*see* **Note 22**).
2. For anesthetic imaging, anaesthetize the mouse using 5% isoflurane until the absence of righting reflex. Then place the mouse on an imaging framework equipped with a heating pad to maintain the normal body temperature. During the whole imaging session, use a rodent gas anesthesia mask to supply the mouse with 1% isoflurane (*see* **Note 23**).
3. Check the integrity of the cranial window coverslip using a stereoscope (Zeiss). Clean the coverslip carefully from dust or water stain with a scalpel tip. If the coverslip partially detaches from the sealing material or the sealing becomes unstable, fix it with Histoacryl and dental acrylic. The coverslip must stay intact during the cleaning or resealing process, and breaking of the coverslip makes the mouse no longer suitable for in vivo imaging.
4. Fix the imaging framework onto the piezo-z stage;
5. Adjust the XY-position of the framework until the cranial window is below the objective.
6. Apply a drop of distilled water to the cranial window, and adjust the z-position of the framework until the objective contacts the water drop. Continue adjusting the z-position with extreme care until the cranial window is at the closest possible position relative to the objective. Turn on the visible fluorescent light source at the wavelength of GFP excitation, and locate the imaging focus using the fine focus knob under the ocular (*see* **Note 24**).
7. According to the experimental plan, use the ocular to locate ~3 regions of interest for subsequent imaging sessions (*see* **Note 25**). Mark all three locations on the "Stage" panel in the Zen software, and take photographs to record the structure of blood vessels of selected regions for their relocation in further imaging sessions.
8. Once all regions of interest are set, switch to "Acquisition" mode in the Zen software.
9. Tune the wavelength of the laser beam to 880 nm to excite eGFP (*see* **Note 26**).
10. Go to position 1 using the "Stage" panel, switch to "Live" mode, and define an appropriate z-stack by rolling up and down along the z-axis. Once the z-stack is defined, take an overview image stack at zoom 1.0 ($424 \times 424 \times 350\ \mu m^3$, 0.83 μm/pixel xy-resolution, and 1 μm/pixel z-resolution). Trace down the dendrites and ensure they are from neurons of the desired location (e.g., pyramidal neurons of layer 5).

11. Select sufficient quantity (~15) of dendrites of required branching hierarchy (e.g., secondary branches) within one region of interest using the "crop" function (70 × 35 × 23 μm^3, 0.138 μm/pixel xy-resolution, and 1 μm/pixel z-resolution). Switch to "live" mode and define an appropriate z-stack by rolling up and down along the z-axis. Once the z-stack is defined, use "Start experiment" to take the image of the dendrite (*see* **Note 27**).
12. In subsequent imaging sessions, the same imaging field is repositioned by orienting at the unique pattern of blood vessels (*see* **Note 28**).
13. Individual dendrites are aligned to the same position according to the acquired images from the first imaging session. To acquire comparable images of the same dendrite during the consecutive imaging sessions, dendrites are scanned in a consistent manner throughout the experiments.

With appropriately done surgery, inflammation and hemorrhage around the implant can be minimized after the operation. The implanted window can then be stable for months during the entire consecutive imaging sessions (*see* **Note 29**). Representative images acquired using the abovementioned procedure can be found in Fig. 3. Based on our experience, with each imaging session less than 1 h, and ~15 dendrites acquired per session (*see* **Note 30**), adequate images of dendrites could be acquired weekly from one mouse for up to ~100 days [16] or, if necessary, considerably longer. However, it should be noted that with higher imaging frequency, multiple factors, e.g., cumulating phototoxicity, repetitive anesthesia, and potential inflammation, might shorten the total period of longitudinal in vivo imaging.

3.2.2 Imaging Dendritic Spines Through the Intracranial Microprism

This section describes steps for longitudinal in vivo two-photon imaging of dendrites and dendritic spines using microprism. The microprism can be used for monitoring brain regions at deep fissures hence difficult to be reached using the regular cranial window, e.g., entorhinal cortex, mPFC, etc. Here we select mPFC as a representative. To image the dendritic spines in the mPFC, the consecutive in vivo two-photon micrographs of eGFP-labelled apical dendritic tufts in *Thy1-eGFP* mice are consecutively imaged once per week through a 16×/0.8 NA water-immersion objective (Nikon) with a working distance of 3 mm (*see* **Note 31**) on an LSM 880 microscope (Zeiss). The detailed imaging procedures are as follows:

1. *Repeat Subheading 3.2.1* **steps 1–5**.
2. Apply a drop of distilled water to the cranial window, and adjust the z-position of the framework until the objective contacts the water drop. Continue adjusting the z-position with extreme

care until the cranial window is at the closest possible position relative to the objective. Turn on the visible fluorescent light source at the wavelength of GFP excitation, locate the microprism first, and then find the imaging focus through the microprism using the fine focus knob under the ocular (*see* **Note 24**).

3. *Repeat Subheading 3.2.1* **steps 7–9**.
4. Once the GFP signal through the microprism can be visualized under the "Acquisition" mode, go to position 1 using the "Stage" panel, switch to "Live" mode, and define an appropriate z-stack by rolling up and down along the z-axis. Once the z-stack is defined, take an overview image stack at zoom 1.0 ($628 \times 628 \times 200\ \mu m^3$, 1.230 μm/pixel xy-resolution, and 1 μm/pixel z-resolution). Trace down the dendrites and ensure they are from neurons of the desired location.
5. Select sufficient quantity (~15) of dendrites of required branching hierarchy (e.g., secondary branches) within one region of interest using the "Crop" function ($105 \times 52.5 \times 52\ \mu m^3$, 0.205 μm/pixel xy-resolution, and 1 μm/pixel z-resolution). Due to possible distortions of microprism-based imaging, acquire as many as dendrites within an acceptable duration of an imaging session. Switch to "Live" mode and define an appropriate z-stack by rolling up and down along the z-axis. Once the z-stack is defined, use "Start experiment" to take the image of the dendrite (*see* **Note 27**).
6. *Repeat Subheading 3.2.1* **steps 12** and **13**.

With well-performed surgery, the implant with microprism could be stable for in vivo imaging as long as the regular cranial window. Representative images acquired using the abovementioned procedure can be found in Fig. 4. However, to monitor the mPFC, the coverslip is located across the frontal and parietal bones of the mouse skull. Therefore, the stability of the microprism implant is not as firm as the regular cranial window for imaging dendritic spines of the superficial cortical layers, making it highly susceptible to mechanical force. To achieve a similar imaging duration as for the superficial cortical layers, more caution should be taken to prevent the objective and the coverslip from crashing during the focus searching and imaging.

3.2.3 Imaging Dendritic Spines Through the Hippocampal Window

This section describes steps for longitudinal in vivo two-photon imaging of dendrites and dendritic spines in the hippocampal region through the hippocampal window. To image the dendritic spines in the hippocampus, the consecutive in vivo two-photon micrographs of eGFP-labelled dendritic tufts in *Thy1-eGFP* mice are imaged once per week through a 16×/0.8 NA water-immersion objective (Nikon) with a working distance of 3 mm on an LSM 880 microscope (Zeiss). A regular session of in vivo imaging takes approximately 1 h. The detailed imaging procedures are as follows:

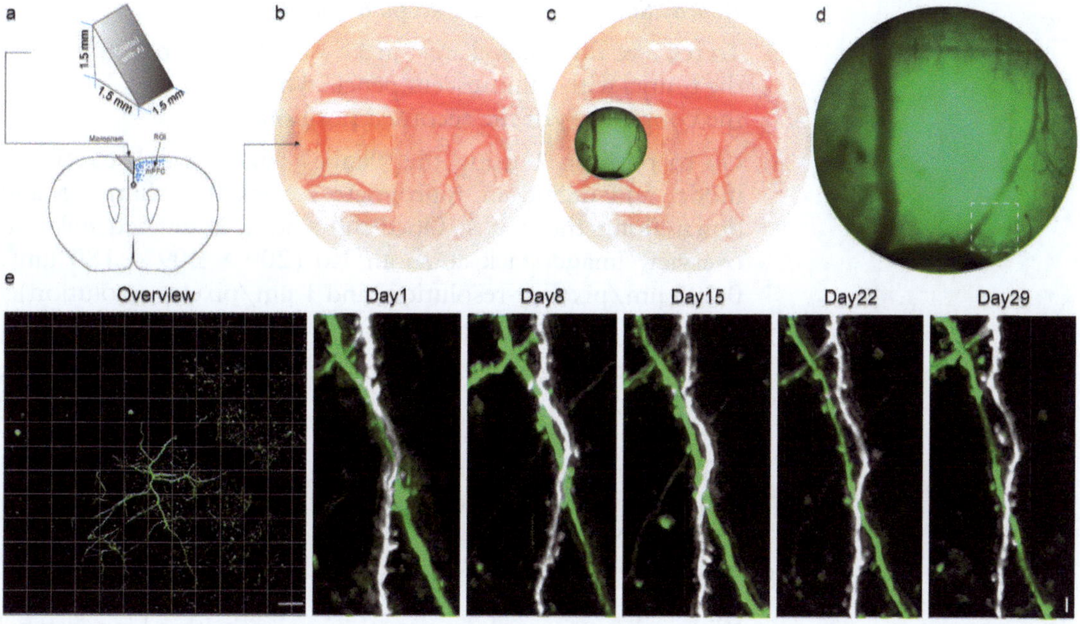

Fig. 4 Representative micrographs showing in vivo imaging area locating and GFP-labelled dendrites in the mPFC. (**a**) Illustration of parameters and installation of the microprism. (**b**) Overview of the window-covered region under the bright field. (**c** and **d**) A smaller region is selected from the window-covered region through the ocular (**c**), and under the GFP exciting fluorescent light source, dendrites in the mPFC are visible (**d**). (**e**) Dendrites in the square with white dash line in **d** are further imaged using the in vivo two-photon microscope, with the selected dendrite imaged longitudinally for several weeks. Scale bars = 50 μm (overview); 5 μm (time series of individual dendrite view). Al aluminum; mPFC medial prefrontal cortex; ROI region of interest

1. *Repeat Subheading 3.2.1* **steps 1–5**.
2. Apply a drop of distilled water into the implanted cannula, and leave no bubbles inside to ensure the optical pathway in a homogenate immersion medium. Adjust the z-position of the framework until the objective contacts the water drop. Continue adjusting the z-position with extreme care until the cranial window is at the closest possible position relative to the objective. Turn on the visible fluorescent light source at the wavelength of GFP excitation. Locate the imaging focus through the coverslip on the bottom of the implant, using the fine focus knob under the ocular (*see* **Note 24**).
3. According to the experimental plan, use the ocular to locate ~3 regions of interest for subsequent imaging sessions (*see* **Note 25**). Before searching for regions of interest, mark a dot using a permanent fluorescent pen on the left edge of the implant, then center the dot in the microscopic field, and set the position on the "Stage" panel in the Zen software as (0, 0, 0) (*see* **Note 28**). The relative positions of the regions of interest should also be marked on the "Stage" panel once they are settled and serve as references for their relocation in subsequent imaging sessions.

4. *Repeat Subheading 3.2.1* **steps 8** and **9**.
5. To retrieve the imaged area, make a marker on the left edge of the tunnel with a fluorescent pencil, which is used to determine the coordinates of the recorded positions.
6. Go to position 1 using the "Stage" panel, switch to "Live" mode, and define an appropriate z-stack by rolling up and down along the z-axis. Once the z-stack is defined, take an overview image stack at zoom 1.0 (209 × 209 × 180 μm^3, 0.205 μm/pixel xy-resolution, and 1 μm/pixel z-resolution).
7. Select dendrites of sufficient quantity (~15) within one region of interest using the "Crop" function (42 × 11 × 28 μm^3, 0.082 μm/pixel xy-resolution, and 1 μm/pixel z-resolution). Switch to "Live" mode and define an appropriate z-stack by rolling up and down along the z-axis. Once the z-stack is defined, use "Start experiment" to take the image of the dendrite (*see* **Note 27**).
8. *Repeat Subheading 3.2.1* **steps 11** and **12**.

With well-performed surgery, the implant with a hippocampal window could be stable for in vivo imaging as long as the regular cranial window. Representative images acquired using the above-mentioned procedure can be found in Fig. 5. However, it should be noted that with higher imaging frequency, the risk of loosening the implant—especially by cleaning the window—will also increase. The loosened implant may suppress the mouse brain and lead to an epileptic seizure.

We have summarized the above imaging procedures for monitoring dendritic spines in different brain regions using the LSM 880 in vivo two-photon microscope. These procedures share

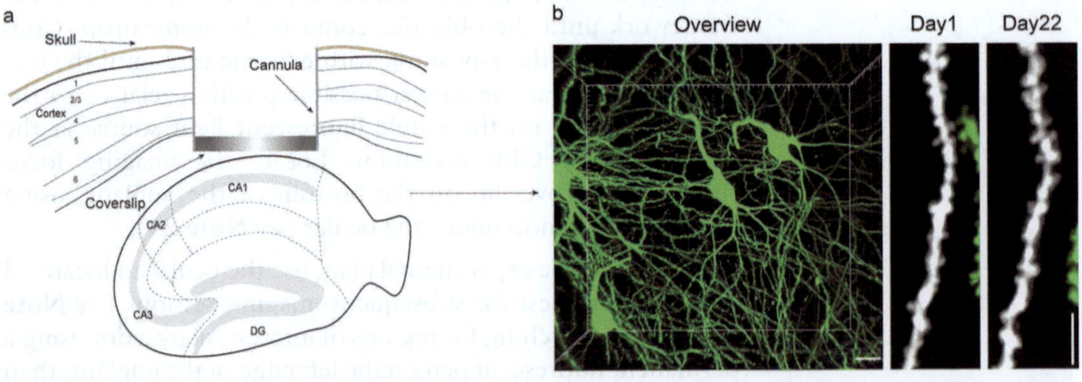

Fig. 5 Illustration of the hippocampal window and representative in vivo two-photon micrograph showing GFP-labelled dendrites in the hippocampus. (**a**) Illustration of the hippocampal window. (**b**) Visualization of GFP-labelled dendrites in the dorsal CA1 region of the mouse hippocampus acquired using the in vivo two-photon microscope, with a selected dendrite within the region for longitudinal imaging. Scale bars = 15 μm (overview); 5 μm (time series of individual dendrite view). DG dentate gyrus

several similar steps; however, due to their imaging location and properties of used optics, procedures have slight differences from each other, e.g., the objective and related parameters of image sizes. On top of this, we described some potential common problems that might be encountered during imaging sessions following the above procedures. The troubleshooting advice can be found in Table 2.

Table 2
Troubleshooting tips

Problem	Possible reason	Solution
No fluorescence can be detected	Excitation and emission settings are wrong	Adjust excitation and emission settings
	Not a mouse with positive fluorescent labelling—due to mistaken genotyping results, or phenotype does not show	Use another mouse with both positive genotype and phenotype
	Loss of water immersion	Make sure the front end of the objective and the coverslip (or the metal cannula for the hippocampal window) is immersed in water drop before and during imaging
Shaking of images	Movement of the head because the head holder is not properly fixed	Fix the clamping of the head holder
	Gasping of the mouse	Make sure that the airflow of the anesthesia unit and isoflurane levels are set properly
	The vibration of components along the optical pathway during imaging	Carefully check the surface on which the microscope system is placed. Remove anything that may cause the vibration
	Window loosening	Make sure the window is sealed properly while cleaning it. If the sealing is partially broken, reseal it using Histoacryl and dental acrylic
Window defects	Inappropriate sealing of the window during the surgery	Apply the Histoacryl and dental acrylic carefully during the surgery, and make sure both are fully cured and the craniotomy is sealed. No cerebrospinal fluid or blood should leak from the implantation area
	The excessive mechanical pressure of the scalpel tip during cleaning	Clean the cranial window before imaging using the scalpel tip with extreme care. If the cranial window partially detaches from the skull/sealing acrylic, use the Histoacryl or dental acrylic to fix the sealing. If the cranial window breaks completely, the mouse is no longer suitable for longitudinal in vivo imaging

(continued)

Table 2 (continued)

Problem	Possible reason	Solution
The cranial window appears somewhat opaque	Dust and water stain	Carefully clean the cranial window before imaging using the scalpel tip
	Contamination of Histoacryl and dental acrylic	Carefully apply the Histoacryl and dental acrylic during surgery. If contamination happens, use the scalpel tip to remove the Histoacryl or dental acrylic. Otherwise, find a spot where dendritic spines are still visible and suitable for longitudinal in vivo imaging. If none of the mentioned methods works, the mouse is no longer suitable for longitudinal in vivo imaging
	Skull regrowth around the craniotomy	Rarely seen, if it happens, the mouse is not suitable for longitudinal in vivo imaging anymore
	Hemorrhage due to physical pressure by the objective during imaging	Once the closest position between the objective and the cranial window is settled before imaging, locate the focal plane with the ocular only by moving the objective and cranial window away from each other using the fine focus knob

3.3 In Vivo Two-Photon Image Processing and Analysis of Dendritic Spines

3.3.1 Manually Identify, Annotate, and Analyze Dendritic Spines

This section describes steps for analyzing images of dendritic spines acquired by longitudinal in vivo two-photon imaging. Some commercial tools, e.g., MicroDynamix (MBF Bioscience), can be used for analyzing dendritic spine images acquired from longitudinal in vivo imaging in a semi-automatic manner. However, due to the inevitable image distortions during the scanning, the occasional fluctuation of signal-to-noise ratio, and the restrained z-resolution, the accuracy of the semi-automatic analysis is sometimes suboptimal. Therefore, to date, analyzing images of dendritic spines from longitudinal in vivo imaging by manually identifying and annotating each one of them individually with colored circles/words remains, in our consideration, an optimal way of handling data on the structural plasticity of dendritic spines. Once spines of all dendrites from one mouse through all time points are identified, readouts can then be summarized as newly gained, lost, or stable dendritic spines.

To identify and annotate in vivo imaged dendritic spines consistently, several criteria should be followed:

1. Only spines with sufficient signal-to-noise ratio and that extend laterally from the dendrite are included for analysis.
2. Along with the longitudinal imaging sessions, newly appeared and disappeared dendritic spines are annotated as gained and lost, respectively.
3. Along with the longitudinal imaging sessions, spines that remain at their original locations (with <0.5 μm displacement) on the dendrite for >= 2 time points are annotated as stable.

Data of annotations are then calculated and analyzed as described below:

1.
$$\text{Dendritic spine density} = \frac{\text{dendritic spines identified at time point } X}{\text{dendrite length } (\mu\text{m}) \text{ counted at time point } X}$$

2. The dendritic spine density acquired from the first imaging session is to be used as a baseline of this dendrite to determine spine density changes in the following time points.

3.
$$\text{Dendritic spine survival rate} * = \frac{\begin{array}{c}\text{dendritic spines identified at both}\\ \text{time point } (X-1) \text{ and time point } X\end{array}}{\text{dendritic spines identified at time point } X}$$

4.
$$\text{Ratio of gained dendritic spines} * = \frac{\text{newly appeared dendritic spines identified at time point } X}{\text{dendritic spines identified at time point } X}$$

5.
$$\text{Survival rate of gained dendritic spines} * = \frac{\begin{array}{c}\text{newly appeared dendritic spines identified at the point}\\ (X-1) \text{ that still exist at time point } X\end{array}}{\begin{array}{c}\text{newly appeared dendritic spines identified}\\ \text{at time point } (X-1)\end{array}}$$

6.
$$\text{Ratio of lost dendritic spines} * = \frac{\begin{array}{c}\text{dendritic spines identified at time point } (X-1)\\ \text{but disappeared at time point } X\end{array}}{\text{dendritic spines identified at time point } X}$$

7. Transient dendritic spines are dendritic spines that are identified in less than n consecutive imaging sessions.

8. Persistent dendritic spines are dendritic spines that are identified in more than n consecutive imaging sessions.
9. The spine density in the first imaging time point was used as a baseline to determine the changes in spine density at the following different time points.

Note: *Parameters are not applicable for the first imaging session.

3.3.2 Imaris-Based 3D Morphometric Analysis of Dendritic Spines

Dendritic spines exhibit a great variety of shapes and sizes, even along a single dendrite. Upon stimuli, dendritic spines change not only by number but also in shape. In general, they are morphologically classified into three major types according to the ratio of the dendritic spine head and neck: mushroom spine, a narrow neck compared to the larger head; stubby spine, no obvious neck between the head and base part; and thin spine, a neck of similar size, but a smaller head [68]. Apart from density, the dendritic spines' shapes are also closely associated with their physiological properties [69] and therefore are one of the important parameters of their structural plasticity. However, due to the abovementioned reasons—inevitable image distortions during the scanning, occasional fluctuation of signal-to-noise ratio, and the restrained z-resolution—images of the dendritic spines acquired by in vivo two-photon imaging are not suitable for sophisticated 3D morphological analysis. Therefore, in our practice, critical time points of the in vivo imaging are usually accompanied by ex vivo confocal imaging of dendritic spines with higher z-resolution for further morphological analysis. This section describes steps for 3D analyzing images of dendrites and dendritic spines acquired by confocal microscopy using Imaris software (Bitplane).

Imaris provides advanced visualization and analysis methods for surface reconstruction, filament tracing, and segmentation in 2D/3D and 4D datasets. It also allows users to measure and plot objects, with group comparisons with statistical tests. Additionally, Imaris is highly customizable in Matlab, Java, or Python with a two-way interface. To conduct the morphometric analysis of dendritic spines in a semi-automatical manner, we employed the Filament Tracer module in Imaris 9.8 to first create the filament 3D reconstruction in the following steps:

1. Load the image (*see* **Note 32**) (*see* Fig. 6a).
2. Add new filaments.
3. Select "Recompute Filament."
4. In our practice, when multiple dendrites are present in one stack, it is more efficient to mark the dendrite of interest manually instead of using automatic detection, which tends to

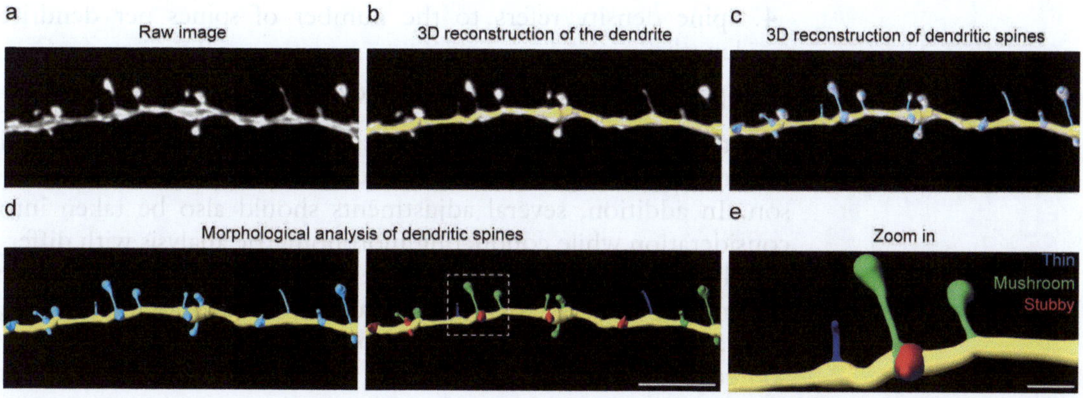

Fig. 6 Analysis procedure of GFP-labelled dendrite and spines in 3D. (**a**) Representative confocal micrograph of GFP-labelled dendrite and its spines in the mouse somatosensory cortex. (**b** and **c**) The dendrite is reconstructed using the Imaris filament module (**b**), following dendritic spine reconstruction around the existing dendrite structure (**c**). (**d**) The reconstructed dendrite and dendritic spines, which are further classified using a Matlab-based algorithm with categories described above. (**e**) The zoom-in image shows classified thin, mushroom, and stubby spines. Scale bars: **a–d** = 5 μm; **e** = 1 μm

reconstruct multiple intertwined dendrites that are difficult to be analyzed. Therefore, select "Skip automatic creation, edit manually."

5. Use "AutoPath" to select and mark the dendrite of interest.
6. Select the "Creation" panel, and then select "Dendrite Diameters" (*see* Fig. 6b).
7. Adjust the threshold to cover all the fluorescent signals of the dendrite, and then click the "Next" button.
8. In the "Spine Points Diameter" view, select "Detect Spines," and then click the "Next" button.
9. Adjust the threshold until all spine heads are marked with seed points (*see* **Notes 33** and **34**), and then click the "Finish" button (*see* Fig. 6c).

Once the dendrite and dendritic spines are reconstructed, a Matlab (The MathWorks)-based Imaris extension is used to classify reconstructed dendritic spines (*see* Fig. 6d and e) in accordance with the below criteria:

1. Stubby spine: length(spine)/mean_width(neck) < 3 and max_width(head)/min_width(neck) < 1.1.
2. Mushroom spine: max_width(head)/min_width(neck) >= 1.1 and max_width(head) > 0.2 μm and min_width(neck) > 0 μm.
3. Thin spine: max_width(head)/min_width(neck) >= 1.1 and max_width(head) <= 0.2 μm or length(spine)/mean_width (neck) >= 3.

4. Spine density refers to the number of spines per dendrite length.
5. Spine head volumes were assessed automatically by Imaris.

It should be noted that consistent criteria are applied to all cohorts in the same experiments to ensure an appropriate comparison. In addition, several adjustments should also be taken into consideration while conducting morphometric analysis with different datasets (*see* **Note 35**). Besides Imaris, there are also other commercial tools for dendritic spine reconstruction and analysis, such as Neurolucida 360 (MBF Bioscience, USA), ImageJ (open-source software), etc., which can be employed in a case-dependent manner.

4 Notes

1. We suggest *Thy1-eGFP* transgenic mice or other transgenic mice with sparsely fluorescently labelled pyramidal neurons for in vivo imaging of dendritic spines. The sparse labelling pattern reduces interference from neighboring GFP-labelled structures, making individual dendrites and dendritic spines more distinguishable; hence, they could be easily imaged and analyzed in a longitudinal manner.
2. Experiments involving animals must be approved by local authorities in terms of experimental design, animal consumption, housing conditions of animals, etc. All animal protocols and procedures described here were conducted in accordance with the regulations of Ludwig-Maximilians-University of Munich and approved by the Government of Upper Bavaria.
3. There are a variety of forms of head holder, for instance, "z"-‐shaped, "T"-shaped, or "-o-"-shaped. They should be chosen based on the window diameter, the location of the window implantation, and the configuration of the imaging setup. The head holders should not restrain the moving range of the objective while imaging and allow it to approach the window to the nearest point, so the working distance of the objective could be mostly utilized. For regular in vivo imaging on the superficial cortical layer, "z"-shaped head holders are usually used.
4. All surgical equipment and the surgical area should be kept clean and sterile during the operation, so the chance of post-op infection and inflammation can be minimized. We use a hot bead sterilizer, which could reach 240 °C to 270 °C within around 20 min and sterilize instruments within 20 s. However, other methods of sterilization could be chosen, for instance, autoclave, or using 70% ethanol.

5. The correct position of the mouthpiece is important for ensuring the air tract is not blocked, and it should be aligned with the mouse maxilla. Additionally, the mouse tongue should be placed appropriately to facilitate the breathing of the mouse during the surgery.
6. Hairs around the wound induce inflammation of the surrounding tissue. The swelling and tissue fluid upon the inflammation weakens the attachment between the mouse skull and the dental cement and subsequently compromises the stability of the cranial window during repetitive imaging sessions.
7. To facilitate long-term repetitive imaging, the attachment of the dental acrylic on the skull should be as firm as possible. Several methods can be used for achieving this purpose. We usually roughen the skull surface using the scalpel tips after the removal of the periosteum. Other than this, while using UV-curable acrylic, UV-curable adhesive like iBond could be used to increase the bond strength of the acrylic to the skull.
8. The mouse skull should always be kept dry during the operation. Residual fluid (blood, tissue fluid, etc.) can potentially on one hand hinder the marking of the coordinates for the craniotomy and on the other hand induce inflammation of surrounding tissues and, when leaking through the craniotomy, brain inflammation. On top of that, a wet skull surface is suboptimal for dental acrylic bonding due to the fluid layer between the bone and the dental acrylic. Collectively it is vital to keep the skull dry during the operation for subsequent longitudinal in vivo imaging.
9. The location and diameter of circular craniotomy can be set using a biopsy punch. After making a mark on the skull, the bone can be removed using a dental drill. Both pieces of equipment should be sterile. The drill bur is non-reusable for each operation. The bur could be worn and heavily contaminated with skull debris during drilling. Poorly performing drill bur is not able to make the clean cut while conducting craniotomy and thus may damage the skull and underneath brain tissue. To remove the marked skull, the drill should gently press the punch mark until a mere leakage of cerebrospinal fluid is visible. Then use forceps to lift the skull piece and detach it from the rest of the skull. It is important that the punch mark around the skull piece is deepened evenly and sufficiently; if the drilling is too shallow, the removal of the skull piece will be difficult and could potentially damage the other part of the mouse skull, thus hindering the installation of the window; if the drilling is too deep, it could cause hemorrhage or damage the underneath brain tissue, thus hindering the subsequent imaging of dendritic spines. Since the skull has varying

thickness—it is thicker near the cranial sutures and thinner elsewhere. Therefore, the buckling of the skull should be checked periodically during drilling by gently pushing it using forceps in case the drill goes too deeply and damages the brain tissue.

10. For routine imaging of dendritic spines, the dura can be left intact after the craniotomy to minimize the inflammation. Optionally, the dura can also be removed using fine forceps for imaging utilizing voltage-sensitive dyes [70] or requiring direct access to the brain tissue, e.g., for tumor implantation and accessing dye uptake [71]. Note that the dura must be removed with extreme care, as it should be repeatedly practiced. The dura around the visible large blood vessels could be preserved in case its removal causes bleeding. If mild bleeding is caused during durotomy, use Gelita-Spon to stop the bleeding and the normal saline to flush the surface. In case of massive bleeding, the mouse should be discarded. However, one should also be aware that for surgeries that involve cortex removal, such as the hippocampal window implantation, some moderate bleeding from the cortex removal may sometimes be unavoidable, because the tissue removal site depends mainly on the coordinates of the hippocampus. Since the imaging area is under the cortex, moderate bleeding from the cortex removal is acceptable.

11. There are different types of coverslips, which should be chosen according to the region of interest of each application. In order to conduct spine imaging on different areas of the brain, the window can be implanted at various locations of the mouse skull. Ideally, the removed skull piece should be relatively flat, so the coverslip can imitate the shape of it and therefore not compress the underneath brain tissue. On top of that, the window should be big enough to allow imaging and acquiring sufficient data but also small enough so the skull structure and head holder attached to it remain stable during longitudinal imaging. These factors require different sizes of window implantation to be performed at different locations of the skull, hence restraining the corresponding sizes of the coverslips. For window implantation on the frontal bone (for spine imaging on the frontal cortex), 3-mm-diameter coverslips can be used due to the relatively small size of the area. For window implantation across hemispheres (medial position implantation), coverslips with a larger diameter can be used and coupled with a "-o-"-shaped head holder. For unilateral window implantation on the parietal bone (for spine imaging of somatosensory cortex), we choose 4-mm-diameter coverslips.

12. Based on the frequently used "z"-shaped head holder, we designed our framework for head-fixed mouse spine imaging. The framework holds the head holder firmly on one side during imaging with an adjustable clamp. The clamp was mounted on a metal plate that fits well on the piezo-microscope stage, which allows to precisely and repeatedly locate the similar region of interest during longitudinal imaging.
13. Currently, there are several types of self-curing and light-curing dental cement for securing the window and mounting the head holder. Irrespective of the types of dental cement, before applying it, the skull periosteum should be removed using a scalpel and covered with a thin layer of superglue. In addition to that, the contact surface must be kept dry while the dental resin is being applied. These ensure the firm bond between the skull and dental cement and keep the window and head holder stable for months during the longitudinal imaging. In terms of the types of dental cement, the self-curing dental cement is more watery when prepared, hence more flexible to be applied. It also polymerizes more evenly. The light-curing resin polymerizes only when exposed to a dental curing light, which falls under the visible blue light spectrum. The light-curing process takes only seconds; however, even polymerization of it requires sufficient light from every angle. Insufficient curing may cause falling off of the head holders during imaging.
14. The microprism can be adhered to the coverslip at different positions, according to the bone structure of the coverslip implantation. For instance, to image the media prefrontal cortex, the microprism should be placed on the anterior area of the mouse brain; however, frontal bones above the region of interest are not wide enough to hold the 3-mm-diameter coverslip. Therefore, the microprism could be placed relative eccentrically on the coverslip—being adhered to the front half of the coverslip—and the coverslip could be implanted relatively posteriorly on wider bone structures.
15. The concrete anatomical location of mouse mPFC is contentious [72]. The term "prefrontal area" was used to describe multiple regions in the mouse brain such as the anterior cingulate cortex, prelimbic cortex, etc., pointing to a long and narrow region underneath the secondary motor cortex. Therefore, for monitoring the structural plasticity of dendritic spines in the mPFC, a 4-mm-diameter microprism-attached coverslip is implanted in the mouse skull with center coordinates, AP 0–2, ML 0 in our practice, so the region being observed coordinates with the current consensus about the mPFC.
16. To implant the microprism, partial frontal bone and parietal bone on both sides will be removed to accommodate the complex of 3-mm-diameter coverslip and the microprism.

This should be conducted with extreme care to avoid damaging the superior sagittal sinus underneath the midline in between the frontal and parietal bones of both sides. It should also be noted that emissary veins inside skull bones could also be damaged during craniotomy. In order to prevent this, the skull could be thinned firstly to visualize the emissary veins. The skull bones can then be broken into small fractures and removed without damaging the veins. In case of mild bleeding, it can be stopped using Gelita-Spon or Histoacryl. In case of massive bleeding, the mouse should be discarded.

17. To insert the microprism in the sagittal fissure, an incision (>1.5 mm) should be made on the dura on the brain surface alongside the superior sagittal sinus. Lower the microprism to the brain surface, and slowly slip it into the incision, and the inclined plane of the microprism should be against the ipsilateral hemisphere. Do not push the microprism into the incision if it cannot smoothly "slip" between hemispheres—remove the microprism, enlarge the incision, and insert the microprism again. It should also be noted that bridging veins from the superior sagittal sinus should not be damaged while making the dura incision; if vessels on the brain surface of one side are too dense, the incision should be made on the other side; otherwise, both sides can be used for microprism insertion. In case the microprism insertion affects the surrounding vessels inevitably, a < 0.5 mm adjustment of vessels is acceptable.
18. If the experimental design does not require a specific hemisphere for imaging, the microprism could be implanted on the side where the durotomy can be conducted more easily and be used for imaging of dendritic spines of contralateral mPFC. If the experimental design requires a specific hemisphere for imaging, e.g., after virus injection in one hemisphere, then mice, in which durotomy cannot be performed on the opposite side of the injected hemisphere, should be discarded.
19. The dental acrylic around the window and on the skull surface should be as thin as possible. The distance between the objective end and imaging plane is relatively long due to the existence of the microprism, and extra distance due to a thick layer of dental acrylic will further shorten the usable working distance of the objective and may make the image depth insufficient to acquire dendritic spines of the required area.
20. To remove the cortex at the implantation site without damaging the surrounding area and underneath the hippocampus, the cortex should be aspirated with gentle suction. Underneath the cortex, the hippocampus is surrounded by a thick myelin layer and a slightly thinner one, both of which have to be carefully removed to reach the alveus CA1 region. Once the

external of the capsule-shaped hippocampus is exposed, stop the aspiration. During the aspiration, bleedings can be stopped using Sugi pointed suction swab and continuous irrigation with normal saline.

21. The cortex aspiration and subsequent cannula implantation are mainly based on a previous study [63]. The optional step of omitting the silicone layer (Kwik-Sil) between the hippocampal surface and the coverslip is meant to stabilize the implant and reduce the movement artifact during the imaging process.
22. The two-photon excitation requires high energy for photon generation. The thermal fluctuation during laser initiation could potentially induce mild deformation and thus compromise the precision of optic and mechanic components of the microscopic system. Besides, the optic pathway, including the electro-optic/acoustic-optic modulator, will also be affected by thermos from the generated laser beam. Therefore, the laser should be allowed to pass the optic pathway until the surrounding components reach a stable state. Additionally, possible electronic drift during the laser initiation could also affect the laser generation. For imaging subcellular structures such as dendritic spines, a 30 min warm-up is a prerequisite.
23. For each step of the imaging process, isoflurane must be controlled with extreme care. The anesthesia induction can be confirmed by the absence of the toe pinch reflex. Once the mouse is put under anesthesia, its vital signs, for instance, breathing and heart rhythm, should be monitored closely to ensure its normal physiological functions are not compromised.
24. Once the cranial window is placed to the closest possible position to the objective, further steps of looking for focus should be done only by moving the cranial window and the objective from each other in order to prevent them from potential crashing.
25. Regions of interest should be selected primarily according to the brightness and integrity of the GFP-positive dendrites. Dendrites with faint GFP signal cannot provide sufficient structural information for further analysis, due to their low signal-to-noise ratio. On top of that, the density of GFP-positive dendrites within the region of interest should be moderate, and overlaying dendrites cannot be separated and analyzed clearly. In addition, despite the wellness of the GFP-positive dendrites in terms of their signal and density, regions of interest should not be selected from margin areas of the cranial window, because, during a consecutive imaging session, due to the slightly different positioning of the mouse on the imaging framework and subsequent alterations of the

relative locations of the objective and the mouse skull, movement of the objective may be interfered by dental acrylic around the window and thus the region imaged during the last session becomes inaccessible. Also, dendrite proximal to big blood vessels should be avoided. The dilation of the vessel due to the blood flow may cause movement of dendrite and thus distortions during imaging.

26. Due to the nonlinear property of the two-photon excitation, the different configurations of the laser sources, and the detectors of the microscope, the optimal excitation wavelength should be tested for each fluorescent label on individual setups to reach the most efficient excitation at the lowest possible laser power, which restrains the optical toxicity and facilitates the longitudinal imaging. On top of that, to maximize the tissue penetration, a longer wavelength is more favorable. When imaging dendritic spines of GFP-M mice, according to our experience, the optimal excitation wavelength falls between 880 nm and 920 nm.
27. The x, y, and z dimensions of the selected dendrite should be defined by the requirements of the experiments. Generally, due to the possible distortion of the in vivo imaging, it is safer to acquire a longer dendrite (1.2–1.5 times longer than the dendrite length planned to be analyzed). To acquire sufficient information on dendritic spine structure in all three dimensions, we keep the pixel size $<=0.15$ μm and the z-step size $<=1$ μm. Smaller pixel size and z-step size provide higher resolution; however, they also cause longer imaging duration and potential phototoxicity; therefore, optimal pixel size and z-step size should be chosen based on multiple factors, e.g., dendrite number per session, imaging frequency, scanning speed of the microscope, etc.
28. To longitudinally monitor multiple imaging areas, their locations should be recordable and traceable in a three-dimensional coordinate system. One of the most widely used ways is to record locations of imaging areas as relative coordinates to a reference point (0, 0, 0 in *x*-, *y*-, and *z*-axis). The reference point can be any top corner the piezo-stage could be moved to, e.g., the (0, 0, 0) could be reached by moving the piezo-stage to the lowest position of the bottom right corner. Locations of the subsequent imaging areas of the specific mouse could then be recorded as relative coordinates to (0, 0, 0). The move-to-top-corner strategy is the simplest way to establish a reference point. However, the long distance between the reference point and the imaging areas (from corner to center) sometimes amplifies the mild but inevitable intersessional deviations during re-mounting mice to the framework on the piezo-stage to large gaps, making relocation of imaging areas difficult. An

alternative to this is to use reference points closer to imaging areas. One of the common methods is to manually make a dot using a permanent marker around the window and keep the reference point (0, 0, 0) in proximity to imaging areas. Instead of using reference points, imaging areas can also be relocated by anatomical landmarks. One of the most unique, recognizable, and relatively stable landmarks on the brain surface is the blood vessel structure. Once the vessel structures of all imaging areas are photographed in the first imaging session, based on their particular pattern, the corresponding imaging area could be easily relocated using the ocular.

29. Several things are affecting the stability of the window, for instance, the mechanical pressure on the window from surrounding tissues by mounting the mouse on the imaging framework. On top of that, the image quality also heavily relies on the general state of the mouse, which can be profoundly affected by the duration of the imaging session. The commonly acceptable duration of in vivo imaging of dendritic spines is less than 1 h.
30. Based on our previous experience in longitudinal in vivo imaging of dendritic spines, the required sample size for statistics of imaged dendrites per mouse is roughly ~8–10. Therefore, the total number of dendrites to be imaged during the initial imaging sessions should exceed ~15, due to the poor resolution or inevitable distortions of certain dendrites at one or more time points along the imaging period, which are to be excluded after the completion of imaging and before analysis.
31. All objectives come with their own working distance, which is defined by the distance between the front lens of the objective and the surface of the specimen under the condition that the specimen is focused. During mPFC imaging through the microprism, both excitation and emission light have to go through the entire prism structure before touching the surface of the brain and reaching the area of interest for spine imaging. Therefore, a long working distance objective is preferred for imaging of this kind.
32. It should be noted that depending on the confocal system used, a deconvolution step might be required. The deconvolution can be conducted using the built-in deconvolution module of Imaris or using commercially available software, e.g., AutoQuant X3 (Media Cybernetics).
33. We usually use the "Slice" view to measure the spine head diameter manually and set it before clicking the "Next."
34. Due to the intensity-dependent auto-detection, autofluorescence acquired during imaging might be marked as "dendritic spines." They could be unchecked by right-clicking while pointing at them.

35. For Imaris-based analysis, it is necessary to adjust the criteria based on the specific images of dendritic spines being analyzed. Some examples of factors that influence the criteria are: the conditions of the subjects that are imaged (e.g., cultured neurons, tissue slices), the species of the animals (e.g., mouse, rat), the methods of labelling (e.g., eGFP transgenic, Golgi staining), and the methods of imaging (e.g., confocal imaging, widefield microscope imaging).

5 Summary and Perspectives

The methods presented above for different types of cranial windows and corresponding imaging procedures allow investigation of dendritic spines across the superficial and deep areas of the brain in live animals in a relatively simple and straight manner. However, several limitations should be considered.

One of the limitations of intracranial microprism implantation is that it requires specific patterns of vasculature—the blood vessels on the brain surface/of the dura around the implantation area must not obstruct the microprism insertion. To work more efficiently, a vasculature screening by thinning the mouse skull and visualizing the blood vessel underneath should be conducted before the implantation. Besides the vasculature pattern, another limiting factor of the microprism application is the restriction of the observation region (~2.25 mm^2), which is far smaller than the regular cranial window (~12.57 mm^2 while using a 4-mm-diameter coverslip). A smaller observation region means fewer opportunities for collecting dendrites suitable for longitudinal monitoring, especially while using mice with sparse labelling fluorescent protein. On the other hand, the hippocampal window implantation is not strictly restricted by the vasculature, which is to be removed with the dura and cortex to expose the underneath hippocampus. Though functionally the removed areas are not directly linked to the hippocampus, it is not easy to accurately estimate the impact of the cortex ablation on dendritic spines within the observed hippocampal region.

Unlike the regular coverslip-based cranial window, for both microprism and hippocampal window, the optical limitations should also be taken into account. Since, for both implantations, the distance between the front end and the intended focal plane is >= 1.5 mm, with the dental acrylic around the coverslip, the effective range of the working distance of any objective employed should be deducted ~2 mm. Therefore, to reach sufficient depth under the brain surface, an objective with sufficient working distance is essential. In our practice, we use a Nikon 16× water immersion objective for both the microprism and hippocampal window. The objective has a 3 mm working distance and can monitor dendritic spines of layer 1 dendrites (mPFC, microprism) and basal dendrites (CA1, hippocampal window). Objectives with longer working distances could also be used. However, they often have bigger front ends and may crash with the head holders or dental acrylic around the coverslip.

On top of these, general limitations of in vivo imaging should also be noticed. For instance, in vivo imaging of dendritic spines is restrained in its z-resolution, making such acquired images not suitable for sophisticated 3D morphological analysis. Therefore, we combine in vivo microscopy with ex vivo confocal microscopy on brain sections to acquire images with fine z-resolution. Together, these methods help to reveal dendritic spine dynamics during spontaneous and stimulus-induced activities. In combination with pharmacological, molecular genetic, opto−/chemogenetic, electrophysiological, and behavioral approaches, the pivotal role of the structural plasticity of dendritic spines in cognition could be further elucidated.

Despite all the abovementioned limitations, the described methods are highly useful for monitoring dendritic spines of areas that regular cranial windows cannot reach; moreover, they can be easily adapted for other applications with minor adjustments. For example, with proper optical settings, emission light at two or more different wavelengths can be detected concomitantly in one imaging session, and therefore, structures in proximity to dendritic spines could also be visualized and analyzed. For instance, by crossing mouse lines with fluorescently labelled dendritic spines (e.g., *Thy1-YFP*) and the ones with fluorescently labelled glial cells (e.g., *Aldh1l1-eGFP* for astrocytes or *CX3CR1-eGFP* for microglia), glial–dendritic spine interactions can be monitored and analyzed in vivo [16]. Dendritic spines can thus be studied with participants of their modulation.

We believe that the methods we described here are capable of helping readers conduct investigations on the structural plasticity of dendritic spines across a variety of brain regions in a longitudinal manner using a set of surgical and imaging procedures that are easy to be put into forms of each other, facilitating studies of multiregional dendritic spine dynamics in a simple to grasp and cost-effective manner.

Acknowledgments

This work was funded by the Deutsche Forschungsgemeinschaft (DFG, German Research Foundation; ID 511905387 to Y.S.; FOR 2858/ID 422181340 to J.H.; EXC 2145/ID 390857198 to J.H.) and the Verein zur Förderung von Wissenschaft und Forschung an der Medizinischen Fakultät der LMU München (WiFoMed; ID 8200676 to Y.S.).

References

1. Crick F (1982) Do dendritic spines twitch? Trends Neurosci 5:44–46
2. García-López P, García-Marín V, Freire M (2007) The discovery of dendritic spines by Cajal in 1888 and its relevance in the present neuroscience. Prog Neurobiol 83:110–130
3. Glickstein M (2006) Golgi and Cajal: The neuron doctrine and the 100th anniversary of the 1906 Nobel Prize. Curr Biol 16:R147–R151
4. Kasai H, Fukuda M, Watanabe S, Hayashi-Takagi A, Noguchi J (2010) Structural dynamics of dendritic spines in memory and cognition. Trends Neurosci 33:121–129
5. Bosch M, Hayashi Y (2012) Structural plasticity of dendritic spines. Curr Opin Neurobiol 22:383–388
6. Zuo Y, Lin A, Chang P, Gan W-B (2005) Development of long-term dendritic spine stability in diverse regions of cerebral cortex. Neuron 46:181–189
7. Lendvai B, Stern EA, Chen B, Svoboda K (2000) Experience-dependent plasticity of dendritic spines in the developing rat barrel cortex in vivo. Nature 404:876–881
8. Gu L et al (2014) Long-term in vivo imaging of dendritic spines in the hippocampus reveals structural plasticity. J Neurosci 34:13948–13953
9. Pfeiffer T et al (2018) Chronic 2P-STED imaging reveals high turnover of dendritic spines in the hippocampus in vivo. elife 7:e34700
10. Grutzendler J, Kasthuri N, Gan W-B (2002) Long-term dendritic spine stability in the adult cortex. Nature 420:812–816
11. Trachtenberg JT et al (2002) Long-term in vivo imaging of experience-dependent synaptic plasticity in adult cortex. Nature 420: 788–794
12. Holtmaat A, Svoboda K (2009) Experience-dependent structural synaptic plasticity in the mammalian brain. Nat Rev Neurosci 10:647–658
13. Cheadle L et al (2020) Sensory experience engages microglia to shape neural connectivity through a non-phagocytic mechanism. Neuron 108:451–468. e459
14. Jung CK, Herms J (2014) Structural dynamics of dendritic spines are influenced by an environmental enrichment: an in vivo imaging study. Cereb Cortex 24:377–384
15. Nithianantharajah J, Hannan AJ (2006) Enriched environments, experience-dependent plasticity and disorders of the nervous system. Nat Rev Neurosci 7:697–709
16. Shi Y et al (2022) Long-term diazepam treatment enhances microglial spine engulfment and impairs cognitive performance via the mitochondrial 18 kDa translocator protein (TSPO). Nat Neurosci 25:317–329
17. Moda-Sava R et al (2019) Sustained rescue of prefrontal circuit dysfunction by antidepressant-induced spine formation. Science 364:eaat8078
18. Segura I et al (2016) The oxygen sensor PHD2 controls dendritic spines and synapses via modification of filamin A. Cell Rep 14:2653–2667
19. Brown CE, Wong C, Murphy TH (2008) Rapid morphologic plasticity of peri-infarct dendritic spines after focal ischemic stroke. Stroke 39:1286–1291
20. Swann JW, Al-Noori S, Jiang M, Lee CL (2000) Spine loss and other dendritic abnormalities in epilepsy. Hippocampus 10:617–625
21. Nimchinsky EA, Oberlander AM, Svoboda K (2001) Abnormal development of dendritic spines in FMR1 knock-out mice. J Neurosci 21:5139–5146
22. Filipello F et al (2018) The microglial innate immune receptor TREM2 is required for synapse elimination and normal brain connectivity. Immunity 48:979–991. e978
23. Penzes P, Cahill ME, Jones KA, VanLeeuwen J-E, Woolfrey KM (2011) Dendritic spine pathology in neuropsychiatric disorders. Nat Neurosci 14:285–293
24. Yilmaz M et al (2021) Overexpression of schizophrenia susceptibility factor human complement C4A promotes excessive synaptic loss and behavioral changes in mice. Nat Neurosci 24:214–224
25. Comer AL et al (2020) Increased expression of schizophrenia-associated gene C4 leads to hypoconnectivity of prefrontal cortex and reduced social interaction. PLoS Biol 18: e3000604
26. Tsai J, Grutzendler J, Duff K, Gan W-B (2004) Fibrillar amyloid deposition leads to local synaptic abnormalities and breakage of neuronal branches. Nat Neurosci 7:1181–1183
27. Zou C et al (2015) Intraneuronal APP and extracellular Aβ independently cause dendritic spine pathology in transgenic mouse models of Alzheimer's disease. Acta Neuropathol 129: 909–920
28. Zou C et al (2016) Neuroinflammation impairs adaptive structural plasticity of dendritic spines in a preclinical model of Alzheimer's disease. Acta Neuropathol 131:235–246

29. Blumenstock S et al (2017) Seeding and transgenic overexpression of alpha-synuclein triggers dendritic spine pathology in the neocortex. EMBO Mol Med 9:716–731
30. Herms J, Dorostkar MM (2016) Dendritic spine pathology in neurodegenerative diseases. Annu Rev Pathol: Mech Dis 11:221–250
31. Helmchen F, Denk W (2005) Deep tissue two-photon microscopy. Nat Methods 2:932–940
32. Denk W, Strickler JH, Webb WW (1990) Two-photon laser scanning fluorescence microscopy. Science 248:73–76
33. Sako Y et al (1997) Comparison of two-photon excitation laser scanning microscopy with UV-confocal laser scanning microscopy in three-dimensional calcium imaging using the fluorescence indicator Indo-1. J Microsc 185: 9–20
34. Denk W et al (1994) Anatomical and functional imaging of neurons using 2-photon laser scanning microscopy. J Neurosci Methods 54:151–162
35. Squirrell JM, Wokosin DL, White JG, Bavister BD (1999) Long-term two-photon fluorescence imaging of mammalian embryos without compromising viability. Nat Biotechnol 17: 763–767
36. Kawauchi T, Chihama K, Nabeshima YI, Hoshino M (2003) The in vivo roles of STEF/Tiam1, Rac1 and JNK in cortical neuronal migration. EMBO J 22:4190–4201
37. Tabata H, Nakajima K (2001) Efficient in utero gene transfer system to the developing mouse brain using electroporation: visualization of neuronal migration in the developing cortex. Neuroscience 103:865–872
38. Saito T, Nakatsuji N (2001) Efficient gene transfer into the embryonic mouse brain using in vivo electroporation. Dev Biol 240:237–246
39. Fiumelli H et al (2013) An ion transport-independent role for the cation-chloride cotransporter KCC2 in dendritic spinogenesis in vivo. Cereb Cortex 23:378–388
40. Peel A, Zolotukhin S, Schrimsher G, Muzyczka N, Reier P (1997) Efficient transduction of green fluorescent protein in spinal cord neurons using adeno-associated virus vectors containing cell type-specific promoters. Gene Ther 4:16–24
41. Flannery JG et al (1997) Efficient photoreceptor-targeted gene expression in vivo by recombinant adeno-associated virus. Proc Natl Acad Sci 94:6916–6921
42. Walther W, Stein U (1996) Cell type specific and inducible promoters for vectors in gene therapy as an approach for cell targeting. J Mol Med 74:379–392
43. Sauer B (1987) Functional expression of the cre-lox site-specific recombination system in the yeast Saccharomyces cerevisiae. Mol Cell Biol 7:2087–2096
44. Sauer B, Henderson N (1988) Site-specific DNA recombination in mammalian cells by the Cre recombinase of bacteriophage P1. Proc Natl Acad Sci 85:5166–5170
45. Akiyoshi R et al (2018) Microglia enhance synapse activity to promote local network synchronization. eneuro 5:ENEURO.0088
46. Feng G et al (2000) Imaging neuronal subsets in transgenic mice expressing multiple spectral variants of GFP. Neuron 28:41–51
47. Spanopoulou E, Giguere V, Grosveld F (1991) The functional domains of the murine Thy-1 gene promoter. Mol Cell Biol 11:2216–2228
48. Spiess E et al (2005) Two-photon excitation and emission spectra of the green fluorescent protein variants ECFP, EGFP and EYFP. J Microsc 217:200–204
49. Porrero C, Rubio-Garrido P, Avendaño C, Clascá F (2010) Mapping of fluorescent protein-expressing neurons and axon pathways in adult and developing Thy1-eYFP-H transgenic mice. Brain Res 1345:59–72
50. Christie R et al (2001) Growth arrest of individual senile plaques in a model of Alzheimer's disease observed by in vivo multiphoton microscopy. J Neurosci 21:858–864
51. Yoder EJ, Kleinfeld D (2002) Cortical imaging through the intact mouse skull using two-photon excitation laser scanning microscopy. Microsc Res Tech 56:304–305
52. Majewska A, Sur M (2003) Motility of dendritic spines in visual cortex in vivo: changes during the critical period and effects of visual deprivation. Proc Natl Acad Sci 100:16024–16029
53. Holtmaat AJ et al (2005) Transient and persistent dendritic spines in the neocortex in vivo. Neuron 45:279–291
54. Levene MJ, Dombeck DA, Kasischke KA, Molloy RP, Webb WW (2004) In vivo multiphoton microscopy of deep brain tissue. J Neurophysiol 91:1908–1912
55. Jung JC, Schnitzer MJ (2003) Multiphoton endoscopy. Opt Lett 28:902–904
56. Barretto RP et al (2011) Time-lapse imaging of disease progression in deep brain areas using fluorescence microendoscopy. Nat Med 17: 223–228

57. Andermann ML et al (2013) Chronic cellular imaging of entire cortical columns in awake mice using microprisms. Neuron 80:900–913
58. Murayama M, Pérez-Garci E, Luscher H-R, Larkum ME (2007) Fiberoptic system for recording dendritic calcium signals in layer 5 neocortical pyramidal cells in freely moving rats. J Neurophysiol 98:1791–1805
59. Pattwell SS et al (2016) Dynamic changes in neural circuitry during adolescence are associated with persistent attenuation of fear memories. Nat Commun 7:1–9
60. Chia TH, Levene MJ (2009) Microprisms for in vivo multilayer cortical imaging. J Neurophysiol 102:1310–1314
61. Jung JC, Mehta AD, Aksay E, Stepnoski R, Schnitzer MJ (2004) In vivo mammalian brain imaging using one-and two-photon fluorescence microendoscopy. J Neurophysiol 92: 3121–3133
62. Meng G et al (2019) High-throughput synapse-resolving two-photon fluorescence microendoscopy for deep-brain volumetric imaging in vivo. elife 8:e40805
63. Dombeck DA, Harvey CD, Tian L, Looger LL, Tank DW (2010) Functional imaging of hippocampal place cells at cellular resolution during virtual navigation. Nat Neurosci 13: 1433–1440
64. Poll S et al (2020) Memory trace interference impairs recall in a mouse model of Alzheimer's disease. Nat Neurosci 23:952–958
65. Dana H et al (2016) Sensitive red protein calcium indicators for imaging neural activity. elife 5:e12727
66. Tischbirek C, Birkner A, Jia H, Sakmann B, Konnerth A (2015) Deep two-photon brain imaging with a red-shifted fluorometric Ca2+ indicator. Proc Natl Acad Sci 112:11377–11382
67. Fuhrmann M, Mitteregger G, Kretzschmar H, Herms J (2007) Dendritic pathology in prion disease starts at the synaptic spine. J Neurosci 27:6224–6233
68. Harris KM, Jensen FE, Tsao B (1992) Three-dimensional structure of dendritic spines and synapses in rat hippocampus (CA1) at postnatal day 15 and adult ages: implications for the maturation of synaptic physiology and long-term potentiation. J Neurosci 12:2685–2705
69. Halpain S, Spencer K, Graber S (2005) Dynamics and pathology of dendritic spines. Prog Brain Res 147:29–37
70. Kunori N, Takashima I (2019) An implantable cranial window using a collagen membrane for chronic voltage-sensitive dye imaging. Micromachines 10:789
71. Orringer DA et al (2010) The brain tumor window model: a combined cranial window and implanted glioma model for evaluating intraoperative contrast agents. Neurosurgery 66:736–743
72. Laubach, M., Amarante, L. M., Swanson, K. & White, S. R. What, if anything, is rodent prefrontal cortex? eneuro 5 (2018)

INDEX

A

Acid-free glyoxal fixation 20, 21
AfterHyperPolarization (AHP) 327–330, 334, 335, 345–350, 356, 357, 360–363, 365
Algorithms 45, 46, 50, 107, 112, 115, 124, 134, 165, 174, 186, 196, 216, 236, 240, 244, 245, 293–295, 301, 309, 323–366, 375, 377, 379, 390, 417
Automated tape-collecting ultramicrotome (ATUM) 98–100, 185, 210
Automation 144
Axon 7, 27, 32, 41, 54, 106, 142, 164, 183, 212, 248, 286, 379

B

Brain connectivity 284
Bursting 163, 223, 323–366

C

Car *vs.* train 9, 10, 16
Chemical stimulation 57, 59
Circuitry 44, 48, 51, 284
Computational models 222, 279–281, 311, 312, 366
Connectivity 41, 42, 46, 117, 161–176, 283–290, 293–295, 302–307, 309–313, 334, 339, 343–345, 355, 356, 358, 395
Connectome 81, 220, 224, 285, 312
Connectomics 41, 184, 186, 224, 248
Correlation 53–56, 196, 197, 243, 248, 265, 289, 294, 296, 304, 309, 327–339, 359
Cranial window 398, 401, 402, 407–410, 412–414, 419, 423, 426, 427
Cryofixation 56, 64, 65, 68, 73, 197

D

Deconvolution 336–338, 389, 425
Deep learning 123–136, 211
Dendritic spine 32, 38, 106, 116–120, 163, 174, 184, 194–197, 221, 249, 294, 336, 379, 388–390, 395–427
Dense segmentation
*d*STORM 141–158

E

Electron microscopy (EM) 19, 31, 41, 54, 81, 105, 123, 142, 184, 201, 234, 264, 284
Electrophysiology 142, 254, 324, 325, 334

F

Focused ion beam scanning electron microscopy (FIB-SEM) 31–39, 82, 83, 105, 106, 109, 118, 201, 205, 209–211, 218, 219, 223, 235
Freeze-fracture replica 123–136
Freeze-substitution (FS) 56, 58–60, 62, 64, 65, 68, 70–73, 264
Functional calcium imaging 384, 385

H

Heterogeneity 107, 183, 261, 266, 281, 284, 301–306, 310
High-pressure freezing (HPF) 56–59, 61, 62, 64–70, 76, 204, 264, 265
Hippocampal window 400, 402, 406–407, 410–413, 420, 426, 427
Holographic projection 374–379, 381, 382, 385, 386, 391
Human brain 3–16, 33, 118, 163, 202, 203, 218, 220, 250, 280
Human brain slices 10
Hypertonic solution 254–256, 258–261

I

Image acquisition 23, 59, 74, 81, 106–107, 112, 127, 143, 193, 194, 223, 244, 246
Imaging 22, 34, 55, 82, 107, 124, 142, 164, 185, 208, 234, 266, 284, 326, 374, 395
Immunogold labeling 25, 28, 29, 123
Immunohistochemistry 148, 158, 167–168, 171, 220

Joachim Lübke and Astrid Rollenhagen (eds.), *New Aspects in Analyzing the Synaptic Organization of the Brain*, Neuromethods, vol. 212, https://doi.org/10.1007/978-1-0716-4019-7,

In vitro slice 163, 169, 176
In vivo two-photon imaging 396, 400, 407–414, 416

L

Labeling 20, 24–26, 28, 29, 31–39, 87, 91, 118, 123, 125, 126, 143, 152–154, 157, 164, 166–168, 171, 173, 174, 184–188, 191, 197, 210, 219, 248, 266–267, 395–397, 413, 418, 426
Light-scattering through tissues 376, 379
Long-distance transport 3–16

M

Microprism 398–402, 404–406, 409–411, 421, 422, 425–427
Modeling 41, 45, 47, 50, 254, 279–313, 324, 332, 334, 337–342, 345–348, 351–354, 357–365
Morphology 28, 32, 42, 51, 54, 56, 154, 162–165, 167, 184, 185, 195, 210, 218, 283–285, 381
Multicolor 23, 25, 28
Multi-modal 162, 164–166, 175, 202
Multineuron patch-clamp 8, 15
Multiplexed imaging 141–158
Multiplexed immunohistochemistry

N

Nano-pipette 336–338
Nanoscopy 211
Neurobiology 279, 283, 301
Neuronal network 54, 224, 260, 283, 284, 287, 288, 290, 302, 303, 305, 306, 308, 310, 311, 323, 324, 326, 334, 339–359, 365, 366
Neurosurgery 8–15, 202, 220, 221
Neurotransmitter release 58, 75, 183, 184, 253, 266, 296, 310, 380

O

Optimization 146, 303, 305–307, 348, 364, 378, 379, 390

P

Parameter estimation 359–360
Phenomenological model 279–313
Physiology 3, 16, 161–176, 285, 358, 359
Plasticity 213, 222, 224, 233–235, 250, 253, 258, 260, 281, 285, 288, 290, 293–307, 309, 311–313, 325, 339–358, 360, 361, 365, 366, 395–427
Pre-embedding immunocytochemistry
Pyramidal neuron 12, 38, 161–164, 166–168, 173–175, 207, 217, 376, 379, 380, 382–384, 386, 388, 389, 407, 408, 418

Q

Quadruple immunofluorescence
Quantitative 3D synaptic models 222
Quantitative immunogold-protein localization 20

R

Readily releasable pool (RRP) 58, 211, 213, 215, 233–235, 239, 240, 244, 246–248, 250, 253–267, 353, 354

S

Serial block face-SEM (SBF-SEM) 82
Serial EM (SEM) 36, 38, 41–44, 46–48, 51, 82–85, 105, 110–112, 185, 189, 209, 211, 223
Serial sectioning 185, 193
Signal segmentation 336
Simulations 45, 80, 124–126, 131, 132, 224, 280, 281, 283, 284, 287–289, 291–293, 295, 300–301, 305–312, 333, 339, 342, 344, 348–350, 352, 353, 358, 359, 362, 363
Single-molecule localization microscopy (SMLM) 142, 143, 149, 154
Slice quality evaluation 3, 8–10, 13, 15
Somatosensory cortex 39, 166, 188, 194, 402, 407, 417, 420
Spiking neuronal networks 283
Sterile preparation and slicing procedure
Structural plasticity of dendritic spines 395–427
Structure-function relationship 184
Super-resolution imaging 143
Synapses 27, 31, 43, 53, 106, 123, 143, 183, 201, 234, 253, 280, 343, 379, 395
Synaptic inputs 41, 162, 165, 172, 257, 289, 310, 341, 376, 379–383, 390
Synaptic organization 105–120, 201–224, 280, 281, 284
Synaptic short-term plasticity 258, 339, 354, 356–358, 361, 365

T

TEM for 3D reconstructions 193, 209
Thalamocortical 183–186, 189, 193, 197, 248, 250
Three-dimensional (3D) reconstruction 41, 81, 82, 84, 106, 118, 167, 193–195, 197, 209, 210, 215, 218, 224, 243–246, 248, 401
3D software 37

Three-dimensional (3D) volume reconstructions 211–212, 215–216, 220
Tissue 4, 32, 42, 54, 82, 105, 143, 163, 187, 202, 235, 264, 293, 325, 373, 396
Tracing............................. 32, 44, 47, 54, 116–117, 164, 167, 189–190, 224, 284, 286, 287, 403, 416
Transcriptomics 162, 164–165
Transmission electron microscopy (TEM) 32, 37, 41, 42, 60, 72, 73, 81, 82, 84, 85, 101, 126, 185, 186, 188, 189, 191–194, 201, 205, 209–211, 216, 219, 220, 223, 234, 237–240, 243, 244, 246–249
Two-photon excitation 381, 385, 390, 397, 408, 423, 424
Two-photon photolysis 380, 381, 388–390

U

Ultrastructural analysis 58, 196, 222, 254

V

Volume electron microscopy 31, 184, 219

W

Wavefront correction 376, 377, 379, 388–390

U

V

W

MIX
Papier aus verantwortungsvollen Quellen
Paper from responsible sources
FSC® C105338

If you have any concerns about our products,
you can contact us on
ProductSafety@springernature.com

In case Publisher is established outside the EU,
the EU authorized representative is:
Springer Nature Customer Service Center GmbH
Europaplatz 3, 69115 Heidelberg, Germany

Printed by Libri Plureos GmbH
in Hamburg, Germany